Landmarks in Gene Regulation

Landmarks in Gene Regulation

Edited by D.S. Latchman

PORTLAND PRESS
London and Miami

Published by Portland Press Ltd, 59 Portland Place,
London W I N 3AJ, U.K.

In North America orders should be sent to Ashgate Publishing
Co., Old Post Road, Brookfield, VT 05036-9704, U.S.A.

British Library Cataloguing-in-Publication Data
A catalogue record for this book is available from the British
Library

Although, at the time of going to press, the information
contained in this publication is believed to be correct, neither
the authors, the editor nor the publisher assumes any
responsibility for any errors or omissions herein contained.
Opinions expressed in this book are those of the authors and
are not necessarily held by the publishers.

Typeset by Portland Press Ltd
Printed in Great Britain by Information Press Ltd, Eynsham, UK

Contents

13 Functional role of transcription factors

Preface

Twenty years ago, when I was an undergraduate at Cambridge University, we were told that there was no known example in eukaryotes of a defined regulatory protein which bound to a specific DNA sequence in a target gene and regulated its expression. Since that time a bewildering array of such regulatory proteins, or transcription factors, have been defined and their roles in regulating the expression of specific genes analysed. When faced with such a vast array of information on individual factors, it is all too easy to forget that the foundations of the study of such factors, and indeed of eukaryotic gene regulation in general, were laid by a relatively small number of seminal papers which established key points in this area.

The aim of the present volume is to present a number of these key papers that have appeared over the last 20 years, together with a commentary which assesses their importance and places them in the context of what was known at the time and what has been established subsequently. This will allow both the student and the more experienced research worker to see how the field has developed over this time and how the key principles were elucidated. It is hoped, therefore, that this collection will serve as a background to the detailed description of these areas contained in my previous books *Gene Regulation: A Eukaryotic Perspective* (2nd edn., Chapman and Hall, 1995) and *Eukaryotic Transcription Factors* (2nd edn., Academic Press, 1995).

The initial section thus contains papers which established that, while the RNA populations of different tissues are different, the DNA content is the same. This indicated, therefore, that gene regulation operates primarily at some point between the DNA and the final production of the mRNA. Indeed, the papers presented in Section 2 indicated that such control operates primarily at the level of transcription, with individual genes being transcribed only in specific tissues or specific cell types. Despite this, some cases of post-transcriptional regulation, for example by alternative splicing, or by regulation of mRNA stability or translation, also exist. Papers describing several examples of such post-transcriptional regulation are therefore included in Sections 3 (alternative splicing) and 4 (RNA stability and translatability).

Having established that regulation is primarily at the level of transcription, the remaining sections of the book present papers that have aided our understanding of the mechanisms of transcriptional regulation. Section 5 presents papers dealing with the changes in chromatin structure that distinguish active or potentially active genes from the remainder of the genome. Such changes function, at least in part, to alter the accessibility of regulatory sequences within the DNA, which can confer a particular pattern of regulation on a gene. Key papers describing the identification of such sequences are presented in Section 6, which deals with promoter elements, and in Section 7, which deals with elements present at greater distances from the promoter, such as enhancers and locus control regions.

In general, such DNA sequences act by binding particular regulatory proteins or transcription factors that either increase or decrease the rate of transcription of their target genes. Section 8 therefore presents papers describing an early example of the purification and cDNA cloning of such a transcription factor. Similarly, subsequent sections present critical papers describing the analysis of such cloned factors that allowed the identification of specific regions within the factors that mediate DNA binding (Section 9), dimerization (Section 10) and the ability to stimulate transcription (Section 11). Clearly, if a particular transcription factor is involved in mediating a particular pattern of gene expression, it is necessary for the activity of this factor to be regulated so that it is only active in the correct cell type or following exposure to the correct stimulus. Section 12 therefore presents examples of papers dealing with the regulation of transcription factor activity which allows this to occur.

Obviously, many transcription factors were identified on the basis of their ability to bind to a specific regulatory sequence and therefore confer a particular pattern of gene expression. In this way their function was evident from the manner of their isolation. However, the central role for transcription factors in specific biological processes is best indicated by the fact that such factors have often been identified during the analysis of genes whose products play a critical role in a particular biological process. When such products were analysed it was found that the genes encoded transcription factors, indicating a critical role for such factors in the particular process under study. Two examples of this are presented in Section 13.

Overall, therefore, it is hoped that a study of these papers will provide a useful overview of the manner in which the highly complex field of eukaryotic gene regulation has gradually been elucidated. As discussed in the concluding section, much remains to be understood. However, immense progress has been made in the last 20 years, and a clear framework exists for the ultimate understanding of how processes as complex as mammalian development are regulated in terms of differential gene expression. The vast amount of information now available in this area has clearly been generated by a number of other publications and laboratories apart from the few selected here. Obviously those that are presented here are a personal choice designed to illustrate the development of the field, and many other excellent papers could have been included. It is hoped, however, that this selection will at least serve as a starting point to provide an overview of the field and perhaps stimulate a detailed reading of other early publications in this area.

Finally, I would like to thank the advisory board of Portland Press Ltd who conceived the idea of the 'Landmarks' series and decided to include a volume on gene regulation. In addition, I am most grateful to Sarah Bell of Portland Press for her enthusiastic and efficient help. I am also most grateful for the enthusiastic support of the authors whose papers have been chosen for this volume, many of whom provided reprints suitable for reproduction. Obviously this work would also not have been possible without the agreement of the publishers of *Cell*, *EMBO Journal*, *Nature*, *Proceedings of the National Academy of Sciences U.S.A.* and *Science*, who generously agreed to the reproduction of papers published in their journals.

D.S. Latchman
Professor of Medical Pathology
University College London Medical School

Acknowledgements

We gratefully acknowledge the kind permission of the various publishers and authors for allowing us to reproduce the original articles in this volume: *Cell* papers are reproduced by permission of Cell Press; the *EMBO J.* paper is reproduced by permission of Oxford University Press; *Nature (London)* papers are reproduced by permission of Macmillan Magazines Ltd; the *Proc. Natl. Acad. Sci. U.S.A.* paper is reproduced by permission of the authors; *Science* papers are reproduced by permission of the American Association for the Advancement of Science.

Different tissues have different RNA populations but generally similar DNA

Following the identification of DNA as the genetic material, it rapidly became clear that the information in the DNA directs the production of specific proteins by means of an intermediate RNA molecule. Thus 'DNA makes RNA and RNA makes protein' became the central dogma of molecular biology.

Obviously this central dogma is of major importance in understanding the processes that regulate gene expression. Thus the use of classical biochemical techniques established that different tissues differ in their protein content. For example, specific enzymes are often made by one cell type and not by others, while secreted factors or hormones are often released only by one tissue and not by others. Obviously such findings beg the question of whether these differences in protein content between different tissues are paralleled at either the RNA or the DNA level.

The two papers presented in this section address this particular question. Thus the paper by Hastie and Bishop represents the first detailed analysis of the differences in RNA populations between different tissues. This analysis was based on an earlier paper from the Bishop laboratory (Bishop et al., 1974) which used the technique of $R_{o}t$ curve analysis. In this method a complementary DNA (cDNA copy) of the RNA is made using radioactive precursor molecules so that the cDNA is radioactively labelled. The cDNA and the RNA are then dissassociated from one another and the rate at which the radioactive cDNA reanneals or hybridizes to the RNA is followed.

Initially, the Bishop laboratory used this method to analyse the different abundances of different RNAs in a single tissue. Thus in the reannealing process a cDNA derived from an abundant RNA will find an RNA partner much more rapidly than one derived from a rare RNA, simply because more potential partners are available (Figure 1.1). Hence this method can be used to examine the abundance of different RNAs within an RNA population. Indeed, using this method the Bishop laboratory established that there were three abundance classes of RNA in eukaryotic tissues, which they termed highly abundant, moderately abundant and rare. This analysis, which was carried out using HeLa cell mRNA in the initial paper (Bishop et al., 1974), was extended in the paper presented here by Hastie and Bishop to show that a similar distribution of mRNA into three abundance classes is also observed in mouse tissues. This result is illustrated in Figure 1 of the paper.

More importantly, in the paper presented here, Hastie and Bishop also took the analysis one stage further by mixing RNA prepared from one tissue with labelled cDNA copied from the RNA of another tissue. The mixture was then allowed to anneal. Obviously the rate of hybridization of the radiolabelled cDNA will be determined by the abundance of its corresponding RNA in the tissue from which the RNA is derived. The more abundant the RNA in this tissue, the more rapidly the cDNA will anneal. In addition, if the RNA is specific to the tissue from which the cDNA

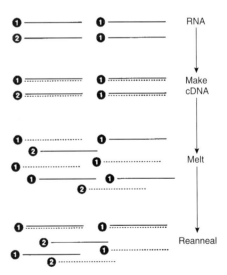

Figure 1.1 Principle of $R_{o}t$ curve analysis. RNA (solid lines) is copied into cDNA (broken lines) and the RNA–cDNA strands are then separated by melting at high temperature. The temperature is then lowered and the reannealing of the complementary strands is followed. An abundant RNA (labelled 1) will find its cDNA partner more rapidly than a rare RNA (labelled 2) simply because more partners are available. In this way the abundance of a particular RNA can readily be measured.

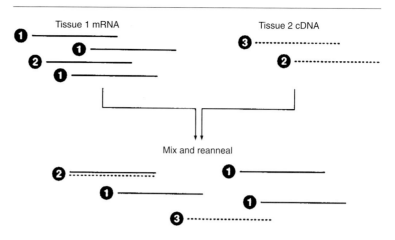

Figure 1.2 R_0t curve analysis using mRNA from tissue 1 and cDNA prepared from the mRNA of tissue 2. Note than an RNA present in both tissues (labelled 2) will find its partner in the reannealing, whereas mRNA species present in only one tissue (labelled 1 and 3) will be unable to find a partner regardless of their abundance in any single tissue.

was derived and is absent from the other tissues, no reannealing will occur (Figure 1.2). Thus Figure 2 of the paper compares the hybridization of liver mRNA with either liver cDNA or cDNA prepared from brain or kidney mRNA. Hybridization was much slower when the cDNA was derived from brain or kidney than when the homologous liver cDNA was used, indicating that there are indeed differences in the mRNA populations of the different tissues. Further analysis of these data, and the results presented later in the paper, showed that some mRNAs which were abundant in one tissue were absent from or present at much lower levels in the other tissues, although there were also some mRNAs common to all three tissues.

This method therefore allowed both qualitative and quantitative differences between the RNA populations of different tissues to be measured at a time when gene cloning to allow the isolation of individual genes was not yet available. The conclusion reached in this way by Hastie and Bishop has, however, been abundantly confirmed in subsequent studies using cloned probes for specific mRNAs (for a review, see Latchman, 1995a). Thus, while some RNAs are present at similar levels in all tissues, both qualitative and quantitative differences between the RNA populations of different tissues are observed. RNA species which are abundant in one cell type can be rare or absent in others. Hence the protein differences between different tissues are indeed paralleled by differences in their RNA content.

Having established that the RNA contents of different tissues are different, the obvious next step is to determine whether this is paralleled at the DNA level. Thus it is perfectly possible that genes whose RNA and protein products are not required in one particular tissue could be lost from that tissue while being retained in another tissue where their corresponding RNA and protein are required. This question was first examined shortly after cloned probes for individual genes became available.

Unlike the study of Hastie and Bishop, which was specifically intended to examine the RNA populations of different tissues, the study of Jeffreys and Flavell presented here was not designed specifically to contribute to the investigation of gene regulation.

Thus, following the initial cloning of the β-globin cDNA (Maniatis et al., 1976), Jeffreys and Flavell used this probe to study the structure of the β-globin gene. They made the, then very surprising, finding that the β-globin gene was interrupted by an intervening sequence (intron) which was not found in the corresponding mRNA. Obviously this finding that eukaryotic genes contain introns has since been replicated in a number of other different genes, and it is now known that these introns are removed from the initial RNA transcript by the process of RNA splicing (for reviews, see Sharp, 1987; Lamond, 1991a). However, as the β-globin cDNA was one of the earliest to be cloned, when Jeffreys and Flavell performed their experiment it was entirely unclear how the intron in the β-globin gene would be removed. Indeed, it was thought possible that this intron was present in the DNA of all tissues which do not express β-globin but was absent from the reticulocyte, which does express β-globin.

To test this hypothesis, Jeffreys and Flavell tested for the presence of the intron in a number of different tissues derived from the rabbit. In the experiment presented in Figure 10 of their paper they detected the same-sized genomic DNA fragments with the β-globin cDNA probe in all the tissues tested, regardless of whether or not the tissue expressed the β-globin gene. As well as eliminating the possibility that the intron is specifically removed from the gene in expressing tissues, this result also indicated that genes are not activated by any other DNA rearrangements (which would generate different-sized DNA fragments) in the tissues where they are expressed. Similarly, since the DNA bands did not disappear in non-expressing tissues or become more intense in expressing tissues, this result also eliminated models in which genes are specifically deleted in tissues where they are not expressed or specifically amplified in tissues where they are expressed. Hence these findings showed that the structure of the gene was identical in the different tissues, regardless of whether the gene was expressed in that tissue or not.

This report is thus of importance for the study of gene regulation, since it establishes the principle that the differences in RNA and protein content of different tissues are not paralleled at the DNA level. Thus, with a few exceptions (such as the immunoglobulin genes; for a review, see Gellert, 1992), the DNA content of different tissues is identical. The different RNA contents of different tissues must therefore be produced from similar DNA. The papers that elucidated how this occurs are discussed in Section 2.

Hastie & Bishop (1976) Cell **9**, 761–774

The Expression of Three Abundance Classes of Messenger RNA in Mouse Tissues

Nicholas D. Hastie* and John O. Bishop
Institute of Animal Genetics
West Mains Road
Edinburgh, Scotland

Summary

Using the technique of mRNA-cDNA hybridization, we have shown that there are between 11,500 and 12,500 different mRNAs in three different mouse tissues: kidneys, brains, and livers. Several experiments suggest that in each tissue the mRNAs are organized into three abundance classes rather than as a continuum with respect to concentration. Cross-hybridization experiments show that the most abundant class of mRNA in each tissue is characteristic, and that a high proportion of the total sequences are common between tissues. For a more complete analysis, cDNA was fractionated into three classes. Studies using isolated abundant cDNA show that some abundant sequences of liver and kidney are present in other tissues, but among the lower frequency classes. Thus tissue-specific differences in mRNA populations may be related to abundance as well as qualitative differences. Using isolated middle frequency cDNA of the kidney, it was shown that of the 550 or so sequences in this class, approximately 500 are shared with the liver. Similarly, between 9,500 and 10,500 of the low frequency kidney cDNAs are shared with the brain and liver, respectively, suggesting that the majority of mRNAs may be involved with "housekeeping" activities. In an attempt to see whether abundance of mRNA is related to repetition of the sequence in the genome, it was shown that abundant and middle frequency cDNA of the liver and kidney contain a component that anneals with DNA repeated approximately 100 fold. However, the low frequency cDNA of the kidney contains no repeated sequences.

Introduction

The mouse genome has a complexity of 1.4×10^{12} daltons. This amount of information is sufficient to encode approximately 1,000,000 average sized genes. Estimates based on genetic considerations suggest that between 50,000 and 100,000 structural genes are expressed in the lifetime of a mammal (Bishop, 1974; Lewin, 1974). It is assumed that different sets of these genes are functional during dif-

ferent stages of development and in different adult tissues. However, the extent of such differential gene expression remains unclear.

We have approached this problem by measuring the complexity of mRNA populations in three different mouse tissues. This was done by analyzing the kinetics of hybridization of mRNA to a cDNA copy, a method that was used initially to characterize HeLa cell mRNA (Bishop et al., 1974). It was concluded from the earlier work that HeLa cell mRNA is distributed in three abundance groups, rather than as a continuum with respect to concentration. We have observed similar distributions in the mouse tissue mRNA populations and have examined the representation of these abundance groups in the different tissues. We also determined whether abundance of mRNA is related to the frequency of the sequence in the genome.

Results

Properties of the Mouse Tissue mRNA

To estimate the complexity of an mRNA population, it is necessary to know the number average molecular weight of the mRNA. This was determined by hybridizing ^{3}H–poly(U) across fractions of a sucrose gradient as described by Rosbash and Ford (1974) and Bishop et al. (1974). As shown in Table 1, the mRNA from each tissue studied has an average molecular weight between 625,000 and 650,000 daltons. A value of between 600,000 and 650,000 daltons was also calculated from the optical density profiles of the mRNA preparations least contaminated by ribosomal RNA. This is close to the size reported for HeLa cell mRNA (Bishop et al., 1974) and sea urchin gastrula mRNA (Galau, Britten, and Davidson, 1974). The mouse tissue mRNA was extracted using conditions that did not degrade highly labeled HeLa cell mRNA upon co-extraction (Experimental Procedures). For hybridization analysis, it is necessary to know the concentration of the mRNA. To do this, the poly(A) content of the mRNA is quantitated by hybridization to ^{3}H–poly(U) (Bishop et al., 1974). Provided the size relationship between the poly(A) and mRNA is known, this provides an estimate of the mRNA concentration. As shown in Table 1, the poly(A) moiety (sized by Dr. Jacques Beckmann) comprises approximately 5% of the length of the mRNA, or 100 nucleotides.

Complexity of the mRNA

The complexity of an mRNA population can be estimated by analyzing the kinetics of hybridization of mRNA to a highly labeled cDNA copy where the mRNA is in vast excess (Bishop et al., 1974). The reaction is followed by expressing the percentage of cDNA made double-stranded as a function of the

*Present address: Department of Medical Viral Oncology, Roswell Park Memorial Institute, 666 Elm Street, Buffalo, New York 14263. Requests for reprints should be addressed to Dr. Hastie at this address.

Table 1. Properties of Mouse Tissue mRNAs

Source of mRNA	Number Average Molecular[a] Weight of mRNA	Size of Poly(A)[b] in Nucleotides	mRNA as % Cytoplasmic RNA	Number of Copies mRNA per Cell
Liver	640,000	95–100	3.6	505,441
Kidney	628,000	95–100	3.3	364,990
Brain	625,000	95–100	5.0	563,505

[a]In each case the mean of three estimates.
[b]Sized on polyacrylamide gels as described by Rosbash and Ford (1974).

logarithm of Rot (RNA concentration in moles per liter × time in sec). Ideally, the Rot value sufficient to hybridize 50% of the cDNA (Rot $_{1/2}$) is proportional to the complexity of the mRNA and inversely proportional to the rate constant of the reaction.

The complexity of an unknown mRNA population is obtained by comparing the hybridization of the unknown mRNA to a cDNA copy with the hybridization of a standard mRNA of known complexity to a cDNA copy. We have analyzed the hybridization kinetics of three such standard mRNAs with their respective cDNAs: rabbit α globin mRNA, a mixture of rabbit α + β globin mRNA, and EMC virus RNA (N. D. Hastie et al., manuscript in preparation). The Rot $_{1/2}$ of these reactions are directly proportional to the complexity of the mRNAs. In addition, all reactions follow ideal first-order kinetics. From these data, it can be calculated that a pure mRNA species of molecular weight 650,000 daltons will hybridize a cDNA copy with a Rot $_{1/2}$ of 8×10^{-4} (when the cDNA is 400 nucleotides long). This value is used for computations of tissue mRNA complexities.

Figure 1A compares the hybridization of rabbit globin mRNA to globin cDNA with the hybridization of mouse liver mRNA to liver cDNA. In both cases, the symbols represent the experimental data, and the solid lines are idealized computer fits to the data (see "Theory"). There are several obvious differences between the curves. The hybridization of liver cDNA to liver mRNA is very much slower than the globin reaction. This means, as expected, that the liver mRNA population is very much more complex than the globin mRNA. In addition, the liver curve is very much broader than the globin curve. If all the different liver mRNA species were present in the same concentration, the reaction would cover approximately two orders of magnitude of Rot (see globin curve). However, the liver curve covers five orders of magnitude. In addition, the liver curve appears to be organized into three separate transitions, suggesting that the mRNAs in the liver are grouped into three abundance classes, rather than as a continuous distribution. The computer fit represents the summing of three first-order curves. If we assume that this interpretation is correct, we can calculate the number of different mRNA species in

each class and the numbers of copies of each mRNA per cell. This is done by relating the Rot $_{1/2}$ of each transition to the Rot $_{1/2}$ value of 8×10^{-4} which, as discussed above, is that expected for a pure mRNA of molecular weight 650,000 daltons. A correction must be made to allow for the fact that each class of mRNA is diluted by the other classes.

As shown in Table 2, liver mRNA contains an abundant group of nine different mRNAs each present 12,000 times per cell, a middle abundance group consisting of about 700 different mRNAs each present several hundred times per cell, and a low abundance, high complexity group consisting of 12,000 mRNAs each present in fifteen copies per cell.

Figures 1B and 1C show the results of a similar analysis for mouse kidney and brain mRNA. In both cases, there also seems to be a grouping of mRNA into three abundance classes. Both kidney and brain have a very obvious abundant class representing approximately 10% of the mRNA.

The numbers of different mRNAs in each class for the brain and kidney are listed in Table 2.

The mRNA populations from the three tissues are similar in several respects. The total number of sequences is almost identical; there appear to be three abundance classes in each tissue, and sequences from corresponding frequency classes are present in similar numbers per cell in the different tissues.

There are, however, differences between tissues in the complexity and magnitude of the most abundant class of mRNA.

Comparison of the mRNA Populations of Different Tissues

To make a preliminary comparison of the mRNA populations of two tissues, the mRNA of one tissue is hybridized to the cDNA of a second tissue.

Figure 2 shows the hybridization of brain cDNA and kidney cDNA to liver mRNA. Approximately 70% and 85% (relative to the homologous hybridization) of the brain and kidney cDNA, respectively, cross-hybridized to the liver mRNA. Close analysis reveals that there is no hybridization of kidney or brain cDNA to the abundant mRNA component of the

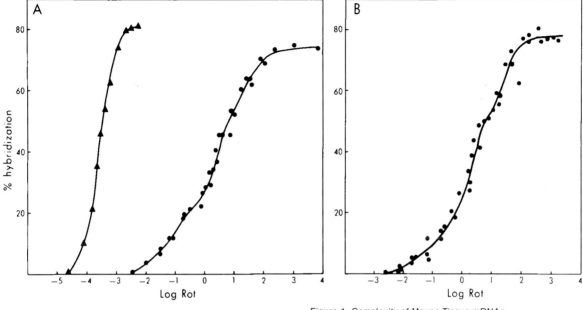

Figure 1. Complexity of Mouse Tissue mRNAs

(A) Hybridization of rabbit globin cDNA to globin mRNA (▲) (the final mRNA concentration was 0.2 μg/ml) and mouse liver cDNA to liver mRNA (●). The data derive from several different reactions in which the mRNA concentrations were 3.6, 41, 60, 408, 863, and 3300 μg/ml.

(B) Hybridization of mouse kidney cDNA to kidney mRNA. mRNA concentrations were 3.3, 33, 63, 332, 720, and 1560 μg/ml.

(C) Hybridization of mouse brain cDNA to brain mRNA. mRNA concentrations were 6, 71, 180, 228, 405, 720, and 1520 μg/ml.

In all cases, the symbols represent experimental data, and the continuous lines are computer fits to the data. All subsequent curves are expressed in a similar way. The curves are corrected for zero time subtractions, which usually varied between 2–5%.

liver. In each case, there is hybridization to the middle and low abundance mRNA of the liver. In fact, each cDNA preparation hybridizes with two transitions. The Rot $_{1/2}$ of the transitions are identical to those calculated for the middle and low frequency mRNA of the liver.

This analysis tells us that the abundant mRNA of the liver is different from the abundant mRNA of the kidney and the brain, and this tends to confirm the notion that the abundant mRNA is a discrete class. In addition, some of the sequences in kidney and brain mRNA are completely absent from the

liver. To determine whether there are differences in all three abundance classes, we have analyzed the situation further by fractionating cDNA into separate abundance groups and examining the relative representation of these in the different tissues.

Fractionation of cDNA into Frequency Classes
Isolation of Abundant cDNA
As described below, we have fractionated kidney cDNA into three frequency classes. To test the fractionation system, abundant liver cDNA was isolated initially. This represents 15% of the total cDNA, so we thought it would be a trivial matter to separate this from the rest of the cDNA by hybridizing to a Rot value of 2×10^{-1} followed by fractionation on a hydroxyapatite (HAP) column. However, this proved to be a difficult task, as described in Experimental Procedures. The abundant cDNA isolated by two cycles of hybridization and HAP fractionation was hybridized to liver mRNA, as shown in Figure 3A. The early part of the curve using unfractionated cDNA (taken from Figure 1A) is included for com-

Table 2. mRNA Complexity Calculated from Hybridization Data

Source of mRNA	Class	% mRNA	Observed Rot$_{1/2}$	Rot$_{1/2}$ if Pure	Number of Different[a] mRNAs	Copies per Cell
Liver	1	22	0.039	0.0086	9	12,444
	2	41	1.64	0.67	754	280
	3	37	30.8	11.4	12,155	15
					Total 12,918	
Kidney	1	10	0.03	0.003	4	12,000
	2	45	1.15	0.52	550	290
	3	45	22	10	10,990	15
					Total 11,544	
Brain	1	7	0.038	0.0036	4	12,444
	2	44	1.00	0.44	480	257
	3	49	21.45	10.5	11,167	27
					Total 11,651	

[a]Slight corrections were made to allow for the difference in cDNA size between the experimental case and the standard of 400 (see text), assuming that rate is proportional to the square root of the length of the cDNA as suggested by Monahan et al. (1976). Since these investigators examined the kinetics with only two different sizes of cDNA, it is still possible that the rate is directly proportional to the length of the cDNA. In practice, since our cDNA preparations are between 300–400 nucleotides long compared with the globin standard of 400 nucleotides, very little adjustment would be necessary in the event that rate is shown to be proportional to the length. We have performed calibration experiments using globin mRNA, EMC virus RNA, and poliovirus RNA with their respective cDNAs. The viral RNAs were about 13 times longer than the globin mRNA, but the cDNAs were approximately the same size. The rate constants of these reactions were directly proportional to the complexities of the RNAs (Bishop et al., 1975a), suggesting that kinetics do not vary with the length of the driver RNA.

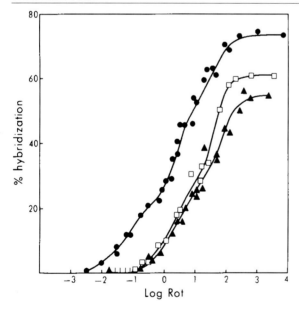

Figure 2. Comparison of Brain and Kidney mRNA with Liver mRNA

The following cDNA preparations were hybridized to liver mRNA: liver cDNA (●); kidney cDNA (□); and brain cDNA (▲).

parison. About 80% of this abundant cDNA fraction had hybridized by a Rot value of 2×10^{-1}. There is very little contamination by less abundant cDNA. This represents a 5–6 fold purification of the abundant cDNA. The kinetics of the reaction are close

to ideal first-order for a single component, suggesting that this is a relatively homogeneous mRNA population. In addition, the Rot$_{1/2}$ is identical to that calculated for the abundant component of the liver from Figure 1A.

Abundant Liver mRNAs Are Present in Other Tissues at Lower Frequencies

We concluded earlier that the abundant mRNA of each tissue is different. It is possible that the abundant sequences of one tissue are present in a second tissue at much lower concentration. This can be tested by hybridizing the purified, abundant cDNA of the liver to kidney and brain mRNA. As shown in Figure 3B, a large proportion of abundant liver cDNA cross-hybridizes to kidney mRNA, but 1000 times slower than to liver mRNA. In fact, the majority of the cDNA cross-hybridizes with a Rot$_{1/2}$ (= 27), close to that expected for the low frequency mRNA of the kidney (= 22; see Table 1). A smaller proportion of the abundant cDNA hybridizes to brain mRNA, but again with kinetics close to those expected for the low frequency mRNA of the brain. No hybridization was observed between abundant liver cDNA and mRNA extracted from a mouse myeloma cell line (a gift from J. Morton). Thus a group of abundant sequences, present in the liver in similar concentration, are expressed to a very different degree in other tissues and a cell line.

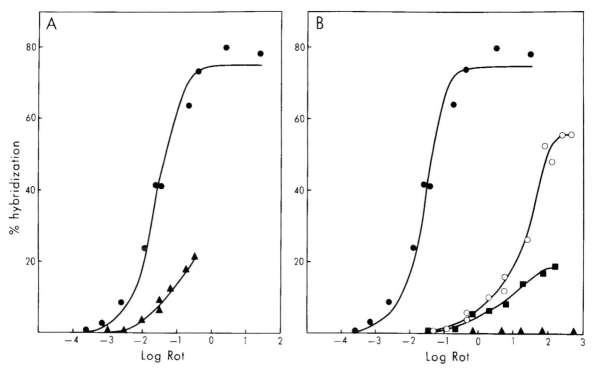

Figure 3. Isolation of the Abundant cDNA of Mouse Liver

(A) The abundant cDNA was purified as described in Experimental Procedures. Hybridization of the abundant liver cDNA to liver mRNA (●); hybridization of unfractionated liver cDNA to liver mRNA (▲) (reproduced from Figures 1A and 2).
(B) Hybridization of abundant liver cDNA to liver mRNA (●); kidney mRNA (○); brain mRNA (■); and to mouse myeloma cell MOPC41 mRNA (▲).

The Abundant Kidney cDNA

The abundant cDNA of the kidney represents 7% of the total cDNA. We attempted to purify this component by hybridization to a Rot value of 3×10^{-1}, followed by HAP fractionation as described above. Three such cycles produced a cDNA fraction that hybridizes back to kidney mRNA as shown in Figure 4A; 58–60% of the cDNA reacts as an obvious component with kinetics close to those expected for the abundant kidney cDNA. There is still, however, contamination by low frequency cDNA. In addition, this method gives low yields of material. As an alternative method for producing abundant cDNA, we hybridized to a Rot value of 3×10^{-1}; the sample was then treated with S1 nuclease after adjustment to the appropriate conditions. To remove RNA and S1 nuclease, the mixture was then boiled in alkali for 3 min, adjusted to neutral pH, and passed over an SP-50-chelex column. Two such cycles of hybridization and S1 treatment produced abundant cDNA which hybridized back to kidney mRNA, as shown in Figure 4A. Between 80% and 85% hybridized as an abundant component; there is still 10–15% contamination by lower frequency cDNA. This method is very much quicker than HAP purification and results in much higher yields of abundant cDNA. In

addition, there is no reduction in size of the cDNA relative to the starting material. The kinetics of the reaction using abundant cDNA prepared by either method suggest that there are six, not four, abundant kidney mRNAs (as calculated earlier from Figure 1B).

Figure 4B shows the hybridization of the abundant kidney cDNA (prepared by S1) to brain and liver mRNA. Between 60% and 65% hybridized to these mRNAs with identical kinetics. It is interesting that the Rot $_{1/2}$ of these reactions are very close to the Rot $_{1/2}$ (identical for the brain and well within a factor of 2 for the liver) of the middle frequency mRNAs of these tissues. The reaction in both cases is an ideal fit to a single first-order component.

Isolation of Middle Frequency cDNA of the Kidney

To isolate middle frequency cDNA, it is necessary to start with material devoid of the abundant cDNA. As shown above, it is possible to remove such material by hybridizing total cDNA to a Rot of 3×10^{-1} and fractionating on HAP. The single-stranded material eluting from HAP should contain a mixture of middle and low frequency cDNA. Figure 5A shows the hybridization of cDNA prepared this way. The

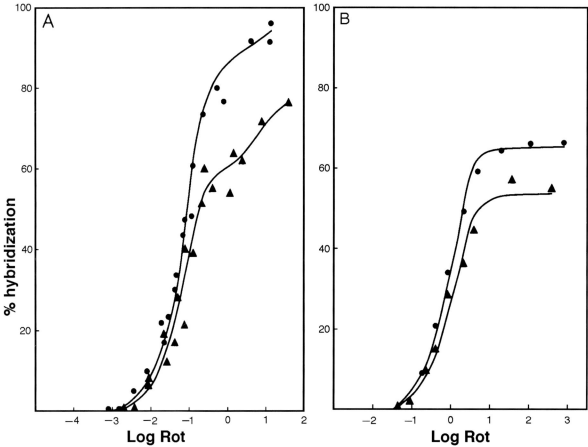

Figure 4. Isolation of Abundant Kidney cDNA

(A) Hybridization of HAP-purified abundant kidney cDNA to kidney mRNA (▲); hybridization of S1 nuclease-purified abundant kidney cDNA to kidney mRNA (●).

(B) Hybridization of abundant kidney cDNA (purified using S1 nuclease) to liver (●) and brain (▲) mRNAs..

abundant cDNA has indeed been removed, and the curve is organized into two obvious transitions representing the middle and low frequency cDNA. To separate the middle from the low frequency sequences, this cDNA fraction was hybridized to a Rot value of 6 (as shown by the upper arrow in Figure 5A). The hybrid fraction was recovered by two cycles through HAP and middle abundance cDNA isolated from mRNA as described above.

Figure 5B shows the hybridization of middle frequency kidney cDNA to kidney mRNA. The kinetics of this reaction are close to ideal first-order for a single component, suggesting that it is a relatively homogeneous mRNA population. The Rot $_{1/2}$ is that calculated earlier for the middle frequency mRNA of the kidney. This middle frequency cDNA was also hybridized to liver mRNA as shown in Figure 5B. Approximately 90% (relative to the homologous control) of the cDNA cross-hybridized to liver mRNA with the expected kinetics. Thus approximately 500 of 550 sequences are common between the tissues

and present in similar concentration in the two tissues.

Is the Middle Frequency mRNA Really a Homogeneous Class of mRNA?

To test whether sequences in the middle frequency class are really homogeneous in terms of concentration, the mixture of middle and low frequency kidney cDNA was hybridized to kidney mRNA to a Rot $_{1/2}$ of 0.3, as shown by the lower arrow in Figure 5A. This mixture was treated with S1 nuclease, alkali-treated, and passed over an SP-50-chelex column. 5% of the material came through in the excluded volume of the column. This cDNA (Rot 0.3 cDNA) was hybridized to kidney mRNA as shown in Figure 5B (left-hand curve). If this cDNA represents a class of mRNA intermediate in concentration between the abundant and middle frequency cDNA, it should hybridize back to kidney mRNA 5–10 times faster than the middle frequency cDNA. Actually, the cDNA hybridized back with a Rot $_{1/2}$ of 0.6. This is within a

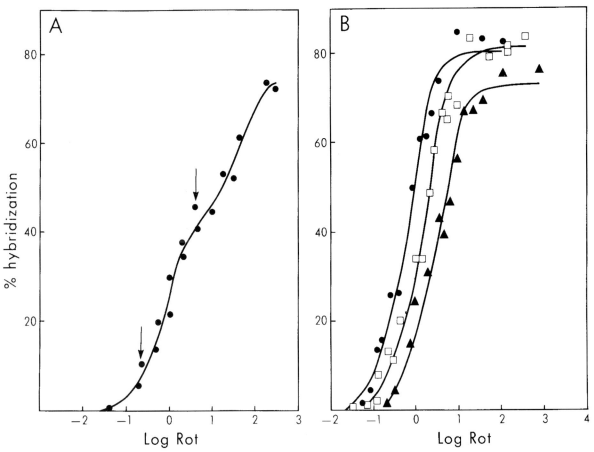

Figure 5. Characterization of Middle Frequency Kidney cDNA

(A) Hybridization of a mixture of middle and low frequency cDNA (the abundant cDNA has been removed on HAP as described in the text) to kidney mRNA. See (B) for explanation of arrows.

(B) An aliquot of the cDNA in (A) was hybridized to kidney mRNA to a Rot of 6 (shown by upper arrow in A) and subsequently fractionated on HAP to produce a middle frequency cDNA which was hybridized to kidney mRNA (□) and to liver mRNA (▲).

A second aliquot of the material shown in (A) was hybridized to a Rot of 0.3 (as shown by the lower arrow in A) and treated with S1 nuclease to produce a fraction which was hybridized to kidney mRNA (●). See text for explanation.

factor of 2 of the Rot ½ for the bulk of the middle abundance cDNA. This suggests that there is little heterogeneity of mRNA concentrations in the middle abundance class. Since the Rot 0.3 cDNA hybridizes with kinetics slightly broader than first-order, it is possible that 25–50% of this material reacts 5 times faster than the Rot 6 cDNA. This would mean, however, that the remaining 50–75% of the Rot 0.3 cDNA reacts with kinetics that are close or identical to those of the Rot 6 cDNA. This would still argue that no more than 2–3% of the middle abundance cDNA is 5 fold higher in concentration than the mean. It is important to add that this Rot 0.3 cDNA was the same size as the starting material.

The Low Frequency cDNA of the Kidney
This fraction is characterized as the single-stranded cDNA which is separated from the middle frequency cDNA-mRNA hybrid on HAP, as discussed above.

This low frequency cDNA hybridized to kidney mRNA as shown in Figure 6. Only 63% of the cDNA hybridized to the homologous mRNA. This is not surprising, since this fraction also contains any cDNA which is incapable of hybridizing to mRNA. Accordingly, we found that there was more small cDNA in this fraction.

The kinetics of this hybridization are slightly broader than ideal first-order; it is difficult to say whether this reflects a true heterogeneity in the population or whether there is contamination by middle frequency cDNA. Again, a large proportion of this 63% cross-hybridized to liver mRNA and slightly less to brain mRNA. Thus of approximately 11,000 sequences in the complex class, 9000–10,000 are common between tissues and in similar concentration in the different tissues.

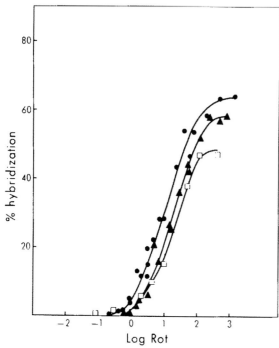

Figure 6. Characterization of Low Frequency Kidney cDNA

Low frequency kidney cDNA (prepared as described in the text) was hybridized to kidney mRNA (●); liver mRNA (▲); and brain mRNA (□).

Is There a Relationship between Abundance of mRNA and Repetition Frequency in the Genome?

To test whether abundance is related to the degree of repetition of the sequence in the genome, the cDNA of each frequency class was annealed to mouse liver DNA. Figure 7A shows the annealing of abundant liver and kidney cDNA to mouse liver DNA. The annealing of ^{3}H single-copy DNA to mouse DNA is included for comparison. The Cot $_{1/2}$ of this reaction is approximately 800. The annealing of abundant kidney cDNA to DNA is very broad. The simplest interpretation is that approximately 20% of the cDNA anneals with a Cot $_{1/2}$ of about 10, and that the remainder anneals with a Cot $_{1/2}$ of about 600. Thus the majority of the sequences are only represented once in the genome, whereas 20% of the sequences are represented approximately 80 times per genome. It seems that 10–15% of the abundant liver cDNA also anneals to sequences repeated 80 fold, but again, the majority of sequences are present once in the DNA. Figure 7B shows the annealing of middle frequency liver and kidney cDNA to DNA. Again, the simplest interpretation for each curve is that 10–15% of the sequences anneal to 80 fold repeated DNA, but the majority anneal to single-copy DNA. By contrast, low frequency kidney cDNA anneals to DNA with identical kinetics to single copy DNA (Figure 7B).

Thus there is no evidence of a repetitive component for this class of sequence. This supports the interpretation that abundant and middle frequency cDNA contain a repetitive component, and also strongly supports the notion that middle and low frequency cDNA are separate classes.

Is There Equal Reverse Transcription of Different mRNAs?

One objection often raised about complexity analysis using cDNA concerns the fact that different mRNAs may be copied by reverse transcriptase with different efficiencies. This becomes a problem only if different classes of mRNA are transcribed with different efficiencies. For example, if the abundant mRNAs as a whole are transcribed per molecule more efficiently than the other classes of mRNA, complexity determinations will be inaccurate.

To test, at a simple level, whether mRNAs are transcribed with different efficiencies, we have copied mixtures of mouse liver and rabbit globin mRNA. Two ratios of liver to globin mRNA were used for reverse transcription, so that globin represented 20% and 8%, respectively, of the total mRNA by weight. Since globin is on average one third the length of liver mRNA, 3 times more globin cDNA should be synthesized than liver cDNA per μg of each respective mRNA, provided the cDNA is the same length (we used conditions which synthesize cDNA 350–400 nucleotides long from each type of mRNA). Thus when globin mRNA represents 20% of the total, globin cDNA should comprise $\left(\frac{20 \times 3}{80 + 60} \right)$ = 42% of the total cDNA. Figure 8A shows the hybridization of cDNA synthesized from such a mixture back to the mixed template mRNAs. As expected, a new abundant component, corresponding to globin, has appeared. This represents 35% of the reactive cDNA. This abundant component hybridized, as predicted, with a Rot $_{1/2}$ of 2.5×10^{-3}, which is 5 times slower than the reaction using pure globin mRNA. 36% (compared to 42% expected) of the cDNA hybridized to globin mRNA, as shown in Figure 8B.

Similarly, a mixture of mRNAs in which globin comprises 8% should provide a cDNA copy which is 20% globin-specific. In this case, 16% was found to hybridize to globin mRNA as shown in Figure 8B. Accordingly, when the mixture of mRNAs was used in the hybridization (Figure 8A), the abundant component had increased from 15–30% (that is, 15% in the reaction using liver cDNA and liver mRNA). Thus in both cases, whether globin is 20% or 8% of the total mRNA, globin sequences are transcribed at 80–85% the efficiency with which liver mRNAs are transcribed.

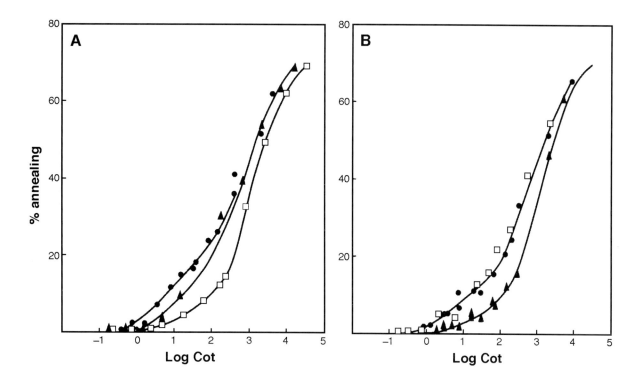

Figure 7. Reiteration Frequency of Various cDNA Classes in Mouse DNA

(A) Annealing of abundant liver cDNA (▲); abundant kidney cDNA (●); and ³H-labeled mouse single-copy DNA (□) to mouse DNA in vast DNA excess.

(B) Annealing of middle frequency liver cDNA (□); middle frequency kidney cDNA (●); and low frequency kidney cDNA (▲) to mouse liver DNA. The continuous line coincident with the low frequency cDNA annealing represents the annealing of single-copy DNA reproduced from (A).

Discussion

Complexity of the mRNA

We conclude that there are between 12,000 and 13,000 different mRNA species present in each of the three tissues studied. This estimate is not dependent upon the ability to identify and quantitate abundance groups, because the total complexity estimate changes very little when an increasing number of components is assigned (data not shown). From a similar series of experiments, Ryffel and McCarthy (1975a) concluded that the mouse liver contains approximately 8000 sequences, but that the mouse brain mRNA population is more complex. It is possible to obtain an independent evaluation of the complexity of mRNA by measuring the percentage of highly labeled single-copy DNA which will hybridize to mRNA at saturation (Galau et al., 1974). Axel, Feigelson, and Schutz (1976) calculated that there are 12,500 mRNA sequences in the avian liver. This complexity figure was obtained both from kinetic studies using cDNA and from experiments using single-copy DNA.

Since we are dealing with mixed populations of cells, it is possible that a rare cell type contains, for example, 50,000 sequences which comprise such a small percentage of the mRNA that the cDNA will be undetectable. Such a complex group

of mRNAs would be picked up more readily using the single-copy DNA method. Axel et al. (1976) observed no such class, even at very high Rot values. However, the brain, for example, is a much more heterogeneous tissue than the liver and could possibly contain such a complex group of sequences representing only a small percentage of the mRNA.

Bantle and Hahn (1976) have recently reported experiments which suggest that the complexity of mouse brain polyadenylated mRNA may be as high as 1.4×10^5 kb, or over 10^5 average sequences, in contrast to our estimate of 12,000 sequences. Bantle and Hahn's estimate is based on the hybridization of single-copy DNA with an excess of poly(A) RNA as described above. The cause of the discrepancy between the two sets of results is not immediately apparent. One partial explanation could be that the single-copy DNA is contaminated by repetitive sequences, some of which will hybridize to repetitive mRNA transcripts shown to comprise 10–20% of mammalian cell mRNA (Spradling et al., 1974; Klein et al., 1974; Campo and Bishop, 1974; Ryffel and McCarthy, 1975b). However, calculations suggest that this could account for only a small part of the difference. Another explanation is that these 100,000 sequences comprise only a small percentage of the mRNA mass, undetectable by hybridization with cDNA. Bantle and Hahn suggested that

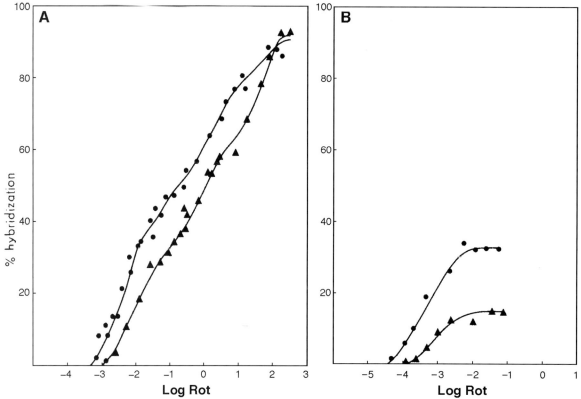

Figure 8. The Relative Efficiency of Copying by Reverse Transcriptase of Rabbit Globin mRNA and Mouse Liver mRNA

(A) A mixture of mouse liver and rabbit globin (at a ratio of 4/1) mRNA was used to synthesize a cDNA copy. This cDNA was hybridized back to the 4/1 mixture of mRNA (●). Similarly, a mixture of mouse liver and rabbit globin (at a ratio of 11.5/1) mRNA was used to synthesize a cDNA copy. This cDNA copy was hybridized back to the 11.5/1 mixture of mRNA (▲).

(B) The cDNA copied from the 4/1 mixture of mRNAs was hybridized to globin mRNA (●); in addition, the cDNA copied from the 11.5/1 mixture was hybridized to globin mRNA (▲).

the bulk of the 10^5 sequences represent 3–5% of the mRNA population. However, our own calculations (Bishop et al., 1975a) and those of other investigators predict that this complex group of mRNAs would occupy 10–16% of the mRNA. Even though there is scatter among the points between a Rot of 100 and 3000, there is no suggestion of a 10–15% increase (Figure 1C).

If our interpretation is correct, there is a total of between 15,000 and 16,000 mRNAs in the three tissues studied. These figures are perfectly comparable with genetically based estimates that between 30,000 and 50,000 different mRNA sequences are expressed during the life cycle of the mouse. Similar experiments with mRNA populations from various stages of the Drosophila life cycle (Bishop et al., 1975a, 1975b) and Drosophila tissue culture cells (Levy and McCarthy 1975; Bishop et al., 1975a, 1975b) support genetic estimates that about 5000 mRNAs are expressed in the Drosophila life cycle. Thus a single mouse tissue appears to contain several fold more sequences than necessary for the Drosophila life cycle.

It was shown several years ago that 4–5% of the single-copy DNA will hybridize to mouse liver and kidney nuclear RNA (Brown and Church, 1971; Hahn and Laird, 1971; Grouse, Chilton, and McCarthy, 1972). Thus by comparison with our data, the nuclear RNA is 5–6 fold more complex than mRNA in these tissues, a ratio similar to that found in the sea urchin gastrula (Hough et al., 1975) and the Friend cell (Getz et al., 1975). However, in view of the fact that experiments with single-copy DNA may provide an overestimate of complexity and experiments with cDNA an underestimate (as discussed above), the complexity ratio between nuclear RNA and mRNA in the mouse tissues may be much lower than 5–6.

Abundance Classes
Several experiments presented in this paper support the idea that there are three major discrete abundance groups in the mouse tissue mRNAs. First, the hybridization of unfractionated cDNA to homologous mRNA shows three transitions. Second, the most abundant class is different in each

tissue, so that the kidney and brain cDNA hybridize to liver mRNA over three, not five, orders of magnitude and with two, not three transitions. This supports the idea that the most abundant class, at least, is relatively discrete. Third, removal of the abundant cDNA leaves cDNA which hybridizes back to kidney mRNA with two obvious transitions (Figure 5A). Fourth, abundant and middle frequency kidney cDNA contains a component that anneals with repetitive cDNA, while the low frequency kidney cDNA does not (Figure 7). This supports the idea that the low frequency mRNA constitutes a distinct class. These observations point to the conclusion that three discrete abundance classes of mRNA exist in each of the tissues. In the experiments with isolated cDNA classes, the homogeneity of each class, as judged by the fit of the data to ideal first-order kinetics, is quite striking. However, the method is not sufficiently sensitive to establish just how homogeneous the classes are.

As shown in Figure 5, the most rapidly reacting 20% of the middle frequency class of kidney cDNA hybridizes twice as rapidly as the middle component cDNA as a whole. In this case, therefore, the mRNA class may be distributed with a variation of 2 or 3 fold on each side of the mean abundance. However, the fact that more or less discrete abundance classes can be detected shows that within each class, the concentration of most sequences is close to the class average.

The reactions between isolated cDNA classes and heterologous mRNA populations provide indications that the existence of abundance classes is significant in terms of differential gene expression. About 75% of the cDNA specific to the most abundant liver mRNA reacts with kidney mRNA, all of it with the least abundant sequences. Again, about 70% and 50%, respectively, of the equivalent fraction of kidney cDNA reacts with liver and brain mRNA, all of it in each case to the middle abundant sequences. This shift in expression of a group of sequences squarely from class to class can scarcely be a chance phenomenon.

In this study, we have measured the steady state levels of different mRNAs. These are determined by the rates of transcription, processing, and degradation, any or all of which may vary from one mRNA to another. At present, our knowledge is insufficient to assess which of these factors is the chief determinant of abundance.

Tissue Specificity

A large proportion of mRNAs are common to the three tissues. It appears that of the 12,000 or so sequences in the kidney, about 9,500 and 10,500 are shared with the brain and liver, respectively. This suggests that the majority of sequences are

concerned with "housekeeping" functions, those required for the growth and maintenance of all cells. This also seems to be the case in the chick, where a large part of the 12,500 mRNAs in the liver are shared with the oviduct (Axel et al., 1976). It is interesting to make a comparison with the situation in adult sea urchin tissues. Galau et al. (1976) have calculated that only about 1500 mRNAs are necessary for housekeeping activities in this organism. This is almost an order of magnitude less than the number of such functions in the mammal or bird, and raises the question whether all the sequences present in low frequency are translated to produce functional proteins. Galau et al. (1976) have calculated that 1–10 mRNAs per cell would be sufficient to maintain the known levels of several specific liver enzymes, each of which has an important physiological function. We have calculated that the lowest frequency mRNAs in the mouse liver are present about 15 times per cell and are therefore in sufficient quantities to specify such low frequency enzymes. The most striking difference between tissues is the abundant sequences. One would predict that the abundant liver mRNAs would not be abundant also in the kidney. The abundant liver mRNAs probably code for albumin, ferritin, transferrin, and macroglobulin, among others. It is known that most of these are not made in high concentration in the kidney. It was shown recently, however, that albumin is synthesized in kidneys at very low levels (Lin and Chang, 1975). As shown in Figure 3, some of the abundant liver mRNAs are present in the kidney at very much lower levels.

As pointed out above, several abundant liver mRNAs are simultaneously expressed in the kidney mRNA at lower frequency but within a narrow concentration range (Figure 3B). This observation suggests coordinate regulation. Thus we can also conclude that tissue-specific differences in mRNA populations may be related to abundance of individual mRNAs as well as qualitative differences.

Repetition Frequency of the mRNA Sequences in the DNA

The major part of the kidney and liver abundant cDNAs anneal to sequences present about once per genome. Thus it appears that the abundance is not due to transcription from repetitive sequences. However, 10–20% of this class is transcribed from sequences represented approximately 100 times per haploid genome. We have not yet distinguished whether this is a discrete class of mRNA completely complementary to repetitive DNA like that observed by Campo and Bishop (1974), Klein et al. (1974), and Ryffel and McCarthy (1975b), or whether a 15% portion of each cDNA is repetitive.

Experimental Procedures

Extraction of Tissue mRNA

Batches of livers from ten mice (all Balb/c mice approximately 3 months old) or kidneys and brains from thirty mice were homogenized in 3 vol of TKM buffer [50 mM Tris–HCl (pH 7.5), 24 mM KCl, 5 mM $MgCl_2$] containing 0.25 M sucrose. The following were added to inhibit RNAase: commercial RNAase inhibitor (5 units per ml) purchased from Searle; polyvinyl sulphate (25 μg/ml), and spermine (25 μg/ml) which stabilized nuclei. Several strokes of a loose-fitting, motor-driven teflon-to-glass homogenizer were used to disrupt the cells. The homogenate was centrifuged at 10,000 × **g** for 10 min, and the resulting supernatant was extracted for RNA by a modification of the method of Parish and Kirby (1968) as described by Campo and Bishop (1974). This procedure would yield about 25 mg of RNA from ten livers, 10 mg of RNA from thirty pairs of kidneys, and 3–4 mg of RNA from thirty brains. To extract mRNA, the RNA was passed over an oligo(dT)-cellulose column as described by Aviv and Leder (1972).

Extraction of Rabbit Globin mRNA

Polysomes were prepared from anemic rabbits as described by Temple and Housman (1972). mRNA was extracted from polysomes using proteinase K treatment, followed by oligo(dT)-cellulose chromatography as described by Macnaughton, Freeman, and Bishop (1974). The globin mRNA was further purified on a 15–30% sucrose gradient as described by Macnaughton et al. (1974).

Synthesis of cDNA

cDNA was synthesized using purified avian myeloblastosis virus reverse transcriptase as described by Bishop and Rosbash (1973). The cDNA was purified as described by Bishop and Rosbash (1973).

Sizing of cDNA

All cDNA preparations were sized on alkaline 5–20% sucrose gradients using conditions described by Waqar and Huberman (1975). Centrifugation for 17 hr at 40,000 rpm in the Beckman SW50 rotor gives very good separation between molecules 300 and 350 nucleotides long. Fragments of SV40 DNA obtained by digestion with restriction enzymes Hind II and Hind III were used as markers. Most cDNA preparations were within the size range 300–400 nucleotides.

Preparation of Mouse Liver DNA

Mouse liver DNA was extracted from mouse liver nuclei as described for rat liver DNA by Campo and Bishop (1974). The DNA was sheared and purified as described by Campo and Bishop (1974).

Preparation of ³H Single-Copy DNA

Mouse L cells were labeled with 10 μCi/ml ^{3}H-thymidine (spec. act. 50 Ci/mmole) for approximately two generations. The DNA was prepared from nuclei as described by Campo and Bishop (1974). Single-copy DNA was prepared by two cycles of annealing to a Cot of 20, followed by HAP fractionation as described by Bishop et al. (1974).

Hybridization and Annealing

Hybridization and annealing were carried out at 70°C in 0.24 M PEB as described by Bishop et al. (1974). Reactions were usually in a total volume of 40–100 μl. The amount of cDNA in hybrid or duplex was assayed using S1 nuclease as described by Bishop et al. (1974). For hybridization, mRNA concentrations were estimated by hybridizing ^{3}H-poly(U) to the mRNA as described by Bishop et al. (1974). This gives an estimate of the amount of poly(A), which, in turn, provides a value for mRNA concentration.

Isolation of Abundant cDNA Using Hydroxyapatite

In control experiments using defined, single-stranded cDNA and globin mRNA/cDNA hybrids, we found that most single-stranded cDNA elutes (from this particular batch of HAP) at 0.08 M phosphate butter (PB), and that all has eluted by 0.1 M PB. Most of the hybrid elutes at 0.14 and 0.16 M PB, although 5–10% has already eluted by 0.12 M PB. A large batch of liver cDNA was hybridized to liver mRNA to a Rot of 2×10^{-1}; the phosphate concentration was adjusted to 0.04 M, and the mixture was loaded onto a 200 mg HAP column. The single-stranded material was eluted with 5 ml washes of 0.12 M PB. Although 45–50% of the cDNA eluted in the first three washes, the counts eluted by the twenty-fifth wash were still well above background. In fact, after wash 20, each traction still contained 0.2–0.5% of the total counts. We decided to increase the PB to 0.4 M after the twenty-fifth wash to elute the hybrid. The hybrid molecules eluted very rapidly within 3×1 ml fractions. This material, however, was only 45% resistant to S1 nuclease, suggesting a high degree of contamination by single-stranded cDNA—not surprising, in view of the elution profile at 0.12 M PB.

The cDNA was isolated from the hybrid by the following series of steps: the mixture was boiled for 5 min in 0.5 N NaOH, neutralized using glacial acetic acid, then passed over a Sephadex SP-50-chelex 100 column, as described by Campo and Bishop (1974), and finally precipitated in the presence of 50 μg yeast carrier RNA. The cDNA was then rehybridized to liver mRNA to a Rot $_{1/2}$ of 2×10^{-1}, and the whole procedure was repeated. The hybrid obtained by this second cycle was 75% resistant to S1 nuclease.

Isolation of Abundant Kidney cDNA Using S1 Nuclease

A large batch of kidney cDNA was hybridized to kidney mRNA to a Rot value of 3×10^{-1}. The reaction mix was then treated with S1 nuclease (purchased from Sigma) in the presence of denatured yeast DNA at 50 μg/ml for 30 min at 37°C. The mixture was then boiled In 0.5 N NaOH for 5 min, neutralized with glacial acetic acid, and passed over an SP-50-chelex 100 column (Campo and Bishop, 1974).10% of the cDNA eluted in this void volume. This cDNA was rehybridized to kidney mRNA to a Rot of 3×10^{-1}, and the whole procedure was repeated. 50–60% of the material eluted in the void volume of the column in the second passage. This cDNA was precipitated in the presence of 50 μg yeast carrier DNA.

Acknowledgments

We thank Lorna Kerr and Carolyn Hejna for excellent technical assistance; Drs. William Hill and Lorne Houten for help with the computer program; Dr. Jacques Beckmann for sizing the poly(A) and for helpful discussions; Dr. Mary Gutai for SV40 restriction fragments; and Drs. Robert Taber and William Held for critical evaluation of the manuscript. We are also very grateful to Drs. J. Beard and M. Chirigos for gifts of purified avian myeloblastosis virus reverse transcriptase. This work was supported by grants from the S.R.C. and C.R.C. and, In part, by a USPHS Center Grant in Viral Chemotherapy from the National Cancer Institute.

Received June 17, 1976; revised August 30, 1976

Theory

In the hybridization experiments described here, the RNA is in excess of the cDNA by several orders of magnitude. Consequently, the reactions, if ideal, will be pseudo-first-order. In fact, the reactions do seem to be ideal, or nearly so, probably because the driver (mRNA) molecules are longer than the tracer (cDNA) molecules, and complement them throughout their entire length. Consequently, every time a nucleation event occurs, an entire cDNA molecule becomes duplexed.

For the simple case in which only one base sequence is represented in the cDNA, and the RNA is also homogeneous:

$$dD/dt = -kRoD$$

where D represents the transient concentration of single-stranded cDNA, Ro the initial (and constant) RNA concentration, and k the rate constant of the reaction. Integrating with the condition $D = Do$ at $t = 0$,

$$\ln (Do/D) = kRot$$

and $k = \ln 2/Rot_{\frac{1}{2}} = 0.691/Rot_{\frac{1}{2}}$ where $Rot_{\frac{1}{2}}$ is the product of RNA concentration and time at which $Do/D = 2$. In addition,

$$d/Do = 1 - (D/D_o) = 1 - e^{-kRot} = 1 - e^{-0.691 \, Rot/Rot_{\frac{1}{2}}}$$

where $d = (Do - D)$, that is, the transient concentration of cDNA in duplex form.

In practice, the reactions appear to go to completion without complete exhaustion of the cDNA. This is at least partly due to some of the cDNA being unreactive, and in that case, it is reasonable to ignore it. Letting P represent the proportion of the cDNA which does react,

$$d/Do = P - e^{-0.691 \, Rot/Rot_{\frac{1}{2}}}$$

In the more complex situation in which more than one base sequence is represented in the cDNA and the RNA, d/Do is the sum of a number of independent reactions, $d_1/Do + d_2/Do + d_3/Do...$, each with its own terminal value P_1 P_2, P_3, and so on. We make the assumption that the relative representation of different sequences is the same in the cDNA and the mRNA populations, and slightly simplify the situation by writing the RNA concentrations P_1Ro, P_2Ro, P_3Ro, and so on. Then

$$d/Do = \Sigma dn/Do = \Sigma Pn - e^{-0.691 \, P_n Rot/Rot_{\frac{1}{2}n}}$$

The data points are estimates of d/Do, and we assume that measurement of Rot is precise. Each component of the reaction mixture carries its own values of P and $Rot_{\frac{1}{2}}$. Curve fitting is carried out by means of a computer program which finds the values of P and $Rot_{\frac{1}{2}}$, which minimize the sum of squares of deviations of the data points from the calculated curve. The program that we use has to be reinitiated for each postulated number of components. It makes no allowance for the fact that d/Do may be positive in the absence of any reaction, and so a zero time control value must be accurately measured and deducted from each datum. To reduce running time, the computer is provided with rough estimates of the correct values from which to start its search. Varying these values does not affect the eventual result. In practice, we obtain these estimates by dividing the range of the data equally. Thus to analyze data ranging up to $d/Do = 0.75$ by postulating three components, we would enter $P_1 = P_2 = P_3 = 0.25$ and let $Rot_{\frac{1}{2}1}$, $Rot_{\frac{1}{2}2}$, $Rot_{\frac{1}{2}3}$, equal the approximate values at which $d/Do = 0.125$, 0.375, and 0.625, respectively.

Choice between different postulated numbers of components is made by comparing the sums of squares of deviations.

In all three cases where unfractionated tissue cDNA was hybridized to the homologous tissue mRNA (Figure 1), there is a dramatic improvement in ssd upon increasing the assigned components from 1 to 2, and a slight but consistent improvement from 2 to 3. However, there is no improvement in fit with higher numbers of components (Figure 9). This says that there are at least three different abundance classes in each tissue.

References

Aviv, H., and Leder, P. (1972). Purification of biologically active globin messenger RNA by chromatography on oligothymidylic acid-cellulose. Proc. Nat. Acad. Sci. USA *69*, 1408–1412.

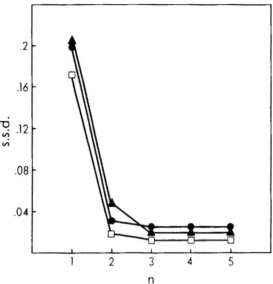

Figure 9. Change in ssd as Number of Postulated Components (n) is Increased

See "Theory" for explanation. (▲) represents hybridization of liver cDNA to liver mRNA; (□) represents hybridization of kidney cDNA to kidney mRNA; (●) represents hybridization of brain cDNA to brain mRNA.

The data used for these analyses are the same as shown in Figures 1A, 1B, and 1C, for the liver, kidney, and brain, respectively.

Axel, R., Feigelson, P., and Schutz, G. (1976). Analysis of the complexity and diversity of mRNA from chicken liver and oviduct. Cell *7*, 247–254.

Bantle, J. A., and Hahn, W. E. (1976). Complexity and characterization of polyadenylated RNA in the mouse brain. Cell *8*, 139–150.

Bishop, J. O. (1974). The gene numbers game. Cell *2*, 81–85.

Bishop, J. O., and Rosbash, M. (1973). Reiteration frequency of duck hemoglobin genes. Nature *241*, 204–207.

Bishop, J. O., Morton, J. G., Rosbash, M., and Richardson, M. (1974). Three abundance classes in HeLa cell messenger RNA. Nature *250*, 199–204.

Bishop, J. O., Beckmann, J. S., Campo, M. S., Hastie, N. D., Izquierdo, M., and Perlman, S. (1975a). DNA–RNA hybridization. Phil. Trans. Roy. Soc. B. *272*, 147–157.

Bishop, J. O., Campo, M. S., Izquierdo, M., Hastie, N. D., Rosbash, M., and Morton, J. G. (1975b). The organization and expression of the eukaryotic genome. FEBS Symposia, *26* (Budapest: Akademia Kiado), pp. 393–402.

Brown, I. R., and Church, R. S. (1971). RNA transcription from nonrepetitive DNA in the mouse. Biochem. Biophys. Res. Commun. *42*, 850–856.

Campo, M. S., and Bishop, J. O. (1974). Two classes of messenger RNA in cultured rat cells: repetitive sequence transcripts and unique sequence transcripts. J. Mol. Biol. *90*, 649–663.

Galau, G. A., Britten, R. J., and Davidson, E. H. (1974). A measurement of the sequence complexity of polysomal messenger RNA in sea urchin embryos. Cell *2*, 9–20.

Galau, G. A., Klein, W. H., Davis, M. M., Wold, B. J., Britten, R. J., and Davidson, E. H. (1976). Structural gene sets active in embryos and adult tissues of the sea urchin. Cell *7*, 487–505.

Getz, M. J., Birnie, G. D., Young, B. D., MacPhail, E., and Paul, J. (1975). A kinetic estimation of base sequence complexity of nuclear poly(A)-containing RNA in mouse friend cells. Cell *4*, 121–129.

Grouse, L., Chilton, M. D., and McCarthy, B. J. (1972). Hybridization of ribonucleic acid with unique sequences of mouse deoxyribonucleic acid. Biochemistry *11*, 798–805.

Hahn, W. E., and Laird, C. D. (1971). Transcription of nonrepeated DNA in mouse brain. Science *173*, 158–161.

Hough, B. R., Smith, M. J., Britten, R. J., and Davidson, E. H. (1975). Sequence complexity of heterogeneous nuclear RNA in sea urchin embryos. Cell *5*, 291–299.

Klein, W. H., Murphy, W., Attardi, G., Britten, R. J., and Davidson, E. H. (1974). Distribution of repetitive and nonrepetitive sequence transcripts in HeLa mRNA. Proc. Nat. Acad. Sci. USA *71*, 1785–1789.

Levy, B. W., and McCarthy, B. J. (1975). Messenger RNA complexity in Drosophila melanogaster. Biochemistry *14*, 2440–2447.

Lewin, B., (1974). Gene Expression–2, Eukaryotic Chromosomes (London and New York: John Wiley and Sons).

Lin, C., and Chang, J., (1975). Electron microscopy of albumin synthesis. Science *190*, 465–467.

Macnaughton, M., Freeman, K. B., and Bishop, J. O. (1974). A precursor to hemoglobin mRNA in nuclei of immature duck red blood cells. Cell *1*, 117–125.

Monahan, J. J., Harris, S. E., Woo, S. L. C., Robbison, D. L., and O'Malley, B. W. (1976). The synthesis and properties of the complete complementary DNA transcript of ovalburnin mRNA. Biochemistry *15*, 223–234.

Parish, J. H., and Kirby, K. S. (1968). Reagents which reduce interactions between ribosomal RNA and rapidly labeled RNA from rat liver. Blochem. Biophys. Acta *129*, 554–562.

Rosbash, M., and Ford, P. J. (1974). Polyadenylic acid-containing RNA in Xenopus laevis oocytes. J. Mol. Biol. *85*, 87–101.

Ryffel, G. U., and McCarthy, B. J. (1975a). Complexity of cytoplasmic RNA in different mouse tissues measured by hybridization of polyadenylated RNA to complementary DNA. Biochemistry *14*, 1379–1385.

Ryffel, G. U., and McCarthy, B. J. (1975b). Polyadenylated RNA complementary to repetitive DNA in mouse L-cells. Biochemistry *14*, 1385–1389.

Spradling, A., Penman, S., Campo, M. S., and Bishop, J. O. (1974). Repetitious and unique sequences in the heterogeneous nuclear and cytoplasmic messenger RNA of mammalian and insect cells. Cell *3*, 23–30.

Temple, G. F., and Housman, D. E. (1972). Separation and translation of the mRNAs coding for α and β chains of rabbit globin. Proc. Nat. Acad. Scl. USA *69*, 1574–1577.

Waqar, M. A., and Huberman, J. A. (1975). Covalent attachment of RNA to nascent DNA in mammalian cells. Cell *6*, 551–557.

Note Added In Proof

Very recently, Young, Birnie, and Paul (Biochemistry *15*, 2823, 1976) published an analysis of polysomal poly(A)$^+$ RNA from mouse, liver, brain, and total embryo, finding total complexities similar to those reported here.

Jeffreys & Flavell (1977) Cell 12, 1097–1108

The Rabbit β–Globin Gene Contains a Large Insert in the Coding Sequence

A. J. Jeffreys* and R. A. Flavell
Section for Medical Enzymology
and Molecular Biology
Laboratory of Biochemistry
University of Amsterdam
Eerste Constantijn Huygensstraat 20
Amsterdam, The Netherlands

Summary

We have used the rabbit β–globin DNA plasmid PβG1 (Maniatis et al., 1976) labeled with ^{32}P as a filter hybridization probe for DNA fragments containing the β–globin gene in restriction endonuclease digests of rabbit liver DNA. The β–globin DNA fragments we detect appear to contain the gene, present in PβG1 DNA, which codes for adult rabbit β–globin. These fragments have been ordered into a physical map of cleavage sites within and neighboring the structural gene in the rabbit genome (Jeffreys and Flavell, 1977). A detailed analysis of β–globin DNA fragments produced by cleavage with restriction endonucleases which are known to cut the β–globin gene has now shown that the β–globin structural gene is not contiguous in rabbit liver DNA, but is interrupted by a 600 base pair DNA segment inserted somewhere within the coding sequence for amino acid residues 101–120 of the 146 residue β–globin chain. Otherwise, the map of cleavage sites within the gene is co-linear with that deduced from the sequence of rabbit β–globin messenger RNA. Preliminary analysis indicates that this insert is also present in the β–globin gene in rabbit brain, kidney, spleen, bone marrow and sperm, and in erythroid cells isolated from the marrow of an anemic rabbit. The insert appears, therefore, to be a general property of the rabbit β–globin gene, even in tissues in which this gene is active, which suggests that the insert is not involved in inactivating the gene in non-erythroid tissues.

Introduction

We have previously shown that it is possible to construct a physical map of restriction endonuclease cleavage sites in the DNA regions flanking a single-copy mammalian structural gene, in our case the rabbit β–globin structural gene (Jeffreys and Flavell, 1977). In brief, our approach is to cleave rabbit liver DNA with a restriction endonuclease and to separate the resulting fragments by electrophoresis in an agarose gel. We then transfer denatured DNA fragments by blotting onto a nitro-

cellulose filter using the method of Southern (1975). The filter is hybridized with denatured β–globin complementary DNA (cDNA) plasmid PβG1 (Maniatis et al., 1976) which has previously been labeled with ^{32}P in vitro by nick translation. After washing the filter, we can detect β–globin DNA fragments as labeled bands by autoradiography. With this method we have been able to map cleavage sites for endonucleases Bgl II, Eco RI, Kpn I and Pst I, and to determine the orientation of the β–globin gene within the physical map.

When we attempted to map the cleavage sites of endonucleases Bam HI and Hae III, which are known to cut the β–globin structural gene, anomalous β–globin DNA fragments were found. We show how these fragments unambiguously lead to a physical map of the rabbit liver β–globin gene in which the 438 base pair (bp) β–globin structural gene is not intact, but is interrupted by a 600 bp DNA segment inserted toward the 3′ end of the β–globin coding sequence.

Results

Construction of a Cleavage Map for Endonucleases Bgl II, Eco RI, Kpn I and Pst I around the Rabbit β–Globin Gene

Rabbit liver DNA fragments containing the β–globin gene present in PβG1 DNA can be ordered into a physical map of Bgl II, Eco RI, Kpn I and Pst I cleavage sites around the rabbit β–globin gene. The mapping of Eco RI, Kpn I and Pst I sites and the determination of the direction of transcription within the physical map are fully described elsewhere (Jeffreys and Flavell, 1977). The β–globin DNA fragments detected and the physical map deduced are shown in Figure 3.

We have previously stated that endonuclease Kpn I does not cut either the 2.6 or 0.8 kb β–globin DNA fragment produced by cleavage of rabbit DNA with endonuclease Eco RI (Jeffreys and Flavell, 1977). More careful analysis has shown, however, that the 2.6 kb Eco RI fragment is reproducibly trimmed by endonuclease Kpn I to produce a new 2.5 kb double digest component (Figure 1). This places one Kpn I site 2.5 kb on the 5′ side of the intragenic Eco RI site. The 6.3 kb Pst I β–globin DNA fragment is cleaved by Kpn I to generate a 3.6 kb double digest fragment (Jeffreys and Flavell, 1977) which places the second Kpn I site on the 3′ side of the gene, near the center of the Pst I fragment. Together these Kpn I sites predict a Kpn I β–globin DNA fragment of 4.8 kb; allowing for experimental error in fragment size determination (see Jeffreys and Flavell, 1977), this is in reasonable agreement with the 5.1 kb fragment found.

We have also localized endonuclease Bgl II

* Present address: Department of Genetics, University of Leicester, University Road, Leicester LE1 7RH, England.

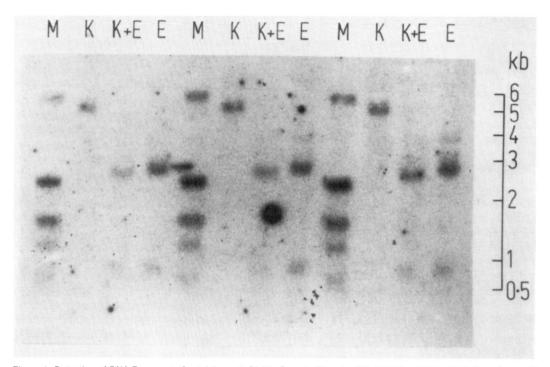

Figure 1. Detection of DNA Fragments Containing a β–Globin Gene in Digests of Rabbit Liver DNA with Endonucleases Kpn I and/or Eco RI

30 μg of rabbit liver DNA cleaved with Kpn I (K) and/or Eco RI (E) were denatured with alkali and electrophoresed in a 1.2% agarose slab gel. Denatured hybridization markers (M) also run were 10 pg PβG1 DNA × Hind III (5.6 kb fragment) plus 25 pg PβG1 DNA × Eco RI × Ava I × Hind III (2.3, 1.5, 1.2, 0.65 kg b fragments; see Jeffreys and Flavell, 1977). Denatured DNA was subsequently transferred by blotting to a nitrocellulose filter. This filter was hybridized with ³²P-labeled PβG1 DNA at 65°C in 3 × SSC [saline sodium citrate, 1 × SSC = 150 mM NaCl, 15 mM sodium citrate (pH 7.0)]. After hybridization, unbound labeled probe was washed from the strips at 65°C in 3 × SSC followed by 0.1 × SSC (first four lanes), 0.3 × SSC (second four lanes) or 1 × SSC (last four lanes). Remaining labeled components on the filter were detected by autoradiography. Full details of hybridization and washing procedures are given elsewhere (see Experimental Procedures and Jeffreys and Flavell, 1977).

cleavage sites within this physical map. Endonuclease Bgl II cleaves the β–globin cDNA insert in PβG1 DNA at a site 80 bp 3′ of the intragenic Eco RI site and just outside the β–globin coding sequence (Efstratiadis, Kafatos and Maniatis, 1977). In rabbit liver DNA cleaved with endonuclease Bgl II, only one prominent β–globin DNA fragment 1.6 kb long was found; the same fragment was found in three different strains of rabbit. This Bgl II fragment was reproducibly trimmed by endonuclease Eco RI to produce a new 1.5 kb double digest fragment (Figure 2). Furthermore, the small (0.8 kb) Eco RI fragment was barely visible after digestion with Bgl II, suggesting the presence of a Bgl II site within this small fragment. Together these observations place one Bgl II site about 0.1 kb 3′ of the intragenic Eco RI site, as found in PβG1 DNA, and the 1.6 kb Bgl II fragment must therefore contain the 5′ region of the β–globin gene (see Figure 3). The other Bgl II β–globin fragment containing the remaining 90 bp of message coding sequence 3′ to the Bgl II site has not yet been identified.

Anomalous Position of an Endonuclease Bam HI Cleavage Site within the Physical Map around the Rabbit β–Globin Gene

The β–globin cDNA insert in PβG1 DNA contains a single endonuclease Bam HI cleavage site 67 bp on the 5′ side of the intragenic Eco RI cleavage site. Digestion of rabbit liver DNA with endonuclease Bam HI should therefore generate two β–globin DNA fragments. To date, only one β–globin fragment 9.9 kb long has been consistently detected in these digests, irrespective of the rabbit strain used (Figure 4). A more detailed analysis of endonuclease Bam HI digests (to be fully described elsewhere) has shown that this 9.9 kb Bam HI fragment contains sequences on the 5′ side of the intragenic Bam HI site; the second (3′) β–globin DNA fragment has not yet been detected in our system. The positions of these extragenic Bam HI sites, however, are not required for the following discussion.

The single 6.3 kb β–globin DNA fragment produced by endonuclease Pst I is cleaved by endonuclease Bam HI to give 5.5 and 0.5 kb double digest fragments, each of which must contain part

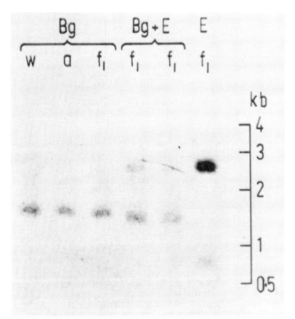

Figure 2. An Analysis of the β–Globin DNA Fragments Produced by Cleavage of Rabbit DNA with Endonuclease Bgl II

DNA prepared from the liver of a single Vienna White (w) or Alaska (a) rabbit or from the F₁ hybrid of these two inbred strains (f₁) was digested with Bgl II (Bg). Digestions with Eco RI (E) and Eco RI + Bgl II were also performed. 30 μg of each DNA digest were denatured with alkali, electrophoresed and transferred to a nitrocellulose filter, and the β–globin DNA fragments were detected by filter hybridization as described in Figure 1. In this experiment, the digestion of Eco RI-cleaved DNA with Bgl II was not complete, and some residual 2.6 kb Eco RI fragment is visible.

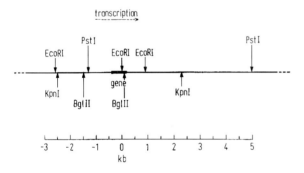

Figure 3. The Physical Map of Restriction Endonuclease Cleavage Sites Flanking the Rabbit β–Globin Gene, as Deduced for Endonucleases Eco RI, Kpn I, Pst I and Bgl II

The map was deduced from the following β–globin DNA fragments detected in rabbit liver DNA digested with various restriction endonucleases (fragment sizes in kb). Eco RI, 2.6, 0.8; Pst I, 6.3; Kpn I, 5.1; Bgl II,1.6; Pst I + Eco RI, 1.3, 0.8; Kpn I + Eco RI, 2.5, 0.8; Kpn I + Pst I, 3.6; Eco RI + Bgl II, 1.5. The fragment sizes are taken from Jeffreys and Flavell (1977) and from Figures 1 and 2. The size of the smaller Eco RI β–globin DNA fragment has been reestimated as 0.8 kb, and not 0.9 kb as previously reported (Jeffreys and Flavell, 1977), by calibration with better DNA markers in this molecular weight range.

The calibration scale (in kb) is centered on the Eco RI cleavage site within the globin gene. It should be stressed that the only extragenic cleavage sites which can be detected for a given enzyme by this analysis are those closest to the gene.

of the β–globin gene (Figure 4). Since the combined molecular weight of these two fragments closely approximates that of the intact Pst 1 fragment, this suggests that the Pst 1 β–globin DNA fragment contains a single Bam HI cleavage site situated within the gene, as predicted from PβG1 DNA. If the Bam HI site, however, is 67 bp on the 5′ side of the intragenic Eco RI site, as in PβG1 DNA, then the Pst β–globin fragment should be cleaved by Bam HI to yield 5.1 ElNd 1.2 kb double digest components. In other words, this Bam HI cleavage site, while still within the β–globin gene, maps 0.7 kb too far to the 5′ side of the intragenic Eco RI Site. The simplest explanation is that between the intragenic Bam HI and Eco RI Sites is an extra 0.7 kb stretch of DNA not present in the β–globin cDNA insert of PβG1 DNA.

We confirmed this strange position of the Bam HI site by mapping it relative to Kpn I, Eco RI and Bgl II cleavage sites. The single 5.1 kb Kpn I β–

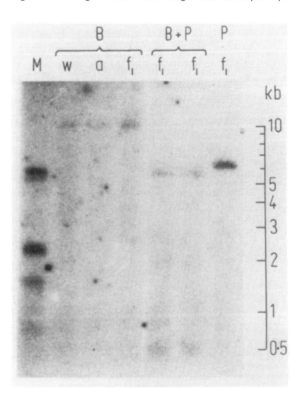

Figure 4. β–Globin DNA Fragments in Digests of Rabbit DNA with Endonucleases Bam HI and/or Pst I

DNA from individual livers of the animals described in Figure 2 was digested with Bam HI (B) and/or Pst I (P). All further electrophoretic and hybridization procedures to detect β–globin DNA fragments were performed as before. Hybridization markers (M) used are those described in Figure 1. Additional faintly labeled components seen in the rabbit DNA digests probably originate from at least two extra β–related globin genes which form relatively poorly matched hybrids with PβG1 DNA; these extra components are described elsewhere (Jeffreys and Flavell, 1977).

globin DNA fragment is cleaved by Bam HI into 3.0 and 1.9 kb fragments (Figure 5), and not the 2.7 and 2.4 kb fragments predicted for a Bam HI site very close to the intragenic Eco RI site. This again places the intragenic Bam HI site some 0.5 kb too far to the 5′ side of the Eco RI site. The alternative location of the Bam HI site predicted from the Kpn I + Bam HI double digest would be close to the 3′ extragenic Eco RI site; this position is inconsistent with all other double digests and with the single Bam HI site on the 5′ side of the Eco RI site in PβG1 DNA. The 2.6 kb Eco RI fragment is shortened by 0.6 kb on cleavage with Bam HI, again consistent with the unusual position of the Bam HI site. Similarly, the 1.6 kb Bgl II fragment is replaced by a new 0.7 kb double digest component after digestion with Bam HI (Figure 6). Thus all four restriction endonucleases predict a Bam HI site separated from the intragenic Eco RI site by an extra 0.5–0.7 kb of DNA inserted into the β–globin coding sequence (Table 1).

In principle, it should be possible to detect the novel (0.7 kb) band containing the insert in an Eco RI + Bam HI double digest; since, however, it contains only 67 bp of globin coding sequence, it might be very difficult to detect by hybridization with labeled PβG1 DNA. The new 0.7 kb insert band in Bgl II + Bam HI double digests should be easier to detect since it contains 147 bp (67 + 80) of globin DNA; unfortunately, this band would co-migrate with the other 0.7 kb Bgl II + Bam HI double digest product. Nevertheless, the absence of a strongly hybridizing "insert band" suggests

that the insert is not comprised of globin coding sequences and is not simply the result of a tandem globin gene duplication. This point is discussed in more detail below.

Confirmation of the β–Globin Gene Insert; Mapping of Endonuclease Hae III Cleavage Sites within the Gene

Endonuclease Hae III introduces two cuts into the β–globin cDNA insert in PβG1 DNA and liberates a 333 bp fragment of β–globin chain coding se-

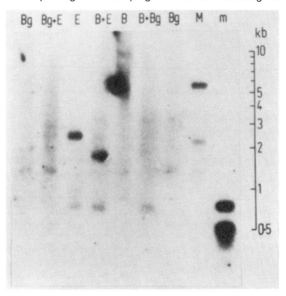

Figure 6. β–Globin DNA Fragments Produced by Cleavage of Rabbit Liver DNA with Various Combinations of Endonucleases Bam HI (B), Bgl II (Bg) and Eco RI (E)

F₁ hybrid rabbit liver DNA was used as in Figure 5. (M) markers described in Figure 1; (m) 200 pg PβG1 DNA × Hae III (see Figure 12). Denatured samples were run in a 1.6% agarose gel.

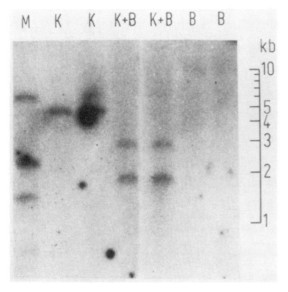

Figure 5. β–Globin DNA Fragments in Rabbit Liver DNA Digested with Endonucleases Kpn I (K) and/or Bam HI (B)

DNA from the liver of an Alaska/Vienna White F₁ hybrid rabbit was used.

Table 1. The Sizes of β–Globin DNA Fragments Produced by Cleavage of Rabbit Liver DNA with Endonuclease Bam HI

Endonucleases	Fragments Predicted for:		Fragments Found
	No Insert	0.6 kb Insert	
Bam HI + Bgl II	1.5	0.9	0.7
Bam HI + Eco RI	2.5; 0.8	1.9; 0.8	1.9; 0.8
Bam HI + Kpn I	2.7; 2.4	3.3; 1.8	3.0; 1.9
Bam HI + Pst I	5.1; 1.2	5.7; 0.6	5.5; 0.5

The sizes (in kb) of β–globin DNA fragments produced by cleaving rabbit liver DNA with Bam HI plus a second endonuclease are taken from Figures 4–6. They are compared with those expected from the physical map in Figure 3, assuming that the β–globin gene contains a single Bam HI site 67 bp 5′ of the intragenic Eco RI site as found in the β–globin cDNA insert in PβG1 DNA (Efstratiadis et al. 1977). They are also compared with fragments expected if the β–globin gene is interrupted by a 0.6 kb DNA segment inserted somewhere between these intragenic Bam HI and Eco RI cleavage sites (see map in Figure 9).

quence containing both the Eco RI and Bam HI cleavage sites (Efstratiadis et al. 1977; see Figure 12). If the rabbit liver β–globin gene does not contain an insert, then digestion of rabbit liver DNA with endonuclease Hae III should produce a 333 bp internal β–globin DNA fragment plus two fragments including each end of the gene. If the 0.6 kb insert is present, the 333 bp fragment should be replaced by a 0.9 kb component, provided that Hae III does not cleave the insert. Analysis of rabbit DNA cleaved with Hae III using ^{32}P-labeled PβG1 DNA as probe, however, could be complicated by the presence of additional 5' and 3' β–globin DNA fragments. We therefore prepared the 333 bp intragenic fragment from a digest of PβG1 DNA × Hae III (see Experimental Procedures and Figure 12), labeled it with ^{32}P by nick translation and used it as a filter hybridization probe for DNA fragments containing sequences coding for amino acid residues 28–137 of the β–globin chain. This probe was capable of detecting both the 2.6 and 0.8 kb β–globin DNA fragments produced by cleavage with Eco RI (Figure 7). In digests of rabbit liver DNA with Hae III, a single prominent 0.8 kb labeled component is seen, with no detectable labeled 333 bp fragment. This fragment fits well with the 0.9 kb component predicted for a β–globin gene containing a 0.6 kb insert. Furthermore, the insert appears either not to be cut by Hae III or cut very asymmetrically toward the 3' side.

We confirmed the map position of this Hae III β–globin fragment by comparing β–globin DNA fragments detected, using labeled PβG1 DNA, in Hae III digests and Hae III + Bgl II or Hae III + Eco RI double digests (Figure 8). Again, a single prominent 0.87 kb β–globin component could be detected in Hae III digests of rabbit liver DNA, this fragment presumably corresponding to the component seen in Figure 7. Other β–globin DNA fragments from each end of the gene were not, in fact, detected, presumably because they contain little β–globin DNA. The 0.87 kb β–globin DNA fragment was reproducibly trimmed by Eco RI to an extent consistent with the 49 bp separation of the Hae III and Eco RI sites in PβG1 DNA. Since Bgl II appears not to cut the Hae III β–globin fragment, yet has a cleavage site on the 3' side of the Eco RI site in the β–globin gene, this places one Hae III site close to the 3' side of the Eco RI site and the second site 0.87 kb 5' of the first site, in the same position relative to the Bam HI site as in PβG1 DNA. This Hae III mapping confirms the presence of the insert in the β–globin gene.

The complete map of restriction endonuclease sites in and around the rabbit liver β–globin gene is shown in Figure 9, together with the position and size of the insert deduced from the preceding analysis.

The β–Globin Gene Insert in Other Rabbit Tissues

The insert within the rabbit liver β–globin gene might be involved in inactivating the gene in this non-erythroid organ and therefore be absent in cells engaged in making hemoglobin. Alternatively, the insert might be an integral part of the globin gene and present in all tissues, irrespective of the activity of the globin gene. We therefore examined other rabbit tissues for the presence of the insert.

Figure 10 shows that DNA isolated from the liver, kidney, spleen, brain or total bone marrow of an adult rabbit in each case contained the normal pattern of 2.6 and 0.8 kb Eco RI β–globin DNA fragments. Since the 0.6 kb insert is carried on the 2.6 kb fragment, this suggests that all these tissues contain the same size insert. In particular, no new 2.0 kb Eco RI fragment containing an intact β–globin gene could be detected in DNA from normal

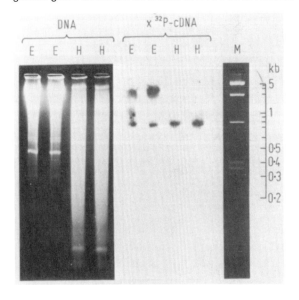

Figure 7. Identification of the Intragenic β–Globin DNA Fragment Produced by Cleaving Rabbit Liver DNA with Endonuclease Hae III

30 μg of duplex F_1 hybrid liver DNA cleaved with Eco RI (E) or Hae III (H) was electrophoresed in a 2.0% agarose slab gel in the presence of ethidium bromide (see Jeffreys and Flavell, 1977). Marker DNA (M, 1 μg phage PM2 DNA × Hind III) was also run. After electrophoresis, the gel was photographed under ultraviolet light, and the ethidium fluorescence patterns are shown above. DNA was then alkali-denatured in situ by the method of Southern (1975) and transferred by blotting onto a nitrocellulose filter. The filter was hybridized under standard conditions with 5 ng/ml heat-denatured, purified 333 bp PβG1 DNA × Hae III globin cDNA fragment which had been labeled with ^{32}P in vitro by nick translation to a specific activity of 6×10^7 cpm/μg DNA. The identification and purification of this cDNA fragment from PβG1 DNA × Hae III are shown in Figure 12. The subsequent washing of the filter and detection of labeled components were performed using standard techniques (Jeffreys and Flavell, 1977). The molecular weight calibration scale was derived from the phage PM2 DNA × Hind III markers, using molecular weights taken from Brack et al. (1976).

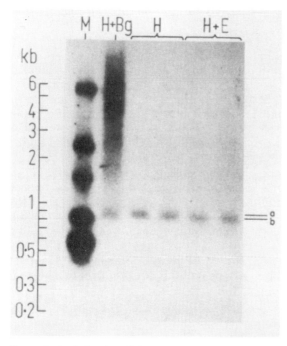

Figure 8. Analysis of β–Globin DNA Fragments Produced by Cleavage of Rabbit Liver DNA with Endonuclease Hae III

F_1 hybrid liver DNA was cleaved with the indicated combinations of Hae III (H), Eco RI (E) and Bgl II (Bg). 30 µg of each digest were alkali-denatured, electrophoresed on a 2.0% agarose gel and transferred to a nitrocellulose filter, and β–globin DNA fragments were detected by hybridization with ^{32}P-labeled PβG1 DNA. Hybridization markers (M) used were 10 pg PβG1 DNA × Hind III plus 25 pg PβG1 DNA × Eco RI × Ava I × Hind III plus 200 pg PβG1 DNA × Hae III. (a) Position of the 0.87 kb Hae III β–globin DNA fragment; (b) position expected for the Hae III + Eco RI double digest component if the Eco RI site lies 49 bp within the Hae III fragment (Efstratiadis et al. 1977). The cause of the high background labeling seen in the Bgl II + Hae III channel is not known.

bone marrow, in which at least 20% of the nucleated cells were erythroid. Since we could detect as little as 0.03 of a gene copy in a reconstitution experiment in which a 1.5 kb PβG1 DNA fragment was added to liver DNA digested with ECO RI (Figure 10), this suggests that the active β–globin gene in bone marrow still contains the insert.

This experiment was repeated with DNA from spleen and bone marrow taken from a Vienna White rabbit which had been made anemic by repeated injections of phenylhydrazine. Both tissues are predominantly erythroid under these conditions (Denton and Arnstein, 1973; Nokin et al., 1975); in this experiment, at least 75% of nucleated bone marrow cells were judged to be erythroid from cytological examination. Again, DNA from those tissues, as well as from brain and blood from the same animal, contained the usual 2.6 and 0.8 kb ECO RI β–globin DNA fragments and 0.8 kb Hae III fragment, with no trace of the new frag-

ments expected for a gene which has lost its insert (Figure 11). This again strongly suggests that the insert is present in tissues expressing the globin gene. DNA isolated from rabbit sperm also showed the usual pattern of Eco RI and Hae III β–globin DNA fragments. From these experiments, we conclude that the β–globin gene contains the insert in all rabbit tissues examined, including the germ line, in a number of tissues not expressing the globin genes and in erythroid tissues. This argues against a role of the insert in the inactivation of globin genes in nonerythroid cells.

Discussion

We have previously shown that DNA fragments containing the β–globin gene homologous to that inserted into PβG1 DNA can readily and specifically be detected in restriction endonuclease digests of rabbit liver DNA, using ^{32}P-labeled PβG1 DNA as a probe in filter hybridizations (Jeffreys and Flavell, 1977). These β–globin DNA fragments can be ordered into a physical map of cleavage sites around what is almost certainly the single-copy structural gene for adult rabbit β–globin. In this paper, we show that this analysis can be extended to map the relative positions of cleavage sites within the structural gene, using endonucleases Eco RI, Bgl II, Bam HI and Hae III which cleave the β–globin gene. Two new enzymes tested – Bam HI and Bgl II – produced β–globin fragments which did not vary between various rabbit strains, reinforcing our earlier conclusion that sequences neighboring (and within) the β–globin gene in the rabbit genome are conserved (Jeffreys and Flavell, 1977). There are limitations to this fine structure mapping. Some enzymes produce β–globin DNA fragments which contain very little message coding sequence and would therefore be difficult to detect by hybridization. Other combinations of restriction endonucleases release small β–globin DNA fragments which may be too small to bind efficiently to nitrocellulose filters (Southern, 1975). Despite these uncertainties, the β–globin DNA fragments which we can detect can be ordered into an unambiguous physical map of intra- and extragenic restriction endonuclease cleavage sites.

This fine structural mapping of the rabbit β–globin gene reveals two properties of the gene. First, the sequence of cleavage sites within the gene is identical to that in the β–globin cDNA insert in PβG1 DNA, with the proviso that the relative order of the Bgl II and Hae III sites on the 3' side of the intragenic Eco RI site in rabbit β–globin DNA cannot as yet be determined. The second and more remarkable property of the gene is a 600 bp insert interrupting the β–globin chain

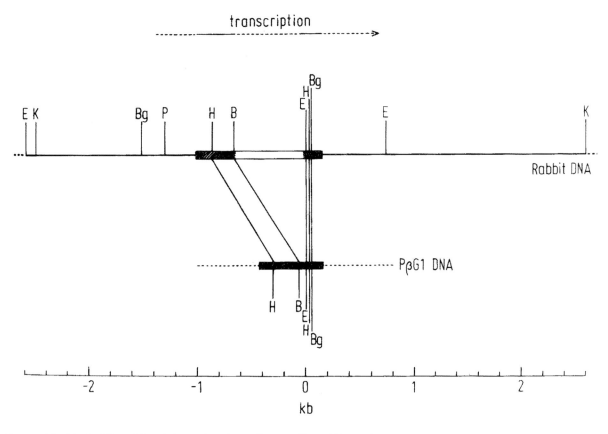

Figure 9. The Physical Map of Restriction Endonuclease Cleavage Sites within and Neighboring the Rabbit β–Globin Gene, Showing the 600 bp Insert within the Structural Gene

The physical map of the β–globin gene in the rabbit genome is compared with the corresponding map of cleavage sites in the β–globin cDNA insert in PβG1 DNA taken from Efstratiadis et al. (1977). Cleavage sites are shown for Bam HI (B), Bgl II (Bgl), Eco RI (E), Hae III (H), Kpn I (K) and Pst I (P). The map rightward of the 3′ Kpn I site is omitted (see Figure 3). The relative order of the intragenic Bgl II and Hae III sites on the 3′ side of the Eco RI site in the rabbit β–globin gene is not known; their order is assumed to be the same as in PβG1 DNA. The probable positions of β–globin chain coding sequences (hatched boxes), transcribed but nontranslated sequences present in mature message (filled boxes) and the insert (open box) are shown, plus flanking sequences in the rabbit genome (solid line) and the PMB9 vector of PβG1 DNA (broken line); restriction endonuclease cleavage sites in PMB9 DNA are not shown. The calibration scale (in kb) is centered on the intragenic Eco RI cleavage site. It must be stressed that this map of the rabbit β–globin gene is the simplest map consistent with our data. The existence of further rearrangements or even absence of certain gene sequences, particularly in regions 5′ of the Bam HI site and 3′ of the intragenic Bgl II site, cannot be excluded.

coding sequence. The presence of this insert was deduced by two independent approaches. First, the intragenic Bam HI site was consistently found to map at a site 700 bp removed from the internal Eco RI site and not at the expected distance of 67 bp; this anomalous position was independently determined relative to known Bgl II, Eco RI, Kpn I and Pst I cleavage sites. The second line of evidence came from using endonuclease Hae III, which cleaves the β–globin cDNA insert in PβG1 DNA to release a 333 bp fragment which contains the Bam HI and Eco RI cleavage sites. This enzyme did not generate this fragment from rabbit DNA, but instead the (0.8 kb) fragment expected if the insert is present between the Bam HI and Eco RI sites. To confirm the map finally, this novel Hae III fragment was independently mapped within the β–globin gene relative to intragenic Bgl II and Eco RI

cleavage sites.

After we deduced the existence of the insert from these data, we learned that the mouse β–related globin genes which have been recently cloned from the mouse plasmacytoma genome by P. Leder and his colleagues (personal communication) have been shown by them to contain a 600 bp DNA segment inserted just 3′ of an intragenic Bam HI site. Our results provide further evidence that this mouse insert is a real component of the mouse genome and is not an artifact, for example, resulting from DNA sequences being scrambled during cloning. Furthermore, the similar position and size of this insert in the globin genes of two species which have been separated in evolution by more than 5×10^7 years (Romer, 1966) suggest that this insert has some general and conserved role in the function of globin genes.

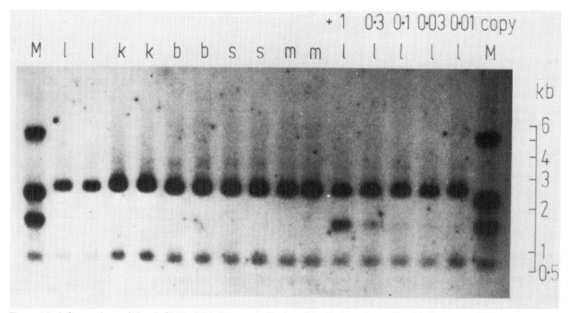

Figure 10. A Comparison of the β–Globin DNA Fragments Produced by Endonuclease Eco RI Cleavage of DNA Isolated from Various Rabbit Tissues

DNA purified from the brain (b), liver (l), kidney (k), spleen (s) or bone marrow (m) of a healthy adult Alaska rabbit was digested with endonuclease Eco RI. 30 μg of each digest were alkali-denatured and electrophoresed, and β–globin DNA fragments were detected as usual. To test the sensitivity of the system, liver DNA digests were mixed prior to electrophoresis with 1, 0.3, 0.1, 0.03 or 0.01 copy of a 1.5 kb PMB9 DNA fragment generated by cleavage of PβG1 DNA with Eco RI + Ava I (see Jeffreys and Flavell, 1977); 1 copy = 15 pg of fragment per 30 μg of rabbit DNA digest. Hybridization markers (M) used were those described in Figure 1.

We know very little about the nature of the β–globin gene insert. It cannot simply be the result of a tandem β–globin gene duplication for three reasons. First, the fragment containing the insert generated by cleavage of rabbit DNA with Bam HI plus Eco RI cannot be detected by hybridization and therefore presumably contains little β–globin DNA sequence. Second, if there were a gene duplication, this insert would be expected to contain one Bam HI site, one Bgl II site, one Eco RI site and two Hae III sites; none of these is detectable. Third, a gene duplication would result in the insert containing message coding sequences from the 3′ side of the Eco RI site in PβG1 DNA; filter hybridizations with a labeled probe for these 3′ sequences, however, have shown that they are absent in the larger (2.6 kb) Eco RI β–globin DNA fragment which contains the insert (see Jeffreys and Flavell, 1977). Similar arguments suggest that the insert is not a partial globin gene duplication. We consider it most probable that this insert does not contain sequences which code for β–globin. Furthermore, the co-linearity of cleavage sites in the β–globin gene and in PβG1 DNA suggests that the additional DNA is a direct insert and is not the result of an inversion of part of the globin gene together with a stretch of neighboring extragenic DNA. Of course, there may be further abnormalities in the rabbit β–globin gene as yet undetected, such as inversions, inserts and even absence of

some sequences present in globin message; this can only be resolved by further mapping of the globin or by direct DNA sequencing of a cloned rabbit β–globin gene.

The presence of this insert in the β–globin gene of the adult rabbit liver, a nonerythroid organ, immediately suggests that this insert might in some way be involved in switching off the expression of this gene and might therefore be absent in cells making hemoglobin. Our data suggest, however, that the same insert is present in the β–globin gene of all rabbit tissues, irrespective of the state of expression of the β–globin gene, and also in germ cell DNA. Thus only the 2.6 and 0.8 kb β–globin fragments were detected in Eco RI digests of DNA from rabbit liver, kidney, brain, spleen and bone marrow, with no trace of the new 2.0 kb component predicted if the β–globin genie were to lose the insert. Similarly, rabbit sperm DNA and DNA prepared from anemic rabbit bone marrow, where at least 75% of the cells were judged to be erythroid, also contained only split β–globin genes, with no detectable 333 bp Hae III fragment or 2.0 kb Eco RI fragment predicted for a β–globin gene lacking the insert. It seems probable, therefore, that all β–globin genes, even those being transcribed, contain the insert as an integral part of the gene. Alternative, less probable explanations include excision of β–globin genes on gene activation, resulting in a small episomal element which

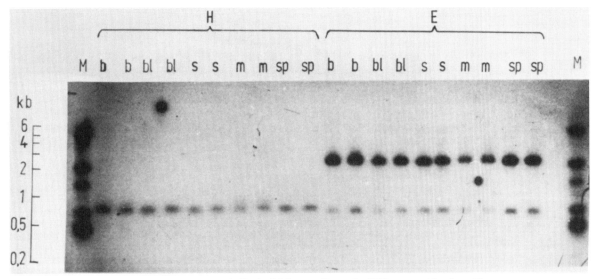

Figure 11. Analysis of the β–Globin DNA Fragments in Erythroid Tissues, Nonerythroid Tissues and the Germ Line

DNA purified from bone marrow (m), spleen (s), brain (b) and blood (bl) from an anemic Vienna White rabbit and DNA purified from sperm (sp) from a Flemish Giant rabbit were digested with endonucleases Eco RI or Hae III. 30 μg of each digest were alkali-denatured and electrophoresed, and β–globin DNA fragments were detected as usual.

is lost during DNA purification, or the presence of very few cells in the erythroid series containing an active β–globin gene (<2% of the erythroid population).

The existence of a split genes [sic] is not entirely without precedent. For example, an insert of various sizes has been found in some 28S ribosomal cistrons of Drosophila (Glover and Hogness, 1977; Wellauer and Dawid, 1977; White and Hogness, 1977; Pellegrini, Manning and Davidson, 1977). Intact 28S ribosomal cistrons, however, are also present in Drosophila, and the split gene might simply be a defective form of this ribosomal cistron. In Xenopus, 5S cistrons are found alternating with DNA stretches containing four fifths of the coding sequence for complete 5S RNA (N. Fedoroff and D. D. Brown, personal communication). A cloned embryonic mouse immunoglobin V gene has been shown to contain a 93 bp insert in the 5′ region of the protein coding sequence (S. Tonegawa personal communication). The existence of the insert reported in the present paper was deduced from unfractionated rabbit DNA, without the possibility of insertions/rearrangements occurring during cloning or the accidental selection of a defective gene present in eucaryotic DNA. More important, it is the first example of a major insertion into the coding sequence of a single-copy structural gene which is known to be translated as an insert-free message.

Whatever the function of this β–globin gene insert, the insert is not shared by large numbers of other genes. We have shown elsewhere that it is possible to locate repetitive DNA sequences in the physical map around the β–globin gene (Flavell, Jeffreys and Grosveld, 1978). We find that the 2.6 kb Eco RI β–globin DNA fragment which carries the insert contains sequences which are repeated at most 15 times elsewhere in the rabbit genome (unpublished results). There cannot, therefore, be more than 15 copies of this insert in the entire rabbit genome.

The problem now is how mature β–globin messenger RNA (mRNA) is produced from an active β–globin gene which appears to contain this 600 bp insert. The first trivial explanation is that this particular β–globin gene is never active, but represents a split and inactive globin gene derived by duplication and divergence from a functional β–globin gene. This explanation is highly improbable, since this rabbit β–globin gene contains the best matched sequences in rabbit DNA which can hybridize with PβG1 DNA; since this β–globin DNA plasmid contains a perfect copy of mature β–globin mRNA, which must have been transcribed from an active globin gene (Efstratiadis et al., 1977), we conclude that this message is probably derived from the split gene detected in our analysis.

Several processes are possible which could be used to generate intact mRNA from a split gene. For example, during transcription, the RNA polymerase plus nascent 5′ mRNA segment could jump across the insert to the beginning of the 3′ coding segment. Alternatively each half-gene could be transcribed separately to produce half-messages which are ligated after transcription. The third possibility is that the entire gene plus insert could be transcribed to produce a precursor mRNA; post-transcriptional processing would then include excising the insert sequence and splicing to produce mRNA. A similar splicing mechanism has been suggested for the adenovirus system, where transcripts from several parts of the adenovirus genome are found covalently linked together to form a single mRNA molecule (Berget, Moore and Sharp, 1977; Chow et al., 1977). It is interesting to note that the 15S (1.5–1.8 kb) mouse β–globin mRNA precursor described by Curtis and Weissmann (1976) and Bastos and Aviv (1977) is only slightly longer than a hypothetical precursor mRNA containing the globin coding sequences plus the insert and a 3′ poly(A) tail (1.2–1.3 kb). As yet, we have no idea of the role of the insert in globin gene expression, or of the reason for placing this insert within the β–globin chain coding sequence. Perhaps the inactive precursor mRNA produced could afford a second site for temporal regulation of the gene's activity during erythroid differentiation by controlling the RNA processing mechanism required to produce mature mRNA.

An RNA splicing mechanism would need to be exceedingly precise, since a splicing error of a single nucleotide would result in a frame shift within the mRNA sequence leading to the production of a defective β–globin chain. Other mutational errors in the splicing mechanism could result in the ligation of two half-message sequences which normally would never be linked; the result would be a fused polypeptide equivalent to those which until now have been interpreted as resulting from unequal crossover between linked loci. For example, the human hemoglobin variant Hb Lepore contains a non-α–globin chain whose N terminal amino acid sequence is δ-type and C terminal is β–type; the origin of this globin has been generally assumed to be the result of unequal crossover of δ– and β–globin genes leading to a fused $\delta\beta$–globin gene (see Lehmann and Huntsman, 1972). The alternative formal possibility now exists that some Hb Lepore individuals might instead possess an altered splicing mechanism required to join globin mRNA segments which erroneously splices a 5′ δ–globin mRNA sequence onto a 3′ β–globin mRNA segment to produce on translation the fused $\delta\beta$–globin chain.

If the existence of split genes and RNA splicing mechanisms in eucaryotes is a general phenomenon, then there are certain implications for recombinant DNA research. For example, attempts to obtain expression of a gene in a cloned chromosomal eucaryotic DNA fragment in a procaryotic host might well prove impossible if the gene is split, since it is improbable that a procaryote would possess the correct RNA splicing mechanisms or, indeed, any splicing mechanism at all. Conversely, this situation would reduce any chance of expression of potentially harmful eucaryotic genes in procaryotic hosts during the cloning of recombinant DNAs.

Elucidation of the role of the insert in the expression of the β–globin gene must await further characterization of the insert and its fate during transcription and post-transcriptional processing. The question remains whether similar inserts are present in all globin genes, and indeed whether split structural genes are a general phenomenon in eucaryotic genomes.

Experimental Procedures

Preparation of DNA

The inbred rabbit strains used and the preparation of DNA from individual livers is described elsewhere (Jeffreys and Flavell, 1977). DNA was prepared from the brain, kidneys, spleen and total femoral bone marrow of a single healthy adult female Alaska rabbit by similar procedures, except that homogenization was performed with a motor-driven teflon-on-glass Potter homogenizer, the homogenate was lysed by adding an equal volume of 2% sodium dodecylsulphate, 8% tri-isopropylnaphthalenesulphonate (Eastman Kodak) dissolved in 12% butan-2-ol, and the phosphate/2-methoxyethanol extraction step to remove glycogen was omitted. DNA was similarly prepared from the liver and from pooled femoral plus humerus bone marrow of a 4 kg adult male Vienna White rabbit which had been made anemic by 6 daily subcutaneous injections of 1.2 ml 2.5% phenylhydrazine (British Drug Houses Ltd.) dissolved in 0.15 M NaCl (pH 7.2).

The cellular composition of rabbit femoral bone marrow was determined by staining dip smears with May-Grünwald Giemsa using a Sörensen buffer (pH 6.8). The cytological preparations were then mounted in Caedax and examined microscopically.

Freshly ejaculated semen obtained from an adult male Flemish Giant was provided by Mr. P. Kruit (Rijksinstitut voor Volksgezondheid, Bilthoven, The Netherlands): the semen was diluted 10 fold into 150 mM NaCl, 100 mM EDTA (pH 8.0), and the sperm was lysed by incubation for 30 min at 37°C in the presence of 2% sodium dodecylsulphate, 1 M 2–mercaptoethanol and 1 mg/ml proteinase K. DNA was prepared as before.

Restriction Endonuclease Digestions

Robbins (1976), was provided by Mr. C. Sol (Laboratorium voor Gezondheidsleer, University of Amsterdam). Endonuclease Bgl II was a gift from Dr. E. Humphries, and endonuclease Hae III was a gift from Dr. R. I. Kamen (both of the Imperial Cancer Research Fund, London). The sources of all other endonucleases used are given elsewhere (Jeffreys and Flavell, 1977). Digestions with the above enzymes were carried out at 37°C for 2 hr in the following incubation mixtures: Bam HI – 6 mM MgCl$_2$, 6 mM 2–mercaptoethanol, 6 mM Tris–HCl (pH 7.5); Bgl II – 7 mM MgCl$_2$, 1 mM dithiothreitol, 50 mM Tris–HCl (pH 7.8); Hae III – 10 mM MgCl$_2$, 7 mM 2–mercaptoethanol, 90 mM Tris–HCl (pH 7.9). Otherwise, digestions and recovery of DNA digests were performed as described by Jeffreys and Flavell (1977).

Detection of β–Globin DNA Fragments

Electrophoresis of alkali-denatured rabbit DNA digests, transfer to nitrocellulose filters, labeling of PβG1 DNA by nick translation, filter hybridizations and post-hybridization washing of filters were performed as described elsewhere (Jeffreys and Flavell, 1977). The success of this procedure has been dependent on several factors.

– Rabbit DNA digests are denatured with alkali prior to electrophoresis; this permits more DNA to be electrophoresed without overloading.

– Filter hybridizations are performed in flat-bottomed, tightly sealed boxes with sufficient hybridization solution to swamp the filters completely; this prevents any chance of filters drying out during the hybridization, which can lead to very high background labeling.

– Filters are hybridized in the presence of a high concentration of ficoll-polyvinylpyrrolidone-bovine serum albumin (Denhardt, 1966) plus carrier salmon sperm DNA to prevent the nonspecific binding of labeled PβG1 DNA.

– After hybridization, filters are washed under hybridization conditions to remove unbound labeled probe. The salt concentration in the washing solution is then decreased to melt out all mismatched hybrids between probe and rabbit DNA, leaving only well matched hybrids between PβG1 DNA and rabbit β–globin DNA fragments. Under less stringent washing conditions, additional labeled components are detected and appear to be relatively poorly matched hybrids between PβG1 DNA and other β–related globin genes in rabbit DNA (Jeffreys and Flavell, 1977).

We have recently introduced the following modifications to the procedure.

—In an attempt to reduce filter backgrounds, DNA was transferred from agarose gels by blotting onto a sandwich of two Sartorius nitrocellulose filters (0.4 μm pore size) instead of the usual single filter. After transfer, the filter in contact with the gel was

kept for hybridization, while the upper filter in contact with the absorbant pad was discarded.

– We have recently noticed that intense spots of background labeling sometimes occur in identical patterns on pairs of filters which lie adjacent during hybridization. It is probable that these spots are a result of air bubbles coming out of solution during hybridization and becoming trapped between adjacent filters; we now successfully avoid such spots by degassing the hybridization solution under vacuum prior to use.

– The volume of hybridization solution was increased from 12 ml to 20 ml per box to minimize the risk of filters drying out during the hybridization.

The Preparation of the 333 bp β–Globin cDNA Fragment from PβG1 DNA Cleaved with Endonuclease Hae III

PβG1 DNA is cleaved by endonuclease Hae III into a large number of fragments, including a 333 bp fragment containing only β–globin chain coding sequences (Efstratiadis et al., 1977). Since this fragment contains the single Eco RI site in PβG1 DNA, it can easily be identified by comparing Hae III and Hae III + Eco RI digests of PβG1 DNA (Figure 12). This fragment was purified by preparative gel electrophoresis as described by Jeffreys and Flavell (1977). The purified fragment was obtained at about 30% yield and was free from contamination with other fragments as judged by gel electrophoresis (Figure 12). The purified fragment was labeled with ^{32}P by nick translation to a specific activity of 6×10^7 cpm/μg DNA, using standard procedures described elsewhere [Jeffreys and Flavell (1977); method based on that of Rigby et al. (1977); Maniatis, Jeffrey and Kleid (1975)].

Acknowledgments

Our thanks to Ernie De Boer and Jan M. Kooter for excellent technical assistance, and to Professor J. James (Histology Department, University of Amsterdam) for valuable help and advice in experiments with bone marrow cells, and for performing the cytological examination of the bone marrow cells. We are indebted to Professor P. Borst for helpful discussions, and to the following individuals for gifts of restriction endonucleases: Mrs. F. Fase-Fowler (Kpn I, Pst I); Professor C. Weissmann, Zurich (Eco RI); Mr. C. Sol, Amsterdam (Bam HI); Dr. R. I. Kamen, London (Hae III); and Dr. E. Humphries, London (Bgl II). A.J.J. is a postdoctoral fellow of the European Molecular Biology Organization. This work was supported in part by a grant to P. Borst from The Netherlands Foundation for Chemical Research (SON), with financial aid from The Netherlands Organization for the Advancement of Pure Research (ZWO).

The costs of publication of this article were defrayed in part by the payment of page charges. This article must therefore be hereby marked "*advertisement*" in accordance with 18 U.S.C. Section 1734 solely to indicate this fact.

Received September 20, 1977; revised October 3, 1977

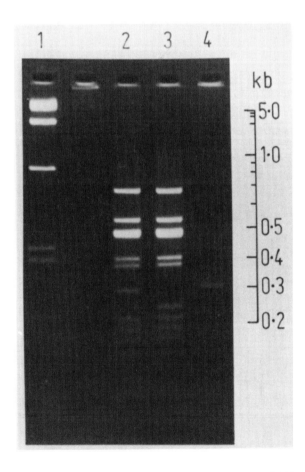

Figure 12. Identification and Purification of the 333 bp β–Globin cDNA Fragment Produced by Cleavage of PβG1 DNA with Endonuclease Hae III

Double-stranded DNA samples were electrophoresed in a 3.0% agarose slab gel, and DNA bands were detected by ethidium fluorescence under ultraviolet light. (1) 1 μo phage PM2 DNA X Hind III; (2) 1 μg PβG1 DNA X Hae III; (3)1 μg PβG1 DNA X Hae III X Eco RI; (4) 40 ng purified 333 bp cDNA fragment.

References

Bastos, R. N. and Aviv, H. (1977). Globin RNA precursor molecules: biosynthesis and processing in erythroid cells. Cell 11, 641–650.

Berget, S. M., Moore, C. and Sharp, P. A. (1977). Spliced segments at the 5' terminus of adenovirus 2 late mRNA. Proc. Nat. Acad. Sci. USA 74, 3171–3175.

Brack, Ch., Eberle, H., Bickle, T. A. and Yuan, R. (1976). A map of the sites on bacteriophage PM2 DNA for the restriction endonucleases Hind III and Hpa II. J. Mol. Biol 104, 305–309.

Chow, L. T., Gelinas, R. E., Broker, T. R. and Roberts, R. J. (1977). An amazing sequence arrangement at the 5' ends of adenovirus 2 messenger RNA. Cell 12, 1–8.

Crawford, L. V. and Robbins, A. K. (1976) . The cleavage of

polyoma virus DNA by restriction enzymes Kpn I and Pst I. J. Gen. Virol. *31*, 315–322.

Curtis, P. J. and Weissmann, C. (1976). Purification of globin messenger RNA from dimethylsulfoxide-induced Friend cells and detection of a putative globin messenger RNA precursor. J. Mol. Biol.*106*, 1061–1075.

Denhardt, D. T. (1966). A membrane-filter technique for the detection of complementary DNA. Biochem. Biophys. Res. Commun. *23*, 641–646.

Denton, M. J. and Arnstein, H. R. V. (1973). Characterization of developing adult mammalian erythroid cells separated by velocity sedimentation. Br. J. Haematol. *24*, 7–17.

Efstratiadis, A., Kafatos, F. C. and Maniatis, T. (1977). The primary structure of rabbit β–globin mRNA as determined from cloned DNA. Cell *10*, 571–585.

Flavell, R. A., Jeffreys, A. J. and Grosveld, G. C. (1978). Physical mapping of repetitive DNA sequences neighboring the rabbit β–globin gene. Cold Spring Harbor Symp. Quant. Biol., in press.

Glover, D. M. and Hogness, D. S. (1977). A novel arrangement of the 18S and 28S sequences in a repeating unit of Drosophila melanogaster rDNA. Cell *10*, 167–176.

Jeffreys, A. J. and Flavell, R. A. (1977). A physical map of the DNA regions flanking the rabbit β–globin gene. Cell *12*, 429–439.

Lehmann, H. and Huntsman, R. G. (1972). The hemoglobinopathies. In The Metabolic Basis of Inherited Disease, J. B. Stanbury, J. B. Wyngaarden and D. S. Fredrickson, eds. (New York: McGraw-Hill), pp. 1398–1431.

Maniatis, T., Jeffrey, A. and Kleid, D. G. (1975). Nucleotide sequence of the rightward operator of phage lambda. Proc. Nat. Acad. Sci. USA *72*, 1184–1188.

Maniatis, T., Kee, S. G., Efstratiadis, A. and Kafatos, F. C. (1976). Amplification and characterization of a β–globin gene synthesized in vitro. Cell *8*, 163–182.

Nokin, P., Burny, A., Cleuter, Y., Huez, G., Marbaix, G. and Chantrenne, H. (1975). Isolation and characterization of highly purified globin messenger RNA from anaemic-rabbit spleen. Eur. J. Biochem. *53*, 83–90.

Pellegrini, M., Manning, J. and Davidson, N. (1977). Sequence arrangement of the rDNA of Drosophila melanogaster. Cell *10*, 213–224.

Rigby, P. W. J., Dieckmann, M., Rhodes, C. and Berg, P. (1977). Labeling deoxyribonucleic acid to high specific activity *in vitro* by nick translation with DNA polymerase I. J. Mol. Biol. *113*, 237–251.

Romer, A. S. (1966). Vertebrate Paleontology, third edition (Chicago: University of Chicago Press).

Southern, E. M. (1975). Detection of specific sequences among DNA fragments separated by gel electrophoresis. J. Mol. Biol. *98*, 503–517.

Wellauer, P. K. and Dawid, I. B. (1977). The structural organization of ribosomal DNA in Drosophila melanogaster. Cell *10*, 193–212.

White, R. L. and Hogness, D. S. (1977). R loop mapping of the 18S and 28S sequences in the long and short repeating units of Drosophila melanogaster rDNA. Cell *10*, 177–192.

Note Added In Proof

We have recently learned that the chick ovalbumin gene also contains a large insert (P. Chambon, personal communication).

Regulation at transcription

2

A number of different stages are required in order to convert the information contained in the DNA into an mRNA molecule (Figure 2.1). In principle, regulation at any one of these stages could be responsible for the observed differences in the mRNA content of different tissues. Thus, for example, regulation may occur at the level of transcription by deciding which genes are actually transcribed into the initial RNA transcript, with the other processes such as splicing and transport to the cytoplasm following. Alternatively, regulation could occur at a post-transcriptional level, for example by transcribing all genes and regulating the splicing process so that only some transcripts have their introns removed and are therefore able to generate a functional mRNA molecule. Similarly it would be possible to regulate gene expression by transcribing and splicing all genes in all tissues but regulating which fully spliced RNA molecules were transported from the nucleus to the cytoplasm, where they can be translated into protein (for a review, see Latchman, 1995a).

This last possibility was addressed by Gilmour et al. (1974) even before it was known that introns existed. They studied the accumulation of the globin RNA in both the cytoplasm and the nucleus of Friend erythroleukaemia cells both prior to and after treatment with dimethyl sulphoxide. They showed that the increase in globin protein production resulting from this treatment was paralleled by an increase in globin RNA in both the cytoplasm and the nucleus. Hence, in this case at least, gene regulation does not occur by regulation of RNA transport, since the globin RNA increased in parallel in both the nucleus and the cytoplasm.

Investigations of this type were greatly extended, however, following the discovery of introns, by using Northern blotting to measure not only the level but also the size of the different RNAs produced from a particular gene. In the paper by Roop et al. which is presented here, the level and nature of the transcripts derived from the ovalbumin gene were studied extensively in the oviduct. More importantly, it was possible to compare the level and nature of these transcripts in the presence or absence of oestrogen stimulation of the oviduct, which results in the 'switching on' of ovalbumin production. Hence the patterns of RNAs could be compared directly in the same tissue when it was or was not producing ovalbumin protein.

As in the experiments of Gilmour et al. (1974) on the globin system, the data shown in Figures 3 and 4 of the paper by Roop et al. indicate that they were unable to detect any significant accumulation of ovalbumin RNA in the nucleus of non-expressing tissues, as would be seen if the ovalbumin RNA was produced in non-expressing tissues but was not transported to the cytoplasm.

In addition, however, due to their use of Northern blotting, Roop et al. were able, in the work illustrated in Figure 7 of their paper, to detect higher-molecular-mass RNA species in the nuclei of expressing tissues that represented unspliced or partially spliced RNAs from which all the introns had not yet been removed. This

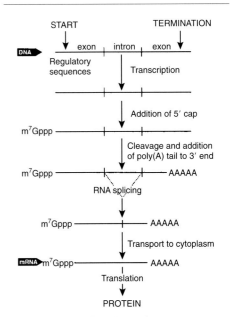

Figure 2.1 Stages in eukaryotic gene expression that could be regulated

allowed them to perform the important experiment illustrated in Figure 8 of the paper, which demonstrated that these higher-molecular-mass RNAs were not present in the oviduct when it was not stimulated with oestrogen and therefore was not producing ovalbumin. This experiment renders unlikely the possibility that the ovalbumin gene is transcribed in the non-expressing condition but that introns are not removed from the primary RNA transcript. In this case it would be expected that the large unspliced or partially spliced RNAs would accumulate in the non-expressing tissue. These experiments therefore support the idea that regulation is at the level of transcription, with the ovalbumin gene only being transcribed following oestrogen stimulation and not in the absence of such stimulation.

It is possible to object, however, that the technique of Northern blotting measures only the steady-state level of a particular RNA and not its rate of production. Thus transcription of the ovalbumin gene could occur in the non-expressing tissue but be followed by the rapid degradation of this RNA so that it was not detectable. In attempting to meet this objection, Roop et al. performed experiments (also illustrated in Figure 8 of the paper) in which material from non-expressing tissues was mixed with the RNA from the expressing tissue in an attempt to see whether degradation of the ovalbumin RNA in the RNA from the expressing tissue would result. No such degradation was observed.

Although this experiment cast doubt on the degradation hypothesis, in order to eliminate it directly it is necessary to measure the actual rate of gene transcription. In the most direct method of doing this, pulse labelling, cells are briefly exposed to radioactively labelled ribonucleotide. The amount of labelled ribonucleotide incorporated into a particular RNA during this brief exposure is then determined and provides a measure of the rate of synthesis of that particular RNA by the process of transcription. Hence the rate of transcription under different circumstances can be determined by carrying out this pulse-labelling experiment under different conditions.

This method was used, for example, by Lowenhaupt et al. (1978) to measure the rate of transcription of the β-globin gene in Friend erythroleukaemia cells both before and after treatment with dimethyl sulphoxide. This allowed these authors to demonstrate that dimethyl sulphoxide treatment produces a dramatic increase in the rate of transcription of the β-globin gene in these cells, paralleling the previously observed increases in the globin mRNA level both in the nucleus and in the cytoplasm, as well the increased production of globin protein (see above). Hence, in this instance, gene regulation occurs at the transcriptional level, with dimethyl sulphoxide inducing an increase in β-globin gene transcription.

Although this experiment was important in that it demonstrated that transcriptional control does occur, the pulse-labelling method used is of relatively limited applicability. This is because the radioactively labelled ribonucleotide is diluted by the unlabelled ribonucleotide already present in the cytoplasm of the cell, and therefore only a small proportion of the radiolabelled molecules are actually incorporated into the nascent RNA. Hence this method has a relatively low sensitivity and can only be used for genes that are transcribed at extremely high levels.

A more generally applicable method which is of higher sensitivity involves the addition of radiolabelled ribonucleotide to isolated nuclei *in vitro*. The isolation of nuclei removes the large pool of unlabelled ribonucleotide present in the cytoplasm of the cell. This results in a greater proportion of the added radiolabelled ribonucleotide being incorporated into the nascent RNA, and therefore produces greater sensitivity.

This method (known as a nuclear run-off assay) was used by Roop et al., and the results are presented in Table 4 of their paper. They demonstrated that significant transcription of the ovalbumin gene could only be detected in oestrogen-stimulated oviduct cells, and not in oviduct cells from which oestrogen had been withdrawn or in other tissues. This conclusion was further substantiated in a subsequent paper from the same group (Swaneck et al., 1979) in which a detailed time course of ovalbumin gene transcription was carried out in oviduct tissue following initial exposure to oestrogen and its subsequent withdrawal.

Hence increased transcription of the ovalbumin gene occurs following oestrogen stimulation, resulting in the previously observed accumulation of nuclear and cytoplasmic RNA for ovalbumin and enhanced production of ovalbumin protein. This finding therefore indicates the importance of transcriptional control in the regulation of ovalbumin gene expression.

Although the nuclear run-off assay thus permitted the demonstration of transcriptional control in the case of the ovalbumin gene, its greatest contribution was to allow the investigation of transcriptional regulation of genes whose corresponding mRNAs are of much lower abundance than those encoding globin or ovalbumin. This was of particular importance at the time because of the existence of the Davidson and Britten model of gene regulation (Davidson and Britten, 1979). This model specifically postulated that the genes encoding high-abundance mRNAs such as globin and ovalbumin are regulated in a different manner to the majority of RNAs which are of much lower abundance. Thus in this model (Figure 2.2) it was postulated that all genes would be transcribed in all tissues at a low basal level and that in most cases gene regulation would operate simply by deciding which of these primary transcripts would be correctly spliced and transported to the cytoplasm. Hence regulation would be mainly at the post-transcriptional level. In contrast, high-abundance RNAs could not be produced in sufficient amounts by this method, and therefore transcriptional control would operate in a particular tissue to boost this low basal level of transcription to a level sufficient to produce the high amounts of RNA required.

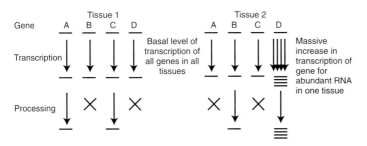

Figure 2.2 Davidson and Britten model of regulation at the post-transcriptional level. In this model all genes are transcribed at a low rate in all tissues, and regulation is achieved by controlling which transcripts are processed to the mature mRNA. Transcriptional control is confined to genes (such as gene D) where the level of RNA required cannot be achieved by processing all of the primary transcripts produced by the low basal rate of transcription.

In view of the existence of this model, the paper by Derman et al. presented here is of particular importance. These authors isolated 16 different cDNA clones corresponding to mRNAs which were present at different abundances in the liver. As indicated in Table 1 of the paper, 11 of these clones were shown to be derived from mRNAs which were specifically expressed in the liver and which were absent from other tissues such as the brain. Nuclear run-off assays were then used to measure the rate of gene transcription in isolated nuclei. As indicated in Table 3 of the paper, it was found that the genes corresponding to these mRNAs were transcribed in the liver, but only at very low levels or not at all in the brain. This paper therefore demonstrated that transcriptional control is not confined to the genes encoding a few RNAs which are expressed at very high abundance in a particular tissue. Rather, such control operates for a wide range of genes that encode mRNAs of widely differing abundances. This indicated that the Davidson and Britten theory is not correct, at least in mammalian cells and tissues.

Thus the papers presented in this section conclusively established the generality of transcriptional control as the major mechanism for regulating gene expression in eukaryotic cells. Post-transcriptional control does occur, however, in some situations, and key papers analysing some cases of this are discussed in the following sections.

Transcription of Structural and Intervening Sequences in the Ovalbumin Gene and Identification of Potential Ovalbumin mRNA Precursors

Dennis R. Roop, Jeffrey L. Nordstrom,
Sophia Y. Tsai, Ming-Jer Tsai
and Bert W. O'Malley
Department of Cell Biology
Baylor College of Medicine
Houston, Texas 77030

Summary

Structural sequences that are extensively separated by nonstructural intervening sequences in the natural ovalbumin gene are coordinately expressed in target and nontarget tissue. The intervening sequences, which consist of unique sequences in the chick genome, are transcribed in their entirety. The amount of nuclear RNA corresponding to these sequences, however, is approximately 10 times less than that observed for structural sequences. The accumulation of RNA corresponding to structural and intervening sequences during acute estrogen stimulation suggests either that there are different rates of transcription for these regions of the ovalbumin gene or that RNA sequences corresponding to the intervening sequences are preferentially processed and degraded. Comparison of the in vitro expression of portions of the ovalbumin gene in nuclei isolated from chronically stimulated oviducts indicates that both structural and intervening sequences are preferentially transcribed in vitro at rates approximately 500 times greater than expected for random transcription of the haploid chick genome. In addition, electrophoresis of oviduct nuclear RNA on agarose gels containing methylmercury hydroxide reveals multiple species of RNA that are from 1.3 to over 4 times larger than ovalbumin mRNA and hybridize to both structural and intervening sequences of the ovalbumin gene. These results are consistent with transcription of the entire ovalbumin gene into a large precursor molecule followed by excision of the intervening sequences and appropriate ligation of the structural sequences to form the mature mRNA.

Introduction

Regulation of the expression of the ovalbumin gene in the chick oviduct appears to be mediated by steroid hormones (O'Malley and Means, 1974; Cox, Haines and Emtage, 1974; McKnight, Pennequin and Schimke, 1975; Harris et al., 1975; Monahan, Harris and O'Malley, 1976a; Palmiter et al., 1976; Hynes et al., 1977). Recent studies of the structure of the natural ovalbumin gene using restriction endonuclease mapping indicate that the structural sequences of this gene are interrupted by regions of nonstructural intervening sequences (Breathnach, Mandel and Chambon, 1977; Doel et al., 1977; Lai et al., 1978; Weinstock et al., 1978). In an attempt to understand the functional importance of these intervening sequences, we have studied the transcription of different portions of the ovalbumin gene in vivo and in vitro. We have prepared probes for different structural regions of the ovalbumin gene that are separated by intervening sequences and have measured their concentration in vivo in target and nontarget cells and during acute estrogen stimulation. We have also measured the concentration of the intervening sequences in vivo using probes prepared to two Eco RI fragments of the natural ovalbumin gene that consist of mostly nonstructural sequence and have recently been cloned in our laboratory (Dugaiczyk et al., 1978; Woo et al., 1978). Comparison of the transcription in vitro of structural and intervening sequences of the ovalbumin gene was performed with nuclei isolated from chronically stimulated oviducts.

To determine whether the intervening and structural sequences of the ovalbumin gene might be transcribed into a larger precursor molecule that is then processed to form mature mRNA, we have searched for the existence of high molecular weight species of ovalbumin sequence-containing RNA. Oviduct nuclear RNA was fractionated by agarose gel electrophoresis in the presence of methylmercury hydroxide, transferred to diazobenzyloxymethyl paper (Alwine, Kemp and Stark, 1977) and hybridized to the probes for different portions of the ovalbumin gene.

Results

Preparation of Probes for Structural and Intervening Sequences within the Ovalbumin Gene

Before the expression of structural and intervening sequences within the ovalbumin gene could be studied, it was necessary to prepare hybridization probes for these sequences. A model of the natural ovalbumin gene generated by restriction endonuclease mapping (Lai et al., 1978) was used as a basis for preparing probes for the structural sequence regions of the gene (we explain below the rationale for preparing probes to these particular regions). Probes were prepared from pOV230, a chimeric plasmid that was previously constructed in our laboratory and contains a full-length oval-bumin cDNA insert (1847 nucleotides in length) lacking only 12 nucleotides at the 5' end (Mc-Reynolds, Catterall and O'Malley, 1977; Mc-Reynolds et al., 1978). Fragments containing DNA sequences corresponding to the 5' terminus (OV_L) and the 3' terminus (OV_R) of ovalbumin mRNA ($mRNA_{ov}$) with respect to the single Hae III cleavage site present in ovalbumin cDNA ($cDNA_{ov}$) (Mc-Reynolds et al., 1977; Monahan et al., 1977) were obtained by simultaneous digestion with endonu-clease Hind III and Hae III and resolution by electro-phoresis in agarose gels (Figure 1). The OV_L frag-ment is 1.15 kb in length (0.95 kb corresponds to the 5' terminus of $mRNA_{ov}$) and the OV_R fragment is 1.45 kb in length (1.1 kb corresponds to the 3' terminus of $mRNA_{ov}$). The OV_L fragment was fur-ther fractionated by digestion with Pst I and reso-lution in agarose gels to yield fragments desig-nated OV_{L1} and OV_{L2}. The OV_{L1} fragment is 0.81 kb in length (0.41 kb corresponds to $mRNA_{ov}$ se-quence) and the OV_{L2} fragment is 0.34 kb long (the whole sequence corresponds to $mRNA_{ov}$). A frag-ment containing essentially the entire structural gene, designated OV, was obtained by digestion of pOV230 with Hha I and resolution in agarose gels (2.95 kb in length with 1.85 kb corresponding to $mRNA_{ov}$ sequence).

The isolated fragments (OV, OV_L, OV_{L1}, OV_{L2}, and OV_R) were labeled with ^{3}H-dCTP by nick translation to specific activities of $4–10 \times 10^6$ cpm/μg as described in Experimental Procedures. Single-stranded hybridization probes were prepared by hybridizing these fragments to completion with a 100 fold excess of ovalbumin mRNA (Rot = 2.5×10^{-2}). The reactions (except the reaction contain-ing OV_{L2}) also contained 20 μg of poly(A) and poly(U) to minimize the effect of hybridization of poly(dA)-poly(dT) linkers present in the fragments.

The reaction mixtures were then treated with S1 nuclease to digest the plasmid pMB9 sequences and anticoding sequences that remain single-stranded under the conditions used. These steps were repeated with the recovered ^{3}H-DNA to insure the complete removal of sequences not contained in $mRNA_{ov}$. Single-stranded DNA probes were then obtained by alkaline hydrolysis of the RNA. The purity of these probes was determined as shown in Table 1. The absence of anticoding sequences is demonstrated by the failure of the probes to self-reassociate and by the ability of the probes to hybridize to completion with $mRNA_{ov}$. The absence of plasmid pMB9 sequences is also demonstrated by the inability of the probes to hybridize to pMB9 DNA. A very small amount of cross-contamination exists between the OV_L and OV_R probes, but the amount observed had little if any effect on the experiments described below. No significant cross-contamination was observed between the OV_{L1} and OV_{L2} probes. A higher percentage of hybridization (75–80%) could be obtained for reactions between probe and homologous DNA if the ratio of cold to labeled DNA was increased.

The rationale for preparing probes for these particular regions of the structural gene is as fol-lows. The natural ovalbumin gene is digested by Eco RI into three major fragments (2.4, 1.8 and 9.5 kb in length) as shown in Figure 1. The Eco RI sites are located in nonstructural intervening sequences which interrupt structural sequences in the oval-bumin gene (Lai et al., 1978). This Pst I site shown on the map of pOV230 is located within the 2.4 kb fragment. The OV_{L2} probe therefore corresponds to most of the structural sequences in the 2.4 kb fragment. Approximately 50 nucleotides corre-sponding to the OV_{L1} probe are located in the 2.4 kb fragment and the remaining sequences are located in the 1.8 kb fragment. The OV_L probe corresponds to structural sequences in both the 2.4 and 1.8 kb fragments. The Hae III site shown on the pOV230 map is located within the 1.8 kb frag-ment, and approximately 15 nucleotides corre-sponding to the OV_R probe are located in this fragment; the remaining OV_R sequences are lo-cated in the 9.5 kb fragment (Dugaiczyk et al., 1978). The OV_{L2} probe thus represents most of the structural sequence within the 1.8 kb fragment, and the OV_R probe represents structural sequences within the 9.5 kb fragment. These probes enable us to study the expression of different structural se-quences that are extensively separated by interven-ing sequences within the ovalbumin gene.

The 2.4 and 1.8 kb Eco RI fragments of the natural ovalbumin gene have recently been cloned in our laboratory (Dugaiczyk et all., 1978; Woo et al., 1978). These fragments can be recovered in

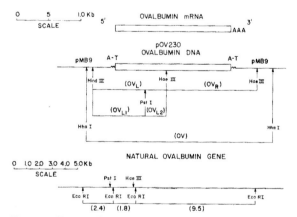

Figure 1. Diagram Illustrating the Position of Restriction Endo-nuclease Sites Used in the Preparation of Probes for Structural Regions of the Ovalbumin Gene

Table 1. Purity of Probes for Structural Sequences in the Ovalbumin Gene

		Hybridization[a] (%)					
	Self-Reassociation[c]	pMB9 DNA	mRNA$_{ov}$	OV$_L$	OV$_R$	OV$_{L1}$	OV$_{L2}$
Labeled Probe[b]							
OV$_L$	2	1	98	52	6		
OV$_R$	2	3	98	3	49		
OV	1	2	97				
OV$_{L1}$	5	3	98			51	1
OV$_{L2}$	3	2	97			1	50
Ratio Cold/Labeled		400	400	20	20	20	20

[a]Reactions carried out to a Cot of 1×10^{-1}.

[b]Specific activities (cpm/μg) of the probes used were as follows: OV$_L$ (5.4×10^6), OV$_R$ (9.1×10^6), OV (7.7×10^6), OV$_{L1}$ (6.1×10^6) and OV$_{L2}$ (4.4×10^6).

[c]Probes were reassociated alone to a Cot of 1×10^{-1}.

pure form from the pOV2.4 and pOV1.8 clones by Eco RI digestion followed by gel electrophoresis. The purified fragments were labeled with ^{3}H-dCTP and ^{3}H-dTTP by nick translation to specific activities of 6×10^7 and 4×10^7 cpm/μg, respectively. These labeled fragments were used as probes for intervening sequences without further purification, since only 7% of the 2.4 kb fragment and 5% of the 1.8 kb fragment were protected from S1 nuclease digestion after hybridization with excess mRNA$_{ov}$.

Hybridization of Probes for Structural and Intervening Sequences within the Ovalbumin Gene to Chick DNA

It was of interest to determine the repetition frequency for the intervening sequences located within the 2.4 and 1.8 kb fragments. The equivalent Cot$_{1/2}$ values obtained for the hybridization of these fragments with excess chick DNA, approximately 1.5×10^3, are essentially identical to that observed for the OV probe of the complete structural gene (Figure 2A). Thus within the limits of detection of this assay, the intervening sequences present in the 2.4 and 1.8 kb fragments are unique sequences in the chick genome. Pure single-stranded probes for the complete structural gene (OV) and the OV$_L$ and OV$_R$ structural regions were also hybridized with excess chick DNA (Figure 2B). Equivalent Cot$_{1/2}$ values of approximately 1×10^3 for the OV and OV$_L$ probes and 4.5×10^7 for the OV$_R$ probe indicate that these are unique sequence DNAs. These results are in agreement with previous studies of the hybridization kinetics of cDNA$_{ov}$ with chick DNA (Harris et al., 1973; Monahan et al., 1976b).

Expression of Structural and Intervening Sequences within the Ovalbumin Gene in Target and Nontarget Cells

Total nuclear RNA from estrogen-stimulated ovi-

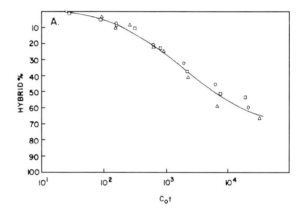

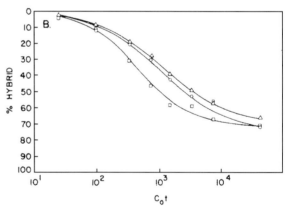

Figure 2. Hybridization of Excess Chick DNA to ^{3}H Probes Corresponding to Structural and Intervening Sequences in the Ovalbumin Gene

(A) (○) OV, (□) OV1.8 and (△) OV2.4. (B) (○) OV$_L$, (□) OV$_R$ and (△) OV. Hybridization was performed as described in Experimental Procedures.

duct, estrogen-withdrawn oviduct, unstimulated liver and unstimulated spleen were hybridized in excess to probes for the OV$_L$ and OV$_R$ regions of the ovalbumin structural gene to measure the concentration of mRNA$_{ov}$ sequences corresponding to these regions (Figure 3). The $Rot_{1/2}$ value obtained for nuclear RNA from estrogen-stimulated oviduct (7.5×10^{-1}) is 1660 times less than that obtained with estrogen-withdrawn oviduct (1.6×10^3). Using these values, the RNA to DNA ratio in the isolated nuclei and the percentage of tubular gland cells in the stimulated oviduct, we have calculated the number of $mRNA_{ov}$ molecules per cell nucleus. There are approximately 3075 molecules of $mRNA_{ov}$ per tubular gland cell nucleus in estrogen-stimulated oviduct, and approximately 2 molecules of $mRNA_{ov}$ per tubular gland cell nucleus in estrogen-withdrawn oviduct (Table 2). The hybridization observed with RNA from estrogen-withdrawn oviduct is indeed due to RNA-DNA hybrids, since the ability

of this preparation to hybridize to the probes is lost if the RNA preparation is subjected to alkaline hydrolysis prior to the hybridization reaction (Figure 3). We have also determined the thermal stability of these hybrids, and it is comparable to that observed for $mRNA_{ov}$-$cDNA_{ov}$ hybrids – a sharp melting curve with a T_m of 84°C in 2 × SSC. Very low levels of structural gene transcripts were detected in nuclear RNA from unstimulated liver and spleen. Since equimolar amounts of RNA corresponding to OV_L and OV_R were detected in all four tissues, it appears that the structural sequence regions of the ovalbumin gene are under coordinate control in these tissues and that the intervening sequences do not function as terminators in

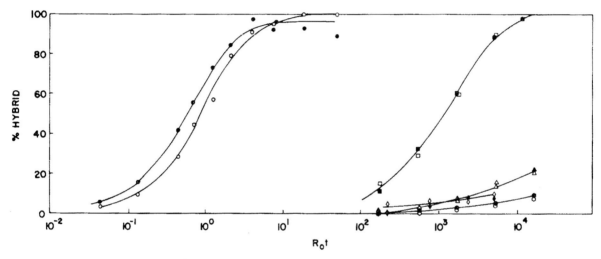

Figure 3. Hybridization of Nuclear RNA to ³H Probes Corresponding to Structural Sequences in the Ovalbumin Gene

The probes used were OV_L (closed symbols) and OV_R (open symbols). The RNAs used were extracted from stimulated oviduct (●, ○), withdrawn oviduct (■, □), liver (▲, △), spleen (●, ○) and withdrawn oviduct (subjected to alkaline hydrolyzed prior to hybridization) (♦, ◇). Hybridization was performed as described in Experimental Procedures.

Table 2. Concentration of $mRNA_{ov}$ Sequences in Total Nuclear RNA

Tissue	$Rot_{1/2}$	Fraction of $mRNA_{ov}$[a]	RNA/DNA	Molecules $mRNA_{ov}$[b] per Cell Nucleus	Tubular Gland Cells per Total Cells (%)	Molecules $mRNA_{ov}$ per Gland Cell Nucleus
Oviduct$_s$	7.5×10^{-1}	4×10^{-3}	0.25	2460	80	3075
Oviduct$_w$	1.6×10^3	1.9×10^{-6}	0.07	0.3	15	2
Oviduct$_w$ + 2 hr DES	7.8×10^2	3.8×10^{-6}	0.09	0.8	15	6
Oviduct$_w$ + 4 hr DES	9.6×10^1	3.1×10^{-5}	0.15	11	16	71
Oviduct$_w$ + 8 hr DES	1.2×10^1	2.5×10^{-4}	0.15	92	20	460
Oviduct$_w$ + 16 hr DES	5.6×10^0	5.4×10^{-4}	0.15	199	20	995
Oviduct$_w$ + 48 hr DES	1.7×10^0	1.8×10^{-3}	0.25	1107	35	3049

[a] Fraction of $mRNA_{ov} = \dfrac{Rot_{1/2} \text{ for pure } mRNA_{ov} \text{ hybridized to } cDNA_{ov}\ (3 \times 10^{-3})}{Rot_{1/2} \text{ obtained for RNA isolated from given tissue (Figure 5)}}$.

[b] Molecules $mRNA_{ov} = $ Fraction of $mRNA_{ov} \times \dfrac{RNA}{DNA} \times 2.6 \times 10^{-12}$ g DNA $\times \dfrac{6.02 \times 10^{23} \text{ molecules}}{6.37 \times 10^6 \text{ g } mRNA_{ov}}$.

estrogen-withdrawn oviduct or nontarget tissues such as liver and spleen.

Nuclear RNA from the tissues described above were also assayed for the presence of transcripts corresponding to the intervening sequences within the ovalbumin gene. This was accomplished by hybridization with labeled 2.4 and 1.8 kb fragments (Figure 4). Transcripts complementary to these fragments were detected in all the RNAs assayed; their concentration, however, was greatly reduced compared to those for structural gene sequences. We have calculated that there are approximately 233 molecules of RNA corresponding to the 2.4 kb fragment and approximately 286 molecules of RNA corresponding to the 1.8 kb fragment per tubular gland cell nucleus in estrogen-stimulated oviduct, and less than one molecule of RNA corresponding to each fragment per tubular gland cell nucleus in estrogen-withdrawn oviduct (Table 3). The low level of hybridization observed with RNA from unstimulated liver and spleen may be due to the detection of structural gene transcripts, since the 2.4 and 1.8 kb fragments contain a small amount of structural gene sequence (7 and 5%, respectively). The hybridization observed for the probes with RNA from stimulated oviduct at a Cot of 1 can also be attributed to the hybridization of structural sequences in these fragments to structural sequence transcripts that are present at high concentrations in this RNA.

It should be noted that the extent of hybridization of both 2.4 and 1.8 kb fragments to RNA from stimulated oviduct was approximately 50% (Figure 4). This is equivalent to 100% hybridization, assuming that both DNA strands of the fragments were labeled to the same degree by nick translation and that only the coding DNA strand was transcribed into RNA in vivo. These findings suggest that all the intervening sequences within the 2.4 and 1.8

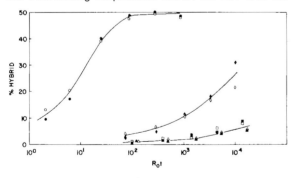

Figure 4. Hybridization of Nuclear RNA to [3]H Probes Corresponding to Intervening Sequences in the Ovalbumin Gene

The probes used were OV1.8 (closed symbols) and OV2.4 (open symbols). The RNAs used were extracted from stimulated oviduct (●,○), withdrawn oviduct (◆,◇), liver (■,□) and spleen (▲,△). Hybridization was performed as described in Experimental Procedures.

kb fragments of the ovalbumin gene are transcribed in vivo.

Expression of Structural and Intervening Sequences within the Ovalbumin Gene during Acute Estrogen Stimulation

Chicks that had been chronically stimulated with estrogen were withdrawn from hormone for 14 days and then given an injection of 2.5 mg of DES on the 14th day of withdrawal and again 24 hr later. Total nuclear RNA was extracted from oviducts which were isolated at 2, 4, 8, 16, 24 and 48 hr after secondary stimulation. These RNAs were hybridized in excess to probes for the OV_{L1} OV_{L2} and OV_R regions of the structural gene (Figure 5). Transcripts corresponding to these regions of the structural gene have accumulated even after the 2 hr interval. Equimolar amounts of RNA corresponding to these regions were detected at all intervals. We have used the data in Figure 5 to calculate the accumulation of $mRNA_{ov}$ molecules per tubular gland cell nucleus during secondary stimulation (Tables 2 and 3). A substantial increase in the concentration of $mRNA_{ov}$ sequences is observed by 4 hr, and this rate of accumulation remains constant for at least 16 hr. After 48 hr of secondary

Table 3. Concentration of Sequences Corresponding to Structural and Intervening Sequences within the Ovalbumin Gene in Total Nuclear RNA

Tissue	Molecules per Tubular Gland Cell Nucleus		
	Structural Sequences[a]	Intervening Sequences[b] 2.4	Intervening Sequences[c] 1.8
Oviduct$_s$	3075	233	286
Oviduct$_w$	2	<1	<1
Oviduct$_w$ + 2 hr DES	6	2	3
Oviduct$_w$ + 4 hr DES	71	–	15
Oviduct$_w$ + 8 hr DES	460	25	33
Oviduct$_w$ + 16 hr DES	995	9	26
Oviduct$_w$ + 24 hr DES	–	5	22[d]
Oviduct$_w$ + 48 hr DES	3049	59	93

[a] Data taken from Table 2.

[b] Calculated as described in Table 2 using the data in Figure 6A and a $Rot_{1/2}$ value for pure 2.4 intervening sequence of 2.98×10^{-3}, calculated as follows:

$$Rot_{1/2ov2.4} = \frac{N_{ov2.4}/\sqrt{N_{ov2.4}}}{N_{ov}/\sqrt{N_{ov}}} \times Rot_{1/2ov}$$

where $N_{ov2.4}$ = intervening sequences in 2.4 fragment = 1900 nucleotides; N_{ov} = 1930 nucleotides; $Rot_{1/2ov}$ = 3×10^{-3}.

[c] Calculated as described above using the data in Figure 6B and $N_{ov1.8}$ = 1500 nucleotides.

[d] Rot curve for this time point not shown in Figure 6B.

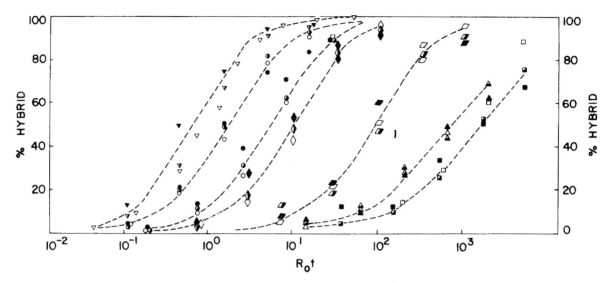

Figure 5. Accumulation of RNA Corresponding to Structural Sequences in the Ovalbumin Gene during Secondary Stimulation with Estrogen

Hybridization was performed as described in Experimental Procedures. The probes used were OV_{L1} (closed symbols;), OV_{L2} (partially closed symbols) and OV_R (open symbols). The RNAs were extracted from withdrawn oviduct (■ , ◨ , □), withdrawn oviduct after secondary stimulation for 2 hr (▲,△,△), 4 hr (◢,◪,◸), 8 hr (◆,◈,◇), 16 hr (●,◑,○) and 48 hr (●,◑,○). RNA from chronically stimulated oviduct was included for comparison (▼,▼,▽).

stimulation. The concentration of mRNA$_{ov}$ has increased approximately 1500 fold to 3049 molecules per tubular gland cell nucleus, a level comparable to that observed in chronically stimulated oviducts.

The accumulation during secondary stimulation of transcripts that correspond to the intervening sequences was determined by hybridizing the labeled 2.4 and 1.8 kb fragments with the nuclear RNAs as described above. The results of this experiment are shown in Figures 6A and 6B. We have used these data to calculate the number of molecules corresponding to these sequences per tubular gland cell nucleus (Table 3). Transcripts representing sequences in the 2.4 and 1.8 kb fragment accumulate up to 8 hr after secondary stimulation to approximately 25–33 copies per tubular gland cell nucleus, and then begin to decrease to 5–22 copies by 24 hr. An increase in [sic] observed 24 hr after a second injection to 59 copies for the 2.4 kb fragment and 93 copies for the 1.8 kb fragment. It is quite evidence [sic] from the data in Table 3 that the accumulation of nuclear transcripts corresponding to intervening sequences is much less, during secondary stimulation, than that observed for structural gene sequences.

In Vitro Transcription in Nuclei of Structural and Intervening Sequences within the Ovalbumin Gene

Labeled RNA was synthesized in nuclei isolated from estrogen-stimulated oviducts as described in Experimental Procedures. The concentration of RNA transcribed from structural sequences in the

ovalbumin gene was determined by hybridizing the ^{3}H–RNA to filters containing pOV230 DNA. Approximately 0.1% of the RNA synthesized in these nuclei corresponds to mRNA$_{ov}$ sequences (Table 4). This value and all other values reported in this section have been corrected for background binding to filters (counts not competitive with the appropriate RNA competitor) and for recovery of the appropriate ^{32}P–RNA internal standard. The endogenous mRNA$_{ov}$ in the RNA isolated from nuclei does not compete significantly with ^{3}H–RNA for DNA on the fitters, since the same hybridization efficiency was observed after increasing the input ^{3}H–RNA 10 fold; nor did placing an additional filter in the hybridization reaction result in an increase in the binding of ^{3}H–RNA. The presence of 5 μg/ml of α-amanitin inhibits the synthesis of mRNA$_{ov}$ sequences by 98%. The specificity of the RNA synthesized in vitro is further demonstrated by the absence of detectable ^{3}H–mRNA$_{ov}$ sequences in RNA synthesized in chick nuclei isolated from estrogen-withdrawn oviduct and hormonally unresponsive tissues such as liver and spleen. In addition, ^{3}H–RNA synthesized in stimulated oviduct nuclei will not hybridize to filters containing the chick β-globin gene, βpHb1001 (Table 4). In the following experiment, we have attempted to demonstrate that transcription of the ovalbumin gene in nuclei is asymmetrical. Filters containing pOV230 DNA were presaturated with mRNA$_{ov}$ prior to hybridization with in vitro synthesized RNA. Under these conditions, only RNA synthesized from the anticoding strand should hybridize to the presaturated filters. As

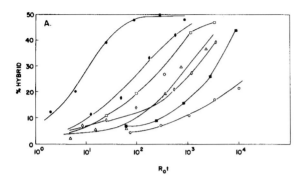

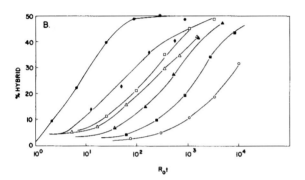

Figure 6. Accumulation of RNA Corresponding to Intervening Sequences in the Ovalbumin Gene during Secondary Stimulation with Estrogen

Hybridization was performed as described in Experimental Procedures. The RNAs used were extracted from withdrawn oviduct (○), withdrawn oviduct after secondary stimulation for 2 hr (■), 4 hr (▲), 8 hr (□), 16 hr (△), 24 hr (◇) and 48 hr (♦). RNA from chronically stimulated oviduct was included for comparison (●). (A) OV2.4, (B) OV1.8.

shown in Table 4, no hybridization was observed to filters treated in this manner. This result suggests that most of the RNA transcribed from the ovalbumin gene in vitro corresponds to coding strand sequences.

The specificity of the filter hybridization assay and the fidelity of in vitro transcription in nuclei is further supported by the following experiment. RNA was synthesized from DNA corresponding to the OV_L and OV_R regions of the structural gene and used as competitor during hybridization reactions. The OV_L RNA should compete out approximately 40% of the hybridizable counts (29% competition was observed) and the OV_R RNA should compete out approximately 60% of the hybridizable counts (60% competition was observed) (Table 5). These results indicate that approximately equimolar amounts of RNA are synthesized from the OV_L and OV_R regions of the structural gene in vitro.

The in vitro synthesis of RNA corresponding to the intervening sequences within the ovalbumin gene was determined by hybridizing labeled RNA synthesized in nuclei to filters containing pOV1.8 and pOV2.4 DNA. The results of a typical experiment and the results averaged from three separate experiments are shown in Table 6. An average of 0.058% of the total synthesized RNA binds to pOV2.4 filters, and competition with $mRNA_{ov}$ suggests that 95% of the counts bound (0.055% of the total counts) correspond to the intervening sequences in the 2.4 kb fragment. An average of 0.057% of the total synthesized RNA binds to pOV1.8 filters, and competition with $mRNA_{ov}$ sug-

Table 4. Specificity of in Vitro Transcription in Nuclei

Filter	Sources of ^3H–RNA	Competitor	^3H–RNA Hybridized[a] (cpm)	^{32}P–cDNA$_{ov}$ Recovery (%)	Hybridizable Gene Sequences (cpm)	Total RNA (%)
pOV230	Oviduct$_{DES}$	–	1041	24.6	3260	0.105
		$mRNA_{ov}$	239			
pOV230	Oviduct$_{DES}$ + α-amanitin	–	52	18.3	32	0.002
		$mRNA_{ov}$	46			
pOV230	Oviduct$_w$	–	19	15.0	20	0.002
		$mRNA_{ov}$	16			
pOV230	Spleen	–	62	18.9	–	–
		$mRNA_{ov}$	87			
pOV230	Liver	–	57	18.1	–	–
		$mRNA_{ov}$	46			
pOV230 (Presaturated with $mRNA_{ov}$)	Oviduct$_{DES}$	–	55	–	–	–
		$mRNA_{ov}$	53			
βpHb1001	Oviduct$_{DES}$	–	43	18.4	–	–
		$mRNA_{ov}$	34			

[a] Input ^3H–RNA: Oviduct$_{DES}$ = 3.1×10^6 cpm, Oviduct$_{DES}$ + α-amantin = 1.63×10^4 cpm, Oviduct$_w$ = 0.93×10^6 cpm, Spleen = 3.1×10^6 cpm, Liver = 2.5×10^6 cpm.

Table 5. In Vitro Transcription of Ovalbumin Gene Structuring Sequences in Nuclei

	^{3}H–RNA Hybridized[a] (cpm)	^{32}P–cRNA$_{ov}$ Recovery (%)	Hybridizable Gene Sequences (cpm)	Total RNA (%)	Competition (%)
	1454	32	3631	0.084	–
mRNA$_{ov}$	292	–	0	–	100
RNA$_L$	1116	–	2575	0.059	29.2
RNA$_R$	757	–	1453	0.033	60.2

[a] Input ^{3}H–RNA = 4.3×10^6 cpm.

Table 6. In Vitro Transcription of Ovalbumin Gene Intervening Sequences in Nuclei

Filter	Competitor	^{3}H–RNA Hybridized[a] (cpm)	^{32}P–cRNA Recovery (%)	Hybridizable Gene Sequences (cpm)	Total RNA (%)	Average[b]
pOV230	–	1009	16.4	4499	0.110	0.10±0.009
	mRNA$_{ov}$	271	–	–	–	–
pOV2.4	–	715	21.7	2295	0.056	0.058±0.004
	mRNA$_{ov}$	712	20.7	2391	0.058	0.055±0.003
	mRNA$_{ov}$ + RNA$_{2.4}$	217	–	–	–	–
pOV1.8	–	624	15.1	2616	0.064	0.057±0.003
	mRNA$_{ov}$	668	18.4	2385	0.058	0.051±0.006
	mRNA$_{ov}$+ RNA$_{1.8}$	229	–	–	–	–

[a] Input ^{3}H–RNA = 4.09×10^6 cpm.
[b] Average and standard error of the mean determined from results of three separate experiments.

gests that 90% of the counts bound (0.051% of the total counts) correspond to the intervening sequences in the 1.8 kb fragment. These values are surprisingly similar to that obtained for the structural sequences (0.1%), if we consider that the accumulation of sequence mass from the intervening and structural regions in total RNA differs substantially – by a factor of more than 100 – under conditions of chronic estrogen stimulation (40,000 versus 300 molecules per tubular gland cell; Harris et al., 1975 and our unpublished observations). The observation that the amount of in vitro synthesized RNA corresponding to structural gene sequences was slightly greater than that for intervening sequences is consistent with the conclusion that some processing of RNA transcribed from the intervening sequences may be occurring in nuclei.

Identification of Ovalbumin Structural Sequences in High Molecular Weight Nuclear RNA

Methylmercury hydroxide prevents base pairing of uridine and guanosine residues and, when used as a denaturing agent during agarose gel electrophoresis, permits excellent fractionation of nucleic acids according to their size (Bailey and Davidson, 1976; Lehrach et al., 1977). To detect species of ovalbumin RNA larger than the mature mRNA$_{ov}$ we subjected nuclear RNA from oviducts of hormone-stimulated chicks to agarose gel electrophoresis in the presence of 4 mM methylmercury hydroxide. After transfer of the RNA from the gel onto diazobenzyloxymethyl paper and hybridization to the ^{32}P-labeled pOV230 probe, we observed multiple discrete bands containing ovalbumin structural sequences (Figure 7). The most intense band of nuclear ovalbumin RNA ran similarly to cytoplasmic ovalbumin mRNA.

At least five bands of ovalbumin RNA that are significantly larger than ovalbumin mRNA (~2000 nucleotides) are readily detected by this method (Figure 7). As shown in Table 7, the sizes of these seven species of ovalbumin RNA range from approximately 1.3 to more than 4 times the size of ovalbumin mRNA. The ovalbumin RNA molecules having mobilities (relative to that of mRNA$_{ov}$) of 0.4 and 0.64 appear to be particularly abundant. The lightest bands of ovalbumin sequence-containing RNA have mobilities (relative to that of mRNA$_{ov}$) of 0.87 and 0.56. These bands (and some of greater molecular weight, which are not listed) are faint and are not observed in every experiment.

The presence in oviduct nuclei of high molecular weight species of RNA that contain ovalbumin sequences is dependent upon estrogen administration and is tissue-specific (Figure 8, slots C and E). Thus both high molecular weight ovalbumin RNA and mature ovalbumin mRNA are found in appreciable amounts only in those tissues of the chick

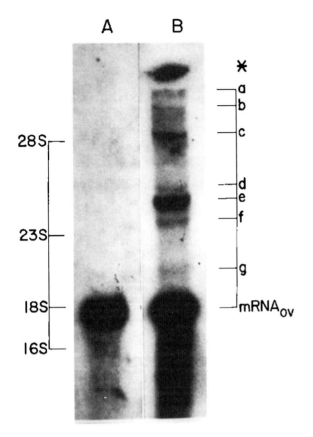

Figure 7. Identification of Ovalbumin RNA Sequences in High Molecular Weight Oviduct Nuclear RNA

Oviduct RNA from chicks stimulated with diethylstilbestrol was subjected to electrophoresis (40 V, 16 hr), transferred to DBM paper and hybridized with the ^{32}P-labeled ovalbumin structural sequence probe pOV230. The mobility of the ribosomal RNA markers was determined by ethidium bromide staining. (A) Purified cytoplasmic mRNA$_{ov}$, 0.1 μg. (B) Oviduct nuclear RNA, 20 μg. The ovalbumin sequence-containing RNA bands were arbitrarily labeled from (a) to (g). The intense spot next to the asterisk is not an actual band but rather an autoradiographic artifact, and is not observed in other gels (Figure 8).

Table 7. Tentative Sizes of the Ovalbumin Sequence-Containing RNA from Chick Oviduct Nuclei

RNA$_{ov}$ Species	Relative Mobility[a] (versus mRNA$_{ov}$)	Tentative Size[b] (Nucleotides)
a	0.22	>7800[c]
b	0.30	>6800[c]
c	0.40	>5600[c]
d	0.56	4350
e	0.64	3900
f	0.71	3400
g	0.87	2600
mRNA$_{ov}$	1.00	2050

[a] Calculated as the average of ten individual experiments.
[b] Sizes were estimated by interpolation of linear plots of log molecular weight versus electrophoretic mobility of RNA standards. Molecular weight values used were 1.9×10^6 for 28S rRNA, 0.71×10^6 for 18S rRNA, 1.07×10^6 for 23S rRNA and 0.53×10^6 for 16S rRNA. The lengths of the RNA sequences in nucleotides were calculated using 346 daltons per nucleotide.
[c] The position on the standard curve of RNA species larger than 28S deviates significantly from linearity (Lehrach et al., 1977). Since appropriate standards to define this region of the standard curve were unavailable at the time of experimentation, values were calculated by linear extrapolation of the standard curve. These values therefore represent minimum estimates of the actual sizes of the RNA species.

that are actively synthesizing ovalbumin.

The denaturing conditions used with the 4 mM methylmercury hydroxide appear to be optimal. Increasing the concentration of methylmercury hydroxide from 4 to 10 mM, or heating (65, 80 or 100°C for 45 sec) and then cooling RNA samples on ice prior to exposure to 4 mM methylmercury hydroxide, failed to alter the mobility of mRNA$_{ov}$ or of the higher molecular weight ovalbumin RNA species (data not shown). Although rigorous denaturing conditions were used, additional experiments were performed to rule out the possibility that the high molecular weight ovalbumin RNA species were the result of aggregation of mRNA$_{ov}$ with itself or other RNA molecules. Compared to the bands observed with nuclear RNA from hormone-stimulated chick oviduct alone (Figure 8, slot

A), none of the higher molecular weight ovalbumin RNA bands increased in intensity when mRNA$_{ov}$ was mixed with nuclear RNA prior to electrophoresis (Figure 8, slot B). Increased intensity was observed only for the mRNA$_{ov}$ band. In addition, neither mixing of mRNA$_{ov}$ with nuclear RNA from chick spleen (data not shown), liver (Figure 8, slot F) or hormone-withdrawn oviduct (Figure 8, slot D), nor extraction of RNA from a mixture of liver nuclei and oviduct cytoplasm prepared from estrogen-stimulated chicks (Figure 8, slot G), generated ovalbumin RNA species that were larger than mRNA$_{ov}$. Finally, the major species of high molecular weight RNA retained the same mobility after being subjected to electrophoresis (data not shown).

Identification of Ovalbumin Intervening Sequences in High Molecular Weight Nuclear RNA

To determine whether the nuclear ovalbumin RNA molecules larger than the mature mRNA$_{ov}$ contain sequences complementary to the nonstructural or intervening sequences of the ovalbumin gene, we used the cloned 2.4 and 1.8 kb fragments of the natural ovalbumin gene as hybridization probes. Most of the major bands of oviduct nuclear RNA that hybridized to the pOV230 DNA probe were again detected when hybridized to either the 2.4 or

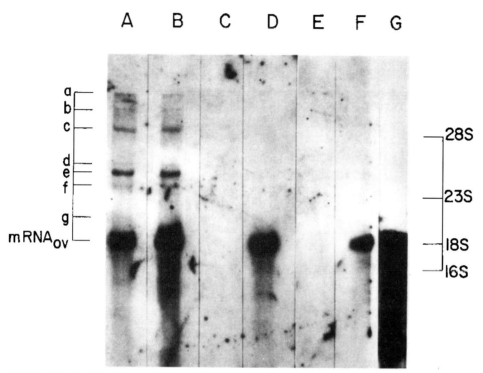

Figure 8. Hormone Dependence and Tissue Specificity of the Presence of Ovalbumin mRNA Sequences in High Molecular Weight Nuclear RNA

Nuclear RNA was isolated from various tissues of the chick, subjected to electrophoresis (40 V, 16 hr), transferred to DBM paper and hybridized with the ^{32}P-pOV230 probe.
(A) Oviduct nuclear RNA from estrogen-stimulated chicks, 20 μg. (B) Oviduct nuclear RNA (20 μg) mixed with mRNA$_{ov}$ (0.1 μg) just prior to electrophoresis. (C) Oviduct nuclear RNA from chicks stimulated with estrogen and then withdrawn from hormone for 14 days, 20 μg. (D) Hormone-withdrawn nuclear RNA (20 μg) mixed with mRNA$_{ov}$ (0.1 μg). (E) Liver nuclear RNA from estrogen-stimulated chicks, 20 μg. (F) Liver nuclear RNA (20 μg) mixed with ovalbumin mRNA (0.1 μg) just prior to electrophoresis. (G) RNA isolated after mixing liver nuclei with oviduct cytoplasm from estrogen-stimulated chicks, 20 μg.

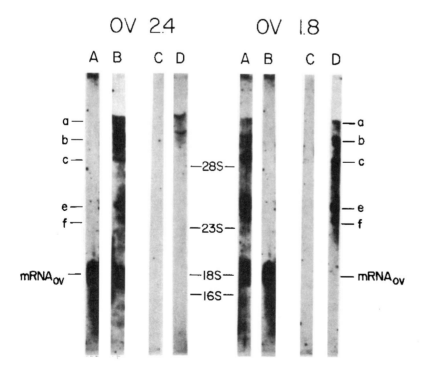

Figure 9. Identification of Sequences Complementary to the Intervening DNA Sequences of the Ovalbumin Gene in Oviduct Nuclear RNA

Oviduct nuclear RNA was subjected to electrophoresis (50 V, 20 hr), transferred to DBM paper and allowed to hybridize to the ^{32}P-labeled 2.4 and 1.8 kb Eco R1 fragments in either the presence or absence of excess unlabeled mRNA$_{OV}$ as a competitor.

(A) Purified cytoplasmic ovalbumin mRNA (0.1 μg). (B) Oviduct nuclear RNA (20 μg). (C) Purified cytoplasmic ovalbumin mRNA (0.1 μg). Unlabeled mRNA$_{ov}$ (20 μg) present during hybridization. (D) Oviduct nuclear RNA (20 μg). Unlabeled mRNA$_{ov}$ (20 μg) present during hybridization.
(Left panel) Hybridization with ^{32}P-2.4 kb Eco R1 fragment of the ovalbumin gene. (Right panel) Hybridization with the ^{32}P-1.8 kb Eco R1 fragment of the ovalbumin gene. Slots (A) and (B) in the right panel have been inadvertently reversed so that slot (A) OV1.8 is oviduct nuclear RNA and slot (B) OV1.8 is cytoplasmic ovalbumin mRNA.

1.8 kb probes (Figure 9, slot B), which was to be expected since both probes contain structural (~20%) as well as intervening sequences. To determine whether the intervening sequences could themselves be detected in the higher molecular weight ovalbumin RNA bands, we added excess unlabeled mRNA$_{ov}$ to the hybridization reaction to act as a competitor for the structural sequences present in the 2.4 and 1.8 kb probes. Under these hybridization conditions, detection of the band corresponding to mRNA$_{ov}$ was eliminated (Figure 9, slot C). Competition for structural sequences thus appears to be complete. Under these conditions, OV2.4 hybridized only to bands a and b and OV1.8 hybridized to all the major bands (a, b, c, e and f in Figure 9, slots D). Thus RNAs corresponding to the intervening sequences of OV1.8 are present in all the prominent high molecular weight ovalbumin species, while those of OV2.4 are present in significant amounts only in the two larger species. These results are consistent with the hypothesis that the intervening sequences of the ovalbumin gene are transcribed into high molecular weight precursors and that the intervening sequences of OV2.4 appear to be removed prior to those of OV1.8.

Discussion

In this investigation, we have addressed the following issues: whether the nonstructural intervening sequences affect the transcription of different structural regions of the ovalbumin gene; whether the intervening sequences are themselves transcribed; whether steroid hormones regulate transcription of structural and intervening sequences in a coordinated fashion; and whether the entire ovalbumin gene is transcribed as a large precursor that is processed to form mature ovalbumin mRNA.

We have determined that the intervening sequences located in the 2.4 and 1.8 kb Eco R1 fragments of the natural ovalbumin gene constitute unique sequence DNA in the chick genome. The expression of different structural regions of the ovalbumin gene that are separated by intervening sequences was studied in estrogen-stimulated chick oviduct that is actively synthesizing large amounts of mRNA$_{ov}$, in estrogen-withdrawn chick oviduct that exhibits a marked decrease in the synthesis of mRNA$_{ov}$ and in unstimulated chick spleen and liver that synthesize very little mRNA$_{ov}$. Although the amounts of mRNA$_{ov}$ synthesized in these tissues are different, equimolar amounts of RNA corresponding to different structural regions of the gene were detected in all tissues. These results suggest that the expression of different

structural regions of the ovalbumin gene is coordinately regulated and that the intervening sequences do not function as terminators or attenuators in estrogen-withdrawn oviduct or nontarget cells.

The intervening sequences in the 2.4 and 1.8 kb fragments of the natural ovalbumin gene are transcribed in approximately equimolar amounts in estrogen-stimulated oviduct and estrogen-withdrawn oviduct, but the amount of RNA corresponding to these sequences is approximately 10 times less than that observed for structural gene sequences. This result suggests either that the transcriptional rate differs markedly for these sequences or that the rate of transcription is the same but the intervening sequence transcripts are preferentially processed and degraded. We should point out that recent analysis of the cloned 2.4 and 1.8 kb fragments by restriction endonuclease mapping has revealed that the sequence arrangement in these fragments is even more complex than initially estimated by restriction mapping of total chick DNA (that is, there are seven intervening sequences rather than two (Dugaiczyk et al., 1978; Lai et al., 1978). These findings do not affect the results presented in this paper, however, since the probes used were specific for structural sequences in the 2.4, 1.8 and 9.5 kb fragments of the natural ovalbumin gene and transcripts corresponding to the entire intervening sequences in the 2.4 and 1.8 kb fragments were detected.

Since the withdrawal of estrogen from chronically stimulated chicks resulted in a marked decrease in the concentration of RNA corresponding to structural sequences (from 3075 molecules to 2 molecules, Table 3) and intervening sequences (from ~250 molecules to <1 molecule, Table 3) in the ovalbumin gene, we believed it would be useful to study the expression of these sequences during acute estrogen stimulation. The accumulation of equimolar amounts of RNA corresponding to different structural regions of the gene begins slowly during the first 2 hr after secondary stimulation. By 4 hr, there is a dramatic increase which continues at a constant rate. This finding is in agreement with previous reports concerning the accumulation of mRNA$_{ov}$ sequences during secondary stimulation (Cox et al., 1974; Harris et al., 1975; McKnight et al., 1975; Palmiter et al., 1976). It is interesting that the accumulation of RNA transcripts corresponding to the intervening sequences is observed for only 8 hr after secondary stimulation, after which the concentration decreases. The rate of accumulation during this time is also significantly less than that observed for structural sequence transcripts. These results again suggest either that differential rates of transcription exist for these sequences or that the RNA corresponding to the intervening sequences is preferentially processed and degraded.

Since the concentration of $mRNA_{ov}$ sequences in the cytoplasm is extremely high [approximately 40,000 molecules per tubular gland cell (Harris et al., 1975)] it is important to demonstrate that the $mRNA_{ov}$ sequences measured in nuclei are not due to cytoplasmic contamination. Although it is difficult to perform experiments that would completely rule out this possibility, we believe that the contribution of cytoplasmic sequences to those observed in nuclei is minimal, for the following reasons. First, similar amounts of $mRNA_{ov}$ were detected in RNA isolated from nuclei not treated with detergent and from nuclei washed repeatedly with 0.5 Triton X-100. Second, the ratio of ovalbumin mRNA to ovomucoid mRNA ($mRNA_{om}$) sequences in isolated nuclei (0.4% $mRNA_{ov}$ versus 0.12% $mRNA_{om}$, Table 2; Tsai et al., 1978) is quite different from that observed in polysomal poly(A) RNA or in total cellular RNA (50% $mRNA_{ov}$ versus 6% $mRNA_{om}$ in polysomal poly(A) RNA and 0.5% $mRNA_{ov}$ versus 0.06% $mRNA_{om}$ in total RNA; Hynes et al., 1977 and our unpublished observations). If most of the $mRNA_{ov}$ sequences in nuclei were the result of contamination by cytoplasmic RNA, we would expect the nuclear ratio to be similar to that observed in the cytoplasm. Most important, during secondary stimulation the kinetics of $mRNA_{ov}$ accumulation in the nucleus and in the total cell are dramatically different. After 48 hr of secondary stimulation, the concentration of $mRNA_{ov}$ per tubular gland cell nucleus has reached a plateau (3000 molecules) while that in the total cell is still accumulating (13,000 molecules). In the chronically stimulated oviduct, the concentration of $mRNA_{ov}$ per tubular gland cell nucleus is 3000 molecules while 40,000 molecules are found in the total cell (Table 2; Harris et al., 1975). These observations suggest that the $mRNA_{ov}$ sequences observed in nuclear RNA are not due to contamination by $mRNA_{ov}$ in the cytoplasm.

The ovalbumin gene is preferentially transcribed in nuclei isolated from estrogen-stimulated oviduct since $mRNA_{ov}$ sequences account for 0.1% of the total RNA synthesized. This value is approximately 500 times greater than would be expected with random transcription of the haploid chick genome. Since the data in Tables 4–6 have been corrected for the hybridization efficiency of ^{32}P internal standards, the absolute value estimated for specific in vitro transcripts could be slightly different. The relative values observed should remain the same, however, and the conclusions drawn from the data will not be affected. A similar value has recently been reported by Nguyen-Huu et al. (1978) using different assay procedures. The present study does not, however, provide information about the fidelity of initiation of RNA synthesis in nuclei in vitro or the existence and definition of eucaryotic promoters. Preliminary results indicate that most of the RNA synthesized in nuclei in vitro results from the elongation of existing chains (our unpublished observations).

The in vitro synthesis in nuclei of RNA from the intervening sequences of the ovalbumin gene does indeed occur. Transcripts corresponding to these sequences were detected using filters containing pOV2.4 and pOV1.8 DNA. The amount of RNA synthesized from the intervening sequences (approximately 0.05% of the total RNA synthesized) was slightly less than that observed for structural sequences (approximately 0.1% of the total RNA synthesized). These results could be explained either by differential rates of transcription for these sequences or by preferential processing and degradation in the nuclei of RNAs transcribed from the intervening sequences. We tend to favor the latter explanation, since the mass ratio of structural versus intervening sequence transcripts (>100:1) present in the total RNA extracted from estrogen-stimulated oviducts differs so greatly. Future pulse-chase experiments will be required to resolve this Issue.

A 15S precursor RNA molecule for β–globin mRNA has been demonstrated by pulse-label and -chase experiments (Ross, 1976; Curtis and Weissman, 1976; Bastos and Aviv, 1977; Kwan, Wood and Lingrel, 1977; Tilghman et al., 1977; Ross and Knecht, 1978). This 15S RNA molecule has recently been shown to consist of a continuous transcript of structural and intervening sequences of the β–globin gene (Tilghman et al., 1978). The results of this study reveal coordinate regulation of structural sequences separated by intervening sequence regions. In addition, the intervening regions are regulated by similar kinetics and appear to be transcribed in their entirety. These results are consistent with the hypothesis that the entire ovalbumin gene is transcribed as a large precursor molecule, after which the intervening sequences are processed. This conclusion is further supported by the demonstration that nuclear RNA from stimulated chick oviduct contains multiple species of RNA that are high in molecular weight and hybridize to structural sequences as well as to the intervening sequences of the 2.4 and 1.8 kb Eco R1 fragments of the ovalbumin gene. The largest of these RNA molecules has more than 7800 nucleotides– >4 times the length of ovalbumin mRNA – and thus is larger than our current estimate of approximately 7000 bp for the ovalbumin gene. Since the exact structure of the 5′ end of the gene is currently unknown, however, (Duzgaiczyk et al., 1978) the full-length size of a transcript may be well over 8000 bases.

An earlier study by McKnight and Schimke (1975) failed to demonstrate the existence of RNA molecules containing sequences complementary to $cDNA_{ov}$ larger than $mRNA_{ov}$. There are at least three reasons for this result. First, total cellular rather than nuclear RNA was used. In oviduct tissue the amount of nuclear RNA is 7% of the total RNA. Thus isolation of nuclear RNA would have meant a 14 fold enrichment of potential precursor molecules. Second, the specific activity of the cDNA hybridization probe was 8×10^6 cpm/μg or 10–30 times less than that of the pure cloned probes used in the present study. Third, resolution of RNA species on sucrose gradients is considerably less than that obtainable by gel electrophoresis. Thus the techniques described in this paper are more than 100 times as sensitive as those used by Mc-Knight and Schimke.

The existence of discrete bands of high molecular weight ovalbumin RNA raises the possibility that ovalbumin gene transcripts may be processed by successive removal of discrete pieces of intervening sequence RNA. This model is supported by the fact that the intervening sequences of OV2.4 or of OV1.8 were detected in all the major bands of ovalbumin RNA higher in molecular weight than mRNA OV. Furthermore, the differential banding pattern of the intervening sequences of OV2.4 with those of OV1.8 suggests that the intervening sequences of OV2.4 may be removed prior to those of OV1.8. While at present it may be premature to draw detailed conclusions on processing, our results are consisent with the idea that the bands represent mRNA precursors which lose intervening sequences as they are processed. Proof that these discrete high molecular weight RNA molecules actually constitute a series of precursors to ovalbumin mRNA requires pulse-label and -chase experiments, however, as well as detailed characterization of the sequences present in each band to determine the relationships between the species.

Experimental Procedures

Hormone Treatment

Stimulated oviducts were obtained from White Leghorn chicks which were implanted weekly with a 20 mg pellet of diethylstilbestrol (DES). This pellet provided continuous release of DES for 8–9 days. Estrogen-withdrawn oviducts were obtained from chicks that had received daily subcutaneous injections of 2.5 mg of DES for 14 days and were subsequently withdrawn from hormone for 14 days. For experiments involving acute stimulation with estrogen, an injection of 2.5 mg of DES was given to chicks on the 14th day of withdrawal and also 24 hr later. Oviducts were collected at the time intervals indicated in the text.

Isolation of Chick Oviduct Nuclei

Chick oviduct nuclei were isolated by a modification of the method described by Busch (1967). Chick oviduct tissue was homogenized in 8 vol of ice-cold 5% citric acid in a tissuemizer at 4°C for 60 sec. The homogenate was filtered through cheesecloth and organza, and nuclei were collected by centrifugation at 600 × g for 10 min. The nuclear pellet was suspended in a solution of 0.25 M sucrose and 1.5% citric acid, layered over a cushion of 0.88 M sucrose in 1.5% citric acid and centrifuged at 600 × g for 20 min. The purified nuclei were washed twice with a solution of 0.5% Triton X-100, 0.25 M sucrose and 1.5% citric acid prior to the preparation of total nuclear RNA. Nuclei were isolated from chick liver and spleen as described above. DNA and RNA determinations were made by diphenylamine and orcinol assays as described previously (Tsai et al., 1975).

Isolation of Nuclear RNA

RNA was isolated from nuclei by a modification cf the procedure of Holmes and Bonner (1973). Purified nuclei were lysed at room temperature in a buffer containing 2% SDS, 7 M urea, 0.35 M NaCl, 1 mM EDTA and 0.01 M Tris–HCl (pH 8.0) by means of a glass teflon homogenizer. Nucleic acids were extracted with an equal volume of phenol mixture (phenol:chloroform:isoamyl alcohol, 25:24:1, v/v). The aqueous layer was removed and the nonaqueous residue was reextracted with $\frac{1}{2}$ vol of lysis buffer. The aqueous phases were pooled and extracted with an equal volume of phenol mixture. The aqueous phase was removed and adjusted to 0.2 M sodium acetate (pH 5) and precipitated with 2 vol of ethanol. The nucleic acid was resuspended in 3 mM $MgCl_2$, 0.1 M NaCl and 10 mM Tris (pH 7.5) and treated with 40 μg/ml of affinity column-purified DNAase I (Maxwell, Maxwell and Hahn, 1977) at room temperature for 1 hr. SDS, proteinase K and EDTA were added to a final concentration of 0.5%, 20 μg/ml and 5 mM, respectively. The mixture was incubated at 37°C for 15 min. extracted twice with phenol mixture and precipitated with ethanol as before. The RNA obtained after this step was treated again with DNAase (as described above except that the DNAase concentration was 20 μg/ml and the incubation time was 40 min), extracted with phenol mixture and passed through a Sephadex G-50 column. The material eluting in the void volume was pooled, lyophilized and stored at -20°C.

In Vitro RNA Synthesis in Isolated Nuclei

Nuclei were isolated from oviducts obtained from chicks chronically stimulated with DES as previously described (Towle et al., 1977) and used for in vitro RNA synthesis. ^{3}H–RNA was synthesized under the following conditions: 80 mM Tris–HCl (pH 7.9); 2.5 mM $MnCl_2$; 3 mM $MgCl_2$; 100 mM $(NH_4)_2SO_4$, 0.6 mM each of ATP, GTP and UTP; 25 mM CTP (20 Ci/mmole); 30 μM EDTA, 6 mM dithiothreitol, 400 μg/ml DNA and 15% glycerol. Reactions were incubated at 37°C for 30 min and then treated with affinity column-purified DNAase I (20 μg/ml) (Maxwell et al., 1977) for 15 min at 37°C. The digestion was terminated by the addition of EDTA, SDS and proteinase K to final concentrations of 5 mM, 0.5% and 20 μg/ml, respectively. The ^{3}H–RNA was then extracted as described above for nuclear RNA, except that there was only one additional treatment with DNAase.

In Vitro Synthesis of RNA_L, RNA_R, $RNA_{1.8}$, $RNA_{2.4}$, ^{32}P–RNA_{OV}, ^{32}P–$RNA_{1.8}$ and ^{32}P–$RNA_{2.4}$

RNAs were transcribed from their corresponding templates [RNA_L (OV_L), RNA_R (OV_R), $RNA_{1.8}$ ($pOV_{1.8}$), $RNA_{2.4}$ ($pOV_{2.4}$) and RNA_{OV} (pOV230)] by E. coli RNA polymerase (isolated as previously described by Towle et al., 1977) using the following conditions: 50 mM Tris–HCl (pH 7.9); 5 mM $MgCl_2$; 100 mM $(NH_4)_2SO_4$; 2 mM 2-mercaptoethanol; 0.8 mM sodium phosphate; 5 mg/ml DNA template; 1 mM each of ATP, GTP, CTP and UTP; and 300 μg/ml E. coli RNA polymerase. For the synthesis of ^{32}P-labeled RNA, ^{32}P–ATP (700 Ci/mmole) at a final concentration of 0.1 mM was substituted for 1 mM ATP. Synthesis was for 2 hr at 37°C. The termination of the reaction and the extraction of RNA were as described above.

Preparation of DNA Filters
Filters containing pOV_{230}, $pOV_{1.8}$ or $pOV_{2.4}$ DNA were prepared as follows. DNA (250 μg) in 4 ml of 10 mM Tris–HCl (pH 7.9) was heat-denatured in a boiling water bath for 10 min and then quickly cooled by the addition of 41 ml of ice-cold 4 × SSC. The denatured DNA was slowly passed through a Millipore filter (3.5 cm in diameter). The filter was then washed with 40 ml of ice-cold 4 × SSC, air-dried for 1 hr and baked in a 60°C vacuum oven overnight. The filters were immersed in Denhart solution (0.002% Ficol, 0.02% BSA and 0.02% polyvinyl pyrrolidone in 4 × SSC) for 1 hr and then baked in a 70°C vacuum oven for 4 hr. Small filter discs (0.35 cm diameter) were cut from the large filter. The amount of DNA on the discs was determined by spectrophotometry after digestion with 0.5 N $HClO_4$ at 100°C for 30 min.

Hybridization of ^{3}H–RNA to DNA Filters
[3]H–RNA (1–6 × 10[6] cpm) along with the appropriate ^{32}P–RNA internal standard (2000 cpm), and in some experiments competitor RNA (7.5 μg of $mRNA_{OV}$ and 9μg of RNA_L, $RNA_{1.8}$ or $RNA_{2.4}$), was heated at 80°C in 40 μl of 50% formamide, 0.21 M NaCl and 0.021 M sodium citrate (pH 7.0) for 5 min to denature any double-stranded sequences. Hybridization reactions were started by the addition of DNA filters (1.2μg per filter) to the mixture, and incubation was for 18 hr at 42°C. Hybridization reactions were covered with paraffin oil to prevent evaporation during the incubation. At the end of the incubation, filters were removed and rinsed with 1.4 × SSC and chloroform to remove the paraffin oil. Filters were subsequently washed twice with 5 ml of 1.4 × SSC at 30°C for 1 hr and twice with 0.1 × SSC at 30°C for 20 min. The filters were incubated in 0.5 ml of 2 × SSC with pancreatic RNAase (20 μg/ml) for 1 hr at room temperature. They were then washed with 5 ml of 2 × SSC at 30°C for 15 min, solubilized in Cellusolve and counted in Aquasol.

Labeling of DNA by Nick Translation
DNA was labeled by nick translation using a modification of the procedure described by Mackey et al. (1977). The reaction was performed in a final volume of 100 μl containing the following: 50 mM Tris (pH 7.8), 5 mM $MgCl_2$, 10 mM 2-mercaptoethanol, 5 μg of BSA and 1.5 μg of DNA. For ^{3}H–DNA the reaction mixture also contained 0.12 mM dATP, dTTP and dGTP; and 0.02 mM ^{3}H–dCTP (25.6 Ci/mmole, evaporated to dryness). For ^{32}P–DNA, the reaction mixture also contained 0.06 mM dATP, dGTP; 0.006 mM ^{32}P–dCTP, dTTP (345 Ci/mmole, each). These components were assembled on ice and then 0.5 ng of DNAase (Worthington, DPFF) was added. The mixture was incubated at room temperature for 1 min and immediately cooled in an ice-water bath. E. coli DNA polymerase I (20 μl, 80 U, Boehringer Mannheim) was added and the mixture was incubated at 14°C. The reaction was stopped after 6 hr of incubation by the addition of 100 μl of 0.2 M EDTA, 100 μg of E. coli DNA and heating at 68°C for 10 min. Unincorporated dNTPs were separated from the labeled DNA by gel filtration on a Sephadex G-50 column using 10 mM Tris (pH 7.5), 1 mM EDTA and 10 mM NaCl. Radioactive fractions eluting in the void volume were pooled, made 0.2 M NaCl and precipitated by the addition of 2 vol of ethanol. The specific activity of ^{3}H–DNA was 4–10 × 10[6] cpm/μg and that of ^{32}P–DNA was 1–3 × 10[8] cpm/μg.

DNA-DNA Hybridization
Hybridizations were performed at an unlabeled DNA to ^{3}H–DNA excess (2 × 10[7]). Unlabeled chick DNA was sonicated so that the average length was approximately 400 nucleotides. Reactions were performed in tapered reaction vials, and contained 0.6 M NaCl, 0.01 M HEPES (pH 7.0), 0.002 M EDTA, 1.3 mg of unlabeled DNA and 0.06 ng of ^{3}H–DNA probe (final volume 200 μl). Samples were denatured at 100°C for 5 min and incubated at 68°C for various time intervals. Following hybridization, the samples were treated with S1 nuclease (Miles, 4800 U) as previously described

(Harris et al., 1976). Equivalent Cot values (those that would obtain at 0.18 M NaCl) have been plotted Britten, Graham and Neufeld, 1974.)

RNA-DNA Hybridization
RNA excess hybridization was performed in tapered reaction vials and contained in a final volume of 100 μl 0.6 M NaCl, 0.01 M HEPES (pH 7.0), 0.002 M EDTA, RNA at concentrations ranging from 3–5000 μg/ml and ^{3}H–DNA probe (0.15 ng for structural sequence probes and 0.01 ng for intervening sequence probes). The samples were denatured at 100°C for 5 min and incubated at 68°C for various time intervals. After hybridization, the samples were treated with S1 nuclease (Miles, 1600 U) as previously described (Harris et al., 1976). Equivalent Rot values (those that would obtain at 0.18 M NaCl) have been plotted (Britten et al., 1974).

Electrophoresis of RNA
RNA samples were made 4 mM in methylmercury hydroxide and then electrophoresed (generally at 40 V for 16 hr at room temperature with electrophoresis buffer recycling between the anode and cathode buffer compartments) on vertical slab gels (16.5 × 15.8 × 0.3 cm) containing 1.5% agarose (Seakem) and 4 mM methylmercury hydroxide (Alfa) as described by Bailey and Davidson (1976). All operations involving methylmercury hydroxide were performed in a fume hood using safeguards recommended for hazardous chemicals.

Transfer of RNA to DBM Paper and Hybridization of DNA Probes
Immediately after electrophoresis, methylmercury hydroxide was removed from the gel, and the RNA was partially degraded by alkali and transferred by blotting to freshly prepared diazobenzyloxymethyl (DBM) paper as described by Alwine et al. (1977). The aminobenzyloxymethyl (ABM) paper used to prepare the DBM paper was obtained from a commercial supplier (Enzobond), and the modified procedure utilizing potassium phosphate buffer (pH 6.5) was used (Alwine et al., 1977). The RNA bound to the DBM paper was hybridized to ^{32}P-labeled DNA probes, extensively washed and autoradiographed as described by Alwine et al. (1977), except that the hybridization buffer was modified to contain 50% formamide (purified through Chelex 100), 0.02% Ficoll, 0.02% bovine serum albumin, 0.02% polyvinyl pyrrolidone, 0.75 M NaCl, 68 mM Na citrate, 7 mM Na_2 EDTA and 29 mM TES (pH 7.0).

Acknowledgments
The authors wish to thank Ms. Melanie Vinion, Ms. Carolyn Engleking and Ms. Valerie McMullian for excellent technical assistance and Dr. Christina Chang for helpful discussions. This work was supported by grants from the NIH.

The costs of publication of this article were defrayed in part by the payment of page charges. This article must therefore be hereby marked "advertisement" in accordance with 18 U.S.C. Section 1734 solely to indicate this fact.

Received July 10, 1978; revised August 7, 1978

References

Alwine, J. C., Kemp, D. J. and Stark, G. R. (1977). Proc. Nat. Acad. Sci. USA 74, 5350–5354.

Bailey, J. M. and Davidson, N. (1976). Anal. Biochem. 70, 75–85.

Bastos, R. N. and Aviv, H. (1977). Cell 11, 641–650.

Breathnach, R., Mandel, J. L. and Chambon, P. (1977). Nature 270, 314–319.

Britten, R. J., Graham, D. E. and Neufeld, B. R. (1974). In Methods in Enzymology, 29, L Grossman and K. Moldave, eds. (New York: Academic Press), pp. 363–418.

Busch, H. (1967). In Methods in Enzymology, 12A, L. Grossman and K. Moldave, eds. (New York: Academic Press), pp. 434–439.

Cox, R. F., Haines, M. E. and Emtage, S. (1974). Eur. J. Biochem. 49, 225–236.

Curtis, P. J. and Weissmann, C. (1976). J. Mol. Biol. 106, 1061–1075.

Doel, M. T., Houghton, M., Cook, E. A. and Carey, N. H. (1977). Nucl. Acids Res.4, 3701–3713.

Dugaiczyk, A., Woo, S. L. C., Lai, E. C., Mace, M. L., McReynolds, L. A. and O'Malley, B. W. (1978). Nature 274, 328–333.

Harris, S. E., Means, A. R., Mitchell, W. M. and O'Malley, B. W. (1973). Proc. Nat. Acad. Sci. USA 70, 3776–3780.

Harris, S. E., Rosen, J. M., Means, A. R. and O'Malley, B. W. (1975). Biochemistry 14, 2072–2081.

Harris, S. E., Schwartz, R. J., Tsai, M.-J., Roy, A. K. and O'Malley, B. W. (1976). J. Biol. Chem. 251, 524–529.

Holmes, D. S. and Bonner, J. (1973). Biochemistry 12, 2330–2338.

Hynes, N. E., Groner, B., Sippel, A. E., Nguyen-Huu, M. C. and Schutz, G. (1977). Cell 11, 923–932.

Kwan, S.-P., Wood, T. G. and Lingrel, J. B. (1977). Proc. Nat. Acad. Sci. USA 74, 178–182.

Lai, E. C., Woo, S. L. C., Dugaiczyk, A., Catterall, J. F. and O'Malley, B. W. (1978). Proc. Nat. Acad. Sci. USA 75, 2205–2209.

Lehrach, H., Diamond, D., Wozney, J. M. and Boedtker, H. (1977). Biochemistry 16, 4743–4751.

Mackey, J. K., Brackmann, K. H., Green, M. R. and Green, M. (1977). Biochemistry 16, 4478–4483.

McKnight, G. S. and Schimke, R. T. (1974). Proc. Nat. Acad. Sci. USA 71, 4327–4331.

McKnight, G. S., Pennequin, P. and Schimke, R. T. (1975). J. Biol. Chem. 250, 8105–8110.

McReynolds, L. A., Catterall, J. and O'Malley, B. W. (1977). Gene 2, 217–230.

McReynolds, L. A., O'Malley, B. W., Nisbet, A. D., Fothergill, J. E., Givol, D., Fields, S., Robertson, M. and Brownlee, G. G. (1978). Nature 273, 723–728.

Maxwell, I. H., Maxwell, F. and Hahn, W. E. (1977). Nucl. Acids Res. 4, 241–246.

Monahan, J. J.. Harris, S. E. and O'Malley, B. W. (1976a). J. Biol. Chem. 251, 3738–3748.

Monahan, J. J., Harris, S. E., Woo, S. L. C., Robberson, D. L. and O'Malley, B. W. (1976b). Biochemistry 15, 223–233.

Monahan, J. J., Woo, S. L. C., Liarakos, C. D. and O'Malley, B. W. (1977). J. Biol. Chem.252, 4722–4728.

Nguyen-Huu, M. C., Sippel, A. E., Hynes, N. E., Groner, B. and Schütz, G. (1978). Proc. Nat. Acad. Sci. USA 75, 686–690.

O'Malley, B. W. and Means, A. R. (1974). Science 182, 610–620.

Palmiter, R. D., Moore, P. B., Mulvihill, E. R. and Emtage, S. (1976). Cell 8, 557–572.

Ross, J. (1976). J. Mol. Biol. 106, 403–420.

Ross, J. and Knecht, D. A. (1978). J. Mol. Biol. 119, 1–20.

Tilghman, S. M.,Tiemeier, D. C., Polsky, F., Edgell, M. H., Seideman, J. G., Leder, A., Enquist, L. W., Norman, B. and Leder P. (1977). Proc. Nat. Acad. Sci. USA 74, 4406–4409.

Tilghman, S. M., Curtis, P. J., Tiemeier, D. C., Leder, P. and Weissmann, C. (1978). Proc. Nat. Acad. Sci. USA 75, 1309–1313.

Towle, H. C., Tsai, M.-J., Tsai, S. Y. and O'Malley, B. W. (1977). J. Biol. Chem. 252, 2396–2404.

Tsai, M.J., Schwartz, R. J., Tsai, S. Y. and O'Malley, B. W. (1975). J. Biol. Chem. 250, 5165–5174.

Tsai, S. Y., Roop, D. R., Tsai, M.J., Stein, J. P., Means, A. R. and O'Malley, B. W. (1978). Biochemistry, in press.

Weinstock, R., Sweet, R., Weise, M., Cedar, H. and Axel, R. (1978). Proc. Nat. Acad. Sci. USA 75, 1299–1303.

Woo, S. L. C., Dugaiczyk, A., Tsai, M.-J., Lai, E. C., Catterall, J. F. and O'Malley, B. W. (1978). Proc. Nat. Acad . Sci. USA 75, 3688–3692.

Derman et al. (1981) Cell **23**, 731–739

Transcriptional Control in the Production of Liver-Specific mRNAs

Eva Derman, Kenneth Krauter, Linda Walling,
Carey Weinberger, Martha Ray and
James E. Darnell, Jr.
The Rockefeller University
New York, New York 10021

Summary

cDNA clones complementary to liver mRNA were prepared and used to determine transcription rates of specific genes in isolated nuclei from liver, brain, and hepatoma cells. The cDNA sequences complementary to mRNA found only (or mainly) in the liver hybridize to labeled nuclear RNA only from liver nuclei. It appears that transcriptional events are primarily responsible for the synthesis of these, and perhaps most, tissue-specific moderately abundant mRNAs.

Introduction

Control of gene expression through the control of DNA transcription was first demonstrated at the molecular level for the lactose operon in E. coli (Hayashi et al., 1963). Those initial experiments involved molecular hybridization of pulse-labeled RNA to purified bacteriophage DNA, into which the lactose operon had been recombined. The formation of β-galactosidase mRNA was shown to be at least 30 times greater in induced compared to uninduced cells. Many other examples of transcriptional control, mediated by both positively acting and negatively acting regulatory proteins, have been demonstrated in bacteria by employing hybridization of labeled RNA to specific DNA (see, for example, Bertrand et al., 1975; Lozeron et al., 1976; Lee, 1978).

Our strategy in studying eucaryotic gene regulation has followed the lead of the successful studies of bacterial gene regulation. No molecular technique other than hybridization of pulse-labeled RNA to specific DNA can answer the question of whether, inside a cell, a gene is controlled by regulating its rate of transcription. Other measurements may be necessary to determine if additional levels of control are important for any particular gene.

Hybridization of pulse-labeled nuclear RNA to restriction fragments of adenovirus (Weber et al., 1977; Evans et al., 1977; Evans and Ziff, 1978) located the initiation sites for a number of adenovirus transcriptional units that are responsible for the production of many different adenovirus mRNAs. Knowing the location of various promoters, the transcriptional control pattern for adenovirus mRNA production could be firmly established (Nevins et al., 1979; Wilson et al., 1979). In addition to various types of transcriptional control, two instances of control at the level of mRNA turnover (M. C. Wilson and J. E. Darnell, Jr., manu-

script submitted) and one probable instance of control at the level of differential processing have been established for adenovirus mRNA production (Chow et al., 1979; Shaw and Ziff; 1980, Nevins and Wilson, 1981).

To study the level at which specific cellular genes are controlled, recombinant DNA segments complementary to specific mRNAs can be used in a manner exactly analogous to the use of recombinant bacteriophage DNAs and adenovirus DNA fragments in earlier work. In this report we describe the preparation of DNA segments complementary to specific mRNAs of mouse liver, and the use of these DNA segments to measure the rate of transcription for liver-specific genes in liver nuclei, brain nuclei and nuclei from cultured hepatoma cells. The results suggest that most moderately abundant tissue-specific genes may be regulated at the level of transcription.

Results

Selection of Recombinant cDNA Plasmids

Mammalian liver is composed of relatively few cell types. The majority of liver cells are the so-called parenchymal cells or hepatocytes, which synthesize many proteins that are not made in other cell types or that are synthesized in other cell types in much smaller amounts. Antibiotic-resistant E. coli plasmid DNA, recombined with individual segments of mouse DNA that was copied from unfractionated mouse liver mRNA "cDNA," would likely contain sequences at fairly high frequency that represent liver-specific mRNAs. We therefore constructed recombinant pBR322 cDNA plasmids containing sequences derived by copying DNA from the total poly(A)$^+$ RNA from the livers of adult male mice, using reverse transcriptase, E. coli DNA polymerase I and the dG:dC tailing procedure (Villa-Komaroff et al., 1978). The mouse sequences were inserted at the single site for the restriction enzyme Pst I (see Experimental Procedures). Colonies containing liver-abundant sequences were detected by colony hybridization screen (Young and Hogness, 1977) of about 1000 bacterial colonies. The individual bacterial colonies were grown on replicate Millipore filters, and were hybridized with labeled poly(A)$^+$ RNA from liver, hepatoma cells and mouse L cells. The RNA was labeled by terminal addition of ^{32}P with polynucleotide kinase after alkali breakage to create 5′ OH ends (Spradling et al., 1980).

The result of one screening test is presented in Figure I. Of the 1000 colonies screened in this fashion, 130 colonies gave unambiguous positive signals only with labeled liver poly(A)$^+$ RNA. Of these, about 80 colonies harbored plasmids complementary to the same, very abundant liver mRNA and the remaining 50 colonies represented sequences from different "liver-specific" mRNAs (see below). Also, 50 colonies were selected that gave positive signals with RNA from all three cell types ("common" clones).

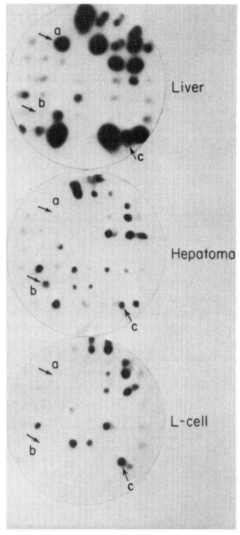

Figure 1. Colony Hybridization with Various Labeled mRNAs

Liver hepatoma cell and L cell poly(A)+ RNA was labeled in vitro (as described in Experimental Procedures) and was used to screen 1000 bacterial colonies harboring recombinant plasmids. An example of such a screen is shown. Arrows: colonies giving autoradiographic signals with liver mRNA only (a), colonies giving the strongest signals with hepatoma cell mRNA (b) and colonies harboring plasmids complementary to mRNAs from all three sources (c).

Characterization of Clones

To ensure that the mouse sequences in the series of individual plasmids represented different mRNA molecules, two tests were carried out. Unlabeled denatured mouse liver mRNA (Rave et al., 1979) was separated by agarose gel electrophoresis and transferred (bound) to Millipore filters (Southern, 1975). These filters were then hybridized to ^{32}P-labeled plasmid DNA, as outlined in Experimental Procedures. Each plasmid complementary to an mRNA of a different size was considered a distinct sequence. In Figure 2 one example is shown, where two different recombinant plasmids hybridized to mRNAs of the same size

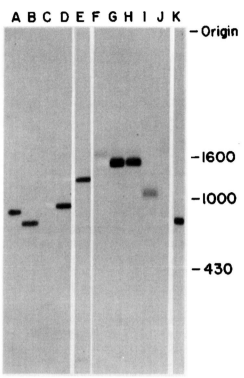

Figure 2. Northern Gel Analysis of Mouse Liver mRNA

Poly(A)-containing mRNA isolated from mouse liver was denatured and subjected to electrophoresis in a 1.25% agarose–formaldehyde gel for 5 hr at 700 V. Following electrophoresis the RNA was transferred to nitrocellulose, baked and cut into individual lanes (A–K) representing about 8 μg of each RNA. The lanes were then hybridized to individual cDNA clones separately by using the two-step hybridization procedure described in Experimental Procedures. At right are the positions of single-stranded DNA markers of 1600, 1000 and 430 nucleotides.

(lanes G and H). All the other lanes show mRNA species of different sizes.

In addition to this "RNA blotting" procedure, different plasmids were also checked for distinct sequence content by the Southern blotting procedure. Purified plasmid DNA was digested with the restriction enzyme Pst I, releasing the inserted mouse DNA sequence (Figure 3A), and the DNA fragments were separated by agarose gel electrophoresis. After transfer of the DNA to nitrocellulose filters, the plasmid DNAs were hybridized to a single ^{32}P-labeled plasmid DNA. Some of the plasmids crossreacted, presumably indicating clones containing portions of the same mouse liver mRNA sequence. In Figure 3B, both the plasmids in lanes I and J appear to have hybridized to the same labeled probe, all the other plasmids did not contain these sequences. Combining the results of the RNA blotting and DNA blotting procedures, some 16 different plasmids with inserts larger than 400 nucleotides were identified.

Table 1 summarizes the properties of the 16 different recombinant plasmids chosen for further studies. ^{32}P-labeled poly(A)+ RNA from liver, brain or cultured

A ABCDEFGHIJKLMNOP

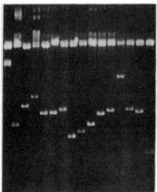

B ABCDEFGHIJKLMNOP

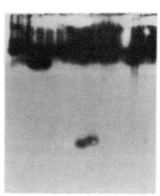

Figure 3. Cross-hybridization Study of Mouse Liver cDNA Clones

(A) Purified plasmids containing double-stranded cDNA segments linked to pBR322 by dG:dC tails were digested with restriction endonuclease Pst I to excise the insert (see Experimental Procedures). Detection of the inserts was accomplished by ultraviolet fluorescence. Approximately 0.5μg of each digest was loaded onto a 1.4% agarose gel and subjected to electrophoresis for 3 hr at 200 V. (B) Following electrophoresis, DNA in the gels was transferred to nitrocellulose, baked and hybridized to nick-translated cDNA clones. The clones in lanes H and I are seen to cross-hybridize to the H probe, but no other clones crosshybridized with this probe.

hepatoma cells was again prepared by alkali breakage and 5′ end-labeling with polynucleotide kinase. Each plasmid DNA bound to a nitrocellulose filter was hybridized to the samples of labeled mRNA, to determine the concentration of specific sequences in the three different mRNA samples. The DNA was in excess of the labeled RNA, as indicated by a failure to detect significant amounts of unbound RNA during a second exposure to filter-bound DNA. Eleven of the plasmids were complementary to liver mRNA but hybridized to less than 10% as much labeled mRNA from brain cells. (These were designated plivS-1 through -11, for plasmids with liver-specific sequences.) Two of the pliv-S plasmids hybridized to about 15–20% as much mRNA from hepatoma as from liver cells; the remaining plasmids hybridized very little hepatoma cell mRNA. Five of the cloned DNAs hybridized to about equal amounts of mRNA from the liver, brain and hepatoma cells. (These plasmids were designated pliv-C, for common sequences cloned from liver.) The concentration of mRNA (Table 1) was recorded as the fraction of poly(A)⁺ RNA that was complementary to a particular clone after pancreatic RNAase digestion. The individual clones hybridized to between 0.015% and 0.4% of the labeled liver mRNA. Because the length of the recombinant mouse DNA segments averages about one half of the average mRNA length, the real abundance of the mRNAs ranged between 0.03 and 0.8% of total mRNA. The one exception is plasmid pliv-S1, which hybridized to as much as 8% of mRNA from male liver. Our current evidence indicates that this plasmid is derived from mRNA for major urinary proteins (E. Derman, unpublished results). Assuming that liver cells contain 5×10^5 mRNA molecules per cell (Hastie & Bishop, 1976), the number of liver mRNA molecules complementary to the eleven clones range between 150 and about 4000. The average abundance classes estimated by Hastie and Bishop were 15 copies per cell for 12,000 different sequences, about 300 copies per cell for about 350 different sequences and 12,000 copies per cell for about 10 different sequences. Thus, the sequences

Table 1. Characterization of liv-pBR Plasmids

	Insert Size (bp)[a]	mRNA Size (bp)[b]	Percent Steady-State Poly(A)⁺ RNA[c]		
			Hepatoma Male Liver Cells		Brain
plivS-1	850	980	8.00	*	*
plivS-2	1350	1700	0.40	*	*
plivS-3	400	1500	0.20	*	*
plivS-4	300	2050	0.17	0.03	*
plivS-5	600	800	0.10	*	0.01
plivS-6	550	2200	0.10	0.02	0.003
plivS-7	1720	1750	0.07	*	*
plivS-8	600	1050	0.07	0.002	*
plivS-9	500	1700	0.05	*	*
plivS-10	700	1650	0.04	0.004	*
plivS-11	500	1600	0.015	*	*
plivC-12	700	940	0.25	0.22	0.32
plivC-13		850	0.17	0.11	0.12
plivC-14	750	750	0.15	0.10	0.13
plivC-15	1100	1250	0.27	0.17	0.12
plivC-16	640	735	0.35	0.30	0.15

[a] The size of the insert was determined by Pst I digestion of the recombinant plasmid and electrophoresis on agarose gels, as described in Experimental Procedures and illustrated in Figure 3.
[b] The size of mRNA was determined by hybridizing plasmid DNA to Northern blots of liver poly(A)⁺ RNA, as described in Experimental Procedures and illustrated in Figure 2.
[c] Poly(A)⁺ RNA was prepared and kinased in vitro, as described in Experimental Procedures, and hybridized to an excess of plasmid DNA bound to Millipore filters. The fraction of total poly(A)⁺ RNA that remains bound to plasmid DNA after pancreatic RNAase digestion is expressed as percentage of steady-state poly(A)⁺ RNA; each figure is the average of at least three measurements. In all cases the variation was less than 0.5–1.5 times the value shown.
* ≤ 0.001%—that is, 10 copies per cell.

we have selected for study cover the moderately abundant range. Because the colony screening procedure with the ³²P-kinased mRNA is based on DNA excess hybridization, the signal-to-noise ratio for mRNAs of very low abundance presumably would not be large enough to allow detection of colonies harboring plasmids complementary to scarce mRNAs.

Transcription in Isolated Nuclei: Comparison of Transcription Rate in Vivo and in Vitro

Having established that eleven of the plasmids contained liver-specific mRNA sequences, we could determine the step or steps during mRNA biosynthesis responsible for this difference. Hypothetically, regulation might occur at three possible levels: transcription, differential processing of primary transcripts or differential mRNA turnover. To be certain which step or steps might participate in gene regulation it is necessary to be able to separately monitor each step leading to the formation of mature mRNA (for review see Darnell, 1979). However, since the first step in the biosynthesis of mRNA is the transcription of the mRNA precursor from the template DNA, the rate of transcription of the liver-specific sequences was examined first. Only if transcriptional differences could not account for the differential abundance of tissue-specific mRNA would it be necessary to investigate further the other levels of regulation.

An ideal measurement of the rate of transcription of a transcriptional unit should be made under conditions that ensure that the synthesis of all sequences is equimolar. This can be accomplished in vivo only in a short pulse, by measuring the rate of transcription of yet unprocessed nascent RNA (Derman et al., 1976; Evans et al., 1977; Evans et al., 1979). It is not possible to obtain nascent labeled RNA of specific activity high enough to measure individual gene transcription rates in intact mouse liver. However, isolated nuclei from tissues can be prepared and pulse-labeled to yield enough radioactive RNA. Available evidence suggests that the majority of RNA synthesis in isolated nuclei represents elongation of already initiated RNA chains (Evans et al., 1977), and that RNA processing in isolated nuclei is only partially carried out even on already completed labeled RNA chains (Blanchard et al., 1978). In addition, Hofer and Darnell (1981) have demonstrated that the proportion of labeled hemoglobin-specific RNA precursor relative to total newly synthesized RNA is similar in intact Friend cells and Friend cell nuclei, and that the synthesis of different segments of the hemoglobin transcriptional unit is equimolar in isolated nuclei. Nascent RNA labeled in isolated nuclei therefore should not be subject to hypothetical regulatory mechanisms involving differential processing of some mRNA precursors and not others. Other transcriptional studies of specific polymerase II products have also suggested that there exists a similarity in the action of polymerase II in vivo and in vitro. For example, the incorporation of ^{3}H–uridine by whole cells, or ^{3}H–UTP by isolated nuclei into adenovirus-specific sequences was similar, both early in infection when virus RNA represents 0.2% of the total and late in infection when it is 30% of the total (Weber et al., 1977; Evans et al., 1977; Weber, 1979).

In order to validate more generally the use of isolated nuclei to measure specific gene transcription rates, we have compared the percentage of pulse-labeled nuclear RNA from isolated nuclei and from intact cultured CHO cells that was complementary to several cloned DNA sequences containing CHO mRNA sequences (Harpold et al., 1979; Table 2). The relative transcription rates were quite similar in whole cells and in isolated nuclei for this set of genes. The percentage of newly labeled RNA complementary to each of the genes was somewhat higher in whole cells. This may reflect a changed balance of polymerase II activity relative to polymerases I and III in isolated nuclei, or the fact that a 10 min pulse in vivo is not an ideal pulse (see above). But we wish to emphasize that the transcription ratio between different mRNA sequences remains the same in isolated nuclei as inside the cell.

Measurement of Specificity of Nuclear Transcription

Isolated nuclei were prepared from mouse liver, brain and from cultured hepatoma cells; the newly synthesized nuclear RNA was labeled by incorporation of

Table 2. Transcription of Five Different mRNA Sequences in Pulse-Labeled Cells and in Isolated Nuclei

| Clone | Pulse-Labeled Cells (10 min) Input: 40×10^6 cpm RNA | | | Isolated Nuclei Input: 90×10^6 cpm RNA | | |
	cpm Hybridized	Fraction of RNA Synthesis $\times 10^{-6}$	Relative Transcription Rate	cpm Hybridized	Fraction of RNA Synthesis $\times 10^{-6}$	Relative Transcription Rate
A	1555	38.9	1.00	1336	14.8	1.00
B	844	21.1	0.54	1007	11.2	0.76
C	539	13.5	0.35	399	4.4	0.30
F	280	7.0	0.18	338	2.4	0.25
D	260	6.5	0.17	218	3.7	0.16

The CHO clones A–D, growth of CHO cells, extraction of pulse-labeled RNA and hybridization conditions are as described in Harpold et al. (1979). In vitro transcription was performed in sucrose-purified nuclei incubated in the presence of 1 mCi α–^{32}P–UTP as described in Experimental Procedures.

α–^{32}P–UTP. The labeled nuclear RNA was then hybridized to excess individual plasmid DNA samples bound to individual nitrocellulose filters, and the RNAase-resistant hybrids were measured.

All eleven of the cloned DNAs designated as liver-specific hybridized to labeled nuclear RNA from liver nuclei (Table 3), and hybridized to less than 10% as much, labeled RNA from brain. With one exception (plivS-6) the same was true for labeled nuclear RNA from hepatoma nuclei. The labeled brain or hepatoma nuclear RNA that did hybridize to liver-specific clones above the background was only in the range of 1–5% of the amount that liver-specific RNA hybridized. Whether this low level of hybridization (only twice background) was significant is not possible to state at present. In contrast, all five of the clones that were complementary to liver, brain and hepatoma mRNAs (plivC12–16) hybridized to about the same proportion of labeled nuclear RNA from liver, brain and hepatoma

nuclei. There was variation from experiment to experiment in the proportion of labeled RNA complementary to some of the clones—for example, clones 1 and 6. These variations were not due to inadequacy of DNA on the filters, because supernatants were checked and they did not contain substantial additional hybridizable RNA. Variations in polymerase I, II and III activity might occur, and true differential transcription in different groups of animals might occur. In any case a general conclusion seems valid: for the eleven liver-specific mRNA sequences studied, differential expression in different cells appears to be accounted for mainly on the basis of differential transcription.

Several interesting points about the transcription rates of the liver-specific and the "common" sequences were apparent. First, there was a general correspondence between the rate of transcription in isolated nuclei and the cell concentration of the liver-specific sequences (plivS-1 through -11). Second,

Table 3. Transcription of Liver mRNA Sequences in Liver, Brain and Hepatoma Nuclei[a] and Hepatoma Cells[b]

Clone	Insert Size	Liver Nuclei — cpm RNA Hybridized — Input: 6×10^7	Input: 9×10^7	Liver — Average Transcription Rate $\times 10^{-6c}$	Brain Nuclei — cpm RNA Hybridized — Input: 8×10^7	Input: 8×10^7	Brain — Average Transcription Rate $\times 10^{-6c}$	Hepatoma — cpm RNA Hybridized — Nuclei Input: 3×10^7	Hepatoma — Pulse-labeled Cells Input: 4×10^7	Hepatoma — Average Transcription Rate $\times 10^{-6c}$
plivS-1	850	6,150	18,300	165.0	10	30	0.2	20	20	0.5
plivS-2	1,350	2,805	2,995	38.6	9	49	0.3	34	7	0.6
plivS-3	400	2,429	1,106	23.5	42	114	1.0	66	77	2.0
plivS-4	300	1,273	952	14.8	71	52	0.8	22	52	1.0
plivS-5	600	1,215	1,007	15.0	20	33	0.3	40	26	0.9
plivS-6	550	1,519	371	12.6	2	44	0.3	66	170	3.3
plivS-7	1,720	573		9.5		30	0.4	30	30	0.8
plivS-8	600	485	278	5.0	0	33	0.3	16	50	0 9
plivS-9	500	1,105	1,257	15.7	6	33	0.3	30	47	1.0
plivS-10	700	547	248	5.3	6	33	0.3	25	37	0.9
plivS-11	500	301	154	3.0	0	18	0.1	20	20	0.5
plivC-12	700	221	470	4.5	223	148	2.3	152	193	5.0
plivC-13		81	115	1.3	213	275	3.0	198	120	4.5
plivC-14	740	112	140	1.8	74	182	1.6	40	60	1.4
plivC-15	1,100	183	320	3.3	269	270	3 4	470	285	10.2
plivC-16	640	167	185	2.4		80	1.0	12	95	3.0
pBR322 (control)		27	47		38	27		30	33	

[a] The nuclei were obtained from fresh tissues or from monolayer hepatoma cells and were incubated in the presence of 0.5–1 mCi α–^{32}P–UTP as described in Experimental Procedures. For one experiment, livers from 6 animals and brains from 50 animals were needed.

[b] Monolayer cells were labeled for 10 min with 10 mCi ^{3}H–uridine. Nuclear RNA was extracted, freed of DNA and hybridized to an excess of plasmid DNA bound to Millipore filters.

[c] Average transcription rate is the fraction of total RNA synthesized in 20 min in isolated nuclei (or in 10 min in pulse-labeled hepatoma cells) that is hybridized to excess plasmid DNA averaged over two experiments. In all experiments transcription of ribosomal RNA constituted 40–50% of total transcription.

some of the common mRNAs (for example, plivC-12, -13, -14, -15 and -16) and the liver-specific mRNAs (plivS-3–6) were present at about the same concentration in the liver. The rates of transcription for the liver-specific group, however, averaged about 3 to 10 fold higher than the common group. The implication of this measurement is that the liver-specific mRNAs either have a shorter half-life or are less efficiently processed. Third, the concentration of the liver-specific sequences in the total labeled RNA from isolated nuclei was about $1/100$ of that in the liver mRNA. Even allowing for possible extra polymerase III action in in vitro nuclei and for about 50% of RNA transcription to be due to polymerase I transcription of rRNA genes, this result emphasizes the earlier conclusion (Harpold et al., 1979) that many RNA sequences in the nucleus are not found among the stable poly(A)$^+$ cytoplasmic mRNA.

Discussion

During animal development a pattern of differential gene expression is established, resulting in mRNA populations that are distinguishable from one cell type to another in both qualitative and quantitative fashion (see, for example, Hastie and Bishop, 1976). The first step in determining how such differential distribution of mRNA is achieved is to prepare pure DNA probes for genes expressed preferentially in one cell type. With the availability of pure DNAs that are complementary to a broad set of tissue-specific mRNAs, it should be possible to identify the regulatory step(s) in mRNA biosynthesis.

Our results, with a group of cloned DNA sequences complementary to moderately abundant liver-specific mRNA, show that the nuclei from brain cells either do not transcribe the liver-specific mRNA sequences or transcribe these sequences at a rate $1/10$ to $1/50$ of that in liver cell nuclei. These results demonstrate that transcriptional events have a primary role in establishing and maintaining the differential gene expression of these two terminally differentiated cell types. That transcriptional events primarily determine mRNA concentration in eucaryotic cells has been suggested since the first description of the Jacob-Monod model (1961). For example, differential puffing patterns in insects (reviewed by Daneholt, 1975) have been interpreted as a demonstration of differential transcription activity of insect genes. More recently, with the ability to measure the transcriptional activity of individual genes, evidence has been reported for increased nuclear RNA synthesis of ovomucoid sequences in oviduct nuclei (Tsai et al., 1978) and for α–2–globulin synthesis in rat liver nuclei (Chan et al., 1978). Until the present work no survey has been available of tissue-specific RNA synthesis rates for a number of different mRNAs that have a broader range of concentrations.

The tissue-specific mRNA pattern, or in fact differentiation itself, is often thought of as infrequently reversible (Mintz, 1974). Possible exceptions to this proposal are malignant cells. Frequently in tumor cells the synthesis of tissue-specific proteins of the presumed cell of origin is reduced or abolished and synthesis of new proteins is observed. At which level does this "deregulation" occur? The experiments presented here show that the decrease in concentration in hepatoma cells of the liver-specific mRNAs is paralleled by a decrease in the transcription of these sequences. The origin of hepatoma cells is uncertain, however. If they arise from "dedifferentiation," our experiments suggest they lose the tissue-specific transcriptional pattern. If hepatoma cells instead arise by proliferation of immature stem cells, then the interpretation of our measurements is that the liver-specific genes are not yet transcriptionally active in stem cells.

A technical limitation of our experiments should be noted. The amount of labeled nuclear RNA from liver cells complementary to most of the liver-specific genes was relatively small. Thus the level of confidence in concluding that brain and hepatoma nuclei do not synthesize these RNAs is limited to a factor between 10 and 50, based on the amount of labeled nuclear RNA hybridized on the DNA filters compared with background. However, the presence of these sequences as mRNAs in brain and hepatoma could be measured with confidence to be $1/100$ to $1/500$ of that in liver. Thus it is possible that some low level of transcription of all genes might proceed in all cells at all times. Even in bacterial cells, where reversible control of gene expression is thought of as "on" or "off," there is a low level of transcription equivalent to 0.1–1% of fully induced levels for most genes (Kumer and Szybalski, 1969). In mammalian cells a low level of nuclear RNA might also be synthesized and remain unprocessed, or a low level of resulting mRNA might not be stabilized. But the rate-limiting step for mRNA production, and therefore the point of control for these liver-specific mRNAs, appears to be transcription.

There have been a number of experiments based on reassociation kinetics with mammalian nuclear RNA (Chikaraishi et al., 1978) and with sea urchin nuclear RNA, in which there is apparent extensive overlap of sequence content of the nuclear RNA sequences. This has been reported particularly in sea urchins for oocyte and blastula tissue compared with specialized cells or gastrula cells (Kleene and Humphrey, 1976; Wold et al., 1978; Sheperd and Nemer, 1980). Several points about these results are worth making in light of the apparent transcriptional control reported here. First, most genes may in fact be transcribed in all cells at very low levels. Second, the scarce mRNAs that are not included in our group of cloned sequences, and which may represent transcripts of the majority of all active genes, might be regulated posttranscriptionally. Also, genes active in early embryonic cells such as those studied in sea urchin development may possibly be regulated by a different mechanism than those active later in development in the adult. However, to distinguish between transcriptional and post-transcriptional events in the regulation of mRNA production, the transcription rate of individual gene sequences must be studied. Only when it is established that the rate of synthesis of RNA from a gene is the same in two tissues but that the mRNA is differentially present can posttranscriptional control be invoked.

We have scored the rate of synthesis of RNA complementary to cDNA segments. The position of these segments within the transcription units relative to the 5′ end of the mRNAs was not established. For this reason these studies are unable to distinguish between two known modes of modulating transcription rates. The mechanisms to be considered are a change in the frequency of initiation of RNA chains, and a change in the proportion of premature termination products relative to complete transcripts (see for example Szybalski et al., 1970; Bertrand et al., 1975).

Experimental Procedures

Animals
Inbred NCS Swiss white male mice (25–30 g) were obtained from the mouse colony maintained at the Rockefeller University.

Cell Lines
Hepatoma cells Hepa I (Darlington et al., 1980), a gift of G. Darlington, were maintained in monolayers in MAB medium supplemented with 10% fetal calf serum. CHO cells were grown in minimal Eagle's medium, 7% calf serum and 1% nonessential amino acids.

Preparation of Poly(A)⁺ RNA from Tissues and Cell Lines
Total cellular RNA was extracted essentially as described by Ullrich et al., (1977), in order to avoid nuclease action. Because in cultured cells most if not all poly(A)⁺ nuclear DNA is represented in the cytoplasmic mRNA (Herman et al., 1976), over 90–95% of the poly(A) is cytoplasmic (Puckett et al., 1975) and finally the size of the poly(A)⁺ molecules that hybridize to our clones is appropriate for mRNA [not poly(A)⁺ hnRNA], we assume we have cloned sequences present in cytoplasmic poly(A)⁺ mRNAs. Tissues were homogenized in 5 M guanidinium thiocyanate (Fluka-Tridom), 0.1% lithium dodecyl sulfate, 50 mM lithium citrate, pH 7.0, and 0.1 m β–mercaptoethanol in a high speed blender. Nucleic acids were precipitated by adjusting the homogenate to 30 mM acetic acid and subsequently adding an equal volume of 95% ethanol. After chilling at −20°C, the precipitate was collected and dissolved in 7.5 M guanidine hydrochloride, 25 mM lithium citrate, 1 mM dithiothreitol. DNA was then precipitated by addition of acetic acid to a concentration of 50 mM and half the volume of 95% ethanol at −20°C. Precipitated RNA was dissolved in 10 mM EDTA, 10 mM Tris, pH 7.4, and freed of impurities by extraction with 4:1 (v/v) chloroform:isobutanol. Poly(A)⁺ RNA was obtained by two passages of RNA through an oligo(dT)–cellulose column (Aviv and Leder, 1972). Poly(A)⁺ RNA from the cytoplasm of hepatoma and CHO cells was prepared as described by Harpold et al. (1979).

Construction of cDNA Recombinant Clones
Reagents
pBR322 DNA was purified from E. coli HB101 according to the method of Norgard et al. (1979). Reverse transcriptase was supplied by J. W Beard. The Klenow fragment of E. coli DNA polymerase I was purchased from New England Biolabs; Pst I restriction enzyme

and terminal transferase were purchased from BRL. S1 nuclease was a gift from J. Logan.

Double-Stranded cDNA Synthesis
Poly(A)⁺ RNA from adult male mouse liver was used as a template for synthesis of double-stranded cDNA essentially as described by Efstradiatis et al. (1976).

cDNA was synthesized in a reaction containing 10 μg poly(A)⁺ RNA, 25 μg / ml oligo(dT), 30 μg /ml actinomycin D, 50 mM Tris, pH 8.3, 10 mM MgCl₂ and 100 U reverse transcriptase in a final volume of 200 μl at 42°C for 1 hr. After digestion of template RNA with heat-inactivated pancreatic RNAase, cDNA served as a template for the second-strand synthesis. The reaction in 1 ml total volume contained 1 μg cDNA, 70 mM KCl, 10 mM MgCl₂, 0.5 ⟩ mM dXTPs, 0.1 M Hepes 6.5, 10 mM DTT and 100 U Klenow fragment of DNA polymerase I, and was carried out at 15°C for 20 hr.

After S1 digestion of hairpin loops the double-stranded cDNA was sedimented through a 15–30% sucrose gradient. Molecules larger than 400 bp were collected. Selected double-stranded DNA was then extended by about 50 nucleotides with dC tails in a 50 μl reaction containing 200 ng double-stranded DNA, 10 U terminal transferase, 100 mM potassium cacodylate, pH 7.4, 2 mM CoCl₂, 1 mM EDTA, 50 μM dCTP at 37°C for 30 min.

Fifty nanograms of dC-tailed double-stranded DNA were annealed to 350 ng of Pst I-cut pBR322 plasmid that had been "tailed" with 20 dG residues. Annealed molecules were used to transfect E. coli X1776 as described by Villa-Komaroff et al. (1978), except that E. coli were heat-shocked at 37°C. Ninety percent of transformants carried recombinant plasmids—that is, their phenotype was Tet⁺ Amp⁻.

Selection of Liver-Specific Recombinant pBR322 Plasmids
About 1000 clones were picked and inoculated onto nitrocellulose filters (Millipore) that were placed on selective nutrient agar plates. After colonies formed, replicates were made as described by Villa-Komaroff et al. (1978). The colonies were screened in duplicate according to the method of Young and Hogness (1977) with in vitro-labeled poly(A)⁺ RNA from male mouse liver, hepatoma cells and L cells.

Preparation of ³²P-Labeled RNA with Polynucleotide Kinase
About 5 μg of poly(A)⁺ RNA was partially hydrolyzed in 100 μl of 50 mM Tris, pH 9.5, 5 mM glycine, 100 μM spermidine and 10 μM EDTA at 90°C for 30 min. Broken RNA was reacted with 150 μCi γ–³²P–ATP (1000–3000 Ci/mmole; New England Nuclear) as substrate and 10 U T4 polynucleotide kinase (P. L. Biochemicals). RNA of specific activity 1–2 × 10⁷ cpm/μg was routinely obtained (Spradling et al., 1980).

Preparation of Nuclei
Nuclei were obtained from mouse liver and brain essentially as described by Tata (1974). One gram of tissue was homogenized in 5 ml of cold 0.32 M sucrose, 3 mM MgCl₂, 1 mM Hepes, pH 6.8, in a motor-driven loose Potter homogenizer. The crude nuclear pellet was resuspended in 2.1 M sucrose, 1 mM MgCl₂, 1 mM Hepes, pH 6.8, and spun for 60 min at 20,000 rpm in an SW40 rotor. The nuclear pellet was washed with 0.25 M sucrose, 1 mM MgCl₂, 1 mM Hepes, pH 6.8, and then with the reaction buffer used for in vitro transcription.

In Vitro Transcription in Isolated Nuclei
We incubated 3×10⁸ nuclei for 20 min at 25°C in 2 ml of total volume With 0.5–1 mCi α–³²P–UTP (400 Ci/mmole; New England Nuclear) in a reaction mixture containing 5 mM MgCl₂, 1 mM MnCl₂, 10 mM Tris, pH 8.0, 0.14 M KCl, 14 mM β-mercaptoethanol, 10% glycerol, 1 mM each of ATP, GTP, CTP. The incorporation for the different cell nuclei was similar. Different preparations from different sources varied over a range of approximately 2 fold.

Nuclear RNA was extracted as described previously (Soeiro and Darnell, 1969), except that proteinase K (200 μg/ml) was added to the lysed nuclei prior to phenol extraction of RNA.

Hybridization of RNA to Excess DNA

Plasmid DNA was boiled for 5 min in 0.1 N NaOH and bound to 0.45 Millipore HA nitrocellulose filters as described by Melli et al. (1975). Hybridization of labeled RNA to at least a 10 fold excess of complementary DNA was performed in $2 \times$ TESS buffer (0.01 M TES, pH 7.4, 0.3 M NaCl, 0.01 M EDTA, 0.2% SDS) at 65°C for 40 hr or in 30% formamide (BRL), 0.1 M Pipes, pH 7.0, 0.3 M NaCl, 10 mM EDTA, 0.2% SDS at 45°C for 72 hr. Carrier yeast RNA (100 μg/ml) and poly(A) (100 μg/ml) were added to the hybridization buffers. Following hybridization, filters were washed extensively at 65°C and then briefly at 70°C in $2 \times$ TESS buffer and treated with pancreatic RNAase in the same buffer minus EDTA and SDS and finally with proteinase K (200 μg/ml). On occasion the supernatants of hybridization were rehybridized to ensure completeness of hybridization.

Gel Electrophoresis of DNA and RNA

Purified cDNA-containing plasmids were digested with restriction endonuclease Pst I according to the instructions of the supplier. Digests were then adjusted to 10% glycerol, 0.001% bromophenol blue, and subjected to agarose gel electrophoresis in 1.25% gels using 0.05 M Tris–borate, pH 8.3, 1 mM EDTA buffer for 3 hr at 200 V.

Poly(A)-containing mRNA was ethanol-precipitated and resuspended at 0.5 mg/ml in 50% formamide, 6%, formaldehyde, 0.02 M borate, pH 8.3, 10% glycerol, 0.2 mM EDTA. Samples were heated to 65°C for 2 min to fully denature the RNA (Rave et al., 1979), then loaded onto 1.25% agarose gels buffered with 0.02 M borate, pH 8.3, 0.2 mM EDTA, 3% formaldehyde as the electrode buffer. Electrophoresis was for 5 hr at 200 V. ^{32}P-labeled DNA restriction fragments were run in a separate well as size markers.

Blotting and Hybridization of DNA and RNA

Size-fractionated DNA was transferred from agarose gels to nitrocellulose sheets as described by Southern (1975). After baking blots at 80°C in vacuo, blots were prehybridized for 6–20 hr in $6 \times$ SSC, $1 \times$ Denhardt's solution (0.02%, BSA, 0.02% Ficoll, 0.02% polyvinyl pyrrolidone), 50 μg/ml denatured salmon sperm DNA, 50 μg/ml poly(A), 50 μg/ml poly(C) at 65°C. Hybridization was performed for 24–36 hr at 65°C in the same buffer plus 50,000 cpm/ml of denatured nick-translated specific plasmid (spec. act. 5×10^7 cpm/μg; Maniatis et al., 1975).

RNA fractionated on formaldehyde gels was transferred to nitrocellulose in the same manner as DNA, except the RNA gels required no alkali pretreatment nor salt equilibration. $10 \times$ SSC was used as the blotting buffer. RNA of sizes ranging from about 100–6000 nucleotides transferred to the nitrocellulose quantitatively, as assayed by the transfer of labeled RNA (data not shown; B. Seed, unpublished observations).

Hybridization of RNA blots to specific cDNA clones was accomplished by a modification of the sandwich hybridization procedure of Dunn and Hassell (1977). Nitrocellulose sheets onto which the denatured RNA transferred were baked for 2 h in vacuo at 80°C. Filters were then prehybridized in 50% formamide, $5 \times$ SSC, $1 \times$ Denhardt's solution (Denhardt, 1966), 50 mM NaPO$_4$, pH 7.4, 50 μg/ml salmon sperm DNA, 50 μg/ml poly(A), 50 μg/ml poly(C) for 4–20 hr at 45°C. After prehybridization, filters were incubated with 1 μg/ml of specific cDNAs cloned in pBR322 in the same buffer at 45°C for 12 hr. Following the first round of hybridization, the blots were washed 3 times in $2 \times$ SSC, 0.2% SDS for 5 min at room temperature, then incubated with 5×10^5 cpm/ml of nick-translated pBR322 (spec. act. 1×10^8 cpm/μg) in the prehybridization buffer at 37°C for 16 hr.

All filters were washed after hybridization 4 times in $2 \times$ SSC, 0.2% SDS for 15 min each at room temperature, followed by two 15 min washes in $0.1 \times$ SSC, 0.2% SDS at 50°C. Damp filters were wrapped in plastic wrap and autoradiographed using DuPont Cronex Lightning-plus intensifying screens. RNA species representing about 0.015% of total poly(A)$^+$ cellular RNA could easily be detected with a few days of exposure under these conditions.

Acknowledgments

E. D. would like to thank Dr. A. Bothwell for his generous advice concerning the constructions of recombinant EDNA clones and Dr. Allan Spradling for suggesting the use of kinased RNA. We would like also to acknowledge the help of Marianne Salditt-Georgieff in performing experiments presented in Table 1. This work was supported by grants from the National Institutes of Health and the American Cancer Society.

Eva Derman was supported by a Helen Hay Whitney Fellowship and an NIH training grant. Ken Krauter is an American Cancer Society Fellow and Linda Bourque is a recipient of an National Institutes of Health fellowship.

The costs of publication of this article were defrayed in part by the payment of page charges. This article must therefore be hereby marked "*advertisement*" in accordance with 18 U.S.C. Section 1734 solely to indicate this fact.

Received October 23, 1980; revised December 9, 1980

References

Aviv, H. and Leder, P. (1972). Purification of biologically active globin messenger RNA by chromatography on oligothymidylic acid cellulose. Proc. Nat. Acac. Sci. USA 69, 1408–1412.

Bertrand, K., Korn, L., Lee, F., Platt, T., Squires, C. L., Squires C. and Yanofsky, C.. (1975). New features of the regulation of the tryptophan operon. Science 189, 22–26.

Blanchard, J. M., Weber, J., Jelinek, W. and Darnell, J. E., Jr. (1978). In vitro RNA–RNA splicing in adenovirus 2 mRNA formation. Proc. Nat. Acad. Sci. USA 75, 5344–5348.

Chan, K. M., Kurtz, D. T. and Feigelson, P. (1978). Transcription of the α2–globulin gene in male rat liver nuclei in vitro. Biochemistry 17, 3092 –3096.

Chikaraishi, D. M., Deeb, S. and Sueoka, N. (1978). Sequence complexity of nuclear RNA in adult rat tissues. Cell 13, 111–120.

Chow, L. T., Lewis, J. B. and Broker, T. R. (1973). RNA transcription and splicing at early and intermediate times after adenovirus-2 infection Cold Spring Harbor Symp. Quant. Biol. 44, 401–414.

Danekolt, B. (1975). Transcription in polytene chromosomes. Cell 4, 1–9.

Darlington, G. J., Bernhard, H. P., Miller, R. A. and Ruddle, F. H. (1980). Expression of liver phenotypes in cultured mouse hepatoma cells. J. Nat. Cancer Inst. 64, 809–819.

Darnell, J. E. (1979). Steps in processing of mRNA: implications for gene regulation. In Gene to Protein: Information Transfer in Normal and Abnormal Cells. T. R. Russell, K. Brew, H. Faber and J Schultz, eds. (New York: Academic Press), pp. 207–208.

Denhardt, D. (1966). A membrane-filter technique for the detection of complementary DNA. Biochem. Biophys. Res. Commun. 23, 641–646.

Dermam, E., Goldberg, S. and Darnell, J. E., Jr. (1976). hnRNA in HeLa cells: distribution of transcript sizes estimated from nascent molecular profile. Cell 9, 465–472.

Dunn, A. R. and Hassel, J. (1977). A novel method to map transcripts: evidence for homology between an adenovirus mRNA and discrete multiple regions of the viral genome. Cell 12, 23–36.

Efstradiatis, A., Kafatos, F. C., Maxam, A. M. and Maniatis, T. (1976). Enzymatic in vitro synthesis of globin genes. Cell 7, 279–288.

Evans, R. M. and Ziff, E. (1978). The promoter and capped 5′ terminus of RNA from the adenovirus-2 major late transcription unit. Cell 15, 1463–1475.

Evans, R. M., Fraser, N., Ziff, E., Weber, J., Wilson, M. and Darnell J. E., Jr. (1977). The initiation sites for RNA transcription in Ad2 DNA. Cell 12, 733–739

Evans, R., Weber, J., Ziff, E. and Darnell, J. E., Jr. (1979). Premature termination during adenovirus transcription. Nature 278, 367–370.

Harpold, M. H., Evans, R. M., Salditt-Georgieff, M. and Darnell, J. E., Jr. (1979). Production of mRNA in Chinese hamster cells: relationship of the rate of synthesis to the cytoplasmic concentration of nine specific mRNA sequences. Cell 17, 1025–1035.

Hastie, N. D. and Bishop, J. O. (1976). The expression of three abundance classes of messenger RNA in mouse tissues. Cell 9, 761–774.

Hayashi, M., Spiegelman, S., Franklin, N. C. and Luria, S. E. (1963). Separation of the RNA message transcribed in response to a specific inducer. Proc. Nat. Acad. Sci. USA 49, 729–736.

Herman, R. C., Williams, J. G. and Penman, S. (1976). Message and non-message sequences adjacent to poly(A) in steady state hnRNA of HeLa cells. Cell 7, 429–438.

Hofer, E. and Darnell, J. E., Jr. (1981). The primary transcription unit of the mouse β-major globin gene. Cell 23, 585–593.

Jacob, F. and Monod, J. (1961). Genetic regulatory mechanisms in the synthesis of proteins. J. Mol. Biol. 3, 318–356.

Kleene, K. C. and Humphreys, T. (1977). Similarity of hnRNA sequences in blastula and pluteus stage sea nuclei embryos. Cell 12, 143–155.

Kumer, S. and Szybalski, W. (1969). Orientation of transcription of the lac operon and its repressor gene in E. coli. J. Mol. Biol. 40, 145–151.

Lee, N. (1978). Molecular aspects of ara regulation. In The Operon, J. H. Miller and W. S. Reznikoff, eds. (New York: Cold Spring Harbor Laboratory Monograph Series), pp. 389–410.

Lozeron, H. A., Dahlberg, J. E. and Szybalski, W. (1976). Processing of the major leftward mRNA of coliphage lambda. Virology 71, 262–277.

Maniatis, T., Jeffrey, A. and Kleid, D. G. (1975). Nucleotide sequence of the rightward operator of phage. Proc. Nat. Acad. Sci. USA 72, 1184–1188.

Melli, M., Ginelli, E., Corneo, G. and diGernia, R. (1975). Clustering of the DNA sequences complementary to repetitive nuclear RNA of HeLa cells. J. Mol. Biol. 93, 23–38.

Mintz, B. (1974). Gene control of mammalian differentiation. Ann. Rev. Genetics 8, 411–470.

Nevins, J. R. and Wilson, M. C. (1981). Expression of the adenovirus-2 major late transcription unit during early infection: regulation at the level of transcription termination and RNA processing. Nature, in press.

Nevins, J. R., Ginsberg, H. S., Blanchard, J. M., Wilson, M. C. and Darnell, J. E., Jr. (1979). Regulation of the primary expression of the early adenovirus transcription units. J. Virol. 32, 727–733.

Norgard, M. V., Emigholz, K. and Monohan, J. J. (1979). Increased amplification of pBR322 plasmid deoxyribonucleic acid in Escherichia coli K-12 strains RR1 and X1776 grown in the presence of high concentration of nucleoside. J. Bacteriol. 138, 270–272.

Puckett, L., Chambers, S. and Darnell, J. E., Jr. (1975). Short-lived mRNA in HeLa cells and its impact on the kinetics of accumulation of cytoplasmic poly(A). Proc. Nat. Acad. Sci. USA 72, 389–392.

Rave, N., Ckvenjakou, R. and Boedtker, H. (1979). Identification of procollagen mRNAs transferred to DBM paper from formaldehyde agarose gels. Nucl. Acids Res. 6, 3559–3567.

Shaw, A. R. and Ziff, E. B. (1980). Transcripts from the adenovirus-2 major late promoter yield a single early family of 3' coterminal mRNAs and five late families. Cell 22, 905–916.

Sheperd, G. W. and Nemer, M. (1980l. Developmental shifts in frequency distribution of polysomal mRNA and their post-transcriptional regulation in sea urchin embryo. Proc. Nat. Acad. Sci. 77, 4653–4654.

Soeiro, R. and Darnell, J. E., Jr. (1969). Competition hybridization by "presaturation" of HeLa cell DNA. J. Mol. Biol. 44, 551–562.

Southern, E. M. (1975). Detection of specific sequences among DNA fragments separated by gel electrophoresis. J. Mol. Biol. 98, 503–517.

Spradlinq, A. C., Bigan, M. E., Mahowald, A. P., Scott, M. and Craig, E. A. (1980). Two clusters of gene for major chorion proteins in Drosophila melanogaster. Cell 19, 905–914.

Swaneck, G. E., Nordstrom, J. L., Kreuzaler, F., Tsai, M.-J. and O'Malley, B. (1979). Effect of estrogen on gene expression in chicken oviduct: evidence for transcriptional control of ovalbumin gene. Proc. Nat. Acad. Sci. USA 76, 1049–1053.

Szybalski, W., Bovre, K., Fiandt, M., Hayes, S., Hradecna, Z., Kumar, S., Lozeron, H. A., Nijkamp, H. J. J. and Stevens, W. F. (1970). Transcriptional units and their controls in Escherichia coli phage: operons and scriptors. Cold Spring Harbor Symp. Quant. Biol. 35, 341–353.

Tata, J. R. (1974). Isolation of nuclei from liver and other tissues. Meth. Enzymol. 31, 253–262.

Tsai, S. Y., Roop, D. R., Tsai, M. J., Stein, P. J., Means, A. R. and O'Malley, B. W. (1978). Effect of estrogen on gene expression in the chick oviduct. Regulation of the ovomucoid gene. Biochemistry 17, 5773–5780.

Ullrich, A., Shine, J., Chirgwin, J., Pictet, R., Rutter, W. J. and Goodman, H. (1977). Insulin genes: construction of plasmids containing the coding sequences. Science 196, 1313–1319.

Villa-Komaroff, L., Efstradiatis, A., Broome, S., Lomdico, P., Tirard, R., Naber, S. P., Chick, W. L. and Gilbert, W. (1978). A bacterial clone synthesizing proinsulin. Proc. Nat. Acad. Sci. USA 75, 3727–3721. [sic]

Weber, J. (1979). Transcription and splicing of nuclear RNA adenovirus-2-infected HeLa cells. Ph.D. thesis. Rockefeller University, New York, New York.

Weber, J., Jelinek, W. and Darnell, J. E., Jr. (1977). The definition of large viral transcription unit late in Ad 2-infection of HeLa cells: mapping of nascent RNA molecules labeled in isolated nuclei. Cell 10, 611–616.

Wilson, M. C., Fraser, N. W. and Darnell, J. E., Jr. (1979). Mapping of RNA initiation sites by high doses of UV irradiation: evidence for three independent promoters within the lefl 11% of the Ad-2 genome. Virology 94, 175–184.

Wold, B. .J., Klein, W. H., Hough-Evans, B. R., Britten, R. J. and Davidson, E. H. (1978). Sea urchin embryo mRNA sequences expressed in the nuclear RNA of adult tissues. Cell 14, 941–950.

·Young, Ivl. W. and Hogness, D. S. (1977). New approach for identifying and mapping structural genes in Drosophila melanogaster. In Eukaryotic Genetics System. ICN-UCLA Symposia on Molecular Biology, G. Wilcox et al., eds. (New York: Academic Press), pp. 315–331.

Post-transcriptional regulation: alternative RNA splicing

3

As discussed in the preceding sections, the finding that most eukaryotic protein-coding genes contain introns raised the possiblity that the removal of such introns might represent a major control point in gene regulation. As discussed in Section 2, however, the major control point of gene regulation is at the level of transcription. Indeed, so-called processing versus discard decisions, in which a transcript is spliced in one tissue and degraded within the nucleus in another tissue, are extremely rare in mammals, although several such cases have been described in *Drosophila* (reviewed by Bingham et al., 1988).

Despite this, regulation of splicing does play a role in regulating gene expression by the use of the process known as alternative splicing (for reviews, see Latchman, 1990; McKeown, 1992). In this process a single primary RNA transcript is spliced differently in different tissues to produce two or more different mRNAs. Often these mRNAs encode different but related proteins with distinct functions. Thus alternative splicing represents a means of producing two or more distinct but related proteins from the same gene.

Moreover, in many situations such alternative splicing is regulated so that one particular pattern of splicing occurs in one particular tissue while in another tissue a distinct pattern of splicing occurs. Thus two or more different mRNAs and proteins are produced in two different tissues from the same gene.

One of the earliest examples of such alternative splicing was reported in the paper by Amara et al. presented here. This group had previously isolated cDNA clones derived from the mRNA encoding the calcium-regulating hormone calcitonin. In using these probes to examine the calcitonin mRNA, they also identified a distinct mRNA which was capable of encoding a protein completely distinct from calcitonin (Rosenfeld et al., 1982). They named the protein encoded by this second mRNA calcitonin-gene-related peptide (CGRP).

In the paper presented here, these authors first showed that both calcitonin and CGRP are encoded by the same gene, as shown by the Southern blotting experiment illustrated in Fig. 1 of the paper. Moreover, detailed comparison of the calcitonin and CGRP cDNA clones (derived from the corresponding mRNAs) with the structure of the gene itself revealed that these two mRNAs were produced by alternative splicing of the gene exons. Thus the calcitonin mRNA is produced by joining the first four exons of the gene by removing the corresponding intervening sequences. In contrast, to produce the CGRP mRNA the first three exons are joined and exon 4 is then omitted, with exon 3 being joined to exon 5 and exon 5 to exon 6. This results in the CGRP mRNA lacking exon 4 and containing exons 1–3, 5 and 6. Thus the two mRNAs contain identical sequences at the 5′ end but differ at the 3′ end. Subsequently, following production of the corresponding proteins from these mRNAs, the two completely distinct peptides calcitonin and CGRP are produced by proteolytic cleavage of the

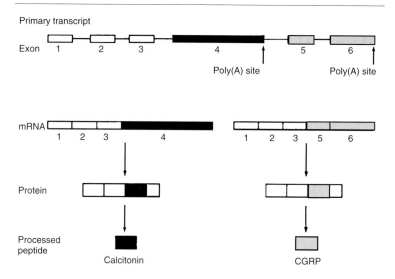

Primary transcript

Exon 1 2 3 4 5 6

Poly(A) site   Poly(A) site

mRNA 1 2 3 4   1 2 3 5 6

Protein

Processed peptide

Calcitonin   CGRP

Figure 3.1 Alternative splicing of the calcitonin/CGRP gene in different cell types. Alternative splicing, followed by cleavage of the protein produced, yields calcitonin in the thyroid gland and CGRP in the brain.

two proteins. The situation as it is currently understood is illustrated in Figure 3.1 (note that Fig. 5 of the original paper is essentially correct in its outline of the processing event, but some details have been corrected since the paper was published).

The most interesting aspect of the paper of Amara et al. from the point of view of gene regulation is that the process of RNA splicing to produce the calcitonin or CGRP mRNAs is clearly regulated in a tissue-specific manner. Thus, as illustrated in Fig. 3(*b*) of the paper, while the thyroid gland produces predominantly calcitonin mRNA, the hypothalamus region of the brain produces predominantly CGRP mRNA. Indeed, subsequent studies indicated that the CGRP mRNA is widely expressed in specific neurons in the central and peripheral nervous systems (Kawai et al., 1985). Hence the alternative splicing event is regulated so that one specific mRNA (encoding calcitonin) is produced in the thyroid gland while another (encoding CGRP) is produced in neuronal cells in the brain. This represents one of the earliest examples of tissue-specific alternative splicing, which has now been shown to play a major role in gene regulation in a number of different tissues (for reviews, see Latchman, 1990; McKeown, 1992).

This conclusion, i.e. that alternative splicing is of major importance in gene regulation, does not conflict with the conclusion in the previous section that transcriptional control is of primary importance. Thus, for example, studies on the calcitonin/CGRP system indicated that the calcitonin/CGRP gene is only transcribed in the thyroid gland and in neuronal cells, and not in a wide variety of other cell types. Hence in this situation the regulation of alternative splicing acts as a supplement to transcriptional control, with transcriptional control restricting expression of the calcitonin/CGRP gene to a few cell types and alternative splicing then determining which mRNA is produced in these cell types. Such a conclusion has subsequently been shown to apply in a number of other systems of alternative splicing so that, for example, the troponin T gene is transcribed only in skeletal muscle cells but its RNA is then spliced in up to 64 different ways in different muscle cell types (Breitbart et al., 1987).

In the second paper from the Rosenfeld group presented in this section, that by Crenshaw et al., the normal restriction of calcitonin/CGRP gene transcription to a few cell types was used in order to throw light on the mechanisms that mediate the tissue-specific alternative splicing event in the calcitonin/CGRP system. Thus in this paper Crenshaw et al. prepared a construct in which the calcitonin/CGRP gene was expressed under the control of the metallothionein promoter, which is active in all cell types. They then constructed a transgenic mouse in which this calcitonin/ CGRP gene would be expressed in all cells of the mouse (Figure 3.2). This allowed Crenshaw et al. to determine whether cells that never normally express the calcitonin/CGRP gene would splice the corresponding RNA to produce either the calcitonin mRNA or the CGRP mRNA. As illustrated in Table 1 of the paper, the vast majority of tissues which do not normally express the gene were able to carry out splicing to produce the calcitonin mRNA but not to produce the CGRP mRNA. Only the heart among the non-expressing tissues was able to splice the primary RNA transcript to produce the CGRP mRNA as well as that encoding calcitonin. As expected the brain (the natural site of CGRP production) was able to carry out the splicing event required to produce the CGRP mRNA. Interestingly, however, this was observed in neurons which do not naturally express the normal calcitonin/CGRP gene as well as in those which do normally express the gene and produce the CGRP mRNA (Figure 3.2).

These findings gave rise to a model in which it was postulated that cells that can carry out CGRP-specific splicing contain an additional factor that is absent from most other cells. The cells that lack this additional splicing protein will only be able to carry out the calcitonin-specific splicing event and hence will produce calcitonin mRNA. This experiment therefore suggested the existence of specific splicing factors, expressed in a tissue-specific manner, which regulate alternative splicing. The factor regulating

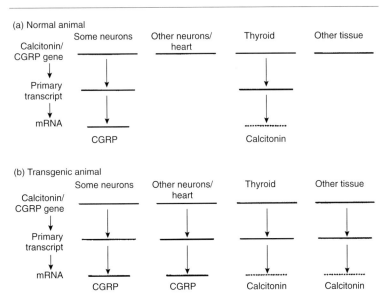

Figure 3.2 Processing of the calcitonin/CGRP transcript in normal animals (a) and in transgenic animals expressing the gene in all cells (b). Note that in normal animals the gene is only transcribed in some neurons, where the transcript is processed to produce CGRP mRNA, and in the thyroid gland where it is processed to produce calcitonin mRNA. In transgenic animals expressing the gene in all cell types, other neurons and the heart also process the primary transcript to produce CGRP mRNA, whereas all other tissues process it to produce calcitonin mRNA.

calcitonin/CGRP splicing would therefore be expressed both in neuronal cells which normally express the calcitonin/CGRP gene as well as in other neuronal cells and in the heart which do not express the endogenous gene. The expression of the factor in cells that never normally transcribe the calcitonin/CGRP gene also suggests that this factor must be involved in regulating other tissue-specific splicing decisions that occur in the heart and in neuronal cells. Thus this experiment not only suggested the existence of such splicing factors but also indicated that they would be likely to regulate more than one splicing decision. The paper by Crenshaw et al. thus represents an important step in the analysis of the regulation of alternative splicing. It should be noted, however, that the putative tissue-specific splicing factor that regulates calcitonin/CGRP splicing has not yet been identified. Moreover, in at least some instances alternative splicing appears to be regulated by differences in the balance between two constitutively expressed factors in different tissues rather than the specific expression of a factor in some tissues and not in others (for a review, see Lamond, 1991b).

Despite this, the calcitonin/CGRP system described here represents one of the best-characterized systems of alternative splicing. Its characterization illustrates the principle that alternative splicing represents a significant supplement to transcriptional regulation. Indeed, it should be noted that CGRP was first described in the papers presented here as an alternatively spliced product of the gene encoding calcitonin, but was subsequently shown to be a molecule of biological importance, being for example the most potent vasodilator that has yet been isolated.

Amara et al. (1982) Nature (London) **298**, 240–244

Alternative RNA processing in calcitonin gene expression generates mRNAs encoding different polypeptide products

Susan G. Amara, Vivian Jonas & Michael G. Rosenfeld

Division of Endocrinology, School of Medicine University of California, San Diego La Jolla, California 92093, USA

Estelita S. Ong, & Ronald M. Evans

Tumor Virology Laboratory, The Salk Institute, San Diego, California 92138, USA

Alternative processing of RNA transcripts from the calcitonin gene results in the production of distinct mRNAs encoding the hormone calcitonin or a predicted product referred to as calcitonin gene-related peptide (CGRP). The calcitonin mRNA predominates in the thyroid while the CGRP-specific mRNA appears to predominate in the hypothalamus. These observations lead us to propose a model in which developmental regulation of RNA processing is used to increase the diversity of the neuroendocrine gene expression.

A LARGE array of hormonal signals provides the specificity of intracellular communication within and between organs. The existence of gene families[1] and the presence of multiple hormones within a single primary translation product[2-4] are two means by which a diversity of peptide hormones can be generated. The discontinuity of genetic regions encoding mature RNAs and the complexity of RNA processing pathways suggest that alternative splicing events could be an additional mechanism for increasing the flexibility of gene expression in the neuroendocrine system. The potential versatility provided by alternative RNA processing events has been used effectively by several eukaryotic viruses to generate multiple protein products from a single transcription unit[5,6]. We show here that multiple mRNAs are generated from the calcitonin gene as a consequence of alternative RNA processing events. These events ultimately produce different protein products and seem to occur in a tissue-specific fashion. Such a mechanism could serve to increase the diversity of proteins encoded by endocrine genes.

The production of multiple calcitonin-related mRNAs was first noted during the spontaneous and permanent switching of serially transplanted rat medullary thyroid carcinoma (MTC) lines from states of 'high' to 'low' calcitonin production[7]. This conversion is associated with the appearance of new calcitonin cDNA-reactive mRNAs, referred to as calcitonin-gene-related product (CGRP) mRNAs, 50–250 nucleotides larger than calcitonin mRNA. These RNAs can function as mRNAs as they are present on polyribosomes and can be translated *in vitro* after being isolated by hybridization to a specific immobilized cDNA clone. A new 16,000 molecular weight protein is the major cell-free translation product directed by the CGRP mRNAs[8]; this protein does not contain immunoreactive calcitonin.

To study the switching events occurring in the tumours, cDNA clones to calcitonin (pCal) and calcitonin gene-related (pCGRP₁, pCGRP₂) mRNAs have been generated and a calcitonin genomic fragment referred to as λ-Cal₁ has been isolated. In this article we examine the structural basis for differences between mature calcitonin mRNA (Cal mRNA) and one of the variant mRNAs (CGRP mRNA) by examination of both cDNA and genomic clones.

Cal and CGRP mRNA sequences are present in the same gene

To determine the relationship of Cal and CGRP sequences within the calcitonin gene, fragments or clones bearing calcitonin-specific or CGRP-specific sequences were used as hybridization probes to locate and orientate exons within the genomic clone, λ-Cal₁. These regions of homology are closely linked as both probes react with a common 8.2-kilobase (kb) *Eco*RI fragment and a common 3.0–kb *Pst*I fragment (Fig. 1*a, b*). Fragments specific to 5′ and 3′ regions of calcitonin mRNA were generated as hybridization probes to determine the transcriptional orientation of the gene. These were prepared by cleaving pCal at the single *Bgl*II site, located 11 nucleotides 5′ to the start of the calcitonin coding sequence[9]. A cDNA clone pCGRP₁ was used as a specific probe for CGRP sequences because it contains a region unique to CGRP mRNAs and does not hybridize Cal mRNA. The pCal- and pCGRP1-reactive *Eco*RI and *Pst*I fragments of λ-Cal₁ correspond precisely to those identified in digests of genomic DNA isolated from rat liver or MTC tumours. Figure 1*c, d* shows the results of this analysis. Comparison of this hybridization with that in *a* and *b* indicates that CGRP sequences map to the 3′ portion of this gene. Further restriction mapping establishes that the 8.2-kb *Bgl*II fragment which hybridizes only with pCGRP₁ (Fig. 1*a*, lane 2) represents the 3′ end of the λ-Cal₁ clonal insert. These results support the notion that the 3′ ends of Cal and CGRP mRNAs are encoded by the same gene.

Comparison of pCGRP₂ and pCal cDNA coding domains

Previously described RNA hybridization analysis[8] and the genomic mapping observations presented above suggest that Cal and CGRP mRNAs contain common or homologous 5′ sequences and nonhomologous 3′ sequences. This prediction was tested by restriction analysis and sequencing of CGRP cDNA clones. One clone, designated as pCGRP₂, containing an insert of >600 nucleotides, was sequenced to identify the

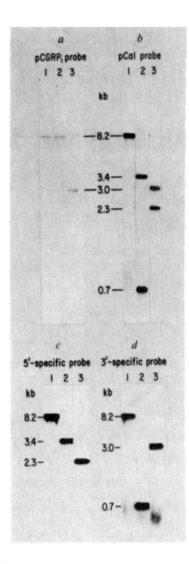

Fig. 1 Localization of pCGRP sequences in the calcitonin genomic clone λ-Cal₁. The rat calcitonin genomic clone (λ-Cal₁) was digested to completion with the restriction enzymes EcoRI, BglII or PstI and electrophoresed on an 0.8% agarose gel. The restriction fragments were transferred to a nitrocellulose filter[27]. Filters were hybridized to ~5 × 10⁶ c.p.m. ml⁻¹ of ³²P-labelled nick-translated probes, washed and autoradiographed as previously described[8]. The clone pCGRP₁ used in this analysis contains a region unique to CGRP mRNA and does not hybridize to Cal mRNA. a, Hybridization of pCGRP₁ to λ-Cal₁ digested with: EcoRI (lane 1), BglII (lane 2), PstI (lane 3). b, Hybridization of pCal to a duplicate blot of a. c, Hybridization of the 5′ fragment of the BglII-digested pCal insert to λ-Cal₁ DNA digested with: EcoRI (lane 1), BglII (lane 2), PstI (lane 3). d, Hybridization of the 3′ fragment of BglII-digested pCal insert to the identical blot shown in c. Sizes are based on the migration of HindIII-digested λ DNA fragments.

point of divergence between Cal and CGRP mRNAs. The sequence of this clone reveals the entire coding region of CGRP mRNA with an open reading frame of 384 nucleotides capable of coding for a primary translation product of 128 amino acids.

The sequences of pCGRP₂ and pCal are compared in Fig. 2a. The 5′ sequences of pCal and pCGRP₂ are identical from the first nucleotide in the pCGRP₂ insert to nucleotide 227 in the coding region of both cDNA clones. As previously reported, the structure of pCal predicts only one open reading frame which encodes a precursor protein whose characteristics correspond to those determined for precalcitonin and its component polypeptides *in vivo* and *in vitro*[9–11]. Thus, the identity of the 5-proximal mRNA sequences including the AUG for both mRNAs suggests that the first 76 amino acids of their translation products must also be identical. The nucleic acid sequence diverges 228 nucleotides into the coding sequence, following an AG, and precisely identifies the junction point that distin-

guishes these two mRNAs. Beyond nucleotide 227, the two mRNAs share no homology and encode completely different amino acid sequences. The junction of shared and divergent sequence in these two mRNAs is detailed in Fig. 2c. These data are most compatible with the interpretation that a single 5′ sequence represents a discrete domain which can be selectively spliced onto Cal- or CGRP-specific exons. In such an instance, the mRNA junction point must correspond to an intervening sequence in which a unique donor site can splice onto one of two alternative acceptor sites. We examined this prediction by a detailed structural analysis of the calcitonin gene.

The point of divergence between Cal and CGRP mRNAs corresponds to a genomic splice junction

Detailed restriction endonuclease mapping using the cDNA clones as probes and partial DNA sequence analysis of the calcitonin genomic clone (λ-Cal₁) confirms the prediction that exons with sequences precisely corresponding to the divergent 3′-coding regions of pCal and pCGRP₂ are present in one genomic transcription unit. As noted in Fig. 2b, all exons identified in the genomic clone have sequences identical to those determined for the two cDNA clones. The 5′ boundaries of the Cal- and CGRP-specific genomic coding sequences establish that splice junctions occur precisely at the point of the divergence between Cal and CGRP₂ mRNAs. Figure 2c (A) and (B) details the sequences at the junctions for the pCal and pCGRP₂ sequences. The observation that an AG precedes both exons in the genomic clone is in accordance with the GT . . . AG rule for splice junctions[12] and is consistent with the prediction that each exon can be alternatively spliced into the mature mRNAs. The AG which appears adjacent to the splice site in the two cDNA clones is presumably contributed by the 3′ end of the preceding exon.

Nuclear RNA transcripts contain both Cal and CGRP sequences

The structure of Cal and CGRP mRNAs and of the calcitonin genomic clone demonstrate that the same region of DNA can be used to generate multiple mature mRNAs. However, there are several possible mechanisms by which each mature mRNA could be produced. Because the sequences unique to CGRP mRNA are located 3′ to Cal mRNA sequences in the gene, it is possible that the calcitonin primary transcript could terminate at a polyadenylation site 5′ with respect to CGRP sequences, while transcription to the next polyadenylation site could give rise to the CGRP primary transcript. This model predicts two different primary transcripts and would parallel the molecular events demonstrated in the case of late adenovirus mRNA production; these events have also been proposed for the alternative generation of membrane and secretory forms of immunoglobulin μ heavy chains[13,14] and the simultaneous synthesis of μ and δ heavy chains[15].

This possibility was investigated by analysing putative nuclear RNA precursors for the presence of both Cal and CGRP mRNA-specific sequences. We have previously identified reactive pCal RNA precursors 6.6, 5.2, 4.2 and 3.8 kb in tumour lines producing CGRP calcitonin mRNA (MTC_L lines)[7]. The comparably sized precursors observed in lines producing predominantly calcitonin mRNA (MTC_H lines) could reflect the production of small amounts of CGRP mRNAs[8], the existence of common precursors for Cal and CGRP mRNAs, or the presence of CGRP RNA precursors not processed to mature CGRP mRNAs in this tissue. RNA from normal thyroid glands was also analysed. Virtually all the pCal-reactive RNA detected in the thyroid gland is the 1,050-nucleotide calcitonin mRNA. The presence of CGRP-specific sequences within the large

Amara et al. (1982) Nature (London) **298**, 240–244

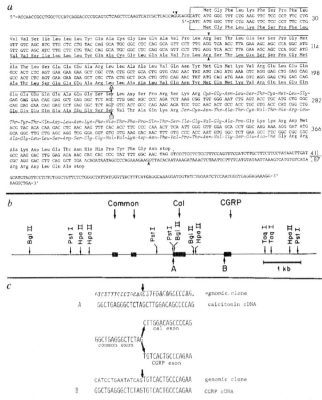

Fig. 2 The nucleotide sequence of pCal and pCGRP2 and the location of corresponding regions within the calcitonin gene. The plasmid pCal, containing DNA complementary to calcitonin mRNA, was cloned and sequenced as described previously[9]. To construct plasmids complementary to CGRP mRNAs, size-fractionated mRNA 800–1,500 nucleotides long was prepared from an MTCL line highly enriched in CGRP mRNAs and used to generate double-stranded DNA using reverse transcriptase. The double-stranded DNAs were inserted into the PstI site of pBR322 by dG-dC tailing and the recombinant plasmid was used to transform an Escherichia coli K-12 host (SF8)[9]. Transformants were identified by their differential antibiotic sensitivity. The plasmid pCGRP2 was selected for further study based on its restriction map and its hybridization to the CGRP-specific 8.2-kb BglII genomic clone fragment. a, The DNA sequence of the 627-base insert in pCal is shown above the sequence of 482 bases of pCGRP2. Amino acid sequences of products encoded by the two clones are indicated above and below. Calcitonin and the putative 37 amino acid cleavage product of pCGRP2 (CGPR) are indicated by italics. The boxed area denotes the region of common nucleotide sequence between the two clones. DNA sequencing has demonstrated that the entire coding sequence of both cDNA clones is present with precise correspondence in the genomic clone λ-Cal1. The genomic DNA sequence corresponding to the first 58 nucleotides of the 5′ noncoding portion of pCal and the last 67 nucleotides of the 3′ noncoding portion of pCGRP2 has not been completed. The sequence for most of pCal and the strategy used has been described previously[10]. To sequence pCGRP2, plasmid was digested with DdeI, HaeII, AluI or RsaI, end-labelled with [γ-32P]ATP and T4 polynucleotide kinase (Bethesda Research Laboratories) and either strand separated or cleaved with a second restriction enzyme before electrophoresis using procedures described by Maxam and Gilbert[28]. The cDNA insert was sequenced in multiple determinations, in both strands and across all sites used for labelling. Arrowheads denote sites of intervening sequences in λ-Cal1 determined by sequencing: A and B are the junctions of the Cal-encoding and CGRP-encoding exons, respectively. Nucleotides are numbered on the right beginning with the first nucleotide by the coding region of both clones. b, Restriction map of λ-Cal1. Endonuclease cleavage sites of λ-Cal1 were determined by analysis of non-limit restriction digests as described by Smith and Birnstiel[29] and/or by comparison of double and single enzyme digestion patterns determined for the clone. Small fragments shown by hybridization to contain regions of pCal and pCGRP2 sequences were isolated and 5′-end labelled for Maxam–Gilbert sequencing as above. The locations of four sequenced exons is represented by black boxes. Fragments generated by RsaI and DdeI cleavage were strand separated and sequenced completely in both strands. Sequence from BglII and HaeIII sites was obtained by multiple determinations sequenced from the same strand. Other sequences were obtained by dideoxy sequencing of sheared genomic fragments cloned into M13[30–32]. Details of the various sequencing strategies are available from the authors on request. Arrows A and B indicate the location of the intron–exon junctions for Cal and CGRP exons, respectively. c, The cDNA sequences at the junction of the region of identity and non-identity in pCal and pCGRP2. The genomic sequences at intron–exon junctions corresponding to those labelled A and B in panels a and b are also labelled A and B in c. Arrows indicate putative splice sites. Intervening sequences are shown using small letters.

nuclear calcitonin mRNA precursors in the thyroid (Fig. 3a, lane T) suggests that sequences unique to CGRP mRNAs may be present in the nuclear RNA transcripts of cells making almost exclusively calcitonin mRNA. In addition, a probe specific for the calcitonin coding region (the 0.7–kb BglII genomic fragment) hybridized to the nuclear RNA species of identical sizes in thyroid, MTCH lines and in MTCL lines. Furthermore, intact nuclear RNA was isolated by hybrid selection of a tumour RNA, using a plasmid containing CGRP-specific sequences. These purified transcripts were size-fractionated and shown to hybridize with Cal-specific probes (data not shown).

Thus, both Cal- and CGRP-reactive regions are present on common nuclear transcripts in cells ultimately producing only Cal or CGRP mRNAs. Although these data suggest that transcription proceeds through both coding regions, irrespective of which mRNA is ultimately produced, it is nevertheless possible that the reactive nuclear species represent large transcripts which are not processed into either mRNA.

pCGRP2 sequence predicts that calcitonin gene encodes an additional peptide product

The splicing of alternative domains generates mRNAs which can encode different polypeptides. The structural relationship of the two polypeptides predicted from the DNA sequence of pCal and pCGRP2 is shown schematically in Fig. 4. For the

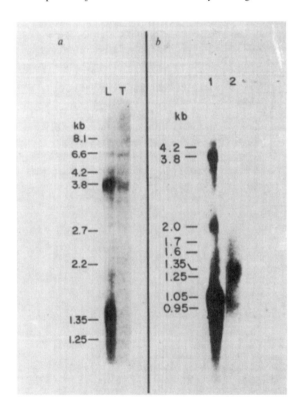

Fig. 3 a, Analysis of calcitonin mRNA precursors for the presence of CGRP regions. Total poly(A)-rich RNAs from an MTCL line and from normal thyroids were denatured and electrophoresed on 1.5% agarose formaldehyde gel as described previously[33]. RNA was transferred to nitrocellulose[27], washed in prehybridization buffer and hybridized to pCGRP1 nick-translated to a specific activity of 8 × 10^8 c.p.m. per μg using [α-32P]dCTP as the labelled nucleotide. The pCGRP1 probe is specific for CGRP sequences. Lane L, 5 μg MTCL tumour mRNA; lane T, 5μg thyroid mRNA. b, Analysis of hypothalamic and thyroid mRNA species hybridizing the calcitonin cDNA clone pCal. Poly(A)-rich mRNA from normal thyroid and gradient-enriched 8–15S RNA from rat hypothalamus were subjected to formaldehyde gel and Northern blotting procedures as in a except that transfer was to diazotysed paper[34]. Immobilized RNA was hybridized with pCal nick-translated to a specific activity of 5 × 10^8 c.p.m. per μg (ref. 33). Lane 1, 5μg poly(A)-rich thyroid mRNA; lane 2, 8μg gradient-enriched poly(A)-rich hypothalamic mRNA. RNA sizes are based on migration of 18S and 28S RNAs.

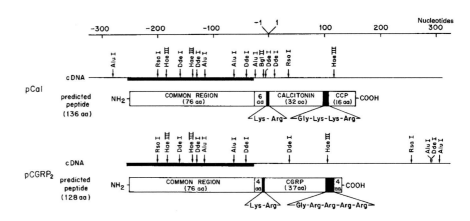

Fig. 4 Schematic representation of pCal and pCGRP₂ DNA inserts and the structure of the precursors they encode. Restriction sites used for DNA sequence analysis are those indicated and regions of identical nucleotide sequence are underlined in black. Nucleotide residues numbered on the scale above begin with the first residue in the coding region for calcitonin. The structure of the predicted protein precursors are represented by the bar below. Potential proteolytic processing sites and the resulting cleavage products are also noted.

calcitonin precursor, the points of excision can be accurately determined from knowledge of the mature calcitonin peptide, as we have previously discussed in detail[9,10]. The excision occurs at paired or multiple basic residues and this processing information is encoded in the calcitonin exon. The DNA sequence of pCGRP₂ predicts a 128 amino acid, 16,000 molecular weight protein which has the same 76 amino acid N-terminal sequence as the calcitonin precursor. The 52 C-terminal residues of the predicted protein contain similar processing signals which would allow excision of a 37 amino acid peptide. The Lys-Arg dipeptide, encoded by nucleotides 241–246 (see Fig. 2a), is proposed to represent the N-terminal cleavage site for excision of this novel product. The presence of glycine followed by three basic amino acid residues is thought to serve as a signal for C-terminal amidation and cleavage in three amidated peptides for which the structure of the precursor is known[4,9,16]. The sequence Gly-Arg-Arg-Arg-Arg in the product encoded by CGRP mRNA may signal the generation of a peptide terminating in a phenylalanine-amide. We refer to this 37 amino acid excised peptide as calcitonin gene-related peptide (CGRP). We note that in parallel with calcitonin, the CGRP exon encodes all the processing signals necessary to generate CGRP.

Expression of Cal and CGRP mRNAs is tissue specific

The demonstration that the 'switching' of MTC tumours from production of Cal to CGRP mRNA is the consequence of alternative RNA splicing events prompted the speculation that such events might occur physiologically during differentiation. This possibility was examined by size-fractionation of RNA from several tissues and challenge with Cal- and CGRP-specific probes. The results of one such study are shown in Fig. 3b. The hypothalamus contains poly(A)-rich RNAs which hybridize specifically with the Cal-specific probe and which are the same size as the CGRP RNA species present in the MTC_L tumours (Fig. 3b, panel 2). These RNA species hybridize to the CGRP-specific probe. A small amount of RNA corresponding to calcitonin mRNA is also present in these cells while virtually all the reactive RNA in the thyroid is calcitonin mRNA (compare Fig. 3a, panel T, with Fig. 3b, lane 1). These data suggest that the alternative RNA splicing events might occur physiologically in the expression of the calcitonin gene. It is tempting to speculate that CGRP, the predicted polypeptide product of CGRP₂ mRNA, is a hypothalamic neuropeptide that may exert hormonal effects of its own.

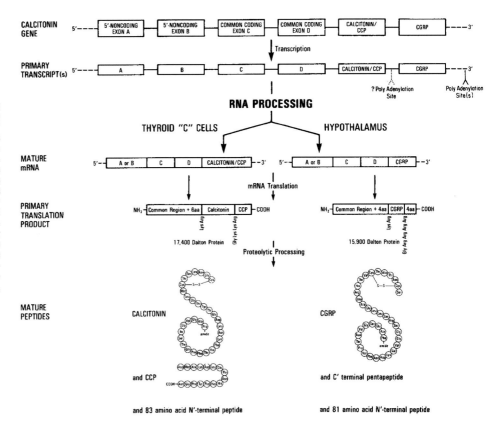

Fig. 5 A model to describe the tissue-specific expression of the calcitonin gene during 'peptide switching' events. The 5' and 3' termini are indicated as dotted lines because their structural organization remains unknown. The structural organization of coding blocks A and B, and the presence of A within CGRP mRNAs are not unequivocally established. Possible sites of polyadenylation of the primary transcript(s) are indicated. The forked arrow indicates tissues in which pCal reactive RNAs predominate (thyroid) or pCGRP reactive RNAs predominate (hypothalamus). The predicted proteolytic products of the primary translation product directed by Cal mRNA are established, those of CGRP mRNA, predicted.

Concluding remarks

The discovery of 'split genes' and definition of hnRNA structure[5,6,17–21] raised the possibility of important regulatory events at the level of processing of initial RNA transcripts. As presented here, such regulation of RNA processing events may be necessary for the expression of certain neuroendocrine genes and may increase the diversity of the resultant peptide products. Genomic mapping data are consistent with the existence of a single calcitonin gene[8]; if a second gene exists it would have to be extremely similar in size and organization. Sequence analysis establishes that Cal and CGRP domains are linked in a single genomic clone whose exons correspond precisely in sequence to the cloned mRNAs. Based on the structure of this gene and its RNA products, we propose a model in which genomic regions can represent discrete hormone encoding domains whose ultimate expression is dependent on differential RNA processing events (Fig. 5). Thus, these domains may be included or excluded during the maturation process to generate multiple mRNAs encoding alternative hormones or neuropeptides. We refer to the consequences of these RNA processing events as 'peptide switching' because the splicing of alternative exons results in the expression of different peptides.

Several explanations could account for the alternative expression of Cal and CGRP mRNAs. First, there could be two overlapping transcription units with different cap and poly(A) sites. Second, there could be a single cap site with two or more poly(A) sites. Multiple RNAs would arise from a choice of poly(A) site either during transcription or due to post-transcriptional cleavage and re-polyadenylation. Third, there could be a small rearrangement or modification of gene structure affecting a splice site or polyadenylation signal. Finally, there could be true regulation of splicing choice for a single primary transcript.

The last possibility would parallel events observed in SV40 and adenovirus gene expression, where several mRNAs can be generated with common 5′ and 3′ termini and yet encode different proteins[5,6]. This process requires the differential use of multiple and optional 5′ and 3′ splicing sites. It has been suggested that the L-1 transcription unit of adenovirus displays splice-choice control[22]. Should such an analogy apply to expression of calcitonin it remains to be defined how such a process is regulated. Splicing regulation could arise from changes in enzyme specificity, small nuclear RNAs and/or other factors directly affecting the splicing process and could account for the accumulation of the larger nuclear RNA transcripts of the calcitonin gene during switching to production of CGRP mRNAs[7].

Certain observations regarding the expression of the calcitonin gene must be accounted for in any model of alternative processing. Both the calcitonin genomic clone and the larger nuclear RNA transcripts contain sequences specific to both the mature calcitonin and CGRP mRNAs. If the observed nuclear transcripts are in fact precursors of both mRNAs, then the expression of either Cal or CGRP mRNAs requires splicing out of an internal exon. This would be in contrast to the RNA processing events proposed in the case of immunoglobulin heavy chain gene expression, where a polyadenylation site intercedes between the exons involved in switching[13–15]. On the other hand, if a polyadenylation site is present between Cal and CGRP exons, the precursors of the two mRNAs will be distinct. In this case, the nuclear CGRP-reactive RNAs in the thyroid would represent species not destined to become either RNA or would require a cleavage and re-polyadenylation to be processed into calcitonin mRNA.

Although it has been proposed that, in the case of liver and salivary gland α-amylase and dihydrofolate reductase gene expression[23], differential 5′ transcription initiation and/or alternative polyadenylation sites[24,25] result in mRNA structural polymorphism, this does not affect the structure of the protein-encoding regions of the mature mRNAs. In fact, 5′ heterogeneity also seems to occur in calcitonin mRNAs, where one splice junction precedes by nine nucleotides the initiator methionine codon (Fig. 2a). The location of this intervening sequence corresponds precisely with the 3′ boundary of a region of 5′ noncoding sequence in pCal that diverges from a sequence reported for another calcitonin clone[26]. The use of 'optional' 5′ exons, in addition to the coding exon switch discussed here, further increases the mRNA polymorphism associated with calcitonin gene expression, but does not alter the coding region of the resultant mRNAs.

The apparent tissue specificity of the observed splicing events implies developmental regulation of these processes. Calcitonin gene expression therefore represents a model system for exploring mechanisms responsible for both transcriptional and post-transcriptional regulation of specific gene expression during development. Whatever the responsible molecular mechanisms, the unexpected consequence of these RNA processing events is that the same gene which generates the hormone calcitonin in thyroid C-cells is apparently responsible for the expression of a new polypeptide product in the hypothalamus.

We thank Drs James E. Darnell, Francis Crick, Geoffrey Wahl, Jean-Jacques Mermod and Geoffrey Murdoch for advice and helpful discussions regarding the manuscript, Arnold Loera and Rodrigo Franco for providing experimental data, our collaborators Drs Bernard A. Roos and Roger Birnbaum for their continued support in these studies, Drs Tom Sargent and James Bonner for providing the rat liver DNA library, Maureen Brennan and Margaret Richards for preparation of the manuscript. This work was supported by grants from the American Cancer Society and the NIH.

Received 9 February; accepted 4 May 1982.

1. Blundell, T. L. & Humbel, R. E. *Nature* **287**, 781–787 (1980).
2. Roberts, J. L., Phillips, M., Rosa, P. A. Herbert, E. *Biochemistry* **17**, 3609–3618 (1978).
3. Eipper, B. A. & Mains, R. E. *J. Biol. Chem.* **253**, 5732–5744 (1978).
4. Nakanishi, S. *et al. Nature* **278**, 423–497 (1979).
5. Ziff, E. B. *Nature* **287**, 491–499 (1980).
6. Darnell, J. E. *Prog. Nucleic Acid Res. molec. Biol.* **22**, 327–353 (1978).
7. Rosenfeld, M. G., Amara, S. G., Roos, B. A., Ong, E. S. & Evans, R. M. *Nature* **290**, 63–65 (1980).
8. Rosenfeld, M. G. *et al. Proc. natn. Acad. Sci. U.S.A.* **79**, 1717–1721 (1982).
9. Amara, S. G., David, D. N., Rosenfeld, M. G., Roos, B. A. & Evans, R. M. *Proc. natn. Acad. Sci. U.S.A.* **77**, 4444–4448 (1980).
10. Amara, S. G. *et al. J. biol. Chem.* **257**, 241–244 (1982).
11. Birnbaum, R. S., O'Neal, J. A., Muszynski, M., Aron, D. C. & Roos, B. A. *J. biol. Chem.* **257**, 241–244 (1982).
12. Breathnach, F., Benoist, C., O'Hare, K., Gannon, F. & Chambon, P. *Proc. natn. Acad. Sci. U.S.A.* **75**, 4853–4857 (1978).
13. Alt, F. W. *et al. Cell* **20**, 293–301 (1980).
14. Early, P. *et al. Cell* **20**, 313–319 (1980).
15. Maki, R. *et al. Cell* **24**, 353–365 (1981).
16. Kreil, G. Suchanek, G. & Kindas-Mugge, I. *Fedn. Proc* **36**, 2081–2086 (1977).
17. Chow, L. T., Gelinas, R. E., Broker, T. R. & Roberts, R. J. *Cell* **12**, 1–8 (1977).
18. Klessig, D. F. *Cell* **12**, 9–21(1977).
19. Sambrook, J. *Nature* **268**, 101–104 (1977).
20. Jeffreys, A. J. & Flavell, R. A. *Cell* **12**, 1097–1108 (1977).
21. Tilghman, S.M. *et al. Proc. natn. Acad. Sci. U.S.A.* **75**, 725–729 (1978).
22. Nevins, J. R. *Cell* **28**, 1–2 (1982).
23. Tosi, M., Young, R. A., Hagenbuchle, O. & Shibler, U. *Nucleic Acids Res.***9**, 2313–2323 (1981).
24. Setzer, D. R., McGrogan, M., Nunberg, J. H. & Schimke, R. T. *Cell* **22**, 361–370 (1980).
25. Young, R. A., Hagenbuchle, O. & Schibler, U. *Cell* **23**, 451–458 (1981).
26. Jacobs, J. W. *et al. Science* **213**, 457–459 (1981).
27. Southern, E. M. *J. molec. Biol.* **98**, 503–517 (1975).
28. Maxam, A. M. & Gilbert, W. *Proc. natn. Acad. Sci. U.S.A.* **74**, 560–564 (1977).
29. Smith, H. O. & Birnstiel, M. L. *Nucleic Acids Res.* **3**, 2387–2398 (1976).
30. Messing, J, Crea, R. & Seeburg, P. M. *Nucleic Acids Res.* **9**, 309–321 (1981).
31. Sanger, F. Nicklen. S. & Coulsen, A. R. *Proc. natn. Acad. Sci. U.S.A.* **74**, 5463–5467 (1977)
32. Fuhrman, S. A., Deininger, P. L., Laporte, P., Friedmann, T. & Geiduschek, E. P. *Nucleic Acids Res.* **9**, 6439–6456 (1981).
33. Potter, E. Nicolaisen, A. K., Ong, E. S., Evans, R. M. & Rosenfeld, M. G. *Proc. natn. Acad. Sci. U.S.A.* **78**, 6662–6666 (1981).
34. Wahl, G. M. Stern. M. & Stark, G. R. *Proc. natn. Acad. Sci. U.S.A.* **76**, 3683–3687 (1979).

Crenshaw et al. (1987) Cell **49**, 389–398

Neuron-Specific Alternative RNA Processing in Transgenic Mice Expressing a Metallothionein-Calcitonin Fusion Gene

E. Bryan Crenshaw III, *† Andrew F. Russo, *
Larry W. Swanson,‡ and Michael G. Rosenfeld *
* Howard Hughes Medical Institute
Eukaryotic Regulatory Biology Program
School of Medicine, M-013
University of California, San Diego
La Jolla, California 92093
† Department of Biology
University of California, San Diego
La Jolla, California 92093
‡ Howard Hughes Medical Institute
Neural Systems Laboratory
The Salk Institute
La Jolla, California 92037

Summary

Alternative RNA processing of the calcitonin/CGRP gene generates transcripts encoding predominantly calcitonin in thyroid C cells or CGRP in the nervous system. To examine the RNA processing choice of this gene in a wide variety of tissues, we created transgenic mice expressing the rat calcitonin/CGRP transcript from the mouse metallothionein-I promoter. Most cells that do not express the endogenous calcitonin/CGRP gene have the capability to make a clear splicing choice for calcitonin or CGRP transcript. In the majority of tissues studied, 90%–97% of the transgene mRNA encodes calcitonin. In contrast, both calcitonin and CGRP mRNAs were detected in the transgenic mice brains. Immunohistochemical and in situ RNA hybridization analyses show that CGRP transcripts are selectively expressed in a wide variety of neurons, while calcitonin is expressed predominantly in nonneuronal structures. Splicing choice operates independently of calcitonin/CGRP gene transcription. The data suggest that a specific regulatory machinery is required for the processing of CGRP transcripts and is restricted primarily to neurons.

Introduction

Alternative RNA processing is a developmental strategy that determines the phenotype of several cell types (for reviews, see Rosenfeld et al., 1984; Padgett et al., 1986; Leff et al., 1986). A clear example of regulated alternative RNA processing is the switch of immunoglobulin M or D from the membrane form to the secreted form during B cell development (Alt et al., 1980; Early et al., 1980; Maki et al., 1981). Like immunoglobulins, many proteins have developmentally regulated isoforms that differ by a single functional domain. These structural changes are often made by the choice of exons included in the RNA transcript. For example, splicing choice is used to produce fetal or adult forms of myosin (Nabeshima et al., 1984; Rozek and Davidson, 1986), cellular or plasma forms of

fibronectin (Schwarzbauer et al., 1983; Tamkun et al., 1984), and several isoforms of troponin T (Breitbart et al., 1985). In the nervous and endocrine system, the functional domain that is included in or excluded from a prohormone can encompass a complete neuropeptide. Examples of neuropeptide splicing choices include calcitonin/CGRP (Amara et al., 1982), pre-kininogen (Kitamura et al., 1983), and substance P/substance K (Nawa et al., 1984). These choices, which increase the diversity of peptides expressed in the neurons and endocrine cells, are undoubtedly crucial events in the development and function of complex systems like the brain.

The peptides calcitonin and calcitonin gene–related peptide (CGRP) are encoded by alternatively spliced transcripts from the same gene (Amara et al., 1982). The calcitonin/CGRP gene contains six exons. There are two polyadenylation sites one following the fourth and one following the sixth exon, used in generating calcitonin and CGRF mRNA, respectively. CGRP mRNA is formed by splicing sequences representing the first three exons (common to calcitonin and CGRP mRNAs) with the fifth and sixth exons (see Figure 1). The calcitonin/CGRP transcription unit uses an identical cap site and termination site approximately 1 kb downstream from the second (exon 6) polyadenylation site, irrespective of the type of mature transcript generated (Amara et al., 1984). This excludes alternative transcription initiation or termination as the regulated event. Therefore, the choice of poly(A) site and the pattern of splicing of exon 3 to either exon 4 (calcitonin-coding) or exon 5 (CGRP-coding) must be regulated. This RNA processing choice is stringently regulated since greater than 95% of the calcitonin/CGRP gene transcripts are processed to mature calcitonin or CGRP mRNAs in their endogenous sites of expression (Amara et al., 1982; Sabate et al., 1985). Calcitonin transcripts are the primary mRNA observed in the thyroid C cells (Jacobs et al., 1981; Sabate et al., 1985). CGRP transcripts, on the other hand, are the primary mRNA observed in specific neurons distributed throughout the central and peripheral nervous systems (Rosenfeld et al., 1983; Kawai et al., 1985).

To determine whether the machinery for alternative splicing of the calcitonin/CGRP transcript can be found in tissues in which the gene is not normally expressed, we have created transgenic mice that widely express this transcript from a heterologous promoter. We chose the metallothionein I (MT-I) promoter because it is expressed in many different tissues, can be regulated by heavy metals, and has previously been shown to direct expression of fusion genes to a number of tissues in transgenic mice (Palmiter et al., 1983), including discrete subsets of neurons (Swanson et al., 1985). Based on the expression pattern of the calcitonin/CGRP fusion gene in transgenic mice, we propose that the ability to splice the calcitonin/CGRP primary transcript to the mature CGRP transcript requires a specific splicing machinery that is expressed in many neurons and that the majority of tissues

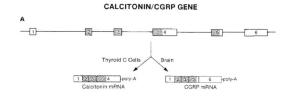

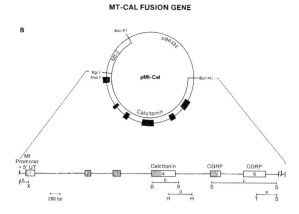

Figure 1. Structure and Expression of the Calcitonin/CGRP and MT–Cal Genes

(A) The endogenous calcitonin/CGRF gene contains six exons that are alternatively processed to produce the calcitonin transcript in thyroid C cells or the CGRP transcript in the nervous system. Coding regions are shown as shaded boxes, and noncoding regions are shown as open boxes.

(B) The MT–Cal fusion gene contains the metallothionein promoter fused to the calcitonin/CGRP gene. The 5′-flanking and a small portion of 5′-untranslated region (BglI at position −185 to XhoI at +62 bases from the MT cap site) of the mouse MT-I gene were fused to the genomic rat calcitonin/CGRP gene at a PstI site, 11 bases upstream of the calcitonin/CGRP CAP site. The coding regions are shown as shaded boxes, the noncoding regions as open boxes, and the MT 5′-untranslated region as a stippled box. The gene and its expected mRNA products are shown to scale. The regions used for hybridization probes (labeled a–e), with the relevant restriction enzyme sites are shown below the gene. Restriction endonucleases are designated by B (BglII), H (HaeIII), S (Sau3A), T (TaqI), and X (XhoI).

express the calcitonin transcript because they lack the requisite CGRP splicing machinery.

Results

Construction and Microinjection of the Metallothionein–Calcitonin Fusion Gene

We constructed a hybrid fusion gene, pMT–Cal, that places the calcitonin/CGRP gene under the transcriptional regulation of the mouse MT-I promoter (Figure 1). The transcript from this fusion gene consists of 62 bp of 5′-untranslated region of MT-I, 11 bp of 5′-flanking sequences from the calcitonin/CGRP gene, and the 6 exons of the calcitonin/CGRP gene. Approximately 200 copies of the BglI–BamHI fragment of MT–Cal were microinjected into the male pronuclei of 210 fertilized eggs, and 115 eggs were selected for transfer into pseudopregnant females. Of 19 mice that developed to term, 7 mice had integrated 1 to 70 copies of the transgene (data not shown). Four

lines were established from the progenitor animals (MC3, MC4, MC5, and MC6). Three of these lines expressed the transgene in the liver, while the MC3 line did not express the transgene in any tissues examined. This result is consistent with the expression pattern seen with other MT fusion genes (Palmiter et al., 1983). The tissue distribution of splicing was examined in greater detail in the three lines that express the transgene.

Tissue-Specific Expression of the MT–Cal Fusion Gene

By using probes specific for calcitonin or CGRP mRNA (see Figure 1, probes b and c), we were able to distinguish the two alternative RNA products, following gel fractionation and RNA blotting analysis. As shown in Figure 2, calcitonin mRNA is the predominant processing choice in the liver and kidney. CGRP probes hybridized to large molecular weight nuclear RNAs (3.6 and 4.2 kb), which contain both the calcitonin and CGRP exons and may represent nonproductive transcripts or precursor RNAs. Only 3%–8% of the mature transcripts of the transgene RNA were expressed as CGRP mRNA (1.2 kb). In contrast, the transgene RNA in the brain was processed to both CGRP and calcitonin RNA (Figure 2). In addition to the liver and kidney, most tissues we tested expressed primarily calcitonin RNA (Table 1). Based on hybridization analysis of size-fractionated total RNA, we found that in skeletal muscle, spleen, gut, and lung greater than 90% of mature transcript was calcitonin RNA. The only non-neuronal location that expressed greater than 10% CGRP mRNA was the heart (Table 1). The relative levels of expression of the transgene (Table 1) parallel that of the endogenous MT-I, as is usually seen with MT fusion genes (Palmiter et al., 1983). The splicing choice is independent of the expression level, because tissues with comparable levels of expression can make very different splicing decisions (Table 1). Furthermore, the MC5 pedigree expressed the transgene at levels 4- to 5-fold higher than the MC4 pedigree in all tissues tested, yet exhibited a virtually identical pattern of relative mRNA processing. The expression level and integration site were not critical determinants for the RNA processing, because the processing choices were qualitatively the same in tissues from all three pedigrees examined, even after induction of liver and kidney MT–Cal RNA levels by zinc treatment. There was no detectable calcitonin or CGRP RNA in these tissues from nontransgenic control mice (with or without zinc), except for relatively low amounts of the endogenous mouse CGRP mRNA in the brain (expressed at roughly 5% the MC5 transgenic CGRP mRNA levels).

Because the nucleic acid probes used for this analysis do not distinguish between RNA from the rat transgene and endogenous mouse calcitonin/CGRP gene, it was necessary to show that the CGRP expression in the transgenic mouse brain was not due to an unexpected induction of the endogenous gene. To differentiate expression of the transgene from the expression of the endogenous gene, we took advantage of the fact that the MT–Cal gene has 62 bases of the MT-I 5′-untranslated region fused to the calcitonin/CGRP gene. Using this region of the trans-

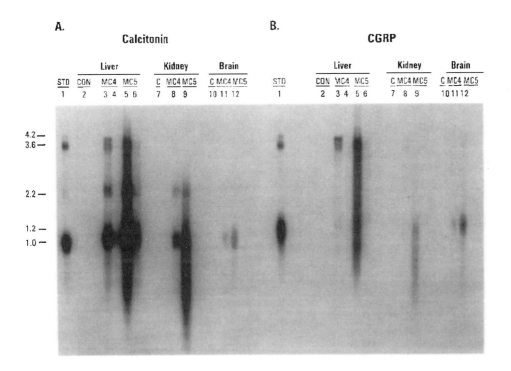

Figure 2. Northern Analysis of Calcitonin and CGRP RNA In the MT–Cal Transgenic Mice Tissues

The RNA processing choice was clearly biased toward calcitonin in the liver and kidney, while the brain produced both calcitonin and CGRP mRNAs. There was no detectable calcitonin or CGRP in the tissues from control mice, except for relatively low amounts of CGRP in the brain (roughly 5% of the transgenic levels). The standard (lane 1) is 50 ng of poly(A)-enriched WA medullary thyroid carcinoma (MTC) RNA, which contains equal amounts of calcitonin and CGR1P RNA. Liver poly(A)-enriched RNA samples are lanes 2, 3, 5 (10 μg), 4, and 6 (1 μg). Kidney poly(A)-enriched samples are lanes 7, 8, and 9 (3 μg). Brain poly(A)-enriched RNA samples are lanes 10, 11, and 12 (1 μg). The 4.2 and 3.6 kb RNA species are nuclear species that contain both the calcitonin and the CGRP exons, and the 2.2 kb nuclear species contains only calcitonin, and not CGRP, exons.

Table 1. Tissue Distribution of Calcitonin and CGRP RNAs in MT–Cal Mice

Tissue	Relative RNA Levels	% Calcitonin RNA
Liver	100	97
Kidney	30	93
Skeletal muscle	0.8	98
Lung	1	>90
Spleen	<0.5	>90
Stomach	3	>90
Submandibular	4	>90
Heart	1	58
Brain	4	42

The RNA processing choice for calcitonin versus CGRP mRNA was determined by scanning densitometry of autoradiographs of size-fractionated RNA from zinc-treated MC5 mice using calcitonin- and CGRP-specific probes, as described in Experimental Procedures. For some tissues there was no detectable CGRP. Based on our limits of detection, we estimate that calcitonin RNA in these tissues represents greater than 90% of the mature transgene mRNA.

gene as a probe (see Figure 1, probe a), there was hybridization to two RNA species, which corresponded in size to CGRP and calcitonin RNA, from the transgenic mice brains, but not from control mice brains (Figure 3). For comparison, transgenic liver contained mostly RNA the size of calcitonin, as well as the endogenous MT-I RNA (Figure 3).

Immunohistochemical Analysis of MT–Cal Expression in the Brain

Because both calcitonin and CGRP RNAs were expressed in the transgenic brain, it was necessary to establish whether discrete neuronal regions selectively express only one mature transcript or whether no clear splicing choice is made. Immunohistochemical analysis showed selective CGRP immunofluorescence in the pyramidal cell layer of hippocampal field CA3 and in the mossy fiber axonal projection to CA3 cells from the granule cells in the dentate gyrus (Figure 4). Colchicine pretreatment, which blocks axonal transport of peptides, diminished CGRP immunoreactivity in the mossy fiber afferents and increased the signal in granule and CA3 pyramidal cell bodies. These cell types were not stained by antisera against calcitonin, and were not stained in normal mice by antisera against either calcitonin or CGRP. Calcitonin, but not CGRP, immunostaining was detected in the livers of these animals, confirming that the antibodies can recognize the calcitonin product of the transgene. Calcitonin and CGRP were also detected in the expected endogenous locations in the thyroid and brain, respectively. These results demonstrate that neurons which do not express the endogenous gene can make an unambiguous choice for CGRP.

Although clear staining for CGRP was observed in hippocampal neurons, we were unable to detect clear immunostaining for calcitonin or CGRP in other regions of

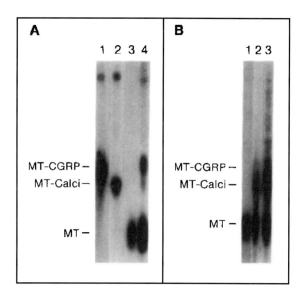

Figure 3. Identification of Calcitonin and CGRP RNAs as Products of the MT–Cal Fusion Gene

Transgene expression in the MT–Cal mice was confirmed by Northern analysis of total RNA hybridized with a probe consisting of metallothionein 5′ region (Figure 1, probe a), except as noted in lanes 1 and 2 of (A). (A) Lanes 1 and 2, WA MTC standard RNA (50 ng poly(A)-enriched) hybridized with either calcitonin (lane 1) or CGRP (lane 2) probes to provide size standards; lane 3, control mouse brain RNA (20 μg); lane 4, MT–Cal brain RNA (20 μg, MC5). (B) Lane 1, control mouse liver RNA (10 μg); lanes 2 and 3, M T–Cal liver RNA (10 μg and 20 μg, MC5). The right panel was exposed for about a 10-fold shorter period to obtain comparable intensities for the endogenous metallothionein.

the brain where CGRP is not normally expressed. The discrepancy between the levels of fusion gene transcripts by RNA blot analysis and the immunohistochemical analysis suggested the necessity of an in situ hybridization histochemical analysis to examine the splicing choices made within the brain.

Analysis of MT–Cal Expression in the Brain by In Situ Hybridization Histochemistry

To localize the MT–Cal RNA products, in situ hybridization of mice brain sections was performed using specific RNA

probes (see Figure 1, probes d and e). We found that the calcitonin RNA was primarily expressed in nonneuronal structures, such as the choroid plexus, ependyma, and pia mater (Figure 5). CGRP RNA was expressed at high levels in the hippocampus (CA3 pyramidal cells and dentate gyrus). It was also the major transcript in several discrete regions, including the reticular nucleus and ventrobasal complex of the thalamus, layer V of the neocortex (and to a lesser extent, layer II), the retrosplenial cortex, and the taenia tecta (Figure 5). A few regions of the brain, such as the granular layer and lateral nuclei of the cerebellum, gave equivalent hybridization signals with both probes. Calcitonin RNA appeared to be the major transcript in only a very few regions that contain neurons, such as the Purkinje layer of the cerebellum and the inferior colliculus (Figure 5).

Because the in situ hybridization technique does not distinguish hybridization to mature mRNA from hybridization to precursor RNAs, it was possible that a portion of calcitonin hybridization seen in the brain was actually due to large molecular weight transcripts containing both the CGRP and calcitonin exons (the 3.6 and 4.2 kb species). To address this possibility, we compared the relative amounts of calcitonin and CGRP RNA in the relatively homogeneous liver and kidney tissues by RNA blotting and in situ hybridization. Typically, the 3.6 and 4.2 kb nuclear RNAs in liver, kidney, and brain represent 10%–20% of the total calcitonin hybridization detected by RNA blot analysis, while mature CGRP mRNA represents 3%–7% of the mature transcripts in liver and kidney (Figure 2). Densitometric scans of the in situ hybridization autoradiographs indicated that liver and kidney contained 80% calcitonin-reactive and 20% CGRP-reactive species. These data are consistent with the interpretation that the in situ technique detects the 10%–20% of nuclear transcripts present in liver, kidney, and brain and underestimates the relative expression of mature calcitonin transcripts in the liver or CGRP transcripts in regions of the brain. Consequently, the results of hybridization histochemical analysis, which show that CGRP mRNA represents approximately 80% of the total hybridization in sites such as the hippocampus and neocortex, are likely to reflect a >90%

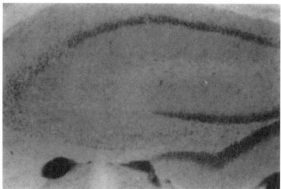

Figure 4. Immunohistochemical Identification of CGRP in the MT–Cal Mouse Hippocampus

The left panel is a bright-field image, and the right panel shows the immunofluorescence seen with anti-CGRP sera (35×) The immunofluorescence is limited to the CA3 pyramidal neurons and mossy fiber projections from the granule cells of the dentate gyrus. No calcitonin immunoreactivity was detected in the hippocampus.

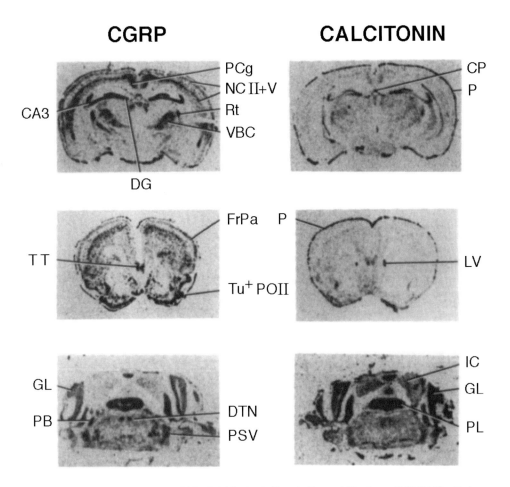

CGRP

CALCITONIN

Figure 5. Localization of Calcitonin and CGRP Hybridization in Discrete Neuronal Structures of MT–Cal Mice Brains

In situ hybridization analysis was performed using ^{35}S-labeled RNA probes to identify the regions expressing the MT–Cal transgene. The results show calcitonin primarily in the pia (P), choroid plexus (CP), cerebellum Purkinje layer (PL), ependyma of the ventricles, e.g., lateral ventricle (LV), and inferior colliculus (IC). CGRP is primarily expressed in the hippocampus (CA3 pyramidal neurons and granule cells of the dentate gyrus [DG]), the retrosplenial (post-cingulate) cortex (PCg), the reticular nucleus (Rt) and ventrobasal complex (VBC) of the thalamus, the neocortex layers II and V (NCII+V), the taenia tecta (TT), the frontoparietal cortex (FrPa), the pyriform cortex layer II (POII), the olfactory tubercle (Tu), and the principal sensory nucleus of the trigeminal (PSV). CGRP RNA was also specifically detected in the parabrachial nucleus (PB) and the dorsal tegmental nucleus (DTN), where CGRP is expressed in control mice. Comparable levels of calcitonin and CGRP RNAs were seen in the cerebellum granular layer (GL). Exposure time was 6 days at 4°C.

processing choice for CGRP mRNA. Conversely, the slight amount of CGRP hybridization in nonneuronal structures, such as the pia mater, is likely to reflect the presence of unprocessed nuclear species.

Within the brain, we find that splicing choice is not correlated with the level of mRNA expression. For example, comparable high levels of expression of mature transcripts are observed in the ependyma of the lateral ventricles, which produces predominantly calcitonin mRNA, and in the taenia tecta, which produces predominantly CGRP mRNA (Figure 5). We also observe differential splicing decisions being made in neurons that express significantly lower mRNA levels (e.g., calcitonin mRNA expression in inferior colliculus is comparable to CGRP mRNA expression in the frontoparietal cortex). Although mRNA level is not solely dependent on the transcription rate, the data suggest that the splicing choice is independent of transcription rate in transgenic animals. Because tissues in the transgenic animals represent a heteroge-

neous population of cells, transcription rates in tissues would not accurately reflect the transcription rate in the cells expressing the transgene. To test directly whether transcription rate is independent of splicing choice, calcitonin/CGRP transcription was determined in a series of medullary thyroid carcinomas expressing either predominantly calcitonin or CGRP transcripts. As shown in Table 2, splicing choice is entirely independent of transcription rate. Therefore, these data show that the splicing decision is independent of mRNA levels and transcription rate in animals and in cell culture, respectively.

Analyses of nontransgenic animals with antisense probes and of transgenic animals with sense strand probes confirmed that the hybridization seen with the transgenic mice was specific (Figure 6). In the control animals, CGRP hybridization was detected only at the expected endogenous locations after long exposures of the autoradiographs. No hybridization was detected either to control or transgenic mice brain sections using sense

Table 2. Calcitonin Gene Transcription in a Series of Rat Medullary Thyroid Carcinoma Tumors

Tumor (Line)	Calcitonin Gene Transcription Rate (ppm/kb Probe)	Type of RNA Produced
WG-1	26	>98% calcitonin mRNA
VE	43	>95% calcitonin mRNA
WG-2	150	>98% calcitonin mRNA
CA-1	98	>95% CGRP mRNA
VF	21	>85% CGRP mRNA
WF	145	>95% CGRP mRNA

A series of rat medullary thyroid carcinoma tumors was analyzed for their content of calcitonin and CGRP mRNAs. Several tumors that contained predominantly calcitonin mRNA or CGRP mRNA were analyzed for the calcitonin gene transcription rate, as described in Experimental Procedures. Simultaneous quantitation of mature calcitonin and CGRP mRNAs was accomplished by S1 nuclease assay. Results are the average of quadruplicate determinations using probes specific for the calcitonin coding region; virtually identical results were obtained when probes specific for the 5′ terminal or CGRP-specific exons were used.

strand RNA probes (Figure 6). The lack of hybridization in control animals or with sense strand probes indicates that the antisense probes are not hybridizing either to DNA or nonspecifically to RNA in regions of dense cell bodies. Thyroid glands from control and transgenic animals gave strong hybridization signals with calcitonin probe and weak signals to CGRP probe, as expected by the relative RNA levels in thyroid (Figure 6).

Cellular Resolution of Hybridization Histochemistry Confirms Neuronal Expression of CGRP

Because of this cellular heterogeneity of the brain, it was important to determine whether neuronal or nonneuronal cell types were expressing the fusion gene. In situ hybridization sections were analyzed by exposure to a thin layer of photographic emulsion and counter-stained with thionin for a Nissl stain. Because most cellular RNA has been removed by RNAase treatment to reduce background hybridization from the RNA probes, only the nuclei (DNA) were Nissl-stained. CGRP mRNA hybridization was detected in the hippocampal field CA3 over cells containing large, pale nuclei, which are characteristic of the pyramidal neurons in this cortical layer (Figures 7A-7C). Cells containing the smaller, darker staining nuclei, characteristic of glial cells, generally had very few or no grains over them. Although the morphology of the stained nuclei is not an absolute criteria of cell type, the distribution of these nuclei in this cortical layer allows unambiguous assignment of the cells as neurons. Calcitonin probes also show the same grain pattern distribution, but at a much lower density, probably reflecting the contribution of precursor RNA to the autoradiographic grain density.

In most regions of the brain, CGRP hybridization occurs in cells that contain neuronal-like nuclei. The ventrobasal

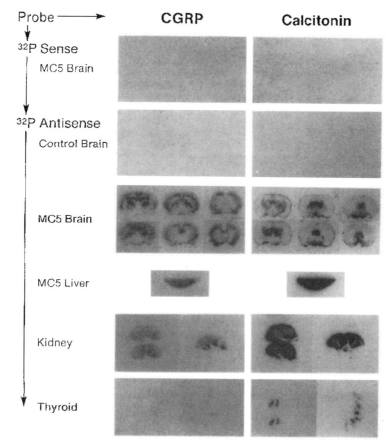

Figure 6. In Situ Hybridization Histochemistry of Calcitonin and CGRP in the Brains of MT–Cal and Control Mice

The specificity and degree of calcitonin and CGRP RNA in situ hybridization analysis were tested by using control mice and both sense and antisense strand ³²P-labeled probes. Brain serial sections, mounted caudal to rostal from brainstem to olfactory bulbs (only six sections shown), were hybridized with antisense and sense CGRP and calcitonin probes, as described in Experimental Procedures. From top to bottom: MT–Cal brain sections (MC5 strain) hybridized using sense strand probes; sections from a control mouse hybridized with antisense strand probes; serial sections from a MT–Cal mouse (MC5) hybridized with antisense strand probes; MC5 liver, kidney, and thyroid hybridized with antisense strand probes. Exposure time was 1 week at 4°C.

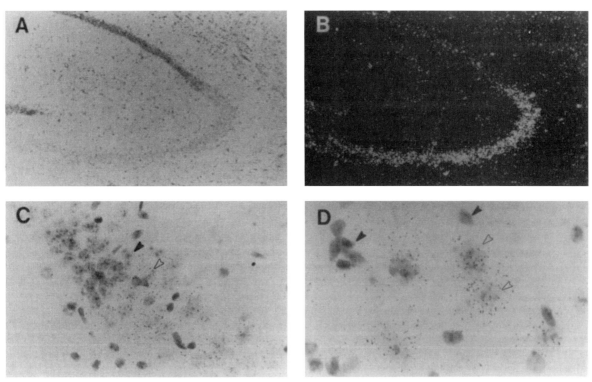

Figure 7. Hybridization Histochemistry of Neurons Expressing CGRP

Sections (20 μm) were hybridized with ^{35}S-labeled CGRP RNA probe (Figure 1, probe e) and stained with thionin (see Experimental Procedures). (A) Bright-field and (B) Dark-field view of MC5 hippocampus section shows autoradiographic siiver grains over field CA3 (50×). (C) Border of fields CA3 and CA1 in hippocampus (190×) Open arrow indicates a CA3 neuron (lighter nucleus) that expresses CGRP from MT–Cal transgene, and solid arrow Indicates a CA1 neuron (darker nucleus) that does not express the transgene. (D) Higher magnification view of the ventrobasal complex of thalamus (300×). Open arrows indicate neurons (lighter nuclei) that express CGRP from the MT–Cal transgene, and solid arrow indicates glial cells (darker nuclei) that do not express the transgene.

complex of the thalamus shows hybridization in cells that contain large, pale nuclei similar to those seen in the hippocampus (Figure 7D). This pattern of hybridization is seen throughout the brain in neuronal structures that express predominantly CGRP transcripts.

While most brain regions that express primarily calcitonin are strictly nonneuronal, the Purkinje layer of the cerebellum is heavily labeled after hybridization with the calcitonin probe, while CGRP hybridization gives a weak signal (Figure 5). The Purkinje layer consists of the Purkinje neurons and specialized types of nonneuronal cells, such as Bergmann glia. Analysis of silver grains over the cerebellum sections shows that calcitonin hybridization is usually excluded from the area around the Purkinje neurons and falls instead over cells containing nuclei with a nonneuronal morphology. This suggests that the calcitonin transcript is being expressed in a cell type other than the Purkinje neurons. Although we cannot unambiguously identify a cell type in this case, it seems likely that the calcitonin transcripts are expressed in specialized glia rather than in neurons.

Although the majority of brain neurons express predominantly CGRP mRNA, the inferior colliculus expresses predominantly calcitonin RNA in cells with neuronal morphology, and the calcitonin transcript is apparently expressed in neurons in the lateral cerebellar nuclei. CGRP mRNA is also neuronally expressed in this region, but whether coexpression of the transcripts occurs at the cellular level is not established.

Discussion

Because transgenic mouse technology allows the expression of alternatively spliced transcript in many cell types, a comprehensive analysis of splicing decisions in vivo provides insights into the organization and relationships of splicing machinery throughout the mouse. Such an analysis has not been previously reported, despite increasing evidence of the widespread use of alternative splicing in a number of genes and cell types. Although alternative RNA processing can be examined in cultured cells, the transgenic mouse model has provided a unique insight into the use of alternative splicing as a developmental strategy. We find that the overwhelming majority of cells throughout the body make unambiguous decisions. Most visceral and muscle tissues express primarily the calcitonin transcript. Tissues, such as liver, lung, and skeletal muscle, are capable of splicing choices that are as definitive as those made in the normal thyroid C cells (Table 1). Predominant calcitonin mRNA expression is found in tissues with ontogenies from endodermal, mesodermal, and ectodermal origin.

In contrast, CGRP mRNA expression is essentially restricted to neurons in MT–CaI mice. Although other tissues express small amounts of the mature CGRP message, we have not detected a subset of cells within these tissues that selectively express CGRP mRNA. These data suggest that cells in which calcitonin mRNA is the predominant product of the transgene may also express CGRP mRNA at low levels. This is analogous to the situation in normal rat thyroid C cells, in which calcitonin mRNA represents >95% of mature transcripts and CGRP has been colocalized to the cells producing calcitonin (Sabate et al., 1985). Within the brain, hybridization histochemistry demonstrates that CGRP expression is localized to neurons. Although most neurons that express MT–CaI generate predominantly CGRP mRNA, there appears to be a very limited set of neurons in which calcitonin mRNA is the predominant transcript.

Immunohistochemical analysis confirms the expression of CGRP in neurons of the hippocampus and dentate gyrus. The reason for the limited detection of CGRP and lack of calcitonin by the immunohistochemical analysis is not known, but could reflect either inadequate sensitivity of the detection method or the inability of these cells to process and store stable peptides that are recognized by our antisera. Because the hippocampus showed the highest level of CGRP expression, it is possible that the other regions of the brain express the transgene at a level below the limits of detection with immunohistochemistry. Alternatively the translation product of these transcripts may not be accurately processed in all neurons. When calcitonin is ectopically expressed in various tumors, the protein precursor is often aberrantly processed (for example, Riley et al., 1986). Low et al. (1985) have demonstrated that the translation product of a metallothionein–somatostatin fusion gene was processed in heterologous pituitary cell types, but not in the liver of transgenic mice. Also, in normal expression of peptide hormones, cell type–specific proteolytic processing of hormone precursors is well documented (for reviews, Liotta and Krieger, 1983; Lynch and Snyder, 1986).

A Model for Calcitonin/CGRP Splice Regulation

We propose that the highly restricted expression of CGRP mRNA is dictated by splice machinery that is restricted to neurons. This would suggest that the calcitonin splice choice is the default or "null" choice when this putative machinery is absent. The model predicts a mechanism that allows for the wide distribution of calcitonin RNA production without invoking the wide distribution of specific factors throughout the animal. Although it is possible that the calcitonin factor is widely distributed and neurons lack this factor, mutational analyses of the calcitonin/CGRP gene argue against this possibility. These analyses show that cells which normally produce calcitonin can express little or no CGRP RNA when the splice site or polyadenylation site of the calcitonin exon is damaged. These mutants are only able to produce mature CGRP transcripts in cells that produce predominantly CGRP (Leff et al., 1987). Furthermore, our results in transgenic animals indicate that the critically regulated step is independent of the level of gene expression. Several additional observations are consistent with this hypothesis. First, in F9 teratocarcinoma cell lines that are transfected with the calcitonin/CGRP gene, the splice choice is independent of gene expression over a 50- to 100-fold range (Leff et al., 1987). Second, the splice choice in medullary thyroid carcinomas is independent of transcription rate. Third, the transcription rate across the calcitonin and CGRP exons is the same in cells making either splicing choice (Amara et al., 1984). Therefore, it seems likely that critical regulation is independent of the extent or the level of gene transcription and that the neuronal pattern of expression is dependent on RNA processing factors specific for the CGRP transcript.

Our results show that neurons, other than those that normally express the CGRP transcript, are capable of making unambiguous splicing choices for CGRP mRNA. These results predict that the CGRP splice machinery is widely expressed in neurons. This raises the possibility that there are a limited number of neuronal factors which dictate alternative splicing and that two or more alternatively spliced transcription units could be regulated by the same machinery. For instance, it is interesting that CGRP is expressed with the alternatively spliced neuropeptide substance P in sensory ganglion cells (Gibson et al., 1984; Weisenfeld-Hallin et al., 1984; Lee et al., 1985). The parallel between these two genes is even more striking when one considers that the second alternatively spliced transcript, encoding substance K, is found in thyroid C cells with calcitonin (Nawa et al., 1984). The colocalization of the transcripts from these genes is consistent with the possibility that they are coregulated by the same splice regulatory machinery. This neuronal splicing phenotype may represent a developmental strategy common to several alternatively spliced gene transcripts that may be regulated by common splice-regulating factors in neurons. Shared splicing factors would allow many transcription units to be alternatively spliced in a wide range of tissues without requiring a large array of splicing factors. Consequently, transcription factors in conjunction with a limited number of splicing factors would increase the potential complexity of developmental strategies.

Experimental Procedures

Plasmid Construction

The calcitonin/CGRP fragment extended from the PstI site, 11 nucleotides 5′ of the calcitonin/CGRP cap site, to the EcoRI site, 1.1 kb 3′ of the poly(A) site in the sixth exon. This fragment was linkered with XhoI at the 5′ PstI site and with BamHI at the 3′ EcoRI site, then inserted into a vector containing the MT–I promoter

Microinjection of Fertilized Eggs and Tail Dot Blot Analysis

Fertilized eggs (C57BL/6J × DBA/2J)F1 or (C57BL/6J × SJL/J)F1 were collected on the morning that a copulation plug was found. Microinjection and transfer to pseudopregnant females was done as described by Costantini and Lacy (1982). Briefly, the BgII–BamHI fragment of MT–CaI was microinjected at approximately 200 copies per cell in a volume of 2 pl. Eggs that survived microinjection were either transferred soon after microinjection or incubated overnight to the two-cell stage and then transferred into (C57BL/6J × DBA/2J)F1 females that had mated with vasectomized males.

Transgenic mice were identified by dot blot analysis of DNA extracted from a segment of the tail as described by Palmiter et al. (1982).

A specific RNA probe was generated by cloning the PstI fragment of the rat calcitonin/CGRP gene that encompasses the first three exons of the gene into pSP65. Run-off transcripts were produced by in vitro transcription using SP6 polymerase as suggested by the manufacturer (BRL). Quantitation of copy number was determined by densitometric scanning of autoradiograms from 3-fold dilution's of tail DNA dot blots. Normal rat DNA was used as a standard.

Northern RNA Analysis

Tissues, which were frozen in liquid nitrogen, were pulverized to a coarse powder, and total nucleic acids were isolated by the method of Shields and Blobel (1977). Total RNA was separated from DNA by precipitation using LiCl. Briefly, total nucleic acids were incubated with 2.5 M LiCl containing 20 mM sodium acetate (pH 5.0) for 3 hr on ice. Aliquots were denatured and subjected to electrophoresis on 1.2% agarose–formaldehyde gels (Maniatis et al., 1982). RNA was transferred to nitrocellulose, washed in prehybridization buffer containing 50% formamide, and hybridized to probes nick-translated with [α-^{32}P]-dCTP as the labeled nucleotide (1×10^8 to 3×10^8 cpm/μg). A Sau3A fragment encompassing the fifth and sixth exons was specific for CGRP transcripts (Figure 1, probe c); a BglII fragment from the fourth exon of the genomic clone was specific for calcitonin containing transcripts (Figure 1, probe b). Size standards were provided by migration of calcitonin and CGRP mRNA species from a rat medullary thyroid carcinoma (MTC) cell line designated WA, which contains equal amounts of the two RNA species (Amara et al., 1982).

Quantitation of relative amounts of calcitonin and CGRP transcripts was obtained by laser densitometric scanning of autoradiographs using Quick Scan R & D densitometer (Helena Laboratories). Scans were standardized using medullary thyroid carcinoma RNA in which calcitonin and CGRP mRNAs are present at equivalent levels (WA MTC).

Immunohistochemistry

Tissue was fixed by vascular perfusion, and frozen sections 20 μm thick were cut on a sliding microtome (Swanson et al., 1983). Immunofluorescence localization of CGRP and calcitonin was carried out as described in detail in Swanson et al. (1983). CGRP serum was used at a dilution of 1:2,000 and calcitonin serum at a dilution of 1:1500.

Hybridization Histochemistry

Tissue was fixed as described above for immunohistochemistry. Sections were mounted on poly-L-lysine-coated slides and air-dried. Pretreatment, hybridizations, and washing conditions have been described (Cox et al., 1984). Briefly, sections were digested with proteinase K (10 μg/ml, 37°C, 30 min), acetylated, and dehydrated. After thorough drying, 50 μl of hybridization mix containing ^{32}P-labeled probe (10^7 cpm/ml) was spotted on each slide. Slides were incubated at 50°C for 16 hr. Slides were rinsed, digested with ribonuclease (RNAase A, a1: 20 μg/ml, 37°C, 30 min), and washed in 0.1 × SSC for 30 min at 55°C. After dehydration, the sections were exposed to Cronex 4 film (DuPont) at 4°C. Hybridizations with ^{35}S-labeled probes were performed in the presence of 10 mM DTT, and 1–10 mM DTT was included in the following washes. Specific RNA probes were generated with in vitro transcription vectors. A CGRP vector was produced by cloning the 450 bp TaqI–Sau3A CGRP genomic fragment (Figure 1, fragment e) into pSP64. This fragment contains 170 bp of single copy 3′-noncoding sequence and 280 bp of single copy 3′-flanking sequence. The calcitonin vector was produced by cloning the 530 bp Hae III genomic fragment (Figure 1, fragment d) into pSP65. This fragment contains 130 bp of 3′-noncoding sequence and 400 bp of 3′-flanking sequence. After the vectors were linearized, run-off transcripts were produced with SP6 polymerase using carrier-free [^{32}P]UTP (800 Ci/mmol, NEN) or [^{35}S]UTP (1200 Ci/mmol, NEN).

Nuclear Run-On Gene Transcription Analysis

A series of independent medullary thyroid carcinoma (MTC) tumor lines, producing either calcitonin or CGRP as the predominant peptide product (referred to as WG-1, WG-2, VE, CA-1, VF and WF), were the generous gifts of Dr. B. Roos (Veterans Administration Medical Center, Tacoma, WA). Cells were washed with ice-cold phosphate-buffered saline and lysed by gentle homogenization in 5 ml of ice-cold buffer containing 150 mM KCl, 5 mM MgCl$_2$, 5 mM DTT, 10 mM Tris (pH 7.8), 0.25 M sucrose, and 0.1% NP40, layered onto a 4 ml cushion of the identical buffer containing 0.5 M sucrose, and centrifuged at $1000 \times$ g for 10 min. The nuclear pellets were washed in the nascent chain labeling buffer (150 mM KCl, 5 mM MgCl$_2$, 10 mM Tris [pH 7.8], 10% glycerol). Nuclei were then incubated (25°C, 10 min) in 100 μl of the same buffer supplemented with 1 mM ATP and GTP, 100 μCi of [^{32}P]UTP and [^{32}P]-CTP (200–400 Ci/mM), and 2 μl of RNAase inhibitor, allowing nascent transcripts to elongate. Labeled RNA was purified and subjected to hybridization analysis, as previously described (Murdoch et al., 1982; McKnight and Palmiter, 1979). Briefly, 8×10^6 to 15×10^6 cpm of ^{32}P-labeled RNA was hybridized to 2 μg of linearized cloned calcitonin cDNA (pCal) or CGRP cDNA (pCGRP$_2$; Amara et al., 1982) bound to a 3 mm disk of nitrocellulose. Immobilized pBR322 was used to determine assay background. Filters were washed with 10 mM Tris–HCl (pH 7.5), 0.1% SDS, 0.2 × SSC at 65°C for 30 min with pancreatic RNAase (1.3 μg/ml) and T1 nuclease (10 U/ml) in 10 mM Tris–HCl (pH 7.4), 0.3 M NaCl at 37°C for 30 min. followed by extensive washing with 10 mM Tris–HCl (pH 7.4), 0.3 M NaCl, 0.1% SDS at 42°C. Each point is the mean +/− SE of the transcription rate (specific cpm bound per 10^6 TCA precipitable cpm added), based on quadruplicate hybridizations. Calcitonin gene transcription is undetectable when α-amanitin (0.8 μg/ml) is included in the incubation. The hybridization is linear with respect to added labeled RNA and constant with hybridization to variable cloned DNA filter (0.5–5.0 μg), confirming that all hybridizations are in DNA excess and that unlabeled RNA does not compete for elongated nascent transcripts.

Acknowledgments

We thank Donna Simmons, Jodi Harrold, and Jeff Arriza for their assistance with the histochemistry; Kelly P. Ahearn for technical assistance; Stacey Dillon for maintaining the animal rooms; Drs. Frank Costantini, Elizabeth Lacy, and Brigid Hogan for instruction on transgenic mouse production; Drs. Ron Evans, Sergio Lira, and Stuart Leff for helpful discussions; Margaret Richards for typing the manuscript. This work was supported by grants from the American Cancer Society and the National Institutes of Health. E. B. C. is supported by training grants to the Department of Biology, UCSD; A. F. R. is supported by a postdoctoral fellowship from the Jane Coffin Childs Memorial Fund for Medical Research.

The costs of publication of this article were defrayed in part by the payment of page charges. This article must therefore be hereby marked "*advertisement*" in accordance with 18 U.S.C. Section 1734 solely to indicate this fact.

Received November 14, 1986; revised January 28, 1987

References

Alt, F. W., Bothwell, A. L. M., Knapp, M., Siden, E., Mather, F., Koshland, M., and Baltimore, D. (1980). Synthesis of secreted and membrane-bound immunoglobulin Mu heavy chains is directed by mRNAs that differ at their 3′ ends. Cell *20*, 293–301.

Amara, S. G., Jonas, V., Rosenfeld, M. G., Ong, E. S., and Evans, R. M. (1982). Alternative RNA processing in calcitonin gene expression generates mRNAs encoding different polypeptide products. Nature *298*, 240–244.

Amara, S. G., Evans, R. M., and Rosenfeld, M. G. (1984). Calcitonin/calcitonin gene-related peptide transcription unit: tissue-specific expression involves selective use of alternative polyadenylation sites. Mol. Cell. Biol. *4*, 2151–2160.

Blattner, F. R., and Tucker, P. W. (1984). The molecular biology of immunoglobulin D. Nature *307*, 417–422.

Breitbart, R. E., Nguyen, H. T., Medford, R. M., Destree, A. T., Mahdavi, V., and Nadal-Ginard, B. (1985). Intricate combinatorial patterns of exon splicing generate multiple regulated troponin T isoforms from a single gene. Cell *41*, 67–82.

Costantini, E, and Lacy, E. (1982). Gene transfer into the mouse germline. J. Cell. Physiol. *1*, 219–226.

Cox, K. H., DeLeon, D. V., Angerer, L. M., and Angerer, R. C. (1984). Detection of mRNAs in sea urchin embryos by *in situ* hybridization using asymmetric RNA probes. Dev. Biol. *101*, 485–502.

Early, R. Rogers, J., Davis, M., Calame, K., Bond, M., Wall, R., and Hood, L. (1980). Two mRNAs can be produced from a single immunoglobulin μ gene by alternative RNA processing pathways. Cell 20, 313–319.

Gibson, S. J., Polak, J. M., Bloom, S. R., Sabate, I. M., Mulderry, P. M., Ghatei, M. A., McGregor, G. P., Morrison, J. F. B., Kelly, J. S., Evans, R. M., and Rosenfeld, M. G. (1984). Calcitonin gene-related peptide immunoreactivity in the spinal cord of man and of eight other species. J. Neurosci. 4, 3101–3111.

Jacobs, J. W., Goodman, R. H., Chin, W. W., Dee, P. C., Habener, J. F., Bell, N. H., and Potts, J. T., Jr. (1981). Calcitonin messenger RNA encodes multiple polypeptides in a single precursor. Science 213, 457–458.

Kawai, Y., Takami, K., Shiosaka, S., Emson, P. C., Hillyard, C. J., Girgis, S., Macintyre, I., and Tohyama, M. (1985). Topographic localization of calcitonin gene-related peptide in the rat brain: an immunohistochemical analysis. Neuroscience 15, 747–763.

Kitamura, N., Takagaki, Y. Furuto, S., Tanaka, T., Nawa, H., and Nakanishi, S. (1983). A single gene for bovine high molecular weight kininogens. Nature 305, 545–549.

Lee, Y., Kawai, Y., Shiosaka, S., Takami, K., Kiyama, H., Hillyard, C. J., Girgis, S., Macintyre, I., Emson, P. C. and Tohyama, M. (1985). Coexistence of calcitonin gene-related peptide and substance P-like peptide in single cells of the trigeminal ganglion of the rat: immunohistochemical analysis. Brain Res. 330, 194–196.

Leff, S. E., Rosenfeld, M. G., and Evans, R. M. (1986). Complex transcriptional units: diversity in gene expression by alternative RNA processing. Ann. Rev. Biochem. 55, 1091–1117

Leff, S. E., Evans, R. M., and Rosenfeld, M. G. (1987). Splice commitment dictates neuron-specific alternative RNA processing in calcitonin/CGRP gene expression. Cell 48, 517–524.

Liotta, A. S., and Krieger, D. T. (1983). Pro-opiomelanocortin-related and other pituitary hormones in the central nervous system. In Brain Peptides, D. T. Krieger, J. B. Martin, and M. Brownstein, eds. (New York: John Wiley and Sons), pp. 613–660.

Low, M. J., Hammer, R. E., Goodman, R. H., Habener, J. F., Palmiter, R. D., and Brinster, R. L. (1985). Tissue-specific posttranslational processing of pre-prosomatostatin encoded by a metallothionein–somatostatin fusion gene in transgenic mice. Cell 41, 211–219.

Lynch, D. R., and Snyder, S. H. (1986). Neuropeptides: multiple molecular forms, metabolic pathways, and receptors. Ann. Rev. Biochem. 55, 773–799

Maki, R., Roeder, W., Traunecker, A., Sidman, C, Wabl, M., Raschke, W., and Tonegawa, S. (1981). The role of DNA rearrangement and alternative RNA processing in the expression of immunoglobulin delta genes. Cell 24, 353–365.

Maniatis, T., Fritsch, E. F., and Sambrook, J. (1982). Molecular Cloning: A Laboratory Manual (Cold Spring Harbor Laboratory, New York: Cold Spring Harbor Laboratory).

McKnight, G. S., and Palmiter, R. D. (1979). Transcriptional regulation of the ovalbumin and conalbumin genes by steroid hormones in chick oviduct. J. Biol. Chem. 254, 9050–9058.

Murdoch, G. H., Potter, E., Nicolaisen, A. K., Evans, R. M., and Rosenfeld, M. G. (1982). Epidermal growth factor rapidly stimulates prolactin gene transcription. Nature 300, 192–194.

Nabeshima, Y. Fujii-Kuriyama, Y., Muramatsu, M., and Ogata, K. (1984). Alternative transcription and two modes of splicing result in two myosin light chains from one gene. Nature 308, 333–338.

Nawa, H., Kotani, H., and Nakanishi, S. (1984). Tissue-specific generation of two preprotachykinin mRNAs from one gene by alternative RNA splicing. Nature 312, 729–734.

Padgett, R. A., Grabowski, P. J., Konarska, M. M., Seiler, S., and Sharp, R. A. (1986). Splicing of messenger RNA precursors. Ann. Rev. Biochem. 55, 1119–1150.

Palmiter, R. D., Chen, H. Y. and Brinster, R. L. (1982). Differential regulation of metallothionein–thymidine kinase fusion genes in transgenic mice and their offspring. Cell 29, 701–710.

Riley, J. H., Edbrooke, M. R., and Craig, R. K. (1986). Ectopic synthesis of high-M$_r$ calcitonin by the BEN lung carcinoma cell line reflects aberrant proteolytic processing. FEBS Lett. 198, 71–79.

Rosenfeld, M. G., Mermod, J.-J., Amara, S. G., Swanson, L. W., Sawchenko, R. E., Rivier, J., Vale, W. W., and Evans, R. M. (1983). Production of a novel neuropeptide encoded by the calcitonin gene via tissue-specific RNA processing. Nature 304, 129–135.

Rosenfeld, M. G., Amara, S. G., and Evans, R. M. (1984). Alternative RNA processing: determining neuronal phenotype. Science 225, 1315–1320.

Rozek, C. E., and Davidson, N. (1986). Differential processing of RNA transcribed from the single-copy Drosophila myosin heavy chain gene produces four mRNAs that encode two polypeptides. Proc. Natl. Acad. Sci. USA 83, 2128–2132.

Sabate, M. I., Stolarsky, L. S., Polak, J. M., Bloom, S. R., Varndell, I. M., Ghatei, M. A., Evans, R. M., and Rosenfeld, M. G. (1985). Regulation of neuroendocrine gene expression by alternative RNA processing: co-localization of calcitonin and calcitonin gene-related peptide (CGRP) in thyroid C-cells. J. Biol. Chem. 260, 2589–2592.

Schwarzbauer, J. E., Tamkun, J. W., Lemischka, I. R., and Hynes, R. O. (1983). Three different fibronectin mRNAs arise by alternative splicing within the coding region. Cell 35, 421–431.

Shields, D., and Blobel, G. (1977). Cell-free synthesis of fish preproinsulin, and processing by heterologous mammalian microsomal membranes. Proc. Natl. Acad. Sci. USA 74, 2059–2063.

Swanson, L. W., Sawchenko, P. E., Rivier, J., and Vale, W. W. (1983). Organization of ovine corticotropin-releasing factor immunoreactive cells and fibers in the rat brain: an immunohistochemical study. Neuroendocrinology 36, 165–186.

Swanson, L. W., Simmons, D. M., Arriza, J., Hammer, R., Brinster, R. L., Rosenfell, M. G., and Evans, R. M. (1985). Novel developmental specificity in the nervous system of transgenic animals expressing growth hormone fusion genes. Nature 317, 363–366.

Tamkun, J. W., Schwarzbauer, J. E., and Hynes, R. O. (1984). A single rat fibronectin gene generates three different mRNAs by alternative splicing of a complex exon. Proc. Natl. Acad. Sci. USA 81, 5140–5144.

Weisenfeld-Hallin, Z., Hokfelt, T., Lundberg, J. M., Forssmann, W. G., Reinecke, M., Tschopp, F. A., and Fischer, J. A. (1984). Immunoreactive calcitonin gene-related peptide and substance P coexist in sensory neurons to the spinal cord and interact in spinal behavioral responses of the rat. Neurosci. Lett. 52, 199–204.

Post-transcriptional regulation: RNA stability and translatability

4

Although transcriptional control and the regulation of alternative splicing represent major control points in the regulation of gene expression, there is also some regulation at subsequent stages in a limited number of cases. Such regulation takes place after the mRNA has been transported to the cytoplasm and can involve regulation of its stability or its translatability. Thus a number of situations where the stability of a specific RNA species is altered in response to a stimulus have been described (for reviews, see Peltz et al., 1991; Sachs, 1993). Obviously the more rapidly an mRNA is degraded the less protein it will produce, so that regulation of RNA stability can represent a means of regulating protein production. Similarly, in other situations the translation of the mRNA can be regulated so that it is translated into protein in one situation and not in another (for a review, see Altmann and Trachsel, 1993).

Both of these cytoplasmic regulatory mechanisms appear to be utilized to supplement transcriptional control in situations where a rapid response to a particular stimulus is required. Thus in such a situation increasing the stability of a pre-existing mRNA or enhancing its translatability allows new protein to be produced more rapidly than if the transcription of the gene had to be induced *de novo*, with all the other stages of splicing, transport etc. having to follow before protein could be produced. In many cases where such cytoplasmic regulation operates, it is paralleled by enhanced transcription of the corresponding gene, so allowing further enhancement of mRNA and protein levels. Hence such cytoplasmic control mechanisms do not controvert our previous conclusion that transcriptional control is the major control point of eukaryotic gene regulation, but should be viewed as a supplement to transcriptional control in situations where a rapid response is required.

The papers presented in this section are of importance in that they define the mechanisms that mediate such post-transcriptional control in particular situations. The paper by Casey et al. defines the mechanism by which decreased stability of the human transferrin receptor mRNA is produced in response to treatment with iron (see Table 4.1). This effect had previously been shown to depend on the 3′ untranslated region of the transferrin receptor mRNA (Owen and Kühn, 1987). In the work illustrated in Fig. 1 of their paper, Casey et al. first showed that transfer of the 3′ untranslated region of the transferrin receptor mRNA to a completely unrelated mRNA was sufficient to render this latter mRNA unstable in the

Table 4.1 Regulation of the transferrin receptor and ferritin genes

Gene	Effect of iron on protein production	Mechanism	Position of stem—loop structure
Ferritin	Increased	Increased mRNA translation	5′ Untranslated region
Transferrin receptor	Decreased	Decreased mRNA stability	3′ Untranslated region

presence of iron. They then followed this up with the data shown in Fig. 2 of the paper by defining a short region between nucleotides 3178 and 3856 of the mRNA which was necessary and sufficient for destabilization of any mRNA containing it in the presence of iron.

Although these studies are of interest in defining the region of the transferrin receptor mRNA responsible for its regulated stability, they have been paralleled in a number of other systems (for a review, see Latchman, 1995a). The real interest of the work performed by Casey et al. came from their analysis of the RNA sequence within this region of the transferrin receptor mRNA, which showed that it could form five stem–loop structures, as illustrated in Fig. 3 of the paper. Most interestingly, comparison of these structures indicated that they were similar to a stem–loop structure in the 5′ untranslated region of the mRNA encoding ferritin. The translatability of this mRNA is known to be increased in response to iron treatment (see Table 4.1). Moreover, when Casey et al. transferred their sequence from the 3′ untranslated region of the transferrin receptor mRNA to the 5′ untranslated region of an unrelated mRNA, they found that the translatability of this latter mRNA was increased in response to iron, exactly paralleling the effect of the similar sequence within the ferritin mRNA itself.

Hence the same sequence taken from the transferrin receptor mRNA can confer enhanced translatability of an mRNA when located in the 5′ untranslated region upstream of the protein coding sequence, and can confer decreased stability of the mRNA when located in the 3′ untranslated region downstream of the protein coding sequence.

This is because the stem–loop structure present in both the transferrin receptor and ferritin mRNAs acts as an iron-response element (IRE) which unfolds in the presence of iron (Figure 4.1). In the ferritin mRNA, where this element is at the 5′ end, such unfolding will allow unimpeded movement of the ribosome along

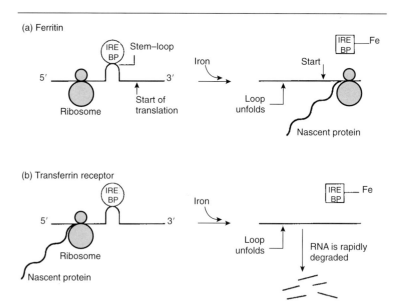

Figure 4.1 Regulated translation of the ferritin mRNA (a) and degradation of the transferrin receptor mRNA in response to iron (b). Note that in each case the presence of iron removes an iron binding protein (IRE BP) from a stem–loop structure in the mRNA. This results in the unfolding of the loop, resulting in increased translation of the ferritin mRNA and increased degradation of the transferrin receptor mRNA.

the mRNA, so enhancing translation. In contrast, in the transferrin receptor mRNA, where this element is in the 3′ untranslated region, its unfolding will result in enhanced degradation of the mRNA. Indeed, a protein which binds to the iron-response element has been identified and shown to increase in activity in cells that have been deprived of iron, indicating that its binding stabilizes the stem–loop structure and prevents unfolding (Figure 4.1) (for a review, see Klausner et al., 1993).

The paper by Casey et al. is of importance, therefore, in that it not only identifies a sequence that mediates degradation of the transferrin receptor mRNA in response to iron but also identifies a general mechanism that can regulate both degradation and translatability in response to such treatment. This indicates that the same sequence can, in some situations, mediate control at the level both of RNA stability and of RNA translatability.

The second paper presented in this section, by Yen et al., is also based on the identification of a sequence that can produce enhanced RNA degradation in response to a specific treatment. Previous work from this group (Yen et al., 1988) had shown that a 13 base sequence at the 5′ end of β-tubulin mRNA was sufficient to cause the degradation of this mRNA in response to the presence of free β-tubulin protein. When transferred to another unrelated mRNA, this sequence was able to render the RNA unstable in the presence of free β-tubulin protein.

The paper presented here focuses on an aspect of RNA degradation which we have not discussed previously. The 13 nucleotides defined initially encode the first four amino acids of the β-tubulin protein. This raises the question of whether the signal that is recognized as the trigger for RNA degradation is actually the nucleotide sequence within the mRNA itself (Figure 4.2a) or the first four amino acids of nascent β-tubulin protein being produced from the corresponding mRNA on the ribosome (Figure 4.2b).

Experiments in the paper by Yen et al. presented here elegantly addressed this question. The authors altered various nucleotides within the first 13 bases of the β-tubulin mRNA and investigated the effect on its regulated degradation. As illustrated in Fig. 1 of the paper, changing the sequence AGG (which encodes an arginine residue) to AGU (which encodes serine) abolished the observed enhanced degradation of the mRNA in response to free tubulin. Interestingly, however, a second mutation which changed the sequence to CGU (which once again encodes arginine) restored the regulation. This finding strongly suggests that the degradation system is recognizing the encoded amino acid sequence rather than the nucleotide sequence itself. Thus a single nucleotide change which alters the encoded amino acid abolishes regulated degradation, whereas a second nucleotide change which results in the original amino acid being restored also restores degradation.

This conclusion was further elegantly substantiated in the data presented in Fig. 3 of the paper, in which the reading frame of the β-tubulin mRNA was changed so that a different protein would be produced while maintaining the identical 13-nucleotide sequence. This had the effect of abolishing the regulated degradation. Thus, for example, introducing four nucleotides containing an initiator methionine codon upstream of the 13-nucleotide sequence abolished regulated degradation, even though the 13-nucleotide sequence remained intact. Hence in this situation regulated degradation appears to proceed via the degradation system recognizing

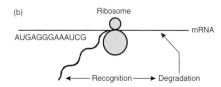

Figure 4.2 Alternative models by which a degradative enzyme could recognize β-tubulin mRNA and degrade it. In (a), the enzyme recognizes the first 13 nucleotides of the tubulin mRNA directly and then degrades the mRNA. In (b), the nuclease recognizes the first four amino acids (MREI) of the nascent β-tubulin protein that is being produced by translation of the β-tubulin mRNA by the ribosome. This recognition then triggers the nuclease to degrade the β-tubulin mRNA attached to the ribosome.

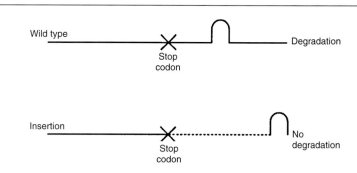

Figure 4.3 Insertion of an extra sequence (broken line) between the stop codon (labelled X) and the stem–loop structure in the histone mRNA prevents its degradation

the first four amino acids of the nascent β-tubulin protein being produced from the ribosome and then degrading the β-tubulin mRNA which is acting as a template for such production (Figure 4.2b).

This paper by Yen et al. is of importance in defining a novel mechanism by which mRNA degradation can take place following recognition of a protein sequence. Its implications are, however, much broader and encompass examples such as that of the transferrin receptor mRNA described above, where RNA degradation is evidently initiated on the basis of recognition of the RNA which is to be degraded rather than of the protein that it encodes. In the case of the histone mRNA, degradation occurs by recognition of a sequence in the 3′ untranslated region of the mRNA which could not encode a protein. However, the insertion of additional sequences between the translation stop codon and this region abolished regulated degradation, suggesting an association between translation and degradation (Graves et al., 1987) (Figure 4.3).

When taken together with the tubulin mRNA observations, these findings suggest a model in which the ribosome carries with it a nuclease capable of degrading specific RNAs in response to a specific signal. In the case of the β-tubulin system this signal is provided by the first four amino acids emerging from the ribosome and results in RNA degradation (Figure 4.4a). In contrast, in the case of the histone mRNA this signal is provided by recognition of a 3′ RNA sequence. However, translation is still required in order to move the ribosome to a position where its associated nuclease is sufficiently close to the target sequence for degradation. If sequences are inserted between the stop codon and the target sequence, the ribosome-associated nuclease does not reach a position where it can carry out degradation of the mRNA (Figure 4.4b).

Interestingly, a number of other situations have been described in which inhibition of translation stabilizes specific mRNAs (e.g. see Kelly et al., 1983). This has been interpreted in many cases as an indication that the nuclease degrading such mRNAs is very short-lived and will be rapidly broken down during a period when protein synthesis is inhibited, so stabilizing the mRNA. It appears much more likely, however, that such mRNAs are also degraded by a mechanism directly requiring translation, as in the case of the histone mRNA (for a review, see Peltz and Jacobson, 1992).

The papers in this section thus elegantly define specific sequences that are associated with regulated stability or translatability of the mRNA. They also point to an association between translation and

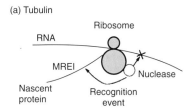

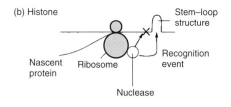

Figure 4.4 Potential mechanisms by which a nuclease associated with the ribosome can recognize nascent tubulin protein (a) or the stem–loop structure in the histone mRNA (b) and degrade the mRNA

the regulation of RNA stability. Thus in some situations, such as the ferritin and transferrin receptor mRNAs, the same sequence can confer regulated translatability or regulated degradation depending on its position in the mRNA molecule. Similarly, in other situations regulated RNA degradation requires translation of the mRNA either because the trigger for degradation involves recognition of the corresponding nascent protein or because translation is necessary to deliver a ribosome-associated nuclease to the region of the RNA that confers regulated instability.

Casey et al. (1988) Science **240**, 924–928

Iron-Responsive Elements: Regulatory RNA Sequences That Control mRNA Levels and Translation

JOHN L. CASEY, MATTHIAS W. HENTZE, DAVID M. KOELLER,
S. WRIGHT CAUGHMAN, TRACEY A. ROUAULT,
RIDHARD D. KLAUSNER, JOE B. HARFORD

The biosynthetic rates for both the transferrin receptor (TfR) and ferritin are regulated by iron. An iron-responsive element (IRE) in the 5′ untranslated portion of the ferritin messenger RNA (mRNA) mediates iron-dependent control of its translation. In this report the 3′ untranslated region of the mRNA for the human TfR was shown to be necessary and sufficient for iron-dependent control of mRNA levels. Deletion studies identified a 678-nucleotide fragment of the TfR complementary DNA that is critical for this iron regulation. Five potential stem-loops that resemble the ferritin IRE are contained within the region critical for TfR regulation. Each of two of the five TfR elements was independently inserted into the 5′ untranslated region of an indicator gene transcript. In this location they conferred iron regulation of translation. Thus, an mRNA element has been implicated in the mediation of distinct regulatory phenomena dependent on the context of the element within the transcript.

IN PROLIFERATING CELLS, THE AVAILability of iron modulates the biosynthetic rates for at least two proteins critical to cellular iron metabolism. The transferrin receptor (TfR) serves as the chief means of iron uptake, and its biosynthesis is decreased when iron is abundant and increased when it is scarce (1). Conversely, ferritin, which sequesters iron in the cytoplasm, is synthesized at a higher rate when iron is abundant than when it is scarce (2). The mechanisms involved in the control of the biosynthetic rates for these two proteins are distinct. For the TfR the regulation of biosynthetic rates can be accounted for by changes in mRNA levels (3). In contrast,

ferritin biosynthesis is altered without a corresponding change in the level of total ferritin mRNA (4) by redistribution of mRNA between polysome and nonpolysome pools (5).

The iron-dependent modulation of the level of TfR mRNA is mediated by two regions of the gene. Together these two regions can produce >20-fold differences in transcript levels between cells treated with an iron chelator and those treated with an iron source. The transcription rate directed by 5′ flanking sequences of the TfR gene is two- to threefold higher in cells treated with the iron chelator desferrioxamine than in cells treated with hemin, an iron source (6). However, deletion of sequences corresponding to the 3′ untranslated region (3′UTR) of the TfR mRNA eliminates most of the

iron regulation of transcript levels regardless of whether the TfR promoter region is present (6, 7).

The 5′ untranslated region (5′UTR) of human ferritin H chain mRNA is both necessary and sufficient to mediate iron-responsive translational regulation of ferritin biosynthesis (8). Moreover, the specific sequence within the ferritin 5′UTR responsible for the modulation of ferritin translation has been identified (9, 10). This RNA sequence, which we have termed an iron-responsive element (IRE), has the potential to form a characteristic stem-loop structure. The 5′UTR of each ferritin gene for which complete mRNA sequence data are available [human (11), rat (12), chicken (13), and frog (14)] contains a sequence capable of forming a similar stem-loop. Here we present a deletion analysis of the TfR 3′UTR that implicates strikingly similar elements in the iron-responsive regulation of TfR mRNA levels. We further show that synthetic oligonucleotides corresponding to elements of the TfR 3′UTR confer ferritin-like translational regulation when inserted in the 5′UTR of an indicator gene transcript.

We (6) and others (7) have shown that the sequences corresponding to the 3′UTR of the TfR mRNA are necessary for the full degree of iron regulation of the level of this transcript. To demonstrate the sufficiency of this region to confer TfR-like iron regulation on another gene, we have inserted a fragment containing most (2.2 kb of 2.5 kb) of the TfR 3′UTR into the 3′UTR of the structural gene for human growth hormone (hGH) and have transformed murine cells with this construct (pSVGH-TR). The hybrid construct gave rise to three major hGH/

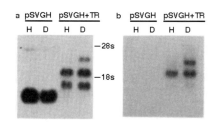

Fig. 1. The 3′UTR of the human transferrin receptor is sufficient to confer iron regulation on the accumulation of a chimeric transcript. Pools of stable transformants were prepared as described (6) after transfection with pSVGH or pSVGH-TR as indicated (17). Cells were treated for 16 hours with 6 mM sodium butyrate plus either 100 μM hemin (H) or 100 μM desferrioxamine (D) prior to isolation of cytoplasmic RNA (6). RNA was separated by electrophoresis in a 1% agarose-formaldehyde gel and analyzed by blot hybridization with an hGH probe consisting of the 650-bp Sty I fragment of the hGH gene (**a**) or the full-length human TfR cDNA (18) (**b**). The relative positions of the 28S and 18S ribosomal RNAs are indicated.

Cell Biology and Metabolism Branch, National Institute of Child Health and Human Development, National Insututes of Health, Bethesda, MD 20892.

TfR hybrid transcripts (Fig. 1). The largest of the chimeric transcripts was regulated by iron in a fashion resembling the native TfR gene, with much higher levels present after desferrioxamine treatment than after treatment with hemin. The levels of the two smaller hGH/TfR transcripts as well as the level of the hGH transcript encoded by the plasmid pSVGH, which lacks TfR sequences, were unaffected by iron availability. The largest of the hGH/TfR transcripts is of the size that would be predicted from the structure of pSVGH-TR. The relative intensities of the hybridization signals when the hGH and TfR probes were compared indicated that the two smaller chimeric transcripts lack portions of the TfR sequences that are part of the plasmid pSVGH-TR.

When hGH protein secretion into the growth medium of the pSVGH-TR transformants was assessed by radioimmunoassay, we detected no appreciable iron regulation of hGH production. This apparent absence of regulation is likely due to the ability of the two smaller hGH/TfR transcripts (which are not iron-regulated because they lack critical TfR sequences) to be used to produce hGH. In our pSVGH-TR transformants, these truncated unregulated transcripts were considerably more abundant than the full-length regulated mRNA. We have also observed multiple human TfR transcripts in murine cells transformed with the full-length human TfR cDNA. The smaller human TfR transcripts in those cells lack some or all of the 3′UTR, and only the full-length transcript was fully iron-regulated (6). The full-length TfR transcript in those cells was the predominant mRNA species, and iron regulation of TfR protein biosynthesis was seen. The appearance of the multiple hGH transcripts in Fig. 1 is clearly a function of the TfR 3′UTR, since the parent plasmid pSVGH that lacks the TfR 3′UTR overwhelmingly gives rise to one hGH transcript. Owen and Kühn (7) transferred the TfR 3′UTR to the 3′ end of the histocompatibility antigen HLA-A2 gene but, on the basis of cell surface expression of the gene's protein product, concluded that there was no iron regulation of this chimeric gene. This observation may also be a reflection of protein production from multiple hybrid transcripts that are not all iron-regulated. Our results (Fig. 1) demonstrate that no portion of the TfR protein-encoding sequences is required for iron regulation of transcript levels. Thus the TfR 3′UTR is both necessary and sufficient to confer the iron regulation of mRNA levels that is characteristic of TfR biosynthetic regulation.

Previously, we prepared stable transformants of murine cells by using a TfR "minigene" containing the full-length TfR cDNA driven by the TfR promoter (Fig. 2a, construct I) and found that expression of the human TfR in those cells was highly regulated by iron. In contrast, cells transformed with a plasmid lacking the bulk of the TfR 3′UTR (construct II) continued to express the normal human TfR protein but had lost the ability to iron-regulate this expression (6). In this report, the sequences within the TfR 3′UTR responsible for iron-dependent regulation are localized further by a restriction fragment deletion analysis. This analysis was first performed with a transient expression assay for the iron regulation of human TfR protein biosynthesis in transfected murine cells (Fig. 2a). Cells transfected with construct I synthesized approximately seven times as much human TfR protein after desferrioxamine treatment as after treatment with hemin. Transfection with constructs IV and V resulted in a similar degree of iron regulation of human TfR biosynthesis. We have tested constructs having smaller deletions 3′ of base 3861, and these too were iron-regulated. There was some residual iron regulation in cells transfected with constructs II and III (a less than twofold difference between desferrioxamine treatment and hemin treatment). Since all of the constructs in Fig. 2 contain 5′ flanking sequences of the human TfR gene that have been shown to mediate a two- to threefold transcriptional effect of iron (6), the residual regulation likely represents a manifestation of this transcriptional element. Nonetheless, the deletion analysis demonstrates that the restriction fragment bounded by TfR cDNA nucleotides 3178 and 3856 contains the region responsible for the majority of TfR iron regulation. When only this region is deleted (construct III), iron regulation is lost. When fragments flanking this region are removed (as in constructs IV and V), iron regulation is retained.

Iron regulation of TfR biosynthesis is due to alteration in TfR mRNA levels (3). We

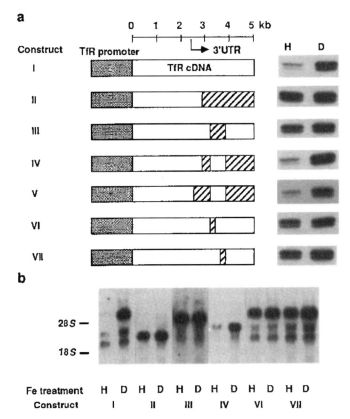

Fig. 2. The sequences responsible for iron regulation of TfR biosynthesis are contained in a 678-nucleotide fragment of the 3′UTR. (**a**) Mouse B6 cells were transfected with calcium phosphate precipitates of plasmids containing the indicated deletion constructs (19) as described previously (6). The shaded region at the 5′ end of the schematic representation of the construct denotes a fragment of genomic DNA containing the TfR promoter, and the hatched regions of each construct represent sequences of the TfR cDNA that have been deleted. After removal of the DNA precipitates, cells were placed in medium containing 6 mM sodium butyrate, and 8 hours later either hemin (H) or desferrioxamine (D) was added to final concentrations of 100 μM. After an additional 16 hours, human TfR biosynthesis was assessed by radiolabeling with [³⁵S]methionine, immunoprecipitation with monoclonal antibody B3/25, SDS–polyacrylamide gel electrophoresis, and autoradiography as previously described (6). The relevant portions of these autoradiographs are shown to the right of each of the constructs tested. (**b**) Mouse B6 cells, which lack thymidine kinase, were stably transformed as described previously (6) by cotransfection with the herpes simplex thymidine kinase gene followed by selection of thymidine kinase-containing colonies. Six of the seven constructs depicted in (a) were used to produce stably transformed populations of murine cells that expressed human TfR. Each transformed cell population represents a pooling of multiple individual colonies of transformants. Cytoplasmic RNA was isolated and analyzed after treatment with hemin (H) or desferrioxamine (D) as in Fig. 1. The full-length human TfR cDNA was used as probe (20). The relative positions of the 28S and 18S ribosomal RNAs are indicated.

prepared stable transformants of murine cells with six of the seven deletion constructs of Fig. 2a and assessed the effect of iron availability on the level of the transcripts encoded by these constructs (Fig. 2b). In several instances multiple transcripts were observed. In each case, the largest human TfR transcript in a given transformant corresponded in size to the full-length transcript expected from the transfected plasmid. The largest transcript in cells transformed with construct I corresponds to the 4.9-kb full-length TfR mRNA that is normally produced in human cells (3). The level of this 4.9-kb transcript in the construct I transformants was highly regulated by iron availability. Deletion of the most 3′ 2.2 kb of the TfR 3′UTR (construct II) resulted in loss of the majority of the iron regulation of TfR transcript level. Deletion of the 3178 to 3856 fragment (construct III) also led to loss of full regulation of transcript level, whereas deletions involving regions flanking this fragment (constructs IV and V) had no effect on regulation of mRNA level. Thus the conclusion regarding the location of the region responsible for iron regulation that was reached on the basis of the transient expression assay for protein synthesis of Fig. 2a was confirmed by the data of Fig. 2b. In addition, the changes in mRNA levels that we observed in the regulated constructs indicated that the regulation of the biosynthesis of TfR encoded by our constructs was occurring in a fashion analogous to iron regulation of a native TfR gene (that is, by alteration of mRNA level).

Computer-aided analysis of the sequence of the human TfR mRNA corresponding to bases 3178 to 3856 of the TfR cDNA revealed that this sequence has potential to form a number of stem-loop structures. Of particular note are the sequences shown in Fig. 3. These TfR 3′UTR sequence elements are capable of folding to form structures that are strikingly similar to that predicted for the IRE located in the 5′UTR of ferritin (9, 10). The stem-loop structures of TfR elements B to E (Fig. 3) have in common with the human ferritin H chain IRE: (i) a loop of CAGUGX, (ii) an "upper stem" of five paired bases, (iii) an unpaired 5′ C residue separated by five bases from the loop, and (iv) a "lower stem" of variable length. This particular consensus of sequence and predicted structure is also contained as a single copy within the 5′UTR of each ferritin whose complete mRNA sequence is known (11–14). The only deviation from this consensus within the sequences of the human TfR 3′UTR is found in TfR element A wherein the putative loop reads CAG<u>A</u>GX. The 3178 to 3856 region of the TfR mRNA represents less than 30%

of the TfR 3′UTR and contains five of these IRE-like elements. No such sequence elements are found in the remainder of the TfR 3′UTR.

The finding of IRE-like sequences within the TfR 3′UTR raised the question of whether these elements were capable of functioning as an IRE in translational regulation. To address this question, oligodeoxyribonucleotides corresponding to either

element B or element C of the TfR 3′UTR were synthesized. Each synthetic TfR element was cloned between the ferritin promoter and the structural gene for hGH such that an mRNA would be produced that contained a single TfR element in the 5′UTR of a chimeric hGH transcript (Fig. 4). A similar approach had been used in the identification and characterization of the human ferritin H chain IRE (9). The regula-

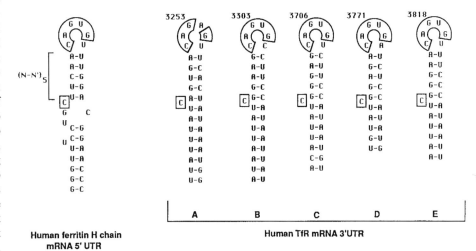

Fig. 3. Similarities between the IRE of the 5′UTR of ferritin mRNA and sequence elements present in the 3′UTR of the transferrin receptor mRNA. Sequence elements from the 5′UTR of the human ferritin H chain mRNA (9) and from the 3′UTR of the human TfR mRNA are depicted in stem-loop configurations (21). With the exception of a one-base deviation in the loop of TfR element A, all structures shown have loops consisting of CAGUGX and an unpaired C in the stem (outlined bases). The relative position of the unpaired C is invariant, in all cases being five paired bases [(N-N′)$_5$] 5′ of the loop. The numbers at the top of each of the TfR elements indicate the position in the human TfR mRNA sequence occupied by the most 5′ G in the loop according to our numbering system (19).

Fig. 4. Elements from the 3′UTR of the TfR mRNA are able to function as an IRE conferring ferritin-like translational regulation when moved to the 5′UTR of a chimeric transcript containing the hGH mRNA. Synthetic deoxyribonucleotides corresponding to the ferritin IRE (**a**), or TfR element B (**b**), or TfR element C (**c**) were cloned between the ferritin promoter and the structural gene encoding hGH (22). (**d**) An analogous plasmid lacking an IRE-like sequence element was similarly tested. These constructs were transfected into mouse B6 cells and, after treatment with either 100 μM hemin (H, ■) or with 100 μM desferrioxamine (D, □), the transient expression of hGH in the medium was assessed as previously described (9). Data shown are representative of

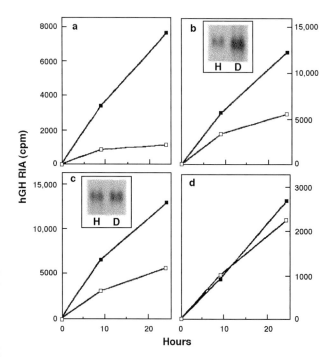

at least three independent transfections with these constructs. Cytoplasmic RNA was isolated from the transfectants involving the TfR elements (23), separated by electrophoresis, and analyzed as described in Fig. 1. The panels (b) and (c) show the blot hybridization analysis of these RNA samples with an hGH probe. RIA, radioimmunoassay.

tion of hGH production seen under control of a synthetic ferritin IRE is shown in Fig. 4a. When in the context of a 5'UTR, each of the TfR elements tested mediated ferritin-like iron regulation of hGH production (Fig. 4, b and c). This regulation is opposite in direction from that normally seen in iron regulation of TfR biosynthesis. The presence of the ferritin promoter alone does not result in iron-dependent control of hGH biosynthesis (Fig. 4d). We have also tested two constructs that contained oligodeoxyribonucleotides with potential to form stem-loop structures not conforming to the above IRE consensus. No iron regulation was observed with these constructs. Thus, TfR elements B and C have the ability to function as translational IREs, and this ability appears to require an IRE consensus of sequence and structure.

A characteristic of the ferritin translational regulatory system is iron-dependent modulation of protein biosynthesis without corresponding alteration in the level of the mRNA (4, 5). To confirm that each of the TfR elements was acting as a translational IRE in the regulation of hGH production seen in Fig. 4, we assessed the level of hGH mRNA in these transfectants. In neither case (Fig. 4, b and c, insets) did we find increased levels of the hGH transcript in the hemintreated cells that would account for their higher rate of hGH protein production. In Fig. 4c the levels of mRNA appeared to be identical in the two samples. In Fig. 4b there appeared to be somewhat more mRNA in the desferrioxamine-treated cells. If the level of mRNA is higher in these desferrioxamine-treated cells, the translational effect of iron availability would have to be more pronounced to yield the higher hGH production rate seen after hemin treatment. Thus, each of the TfR elements tested in the context of a 5'UTR was capable of mediating alterations in protein production without corresponding changes in mRNA levels.

The magnitude of the translational iron regulation mediated by each of the TfR elements (Fig. 4, b and c) appeared to be somewhat lower than that seen in the same experiment when a plasmid containing the ferritin IRE was used (Fig. 4a), but their ability to function as IREs was clear in comparison to the unregulated plasmid lacking an IRE-like sequence element (Fig. 4d). The ability of TfR element B or C to substitute for the ferritin IRE conveys information regarding the requirements for IRE function in translational regulation. Between the two synthetic oligomers corresponding to the TfR elements tested, 22 of 32 bases that are contained in the synthetic ferritin IRE oligomer have been altered and yet IRE function has been retained. It has

been shown that a single base deletion in the consensus loop or a deletion that would disrupt base pairing in the upper stem each abolished iron regulation (9). In the proposed TfR structures depicted in Fig. 3, the nucleotides of the stems are significantly different from the ferritin IRE structure and different to a lesser extent from each other. However, in all instances where a difference in one half of the stem would prevent the base pairing needed to preserve these structures, a complementary difference exists in the other half of the stem. If the multiple elements of the TfR 3'UTR arose by duplication of a single ancestral element, it would appear that their potential to form an IRElike stem-loop has been a more dominant feature to be conserved than the exact nature of the bases that make up the stems of the putative structures.

Although similarities exist between the ferritin and TfR elements depicted in Fig. 3, the elements function in distinct ways in their respective native contexts. One aspect of these differences relates to the number of IRE-like sequences present in the two transcripts. The IRE occurs as a single copy within the ferritin 5'UTR, whereas there are five similar elements within the human TfR 3'UTR. We have shown that a single TfR element is sufficient when moved to a 5'UTR to confer the translational regulation that is normally mediated by the ferritin IRE. However, two of our constructs having restriction fragment deletions within the critical 3178 to 3856 region indicate that a single element is not sufficient to accomplish regulation of TfR mRNA levels (Fig. 2). In construct VI, sequence elements A and B have been deleted and elements C to E are present. Conversely, construct VII contains elements A and B but not C to E. Thus each of these constructs contains an element that has demonstrated IRE function when in a 5'UTR (Fig. 4). However, after treatment with hemin or desferrioxamine, both construct VI and VII give rise to less than twofold differences between human TfR biosynthetic rates in the transient protein synthesis assay (Fig. 2a) and in the RNA levels of stable transformants (Fig. 2b). This low degree of iron regulation was not significantly greater than that seen after deletion of the entire TfR 3'UTR and is likely a reflection of the transcriptional effect described above. The insufficiency of a single element within the 3'UTR of TfR suggests that iron regulation of the TfR mRNA may involve the cooperative influence of more than one element within the TfR 3'UTR.

A portion of the chicken TfR cDNA corresponding to the 3'UTR has recently been cloned and sequenced (15). Our examination of the unpublished chicken se-

quence (16) revealed that the chicken TfR mRNA also contains five IRE-like sequences that correspond to elements A to E of the human TfR mRNA. These sequences are strikingly similar in the human and chicken genes; indeed, 49 consecutive bases encompassing element B are identical in the two genes. In portions of the chicken 3'UTR outside of the region corresponding to that implicated by our deletion analysis, sequence similarity to the human gene is markedly reduced. Four of the five potential stem-loops of the chicken TfR sequence also have loops of CAGUGX; the fifth (corresponding to human TfR element A) has a loop of CAG$\underline{C}$GX. All of the potential chicken structures have five paired bases in an "upper stem" followed by a 5' unpaired C and a lower stem of variable length. We believe that the similarity between these human and chicken TfR sequences supports our contention that these are critical elements involved in iron regulation of TfR expression.

Thus, the IRE is an RNA element capable of mediating two distinct regulatory events depending on its context within the transcript. It is conceivable that a common (or similar) regulatory molecule binds to these RNA structures such that, when in the context of a 5'UTR, translation of the transcript is attenuated and, when in the context of a 3'UTR, the transcript is protected from degradation. The putative regulatory molecule would be associated with the IRE elements when iron is scarce and dissociated when iron is abundant. Such a model would account for the opposite responses of the biosyntheses of ferritin and the TfR to a common primary stimulus (iron). Given the relatively short half-life of the TfR mRNA (3), this model would allow rapid changes in the biosynthetic rates for both ferritin and the TfR in response to changes in iron availability. The difference in the mechanisms of these regulatory phenomena that is apparent from assessing the effect of iron on the level of the respective transcripts would also be consistent with this model.

REFERENCES AND NOTES

1. J. H. Ward, I. P. Kushner, J. Kaplan, J. Biol. Chem. 257, 10317 (1982); P. G. Pelicci et al., FEBS Lett. 145, 350 (1982); E. Mattia, K. Rao, D. S. Shapiro, H. H. Sussman, R. D. Klausner, J. Biol. Chem. 259, 2689 (1984); K. R. Bridges and A. Cudkowicz, ibid., p. 12970.
2. H. N. Munro and M. C. Linder, Physiol. Rev. 58, 317 (1978).
3. K. K. Rao, D. Shapiro, E. Mattia, K. Bridges, R. D. Klausner, Mol. Cell. Biol. 5, 595 (1985); K. Rao et al., ibid. 6, 236 (1986).
4. J. Zähringer, B. S. Baliga, H. N. Munro, Proc. Natl. Acad. Sci. U.S.A. 73, 857 (1976); G. E. Shull and E. C. Theil, J. Biol. Chem. 257, 14187 (1982); ibid. 258, 7921 (1984); T. A. Rouault et al., Proc. Natl. Acad. Sci. U.S.A. 84, 6335 (1987).

5. N. Aziz and H. Munro, *Nucleic Acids Res.* **14**, 915 (1986), J. Rogers and H. Munro, *Proc. Natl. Acad. Sci. U.S.A.* **84**, 2277 (1987).

6. J. L. Casey *et al., Proc. Natl. Acad. Sci. U.S.A.* **85**, 1787 (1988).

7. D. Owen and L. Kühn, *EMBO J.* **6**, 1287 (1987).

8. M. W. Hentze *et al., Proc. Natl. Acad. Sci. U.S.A.* **84**, 6730 (1987).

9. M. W. Hentze *et al., Science* **238**, 1570 (1987).

10. N. Aziz and H. N. Munro, *Proc. Natl. Acad. Sci. U.S.A.* **84**, 8478 (1987).

11. F. Costanzo *et al., Nucleic Acids Res.* **14**, 721 (1986); C. Santoro *et al., ibid.,* p. 2863; M. W. Hentze *et al., Proc. Natl. Acad. Sci. U.S.A.* **83**, 7226 (1986).

12. E. Leibold and H. Munro, *J. Biol. Chem.* **262**, 7335 (1987); M. T. Murray, K. White, H. N. Munro, *Proc. Natl. Acad. Sci. U.S.A.* **84**, 7438 (1987).

13. P. W. Stevens, J. D. Dodgson, J. D. Engel, *Mol. Cell. Biol.* **7**, 1751 (1987).

14. J. Didsbury *et al., J. Biol. Chem.* **261**, 949 (1986).

15. L.-N. Chan, N. Grammatikakas, J. M. Banks, E. M. Gerhardt, *J. Cell Biol.* **105**, pt. 2, 154a (1987).

16. L.-N. Chan, personal communication.

17. The plasmid pSVGH was constructed by ligation of the 2.1-kb Hind III-Eco RI hGH fragment of p0GH [R. F. Selden, K. B. Howie, M. E. Rowe, H. M. Goodman, D. D. Moore, *Mol. Cell. Biol.* **6**, 3173 (1986)] and the 2.6-kb Hind III-Eco RI fragment of pSV2CAT [C. M. Gorman, L. F. Moffat, B. H. Howard, *ibid.* **2**, 1044 (1982)]. The plasmid pSVGH-TR was created by insertion via blunt-end ligation of the 2.2-kb Bgl II-Bam HI fragment of pcDTR1 (*18*) into an Ava I site in the 3′UTR of the hGH gene. This Bgl II-Bam HI fragment represents most of the 2.5-kb 3′UTR of the TfR cDNA and contains no TfR protein-coding sequences.

18. L. C. Kühn, A. McClelland, F. H. Ruddle, *Cell* **37**, 95 (1984); A. McClelland, L. C. Kühn, F. H. Ruddle, *ibid.* **39**, 267 (1984).

19. Construct I (formerly referred to as TRmg2) and construct II (formerly referred to as TRmg3) have been described (*6*). The other constructs were prepared by restriction fragment deletion and standard cloning procedures [T. Maniatis, E. F. Fritsch, J. Sambrook, *Molecular Cloning: A Laboratory Manual* (Cold Spring Harbor Laboratory, Cold Spring Harbor, NY, 1982)]. We have numbered the bases of the TfR cDNA using the sequence data of Schneider *et al.* [C. Schneider, M. Kurkinen, M. Greaves, *EMBO J.* **2**, 2259 (1983); C. Schneider, M. J. Oven D.Banville J.G.Williams, *Nature* **311**, 675 (1984) with base 1 being the transcription start site that they determined by primer extension. However, we have subtracted the 192 bases that are present near the 5′ end of their sequence but which are not in pcDTR1 (*18*) from which our constructs were derived. Using this numbering system, our deletion constructs correspond to deletions of bases: none (construct I); 2802 to the end of the cDNA (construct II); 3178 to 3856 (construct III); 2802 to 3177 plus 3861 to the end of cDNA (construct IV); 2405 to 3177 plus 3861 to the end of cDNA (construct V); 3178 to 3405 (construct VI); and 3636 to 3856 (construct VII).

20. Absolute levels of human TfR transcripts in the pools of stable transformants differed somewhat (less than fivefold) with no apparent pattern relating to the nature of the deletion. It is assumed that this represents a combination of transfection efficiency, gene dosage, and relative growth rate among members of our pools of stable transformants. For purposes of presentation, an autoradiographic exposure was selected for each transformant such that the intensities of the largest transcript in desferrioxamine-treated lanes were approximately equal.

21. We analyzed sequences of the 3′UTR of the human TfR, using an algorithm that accommodates the potential of G-U pairing in RNA [M. Zuker and P. Stiegler, *Nucleic Acids Res.* **9**, 133 (1981)].

22. Plasmid LS(+26mer)-GH (used in Fig. 4a) and its parent plasmid L5-GH (used in Fig. 4d) have been described (*9*). L5-GH contains the most 5′ six nucleotides of the human ferritin H chain 5′UTR. Two pairs of complementary deoxyribonucleotides corresponding to each of TfR elements B and C were synthesized with an Applied Biosystems DNA synthesizer. Plasmids containing these elements were prepared by cloning the double-stranded DNA between the Bam HI and Xba I sites of a plasmid derived from pUC18 with the hGH gene in its Hinc II site and the ferritin pFP$_1$ promoter (*8*) between its Eco RI and Bam HI sites. The correct nucleotide sequences of the insertions in the resultant plasmids were confirmed by DNA sequencing [F. Sanger, S. Nicklen, A. R. Coulson, *Proc. Natl. Acad. Sci. U.S.A.* **74**, 5463 (1977)] to be: 5′-GGATCCAT-TATCGGAAG<u>CAGTGC</u>CTTCCATAATTC TAGA-3′ (for the plasmid used in Fig. 4b) and 5′-G GATCCATTATCGGGAG<u>CAGGT</u>CTTCCA TAATTCTAGA-3′ (for the plasmid used in Fig. 4c). The nucleotides corresponding to the six-membered loops in the stem-loop structures proposed for mRNA arising from these constructs are underlined for purposes of orientation.

23. To increase transient expression of hGH mRNA, transfected cells were treated with 6 m*M* sodium butyrate [T. A. Gottlieb *et al., Proc. Natl. Acad. Sci. U.S.A.* **83**, 2100 (1986)] in addition to either 100 μ*M* hemin or 100 μ*M* desferrioxamine beginning 24 hours after removal of precipitated plasmid DNA. After an additional 24 hours, cytoplasmic RNA was isolated (*6*). The RNA samples were treated with ribonuclease-free deoxyribonuclease for 15 minutes at room temperarure to reduce the content of plasmid DNA before separation by electrophoresis and analysis by blot hybridization. Radioimmunoassay of hGH production in the cells used for RNA preparation indicated that butyrate-treated cells displayed iron regulation comparable to that seen in Fig. 4, a and b.

24. We thank L.-N. Chan for making available to us the sequence of the chicken TfR 3′UTR prior to publication. We also acknowledge the contributions of B. Di Jeso, K. Rao, J. G. Barriocanal, A. Dancis, and V. Ramin to the studies on which this work is based. S.W.C. was supported by the Dermatology Branch, National Cancer Institute, NIH.

26 January 1988; accepted 25 March 1988

Yen et al. (1988) Nature (London) **334**, 580–585

Autoregulated instability of β-tubulin mRNAs by recognition of the nascent amino terminus of β-tubulin

Tim J. Yen, Paula S. Machlin & Don W. Cleveland

Department of Biological Chemistry, Johns Hopkins University School of Medicine, Baltimore, Maryland 21205, USA

Tubulin synthesis in animal cells is controlled by an autoregulatory mechanism that modulates the stability of polysome-bound tubulin messenger RNAs. The β-tubulin RNAs are selectively targeted as substrates for destabilization not through the recognition of specific RNA sequences, but rather through co-translational recognition of the amino-terminal β-tubulin tetrapeptide after its emergence from the ribosome. This motif is likely to be used in other systems where RNA degradation is coupled to ribosome attachment and translation.

EXPRESSION of α- and β-tubulins, the two principal constituents of microtubules, is regulated in higher eukaryotes by molecular processes that operate at two levels. The first of these is the transcriptional activation during cell differentiation of one or more members of the small multigene families comprised of about 6 or 7 functional genes that encode either subunit (see ref. 1 for review). In most animal cells, a second regulatory event establishes the appropriate quantitative level of tubulin expression through an autoregulatory pathway in which the apparent intracellular concentration of tubulin heterodimers (comprised of one α- and one β-tubulin polypeptide) modulates the stability of tubulin messenger RNAs[2–7].

The determination of the molecular mechanism through which autoregulated destabilization of tubulin mRNAs is achieved is of general interest for two reasons. First, although autoregulation is well known in prokaryotes, only a few examples have been found in eukaryotes. Tubulin provides the best-known example of how autoregulation can be achieved for normal cellular genes in higher organisms. Second, it is becoming increasingly apparent that for many genes (for example, histones[8,9] and a series of lymphokine, cytokine and proto-oncogene mRNAs[10–15]), the level of gene expression is strongly dependent not only on the rate of gene transcription but also on the stability of the mature mRNAs. Despite this, essentially nothing is known of how mRNA half-life is established for almost all eukaryotic mRNAs.

As part of our goal to elucidate the mechanism of post-transcriptional regulation of tubulin gene expression, we have recently determined that the first 13 translated nucleotides of a β-tubulin mRNA (which encode the first four amino acids, Met-Arg-Glu-Ile, of β-tubulin) are sufficient to confer autoregulated instability onto heterologous RNAs if present as the first translated nucleotides In addition, as deletions within this domain (either codon 2 or codons 3 and 4) of an authentic β-tubulin gene yield RNAs insensitive to autoregulation, we have concluded this 13-nucleotide segment is both necessary and sufficient to specify RNAs as substrates for autoregulated degradation[7]. Furthermore, using various inhibitors of protein synthesis[16] and by transfection of β-tubulin genes with premature translation termination codons[7,16], we have shown that for β-tubulin mRNAs to be substrates for autoregulation, they must be associated with polysomes and that translation of them must proceed past 41 codons.

Taken together, these data suggest two possible models for the mechanism that establishes autoregulated instability of polyribosome-bound β-tubulin RNAs. The first model proposes that RNA degradation is facilitated by unpolymerized tubulin subunits binding (either directly or indirectly) to the 13-nucleotide recognition sequence present in β-tubulin mRNAs. That only ribosome-bound RNAs are substrates for degradation can be explained if translation is required to make the recognition domain accessible for binding. The second model arises from

the realization that the sequence conferring autoregulated instability actually encodes the four amino-terminal β-tubulin amino acids, which suggests that the true recognition event is a protein–protein interaction in which free subunits interact (directly or indirectly) with the nascent tubulin polypeptide immediately after it emerges from the large ribosomal subunit. This putative binding event must somehow be transduced through the adjacent ribosome to signal the degradation of the RNA being translated.

To distinguish between these two models (recognition of the nucleotide sequence or recognition of the encoded polypeptide), we have systematically introduced 25 different nucleotide-base substitutions into the 13-base regulatory element of a human β-tubulin gene[17]. Some of these mutations alter the amino acids encoded by codons 2 or 3, whereas others leave the polypeptide unchanged (because of the redundancy of the genetic code). We have also changed the translational reading frame of the 13-nucleotide 'identifier' sequence such that its position within the mRNA remains virtually unchanged but it is now translated in an inappropriate reading frame (thereby producing a polypeptide with a different amino-acid sequence). We determined whether RNAs derived from each of these mutant genes remain substrates for autoregulated instability after their re-introduction into animal cells by DNA transfection. Our results show unambiguously that the initial step in the pathway for autoregulated degradation of polysome-bound β-tubulin mRNAs is the recognition by the autoregulatory factor(s) of the nascent, amino terminus (Met-Arg-Glu-Ile) of β-tubulin.

Any Arg codon at aa2 maintains autoregulation

To determine whether autoregulated tubulin mRNA instability is specified by recognition of a nucleotide sequence or of the encoded polypeptide, we used site-directed mutagenesis[18] to introduce one, two or three base substitutions within the 13-nucleotide identifier sequence of the human M40 β-tubulin gene[17]. Initially, we introduced all six possible single base substitutions at the first and third positions of the second codon (AGG). We also introduced nine double and one triple base substitution. Each mutant β-tubulin gene contains all the flanking and intervening sequences necessary for expression following transient transfection into mouse L cells. We then assayed cytoplasmic RNAs isolated from parallel dishes of transfected cells that contain either a normal level of unassembled tubulin subunits or an elevated concentration of free subunits (as a consequence of colchicine-induced microtubule depolymerization) to determine the concentration of mutant β-tubulin transcripts and of the endogenous mouse mβ5-tubulin RNA (Fig. 1).

As previously reported[19], RNAs transcribed from the wild-type M40 gene (which carries codon AGG (encoding arginine) at the second amino-acid position) are destabilized (about five- to sixfold) as a consequence of colchicine-induced elevation of the tubulin subunit levels. Except for mutant codons CGG and AGA (both of which are alternative arginine codons), the

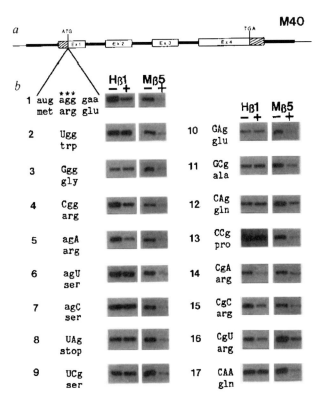

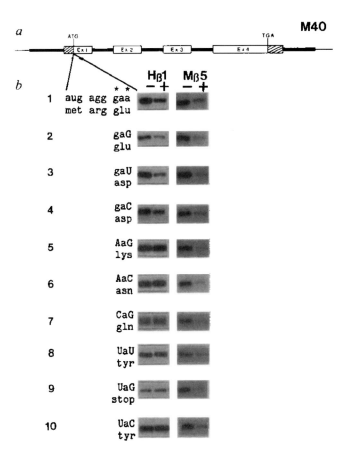

Fig. 1 Autoregulated instability of β-tubulin RNAs with mutations at codon 2. *a*, Schematic diagram of the human β-tubulin gene M40[17], for which we used site-directed mutagenesis to substitute nucleotides at codon 2. Fine line, pUC sequences; bold line, M40 5′ and 3′ or intron sequences; open boxes, exon (Ex) sequences corresponding to coding regions; hatched boxes, exon sequences corresponding to 5′ and 3′ untranslated regions. *b*, Quantitative analysis of levels of cytoplasmic RNAs extracted from mouse Ltk⁻ cells transiently transfected with various mutant β-tubulin genes. Wild-type sequence of the first three codons is shown, followed by the various nucleotide substitutions (capitals) introduced at codon 2. Encoded amino acids are indicated below the nucleic acid sequences. Equal amounts of cytoplasmic RNA from cells with normal (−) or elevated (+) unpolymerized tubulin subunits were analysed for mβ5 (an endogenous mouse β-tubulin mRNA) and for RNAs from each transfected tubulin gene.

Methods. Mixed oligonucleotides (19–21mers) were synthesized using an Applied Biosystems DNA Synthesizer and purified by HPLC. Site-directed mutagenesis was performed as described with the following modifications: T4 DNA polymerase was substituted for Klenow and the *in vitro* synthesis reaction was performed at 37°C for 90 min followed by digestion with *Eco*RI and *Hin*dIII to release a 900-base-pair tubulin insert containing the DNA modification. This heteroduplex fragment was isolated from low-melting agarose, ligated into M13mp19 and transformed into *Escherichia coli*. Positive plaques were identified by hybridization to ³²P-labelled oligonucleotides and confirmed by sequencing the DNA[22]. The double-stranded form of the inserts containing the desired mutations were isolated from the recombinant M13 replicative form DNAs and recloned into the human tubulin gene M40 by insertion between the *Hin*dIII and *Nhe*I sites. Transient transfections using DEAE-dextran were performed as described[7]. Unpolymerized tubulin subunit concentrations were elevated in some samples (+) by treatment of transfected cells with 10 μM colchicine for the final 5 h of culture. RNA levels determined by S1 analysis[7].

Fig. 2 Autoregulated instability of tubulin RNAs with mutations in the third codon. *a*, Schematic diagram of the human β-tubulin gene M40 and of the mutagenesis at the first and third positions of codon 3. *b*, Quantitative analysis of cytoplasmic RNAs isolated from aliquots of cells transfected with each of nine β-tubulin genes that carry mutations in the third codon (see Fig. 1 for methods).

mutants in which autoregulation is still maintained, the sensitivity of each RNA to autoregulated degradation is quantitatively retained (determined by densitometry of the appropriate autoradiographs). In all cases, the autoregulated instability of endogenous mβ5-tubulin RNA demonstrates retention of the autoregulatory response in all transfected cells, whereas levels of RNAs encoding ribosomal protein L16 are completely insensitive to colchicine-induced microtubule depolymerization, thus ensuring that comparable levels of RNA are assayed in each sample (data not shown).

Inspection of the nucleotide substitutions of the five mutants that retain autoregulated RNA instability and of the 11 whose RNAs are not regulated correctly does not identify an obvious consensus nucleotide sequence that specifies autoregulation. For example, two double mutants of the initial AGG codon (CGC and CGU) restore autoregulation to single base substitutions (AGC and AGU) that disrupt it (compare rows 15, 16 with rows 6, 7, respectively, of Fig. 1). The most notable feature among the five mutant RNAs which remain substrates for autoregulation is that, like the wild-type gene, all of them encode arginine at codon 2. In fact, these five mutants and the wild type comprise the entire repertoire of arginine codons.

Mutagenesis at codon 3

The mutagenesis at codon 2 clearly supports a model for tubulin RNA instability in which the encoded tubulin polypeptide is the element that targets an RNA as a substrate. To test this hypothesis further, we mutagenized the next three nucleotides of the 13-nucleotide sequence that is necessary and sufficient to confer autoregulation. We constructed nine mutants of the GAA triplet that normally encode glutamic acid at the third residue of β-tubulin and we tested RNAs transcribed from each for

remaining four single-base substitution mutants at codon 2, UGG (Trp), GGG (Gly), AGU (Ser), and AGC (Ser), yield RNAs that are insensitive to autoregulated instability (see rows 4, 5 and 2, 3, 5, 7, respectively, of Fig. 1). Similarly, analysis of 10 additional mutants (nine 2-base and one 3-base substitution) reveals that RNAs from only three of these, CGA (Arg), CGC (Arg), and CGU (Arg), remain substrates for regulated RNA instability (see rows 14, 15, 16). Moreover, for each of the

autoregulated instability. The results (Fig. 2) show that three of the nine mutant RNAs (with codons GAG (Glu), GAU (Asp) and GAC (Asp)) are still substrates for regulated instability, whereas the remaining six mutants, AAG (Lys), ACC (Asn), CAG (Gln), UAU (Tyr), UAG (stop) and UAC (Tyr) are insensitive (see rows 2–4 and 5–10, respectively, of Fig. 2). Comparison of the nucleotide sequences of the nine mutants shows that whereas all four nucleotides are permitted in the third position, additional base substitutions introduced at the first position of the codon invariably abolish recognition of the RNA. This observation can be interpreted in two ways. First, if the recognition sequence is at the nucleic-acid level, the ninth nucleotide (the third position in the codon) within the 13-nucleotide domain does not comprise an essential part of the regulatory element. Alternatively, consistent with our previous findings, the mutant RNAs which remain substrates for autoregulation either retain an encoded glutamic acid or the conservative amino-acid substitution, aspartic acid.

Autoregulated mRNAs must encode MREI

Although the results obtained from analysing tubulin RNAs with mutations at codons 2 and 3 strongly support a model in which the nascent tubulin polypeptide is the element recognized by the autoregulatory factor(s), mutations that alter the encoded amino acids by necessity also alter the nucleic-acid sequence.

Thus, it remains possible that the loss of recognition of some mutant mRNAs is the consequence of disruption of a nucleotide-recognition sequence rather than the encoded polypeptide. To resolve this ambiguity, we shifted the translational reading frame of the 13-nucleotide identifier sequence but kept it in essentially the same position within the mRNA. To do this, we exploited an existing hybrid gene (MT-M17-tk_{out}; see ref.7) comprised of a metallothionein promoter and 5′ untranslated region; the first 17 codons from a mouse β-tubulin gene; a six-nucleotide linker; and 278 codons plus 3′ flanking sequences from the herpes virus thymidine kinase (*tk*) gene. The translation of this mRNA begins at the authentic tubulin AUG translation initiation codon but the tubulin–tk junction is joined such that translation reads into the wrong reading frame of *tk*, thereby leading to premature translation termination after the 41st codon (17 amino acids from tubulin, 2 from the linker and 22 from the alternate reading frame of *tk*). We have previously used this construct to demonstrate that RNAs yielding such premature translation termination products are not substrates for autoregulation, whereas analogous RNAs that are joined in-frame (and yield long translation products) are efficient substrates[7].

Because some feature of the autoregulatory process requires translation beyond codon 41, autoregulation should be restored to RNAs from MT-M17-tk_{out} by one-base insertions that restore translation to the correct reading frame of *tk*. We strategically

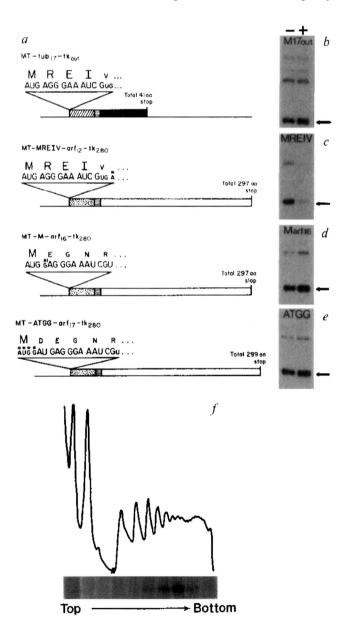

Fig. 3 Alteration of the translational reading frame of the β-tubulin RNA sequences abolishes autoregulation. *a*, Schematic diagram of MT–tubulin–*tk* genes into which nucleotide insertions (asterisks) were introduced to restore the proper reading frame to MT–M$_{17}$–tk_{out}. Fine line, untranslated sequences; hatched boxes, tubulin amino acids; open boxes, *tk* amino acids; lined box, two-amino-acid linker; stippled box, amino acids derived from an alternate reading frame of tubulin sequences; closed box, amino acids from an alternative reading frame of *tk*. Nucleotides and amino acids which comprise the regulatory element are represented by large letters. Smaller letters denote sequences (nucleotide or amino acid) not essential for autoregulation. MT, mouse metallothionein promoter and 5′ untranslated sequences; M17, mouse β-tubulin encoding the amino-terminal 17 β-tubulin amino acids; arf, alternate reading frame of β-tubulin as a result of nucleotide insertions; *tk* thymidine kinase. The numbers in subscripts denote numbers of amino acids or number of codons. Single letter designations for amino acids are presented. *b–e*, Mouse Ltk⁻ cells were transfected with MT–M$_{17out}$–*tk* (M$_{17out}$—*b*), MT–MREIV–arf$_{12}$–tk_{280} (MREIV—*c*), MT–M–arf$_{16}$–tk_{280} (Marf—*d*), or MT–ATGG–arf$_{18}$–tk_{280} (ATGG—*e*). Equal amounts of cytoplasmic RNA from cells with normal (−) or elevated (+) unpolymerized tubulin subunits were analysed for endogenous mβ5 RNAs and for RNAs from each transfected tubulin gene. (The faint larger species seen in each lane represents RNAs initiated at minor upstream transcriptional initiation sites of the metallothionein (MT) promoter.) Arrows mark the positions for fragments protected by RNAs initiated from the major transcriptional start site in MT. *f*, The distribution of RNAs derived from MT–ATGG–arf$_{18}$–tk_{280} within a polysome profile. Polysomes from cells transfected with MT–ATGG–arf$_{18}$–tk_{280} were analysed on a 10–40% sucrose gradient[16]. The gradient was divided into 10 fractions and RNA isolated from each and analysed by S1 nuclease analysis. Polysome isolation was performed as described[16] with the following modifications: 100-mm diameter dishes of mouse L cells transfected with MT–ATGG–arf$_{18}$–tk_{280} were lysed using NP40-lysis buffer (10 mM Tris-HCl, *p*H 8.6, 100 mM NaCl, 10 mM MgCl$_2$, 0.5% NP40, 10 mM vanadyl-ribonucleoside complexes and 100 μg ml⁻¹ of cycloheximide). Gradients (10–40% sucrose) in lysis buffer without NP40 and cycloheximide were centrifuged for 2.5 h at 27,000 r.p.m. in an SW41 rotor[16].

introduced single nucleotide insertions into MT-M17-tk_{out}, immediately after the initiating AUG of tubulin or immediately after the fifth tubulin codon. In both cases, the reading frame of tk is restored such that a 297-amino-acid fusion polypeptide is produced. Although the nucleotide sequences of these two hybrid genes are virtually identical, their translation products are different (see Fig. 3a). An insertion immediately following the AUG (to produce MT–M–arf_{16}–tk_{280}) will alter the translation reading frame of the remaining 48 tubulin-coding nucleotides to produce a 16-amino-acid polypeptide (encoded by the wrong frame of β-tubulin) which its fused with the remaining 280 amino acids of tk. As a control, insertions introduced after the fifth tubulin codon (MT-MREIV-arf_{12}–tk_{280}) produce a fusion polypeptide comprised of the amino-terminal five β-tubulin amino acids (the shortest sequence required for autoregulation contains the first four amino acids) fused to a 12-amino-acid polypeptide derived from the translation of the alternate reading frame (arf) of the remaining portion of the 36-nucleotide tubulin-coding sequence which in turn is joined in-frame to tk.

If the recognition element is indeed the amino-terminal tubulin peptide, we would predict that RNAs with an insertion altering the reading frame of the amino-terminal tubulin-coding sequence will not be substrates for autoregulation, whereas those RNAs that encode the amino-terminal five tubulin amino acids would be substrates. (Before the introduction of these nucleotide insertions, inspection of the tubulin nucleotide sequence within MT–M17$_{out}$–tk revealed that a translation termination codon would be encountered in the inappropriate tubulin reading frame. To overcome this, we substituted by site-directed mutagenesis the T at position 51 of the tubulin sequence to a C before additional nucleotide insertions were introduced.)

We transfected the constructs MT-M-arf_{16}-tk_{280} and MT–MREIV-arf_{12}-tk_{280} into mouse cells and tested their encoded RNAs for autoregulated degradation after colchicine-induced elevation of tubulin subunit concentration. Although transcripts derived from MT–M17$_{out}$-tk are unaffected by increases in tubulin subunit concentrations (Fig. 3b), a single base insertion (A) after the fifth tubulin codon in MT–M17$_{out}$-tk restores autoregulated instability of RNAs transcribed from this gene (MT-MREIV-arf_{12}-tk_{280}; Fig. 3c). (Similarly, introduction of a C after the fifth tubulin codon also restored autoregulation; data not shown.) In contrast, transcripts with an insertion positioned immediately after the initiating AUG are not autoregulated (Fig. 3d). We eliminated the possibility that the failure of these latter RNAs to be recognized stems from their inability to be translated into a long polypeptide by demonstrating that they are present in the fast-sedimenting region of a polysome profile (data not shown).

Although these results further support the notion that the recognition element for tubulin autoregulation is the amino-terminal peptide, it still remained possible that the single base insertion introduced after the initiating AUG disrupts the putative RNA recognition sequence. To settle this question unambiguously, we constructed one additional mutant gene (MT–ATGG–arf_{17}-tk_{280}), again starting with MT–M17-tk_{out}. This time, however, we restored the tk sequences to the proper translational reading frame not by insertion of a single nucleotide but rather by a four-base insertion (ATGG) just before the authentic β-tubulin translation initiation codon. In the final gene (Fig. 3a), the 13-nucleotide sequence capable of conferring autoregulation remains intact, although its position in the mRNA has been moved four bases. Assuming that translation initiates at the first AUG (that provided by the inserted ATGG), a 299-amino-acid polypeptide will be produced that consists of a methionine followed by 18 amino acids encoded by an alternative translation frame of the tubulin sequence, a 2-amino-acid linker, and the carboxy-terminal 278 amino acids of TK. The levels and polysomal distribution of RNA accumulated in cells transfected with MT–ATGG–arf_{17}–tk are shown in Fig. 3e, f. Clearly, these RNAs are not substrates for facilitated degrada-

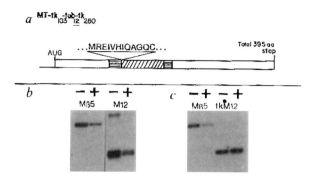

Fig. 4 Translocation of the β-tubulin tetrapeptide recognition sequence within tk prevents its recognition by the autoregulatory factor(s). a, Schematic diagram of MT–rk_{103}–tub_{12}–tk_{280}, a construct that internalizes the amino-terminal 12 β-tubulin amino acids within the tk polypeptide. The actual encoded polypeptide comprises 97 aa of tk followed by the 12 amino-terminal β-tubulin residues, a 6 aa linker and 280 additional aa of tk. See legend to Fig. 3a for explanation of symbols. b, c, S1 nuclease analysis of cytplasmic RNA isolated from cells transfected with (b) MT–tub_{12}–tk_{280} or (c) MT–tk_{103}–tub_{12}–tk_{280}. Mβ5: RNAs from an endogenous mouse β-tubulin gene; M12: RNAs from MT–tub_{12}–tk_{280}; tkM12: RNAs from MT–tk_{103}–tub_{12}–tk_{280}. Equal amounts of cytoplasmic RNA from cells with normal (−) or elevated (+) unpolymerized tubulin subunits were probed for Mβ5 and for RNAs from each transfected gene.

tion after elevation of tubulin subunit concentrations. Further, because most MT–ATGG–arf_{17}–tk_{280} RNAs are in the fast-sedimenting polysomal fractions (Fig. 3f), translation of these mRNAs must initiate at the 5′-most AUG codon, as initiation at the second methionine codon (that specified by the intact 13-base tubulin recognition sequence) can yield only a 41-amino-acid translation unit which is too short to attach simultaneously to more than 2–3 ribosomes[7]. Thus, the disruption of autoregulation in MT–ATGG–arf_{17}–tk_{280} RNAs must be caused by the failure to translate the β-tubulin amino-terminal tetrapeptide (Met-Arg-Glu-Ile), as the 13-nucleotide RNA sequence is uninterrupted and resides in essentially the same position in the mRNA.

Internalization of MREI blocks recognition

From the cumulative results of the site-directed mutagenesis experiments, we conclude that the recognition of polysome-bound β-tubulin mRNAs by the autoregulatory factor(s) occurs via direct protein–protein interactions involving the nascent β-tubulin polypeptide. To examine whether the position of this polypeptide sequence relative to the remaining portion of the protein is important for proper recognition by the autoregulatory mechanism, we constructed a gene (derived from MT–M12–tk_{in}[7]), in which we placed the amino-terminal 12 tubulin amino acids (in the correct translational reading frame) within a tk polypeptide. This construct, MT–tk_{103}–tub_{12}–tk_{280} (see Fig. 4), encodes a polypeptide comprised of 395 amino acids (97 from tk, 6 encoded by a linker, 12 from the amino terminus of β-tubulin and 280 corresponding to the carboxy terminus of tk). We transfected this construct and examined RNAs derived from it for autoregulated instability. As is shown in Fig. 4b, RNAs which are translated into a fusion polypeptide that carries 12 β-tubulin amino acids at its amino terminus are sensitive to autoregulation (M12), whereas those RNAs that encode the identical 12 tubulin amino acids but located within the tk polypeptide are insensitive to the tubulin subunit concentration (Fig. 4c). Thus, an autoregulated RNA must encode the amino-terminal β-tubulin amino acids; it must be translated to yield a long polypeptide; and the position of the tubulin residues within the polypeptide is important.

A model for autoregulated mRNA instability

The transfection experiments with gene constructs carrying

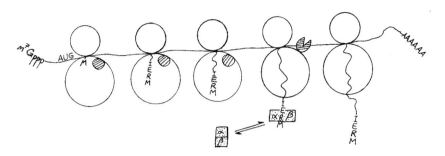

Fig. 5 Proposed model for autoregulated instability of β-tubulin mRNA. Unpolymerized tubulin subunits bind directly (or activate a factor(s) which binds) to the nascent amino-terminal tetrapeptide (Met-Arg-Glu-Ile) of β-tubulin. This binding is transduced through the adjacent ribosome to activate an RNase which degrades the ribosome-bound mRNA. The RNase has been drawn (shaded areas) to be ribosome-associated, although this has not yet been demonstrated. MREI: the amino-terminal β-tubulin polypeptide.

altered nucleotide sequences and/or translation frames lead us to conclude that the recognition element for autoregulated RNA instability is actually the encoded amino-terminal tetrapeptide (Met-Arg-Glu-Ile). We have derived a model for autoregulated instability from these and from our previous work (Fig. 5). As the unpolymerized tubulin subunit pool is elevated, the initial event that leads ultimately to regulated degradation of β-tubulin RNAs is the interaction between the autoregulatory factor(s), possibly tubulin itself (as shown in Fig. 5) and the nascent amino-terminal tubulin tetrapeptide just after it emerges from the ribosome. How this protein–protein interaction is transduced through the ribosome to yield degradation of the corresponding RNA is not yet known, although there are two general possibilities. First, the binding event could activate a cellular RNase (which itself might be a peripheral ribosomal component). Alternatively, binding could induce a transient stalling of the ribosome that leaves the RNA in an exposed conformation that is a better substrate for non-specific RNases. We cannot yet distinguish between these two possibilities. Further, although we show the binding event occurring immediately after the amino terminus of β-tubulin emerges from the ribosome, the actual position along the polysome profile where this interaction occurs has not been determined experimentally.

That so short a sequence as a tetrapeptide can be specifically recognized may at first seem surprising. But inspection of the predicted amino-terminal four amino acids among all β-tubulin genes sequenced to date reveals that the tetrapeptide Met-Arg-Glu-Ile is absolutely conserved at the amino terminus. Furthermore, a search of the protein database for other proteins that contain this tetrapeptide sequence identifies only β-tubulins. Even considering the possibility that the true minimum sequence is only Met-Arg-Glu (we have not obtained and tested a construct with only the first three amino-terminal β-tubulin codons), a search of the protein-sequence library shows that although several known proteins carry this tripeptide sequence internally, only tubulins (α and β) have it at their amino termini. Further, as translocation of the first 12 β-tubulin amino acids internally within a *tk* polypeptide disrupts its recognition by the autoregu-

latory factor(s), the context in which Met-Arg-Glu-Ile appears is important. It seems plausible that an amino-terminal localization may be necessary for autorecognition.

The model easily explains why RNAs that yield only short translation products fail to be autoregulated even though they are efficiently associated with ribosomes[7]. Because it is known for bacterial ribosomes that 30 to 40 amino acids are contained in a tunnel in the large ribosomal subunit[20], at least this many amino acids must be translated before the nascent Met-Arg-Glu-Ile peptide would be accessible for binding by the autoregulatory factor(s). Obviously, if translation terminates before this point, it would not be possible to link tubulin RNA levels to subunit concentration. Although we have not formally determined the minimum number of codons through which translation must proceed so that the amino terminus of β-tubulin emerges from the large ribosomal subunit, translation of as few as 90 codons is sufficient for an RNA to be sensitive to autoregulated destabilization (T.J.Y. and D.W.C., unpublished).

In addition to the tubulin example reported here, an increasing number of eukaryotic proteins are regulated at least in part by RNA stability. Examples include many proto-oncogene, lymphokine and cytokine mRNAs that contain within their 3′ untranslated regions a ~50-base AU-rich sequence that destabilizes the mRNA[14]. Cell-cycle-dependent expression of histone mRNAs is also established in part through changes in mRNA stability specified by sequences carried in the 3′-untranslated region[9,21]. Although the mechanisms of degradation must differ for each of these RNAs, a striking parallel does emerge: RNA instability is linked to ongoing protein synthesis (tubulin (ref. 16 and this paper); c-*myc*[10]; c-*fos*[11–13]; and histones[9]). Recognizing this common linkage of mRNA stability to translation, we propose the existence of a widely used motif in which the degradation of RNAs is integrally linked to ribosome attachment and translation. We also make the appealing suggestion that the RNase(s) are localized to actively translating ribosomes.

This work has been supported by a NIH grant to D.W.C. T.J.Y. has been supported by an American Cancer Society Postdoctoral Fellowship.

Received 14 April; accepted 7 June 1988.

1. Cleveland, D. W. *J. Cell Biol.* **104**, 381–383 (1987).
2. Cleveland, D. W., Lopata, M. A., Sherline, P. & Kirschner, M. W. *Cell* **25**, 537–546 (1981).
3. Caron, J. M., Jones, A. L., Rall, L. B. & Kirschner, M. W. *Nature* **317**, 648–650 (1985).
4. Caron, J. M., Jones, A. L. & Kirschner, M. W. *J. Cell. Biol* **101**, 1763–1772 (1985).
5. Pittenger, M. F. & Cleveland, D. W. *J. Cell Biol.* **101**, 1941–1952 (1985).
6. Gay, D. A., Yen, T. J, Lau, J. T. Y. & Cleveland, D. W. *Cell* **50**, 671–679 (1987).
7. Yen, T. J., Gay, D. A., Pachter, J. S. & Cleveland, D. W. *Molec. Cell. Biol.* **8**, 1224–1235 (1988).
8. Bird, R. C., Jacobs, F. A. & Sells, B. H. *Biochem. Cell Biol.* **64**, 99–105 (1986).
9. Graves, R. A., Pandey, N. B. Chodchoy, N. & Marzluff, W. F. *Cell* **48**, 795–804 (1987).
10. Linial, M., Gunderson, N. & Groudine, M. *Science* **230**, 1126–1131 (1985).
11. Kruijer, W., Cooper, J. A., Hunter, T. & Verma, I. *Nature* **312**, 711–716 (1984).
12. Miller, A. D., Curran, T & Verma, I. *Cell* **36**, 51–60 (1984)
13. Muller, R., Bravo, R., Burckhardt, J. & Curran, T. *Nature* **312**, 716–720 (1984).
14. Shaw, G. & Kamen, R. *Cell* **46**, 659–667 (1986).
15. Treisman, R. *Cell* **42**, 889–902 (1985).
16. Pachter, J. S., Yen, T. J. & Cleveland, D. W. *Cell* **51**, 283–292 (1987).
17. Lee, M. G.-S., Lewis, S. A., Wilde, C. D. & Cowan, N. J. *Cell* **33**, 477–486 (1983).
18. Zoller, M. J. & Smith, M. *Meth. Enzym.* **100**, 468–500 (1983).
19. Lau, J. T. Y., Pittenger, M. F. & Cleveland, D. W. *Molec. Cell. Biol.* **5**, 1611–1620 (1985).
20. Yonath, A., Leonard, K. R. & Wittman, H. G. *Science* **236**, 813–816 (1987).
21. Capasso, O., Bleeker, G. C. & Heintz, N. *EMBO J.* **6**, 1825–1831 (1987).
22. Sanger, F., Nicklen, & S. Coulson, A. R. *Proc. natn. Acad. Sci. U.S.A.* **74**, 5463–5467 (1977).

Chromatin structure

5

In bacteria, the regulation of gene transcription is achieved by the binding of specific proteins to the regulatory regions of the gene. Such proteins then either activate or repress gene expression (for a review, see Travers, 1993). As will be illustrated in subsequent sections, similar proteins, known as transcription factors, also produce similar effects in eukaryotes.

However, in eukaryotes such factors act against a background of long-term differences between different tissues in the manner in which they respond to specific inducers of gene expression. Thus, as described in Section 2, the ovalbumin gene is activated by the steroid hormone oestrogen in the oviduct. However, similar treatment of the liver with oestrogen results in the activation of the vitellogenin gene while having no effect on the ovalbumin gene.

Hence the regulation of gene transcription involves an interaction between the relatively short-term processes mediated by specific transcription factors and longer-term regulatory processes which establish and maintain the stable differences between different tissues that are characteristic of eukaryotes.

A considerable amount of evidence suggests that such differences involve alterations in the association of particular regions of the DNA with specific nuclear proteins such as the histones. The structure formed by this association of DNA with specific nuclear proteins is known as chromatin (for reviews, see Igo-Kemenes et al., 1982; Morse and Simpson, 1988).

The first indication that genes that are active in a particular tissue are packaged differently in the chromatin was provided by the paper by Weintraub and Groudine which is presented here. These authors studied the pattern of digestion of chicken erythrocyte DNA with the enzyme deoxyribonuclease I (DNase I). This experiment was carried out before the availability of cloned probes for the globin gene or of the Southern blotting technique (used for example by Jeffreys and Flavell in the paper presented in Section 1).

To carry out their experiment, therefore, Weintraub and Groudine partially digested chromatin within nuclei prepared from various tissues using DNase I so that only about 10–20% of the DNA was digested. They then hybridized this partially digested DNA with a globin cDNA probe prepared by copying the globin mRNA isolated from adult reticulocytes. The amount of undigested globin DNA in each nuclear sample was then quantified by determining the rate of hybridization of the cDNA probe with this partially digested total nuclear DNA (Figure 5.1). This method of determining the DNA concentration by the rate of hybridization is known as a $C_o t$ curve, and is comparable with the use of the $R_o t$ curve to determine RNA levels as utilized by Hastie and Bishop in their paper described in Section 1.

The data obtained in this way are illustrated in Fig. 2 in the paper by Weintraub and Groudine. This shows that the globin genes were rapidly digested by DNase I in nuclei derived from 5-day or 18-day chicken embryo red blood cells which actively express these

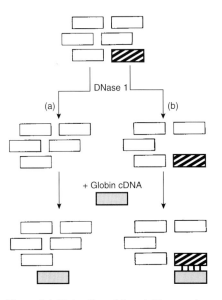

Figure 5.1 Detection of the globin gene in DNA partially digested with DNase I. If the globin DNA (hatched box) is degraded by partial digestion with DNase I (a), no hybridization will occur when the partially digested DNA is mixed with globin cDNA (stippled box). Alternatively, if the globin DNA has not been digested (b), hybridization with the cDNA will occur.

genes. In contrast, the ovalbumin gene, which is not expressed in these cells, was not digested under these conditions. Moreover, such rapid digestion of the globin genes is a tissue-specific phenomenon and does not simply represent a high level of sensitivity of these genes to digestion within the chromatin of all tissues. Thus in Figure 3 (left) of the paper, it is shown that the globin genes were not rapidly digested in nuclei derived from fibroblast or brain cells in which they are not expressed.

This experiment thus described for the first time a tissue-specific difference in the chromatin structure of a particular gene that results in the gene being more accessible to digestion by DNase I in a tissue where it is expressed than in a tissue where it is not expressed (Figure 5.2). As discussed previously in Section 2, however, the globin genes are expressed at very high levels in the red blood cell lineage, and this has led to suggestions that they may be regulated differently to the bulk of cellular genes whose expression never reaches such high levels. It was therefore of importance for Weintraub and Groudine to show that other genes, which are transcribed at lower levels than the globin genes, also exhibit this preferential sensitivity to digestion by DNase I. This obviously posed some difficulties at a time when cloned gene probes were not available. However, it was achieved in two ways.

Firstly, Weintraub and Groudine studied the expression of endogenous avian tumour virus genes, which are present within the genome of chicken cells and for which a suitable probe could be prepared from purified virus. As illustrated in Fig. 5 of the paper, they showed that these genes exhibit preferential sensitivity to digestion compared with the bulk of the DNA present in erythroblast nuclei. As the avian tumour virus genes are expressed at a much lower level than the globin genes in erythroblast cells, this finding indicates that the preferential sensitivity of expressed genes is not confined to genes transcribed at very high levels.

This conclusion was also reinforced in a second type of experiment, in which Weintraub and Groudine used total nuclear RNA as a probe rather than purified globin or avian tumour virus probes. In the hybridization experiments illustrated in Fig. 6 of the paper, they showed that the DNA sequences within the chromatin that are rapidly digested by DNase I are homologous to those that are contained in the nuclear RNA probe. As the nuclear RNA probe is obviously derived from the genes transcribed into RNA in a particular tissue, this finding establishes the generality of the observations of Weintraub and Groudine, namely that expressed genes are preferentially sensitive to DNase I in the tissues in which they are expressed.

The importance of the finding made by Weintraub and Groudine is greatly enhanced by the difficulties under which their experiments were conducted. Southern blotting (as used by Jeffreys and Flavell in their paper presented in Section 1) was not available. Similarly, the globin cDNA probe which they used was not a pure probe but contained sequences derived, for example, from both α- and β-globin mRNAs, and this greatly complicated their analysis. Nonetheless the conclusions which they presented have been confirmed by numerous subsequent experiments using cloned probes and Southern blotting analysis (for a review, see Latchman, 1995a).

As well as confirming the conclusions of Weintraub and Groudine, these subsequent studies also allowed a more precise determination of the relationship between the appearance of enhanced sensitivity

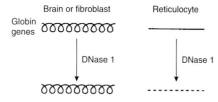

Figure 5.2. Partial digestion with DNase I results in the rapid digestion of the globin gene in reticulocytes, which express the gene, but not in other cell types, such as brain or fibroblasts, which do not. Thus the globin gene is present in a different structural configuration in the reticulocytes compared with other cell types.

to DNase I and the actual expression of the gene. Thus in a paper from the Weintraub group (Stalder et al., 1980) it was shown that the adult β-globin gene exhibited enhanced sensitivity to DNase I in both embryonic and adult red blood cells, even though it was only expressed in the adult red blood cells. Similarly, the embryonic β-globin gene remained sensitive to DNase I even in mature chick red blood cells, where it is not actively transcribed (Figure 5.3). This leads to the important conclusion that enhanced sensitivity to DNase I appears prior to the onset of gene expression and persists after gene expression has ceased. Hence, rather than reflecting gene transcription itself, it appears to be a reflection of cellular commitment in which a cell commits itself to a particular lineage before the genes whose expression is associated with that lineage are actually switched on.

Thus cellular commitment would involve the opening up of the chromatin structure of a particular gene so that it is accessible to subsequent activation by transcription factors binding to it. An excellent example of this is seen in the Friend erythroleukaemia cell system. As discussed in Section 2, the globin gene in these cells can be activated by treatment with the inducing molecule dimethyl sulphoxide. Interestingly, enhanced sensitivity of the globin gene to DNase I digestion can be observed both prior to as well as after dimethyl sulphoxide stimulation (Miller et al., 1978). Hence the Friend erythroleukaemia cells were committed to a lineage allowing globin production but required exposure to the inducer dimethyl sulphoxide to allow actual gene activation to occur. In agreement with this idea, the globin gene cannot be activated by dimethyl sulphoxide treatment in other cell types where the gene is not in a DNase I-sensitive state.

The paper presented here by Stalder et al. (also from the Weintraub laboratory) also extends the original Weintraub and Groudine paper, but in another manner. These authors focused on the extent of the region around the globin gene that is preferentially sensitive to DNase I. In the data illustrated in Figure 2 of the paper, for example, they showed that this preferential sensitivity applies not only to the regions of the gene that actually encode the β-globin protein but also to the adjacent non-coding regions. Hence the preferential sensitivity extends over the entire region that is transcribed into RNA.

In the most important aspect of their paper, Stalder et al. also focused on the distribution of sensitivity within the chromatin containing the transcribed region. Thus in the data illustrated in Figure 6 of the paper they showed that, when chromatin was treated with very low levels of DNase I, specific bands of partially digested DNA could be identified in Southern blotting experiments. This indicates that, within the transcribed region, specific sites are exquisitely sensitive to DNase I. Cutting at these sites in the DNA within the chromatin structure therefore produces specific bands when the DNA is subsequently purified, cut with a restriction enzyme and Southern blotted. These bands contain a cutting site for DNase I at one end of the fragment and a restriction enzyme digestion site at the other end (Figure 5.4).

Hence, while the transcribed DNA is generally more sensitive to digestion by DNase I than the bulk of DNA, it also contains regions within it that are exquisitely sensitive to DNase I and which represent the first sites at which the enzyme cuts. Such so-called 'DNase I-hypersensitive sites' had previously been described in *Drosophila* (Wu et al., 1979). The importance of the paper by

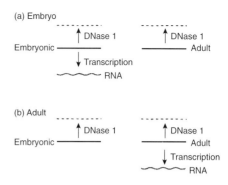

Figure 5.3. Both the embryonic and adult globin genes are sensitive to digestion with DNase I in embryonic reticulocytes, even though only the embryonic gene is transcribed. Similarly, both genes are sensitive in adult reticulocytes, even though only the adult gene is transcribed.

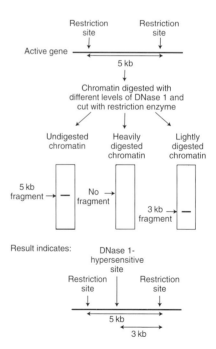

Figure 5.4 Detection of DNase I-hypersensitive sites by mild digestion of chromatin that has already been cut with a restriction enzyme. Specific products are obtained which contain a restriction enzyme site at one end and a DNase I-hypersensitive site at the other end (right-hand panel). More extensive digestion will result in the disappearance of the band.

Stalder et al. presented here, however, is that these authors demonstrated that, like overall DNase I sensitivity, such DNase I-hypersensitive sites are tissue-specific. The hypersensitive sites observed in the globin genes in adult and embryonic red cell nuclei are not observed, for example, in chromatin derived from the brain, where the globin gene is not expressed. This finding has subsequently been confirmed for a variety of other genes for which the DNase I-hypersensitive sites appear in tissues where the gene is active or about to become active (e.g. see Kaye et al., 1984; for a review, see Gross and Garrard, 1988).

Hence, as with overall sensitivity to DNase I, the appearance of DNase I-hypersensitive sites appears to be important in gene regulation. This conclusion is reinforced by the location of these hypersensitive sites, which can be mapped precisely relative to specific sites of restriction enzyme digestion, as illustrated in Figure 5.4. Many of these sites are located at the 5′ ends of genes in positions corresponding to DNA sequences that are important in regulating transcription. It is likely, therefore, that they represent sites in which the chromatin structure has been altered so as to facilitate the binding of regulatory transcription factors.

Thus, in a specific cell lineage in which a gene will be expressed, the chromatin structure of that gene is altered in two ways. First, as illustrated in the paper by Weintraub and Groudine, the entire transcribed region becomes preferentially sensitive to digestion with DNase I. This is likely to involve a change in the association of DNA with the histones or other proteins in this region so that a more open chromatin structure is formed which is more accessible to digestion by DNase I. Secondly, as described in the paper by Stalder et al., short regions within this area become exquisitely sensitive to DNase I. This is likely to involve the removal of histone proteins from these sites, allowing them to serve as binding sites for activating transcription factors.

Obviously, the existence of such changes leads to the question of the mechanism by which they arise. Clearly some process or processes must exist which identifies these regions within the vast bulk of cellular DNA and then results in a structural alteration of these regions which is detected by DNase I. Such processes may involve changes in the DNA within this region itself or in its association with other proteins such as the histones. Numerous changes of this type within active or potentially active genes have been described. These include alterations in the histone proteins themselves (for a review, see Turner, 1993). For example, it has been shown that increased acetylation of specific histones occurs in regions of DNase I sensitivity, and artifical increases in histone acetylation have been shown to result in the activation of some previously silent cellular genes (Reeves and Cserjesi, 1979).

However, the most consistent difference that has been described between active or potentially active DNA and the bulk of cellular DNA involves the level of cytosine methylation. For this reason, the paper by McGhee and Ginder is presented here. Thus, as indicated in the introduction to the paper, the methylation of the cytosine base in DNA is the most common modification detected in the DNA of higher eukaryotes (for reviews, see Razin and Cedar, 1991; Bestor and Coxon, 1993).

The paper by McGhee and Ginder looks for the existence of such methylated cytosine residues at specific sites in the vicinity of the β-globin gene. It takes advantage of the fact that approx. 90% of the methylated cytosine residues occur in the dinucleotide CG, where the methylated C is followed on its 3′ side by a G residue. In a subset of such sites, the CG pair will form part of the sequence CCGG.

This sequence CCGG forms the recognition site for cuttting of the DNA by the restriction endonucleases *Hpa*II and *Msp*I. Most importantly, however, while *Msp*I will cut this sequence whether the central cytosine is methylated or not, *Hpa*II will only cut the DNA when the central cytosine is unmethylated and not when it is methylated. Hence digestion of the DNA from a particular tissue with each of these enzymes followed by Southern blotting with a specific probe will allow differences in the methylation of specific cytosine residues between different tissues to be identified (Figure 5.5).

This method was used by McGhee and Ginder in the paper presented here to study the methylation of specific C residues in the DNA region containing the β-globin gene. Thus in Fig. 1 of their paper they showed that the *Hpa*II digest of adult chicken erythrocyte DNA produced different bands when probed with a β-globin cDNA clone compared with those observed upon probing an *Msp*I digest of the same DNA. This indicates that some sites within the β-globin locus are methylated in adult chicken erythrocyte DNA, whereas others are unmethylated. Such detection of methylated cytosine residues had previously been performed by other workers (e.g. see Waalwijk and Flavell, 1978). The significance of the work presented by McGhee and Ginder lies in the fact that they also used *Hpa*II and *Msp*I to digest DNA from the oviduct, which does not express the globin genes. As illustrated in Fig. 1 of the paper, *Hpa*II digestion of oviduct DNA gave a different pattern of bands hybridizing to the globin probe compared with that produced on similar digestion of erythrocyte DNA. In data mentioned in the paper but not illustrated, it is also stated that digestion of brain DNA (which also does not express the globin genes) gave a similar pattern to that produced by oviduct DNA.

This paper by McGhee and Ginder therefore represented the first report of tissue-specific differences in the methylation pattern of a gene in different tissues, and was the first indication that such differences might be involved in gene regulation. It has subsequently become clear that many sites within a particular gene are unmethylated in tissues where the gene is active or potentially active, whereas they are methylated in other tissues (for reviews, see Cedar, 1988; Bird, 1992). As in the case of sensitivity to DNase I, undermethylation is observed prior to the onset of transcription and persists after its cessation. Moreover, the region in which unmethylated C residues are found correlates with that exhibiting enhanced sensitivity to DNase I (Weintraub et al., 1981).

Thus, as with DNase I sensitivity, undermethylation is a consequence of commitment to a particular pattern of gene expression and is associated with the change in chromatin structure observed in active or potentially active genes. Indeed, in some cases treating cells with the cytidine analogue 5-azacytidine, which cannot be methylated, results in the activation of genes which were previously silent. In the most dramatic of these cases, treatment of the undifferentiated $10T_{1/2}$ cell line with 5-azacytidine results in the activation of a key regulatory gene and leads to their differentiation into multinucleate, twitching, striated muscle cells (Constantinides et al., 1977) (see also Section 13).

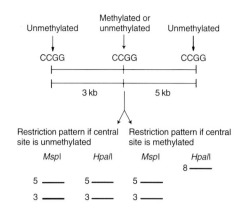

Figure 5.5 Detection of differences in DNA methylation between different tissues using the restriction enzymes *Msp*I and *Hpa*II

Although this is an extreme example, it is clear that the phenomenon of tissue-specific methylation of cytosine residues discovered by McGhee and Ginder plays an important role in the process of commitment to a particular cell lineage and its associated pattern of gene expression. Indeed, the inactivation of the gene encoding the enzyme DNA methyltransferase, which methylates DNA, results in the death of mouse embryos prior to birth (Li et al., 1992), indicating that, at least in mammals, DNA methylation is essential for normal embryonic development.

The papers presented in this section thus describe three key changes, namely enhanced sensitivity to DNase I, appearance of DNase I-hypersensitive sites and undermethylation, that occur in genes that are active or potentially active in a particular lineage. These changes prepare the gene to become activated in response to an appropriate stimulus. Thus, in the example quoted at the beginning of this section, the ovalbumin gene has undergone these changes in oviduct tissue and is therefore activated in response to subsequent oestrogen treatment of the tissue. However, these changes have not occurred in the ovalbumin gene within the liver, and this gene therefore is not activated in response to oestrogen in this tissue. In contrast, the vitellogenin gene has undergone these changes in its chromatin structure in the liver and is therefore activated in response to oestrogen in this tissue but not in the oviduct. Thus the chromatin changes described in this section play an essential role in determining which genes will respond to specific stimuli in a given tissue.

Chromosomal Subunits in Active Genes Have an Altered Conformation

Globin genes are digested by deoxyribonuclease I
in red blood cell nuclei but not in fibroblast nuclei.

Harold Weintraub and Mark Groudine

Knowledge of the structure of DNA has provided many insights into its biological function (*1*). In higher cells, a detailed understanding of the structure of chromatin will probably provide analogous insights into how genes are regulated. Already, there are a number of important observations demonstrating a rela-tion between the structure of chromatin and its biological activity (*2, 3*).

The packaging of most of the nuclear DNA is now thought to be based on repeating units of about 180 to 200 base pairs of DNA associated with specific complexes of histones (*4, 5*), possibly two self-complementary tetramers each containing one of the four major histones (*6*). These two tetramers could define the twofold axis of symmetry within the nucleosome. These complexes interact through 70 to 90 amino acid residues at their carboxyl terminal end to produce a tight, trypsin-resistant core (*7*). The posi-tively charged histone amino terminal residues extend outward from this core and define what may prove to be a "kinked" or "coiled" pathway for the DNA (*5, 8*) about the histone complexes. These so-called "particles-on-a-string" or "nu" bodies constitute the primary level of folding for the bulk of the chromosome. Through their mutual inter-actions higher levels of DNA packaging can be achieved, although details of this organization are not known. At present there is no proof that nu bodies are homo-

Dr. Weintraub is an assistant professor in the Department of Biochemical Sciences, Frick Labora-tories, Princeton University, Princeton, New Jersey 08540. Dr. Groudine was a visiting fellow in the same department and is now at the Department of Radiation Oncology, University in Washington Hospital, Seattle 98105.

genous (9) either in composition or in conformation. Also lacking is concrete evidence that nu bodies are associated with active genes (10) and, if so, whether they are in the same conformation as those nu bodies associated with inactive genes. The important finding that *Escherichia coli* RNA polymerase preferentially transcribes globin genes from reticulocyte chromatin but not from liver or brain chromatin (11) reflects the fact that some structural aspect of an activated gene is different in different tissues. However, whether this difference occurs only at the 5′ end of the gene, perhaps at a promoter region, or throughout the entire length of the gene is not known. Similarly, it is not clear whether the tissue-specific differences in accessibility revealed by RNA polymerase arise from differences in the basic nu body configuration or in the way some or all of the nu bodies in a transcription unit are packaged into higher levels of organization.

In this article, we show that active genes are likely to be packaged by histones but that these histones are in an altered conformation, one that renders the associated DNA extremely sensitive to digestion by pancreatic deoxyribonuclease I.

Digestion of Chicken Erythrocyte Nuclei by Deoxyribonuclease I

The kinetics of digestion of chicken erythrocyte nuclei by pancreatic deoxyribonuclease I (12) are shown in Fig. 1. The bottom insets show the corresponding patterns of resistant single-stranded

DNA fragments displayed on a denaturing acrylamide gel. The kinetics reveal a rapid initial digestion of about 15 percent of the DNA followed by a slower digestion that levels off at about 50 percent, and an even slower process that leads to digestion beyond 50 percent. The characteristic pattern of resistant products (between 20 and 160 bases in our standard gels) appears very early in digestion and persists through 50 percent digestion with no apparent increase or decrease in the intensity of any one particular band. Beyond 50 percent digestion, the larger bands disappear before the smaller bands do. This disappearance is accompanied by an accumulation of DNA between the usually well defined DNA peaks. We interpret this to mean that the enzyme is sequentially digesting a base or two at a time when digestion proceeds beyond 50 percent, and that this phase of the digestion process probably represents "nibbling" (digestion of a base at a time) that occurs in DNA regions that are intimately protected by proteins, presumably histones. In contrast, the repeating pattern of denatured DNA fragments between 20 and 160 bases probably represents the cutting at very accessible regions between and within individual nu bodies. It is not clear to us why larger fragments fail to break down into the smaller ones before digestion reaches 50 percent and nibbling ensues. The conversion of larger fragments into smaller ones would have been predicted if the fragments arose from the statistical cleavage of a number of accessible sites within a homogeneous population of nu bodies.

Tissue-specific, Preferential Digestion of Active Genes

Staphylococcal nuclease shows no preferential digestion of specific nuclear DNA sequences, in particular sequences coding for active genes (13). Since it is clear that pancreatic deoxyribonuclease has a much higher affinity for sites within nu bodies, and hence might be expected to differentiate between similar nu bodies, we decided to investigate whether pancreatic deoxyribonuclease preferentially digested the DNA coding for active genes. There are a number of indications that this might be occurring during the digestion of nuclei with this enzyme. We have shown that ribosomal DNA in nuclei is especially sensitive to deoxyribonuclease (14); in addition, Billing and Bonner (15) have shown that RNA labeled for a short period is rapidly released on mild digestion of nuclei with either deoxyribonuclease I or II. Finally Berkowitz and Doty (16) have shown that a putative active fraction of sheared chromatin (isolated as a slow-sedimenting fraction on sucrose gradients) is much more sensitive to deoxyribonuclease I than is bulk chromatin.

To investigate this question further, we have prepared a complementary DNA (cDNA) probe to globin messenger RNA (mRNA) isolated from the reticulocytes of the adult chicken. The details of the purification and analysis and the characteristics of the cDNA have previously been described (17); in particular it has been demonstrated that the cDNA made to adult globin mRNA can be used to detect globin mRNA coding for embryon-

Fig. 1. Digestion of chick erythrocyte nuclei with pancreatic deoxyribonuclease I. (Curves) Avian red blood cells were isolated as described (17). The cells were washed twice in phosphate-buffered saline (PBS) (Grand Island), and the nuclei were isolated by suspension in reticulocyte standard buffer (RSB) (0.01M tris-HCl, pH 7.4; 0.01M NaCl; 3 mM MgCl₂) containing 0.5 percent NP-40 (British Drug House). The nuclei were washed several times in RSB and then digested at a DNA concentration of 1 mg/ml at 37°C with pancreatic deoxyribonuclease I (Sigma) (20 μg/ml) for increasing periods of time. The percentage of DNA remaining (● —— ●) was determined either by precipitation with cold 7 percent perchloric acid and subsequent measurement of the absorbancy at 260 nm of the acid soluble and insoluble fraction or by sedimenting the nuclei at low speed and measuring the release of material absorbing at 260 nm. Either method gave essentially the same results. Addition of fresh deoxyribonuclease (20 μg/ml) when the digesion begins to level off (at about 50 percent) does not affect the course of subsequent digestion. (▲ —— ▲) The time course of digestion and the pattern of resistant fragments (see below) is the same in chromatin isolated either by sonication or by mild treatment with nuclease. Substitution of CaCl₂ for MgCl₂ also did not affect the course of digesion or pattern of resistant frag-

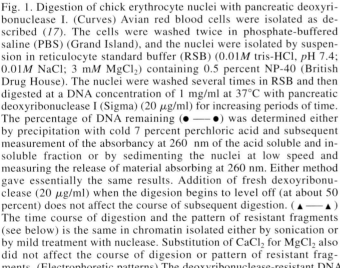
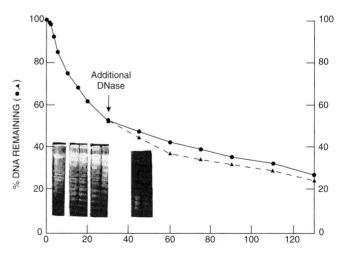

ments. (Electrophoretic patterns) The deoxyribonuclease-resistant DNA was obtained from the sedimented nuclei at intervals during the digestion. The following were added to the nuclear pellet (final concentration) EDTA (10 mM), sodium dodecyl sulfate (SDS) (0.1 percent), and protease (Aldrich) (500 μg/ml). The mixture was incubated for 60 minutes at 37°C, boiled for 5 minutes, added directly to sample buffer (50 percent glycerol, 0.01 percent bromophenol blue), and loaded onto a 6 percent acrylamide slab gel containing 98 percent formamide and 20 mM sodium phosphate at pH 7.0. Electrophoresis was conducted for 5 hours at 160 volts in a running buffer containing 20 mM sodium phosphate, pH 7.0. The gels were stained with ethidium bromide (2 μg/ml) and the denatured DNA bands (bottom insets) were photographed through a red filter after illumination with ultraviolet light. *DNase*, deoxyribonuclease.

ic globin polypeptide chains (*17*). Thus, embryonic red cell RNA saturates about 70 percent of the cDNA probe prepared from adult red cell globin mRNA, while adult red cell RNA saturates the labeled probe at about 95 percent (*17*). The cross reaction of the adult probe with embryonic RNA is qualitatively attributable to the fact that the adult line of red cells synthesize three main types of globin chains, two of which are identical to chains synthesized by the embryonic line of red cells and the third bears about a 50 percent homology by tryptic peptide analysis (*18*) to two other embryonic globin chains.

When the DNA in nuclei from adult chick erythrocytes—this line of red blood cells contains the adult globin polypeptide chains—is digested with pancreatic deoxyribonuclease I so that 10 percent is acid soluble, approximately 75 percent of the globin cDNA probe fails to hybridize with the remaining 90 percent of the purified erthyrocyte DNA (Fig. 2). Similarly, after the same amount of digestion, DNA isolated from the embryonic line of erythroblasts protects only about 50 percent of the cDNA at saturation. In contrast, undigested DNA isolated from either embryonic or adult red cells saturates the probe at more than 94 percent. These experiments show that, in the different red cell lines, specific globin se-

quences are particularly sensitive to digestion. We present evidence below that embryonic but not adult-specific globin sequences and digested in embryonic red cells and adult but not embryo-specific sequences are digested in adult red cells. The same preparations of digested nuclei from either the adult or embryonic line of red cells contains the DNA sequences coding for ovalbumin mRNA (Fig. 2). Most importantly, nuclei from cultured chick fibroblasts or freshly isolated chick brain, after 10 to 20 percent digestion (more extensive amounts of digestion were not tested), retain the DNA sequences coding for both adult and embryonic globin (Fig. 3). In addition, the fibroblast nuclei retain the DNA sequences coding for ovalbumin (Fog. 2). In contrast, digestion of red cell nuclei to DNA fragments of approximately the same size by a different nuclease, staphylococcal nuclease, results in no preferential digestion of any globin sequences (Fig. 3). This was first shown by Axel *et al.* (*13*) and recently extended by Lacy and Axel (*19*).

In summary, digestion of nuclei with pancreatic deoxyribonuclease reveals that specific globin sequences are preferentially degraded in erythroid cells, but not in nonerythroid cells. Similarly, the ovalbumin gene is not preferentially digested in cells that do not produce ovalbumin.

Fig. 2. Preferential digesion of active genes by pancreatic deoxyribonuclease I. Red blood cells (RBC) were obtained by vein puncture from 18-day (containing adult-type globin chains) and 5-day (containing embryo-type globin chains) chick embryos. Fibroblasts were dissected from the region of the developing breast muscle of 11-day chick embryos and grown in culture (*17*). Nuclei from 18-day embryo RBC's, 5-day embryo RBC's, and 11-day cultured chick embryo fibroblasts are isolated in RSB containing 0.5 percent NP-40 and digested with pancreatic deoxyribonuclease I until 10 to 20 percent of the DNA was soluble in acid (legend to Fig. 1). DNA was prepared as follows. Nuclei were centrifuged and suspended in 0.1 percent SDS, 100 μg of pronase per milliliter, and 5 mM EDTA overnight at 37°C. The sample was extracted several times with equal volumes of a mixture of phenol and chloroform (1 : 1), and several times with a mixture of

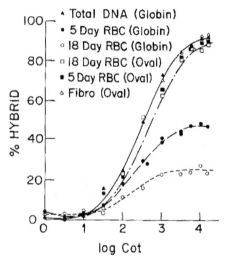

chloroform and isoamyl alcohol (24 : 1). The resultant aqueos phase was made 0.1M with respect to NaCl, and the nucleic acid was precipitated overnight at -20°C with two volumes of 95 percent ethanol. The nucleic acid was recovered by centrifugation for 30 minutes at 10,000 rev/min (HB-4 head of a Sorvall RC-5 centrifuge), suspended in 10 mM NaCl, 10 mM tris-HCL (pH 7.4), and incubated for 30 minutes at 37°C with ribonuclease A (20 μg/ml) (Worthington) that had been boiled for 30 minutes. The preparation was again extracted with the phenol-chloroform and then chloroform–isoamyl alcohol mixtures and the extract was precipitated with ethanol. The DNA concentration was determined by absorbancy at 260 nm in a Zeiss spectrophotometer (DNA at 1 mg/ml gave 20 A_{260}). Total DNA was prepared directly from 18-day red cell nuclei and sonicated so that the average length corresponded to 500 nucleotides. The cDNA complementary to globin mRNA from adult reticulocytes was prepared as described (*17*). Hybridizations were conducted with an excess of DNA to cDNA (1 $\times$ 10^7 to 2 $\times$ 10^7 : 1) and analyzed (*17*). The DNA samples suspended in a mixture of 0.3 M NaCl, 50 mM tris-HCl (pH7.4), and 0.1 percent SDS and ranging in concentration from 1 to 20 mg/ml were denatured by heat and annealed at 65°C (at 1000 count/min per 5 μl of reaction mixture) to [^{3}H]deoxycytidine- and [^{3}H]thymidine-labeled cDNA (5 $\times$ 10^7 to 8 $\times$ 10^{7}count/min per microgram). Under these conditions the calculated ratio of globin DNA to globin cDNA was 10 : 1 to 15 : 1. Polypropylene tubes overlaid with paraffin oil were used for the hybridizations. At intervals (from 5.7 minutes to 96 hours), 10-μl samples of reaction mixtures were pipetted into 400 μl of a mixture of 30 mM sodium acetate (pH 4.5), 0.15M NaCl, 1mM ZnSO$_4$, and 10 μg of denatured DNA from salmon sperm. Half (200 μl) of the above mixture was immediately precipitated with trichloroacetic acid (TCA) and the other half was incubated with partially purified S1 nuclease at 45°C for 40 minutes. The percentage of hybridization was determined by comparison of the TCA-precipitable radioactivity remaining after S1 digestion to that precipitable in the undigested samples. The S1 background radioactivity ranged from 2 to 6 percent. The percentage of hybridized cDNA is plotted as a function of the C_0t, which represents the concentration of deoxyribonucleotide (moles) times the time of digestion (seconds) per liter (*28*). The same plateaus were obtained at concentrations of driver DNA from 2 to 20 mg/ml in a reaction volume of 50 μl, to which cDNA (5000 count/min) was added. ($\blacktriangle$——$\blacktriangle$) Total DNA hybridized to globin cDNA; ($\bullet$——$\bullet$) 5-day red cell DNA treated with deoxyribonuclease I and hybridized to globin cDNA; ($\circ$——$\circ$) 18-day red cell DNA treated with deoxyribonuclease I and hybridized to globin cDNA; ($\square$——$\square$) 18-day red cell DNA treated with deoxyribonuclease I and hybridized to ovalbumin cDNA; ($\blacksquare$——$\blacksquare$) 5-day red cell DNA treated with deoxyribonuclease I and hybridized to ovalbumin cDNA; ($\triangle$——$\triangle$) fibroblast DNA treated with deoxyribonuclease I and hybridized to ovalbumin cDNA.

Identification of Globin Genes Digested in Adult and Embryonic Nuclei

The failure of DNA isolated from red cell nuclei treated with deoxyribonuclease I to saturate the cDNA probe (Fig. 2) could be due to (i) to the specific degradation of a unique subset of globin genes, as we have suggested, or (ii) to an overall reduction in DNA sequences complementary to all sequences present in the cDNA, resulting in a situation where the cDNA is in excess. To exclude the second possibility, we have performed several of the reannealing experiments shown in Fig. 2 with one-tenth the amount of driver DNA (that is, the DNA which determines the rate of the reaction) and the same amount of cDNA. Under these conditions, no change is observed in the level of saturation of the cDNA. This demonstrates that specific globin DNA sequences are digested by pancreatic deoxyribonuclease and also that specific sequences are resistant.

As was mentioned previously, our cDNA probe contains sequences complementary to three adult globin mRNA molecules. The resistant DNA obtained after embryonic erythroblast nuclei are mildly treated with deoxyribonuclease I saturates the cDNA probe at about 50 percent. Our hypothesis is that only actively transcribing globin genes are digested. Since two of the three adult globin genes are also active in the embryonic red cell line, we predict that two adult globin genes are digested in embryonic cells and the third is resistant. To test this, polysomal RNA from embryonic red cells was added to the hybridization mixture which also contained ^{3}H-labeled cDNA from adult globin mRNA and partially digested DNA from embryonic red cell nuclei. Under these conditions, the cDNA is fully saturated at about 98 percent (Fig. 4). Thus, embryonic red cell RNA fully complements the deficiency in globin DNA sequences resulting from the digestion. This implies that embryonic sequences are absent and the adult-specific sequences are present in the DNA from the embryonic red cell nuclei treated with deoxyribonuclease I.

A similar type of analysis suggests that adult-specific sequences are preferentially digested in the adult line of red cell nuclei. Figure 2 shows that the cDNA is saturated at about 25 percent by DNA isolated from adult red cell nuclei treated with pancreatic deoxyribonuclease. Our hypothesis in this case is that only adult-specific globin genes are digested. In principle, if this is true, then none of the probe should be protected from S1 nuclease. We believe that the partial protection of the probe by DNA obtained from adult nuclei treated with deoxynuclease I can be explained by two embryonic globin genes that bear a marked homology to the β-globin gene in the adult (18). These embryo-specific globin genes (named ε and ρ) would be present in the deoxyribonuclease I-resistant fraction from adult red cell nuclei and, by virtue of their homology to the adult β-globin, would be expected to partially protect some of the cDNA. Since the adult β-gene codes for about 50 percent of the adult globin chains, it is not unreasonable that it constitutes about 50 percent of our cDNA population. In addition, peptide analysis has shown about a 50 percent identity in tryptic peptides between the β-globin in adults and the ε- and ρ-globin chains in the embryo (18). Thus, if our logic is correct, we predict that the digested adult nuclei should saturate about 25 percent of the cDNA probe (50 percent × 50 percent = 25 percent). To test this, embryonic red cell RNA was added to the hybridization mixture containing the cDNA and DNA from treated adult nuclei. Under these condi-

tions, the probe is saturatd at about 70 percent (Fig. 4). This is precisely the saturation achieved by pure embryonic red cell RNA alone (17), an indication that the deoxyribonuclease I-resistant DNA from adult red cells contains no adult-specific globin sequences (Table 1).

The sensitivity of the globin genes to deoxyribonuclease I could be related to the fact that they are extremely active in transcription and are therefore not typical of most of the genes actively being transcribed. The chicken genome contains endogenous DNA sequences com-

plementary to those of avian tumor viruses (ATV). A cDNA probe made against a specific type of avian tumor virus, avian myeloblastosis virus RNA (AMV), hybridizes with chicken DNA with a sequence homology of about 60 percent and a log $C_0t_{1/2}$ of about 1.75. This corresponds to about eight to ten copies of endogenous ATV DNA per cell. In separate experiments the total RNA from embryonic chick erythroblasts was hybridized to the ^{3}H-labeled AMV cDNA. The kinetics of hybridization indicated that from one-tenth to two copies of endoge-

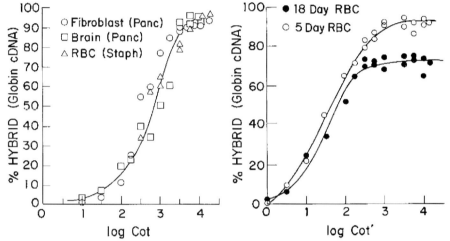

Fig. 3 (left). Retention of inactive genes after treatement of nuclei with pancreatic deoxyribonuclease. As was described (legend to Fig. 2), 18-day RBC nuclei and 11-day fibroblast nuclei were isolated from chick embryos. Brains were dissected from 11-day chick embryos, incubated for 30 minutes in PBS (Gibco) with 0.1 percent trypsin, washed repeatedly with PBS to eliminate contaminating RBC's, and lysed by homogenization in RSB with 0.5 percent NP-40. Fibroblast and brain nuclei were digested so that 20 percent of the DNA was acid soluble (legend to Fig. 1). The 18-day RBC's were digested until 50 percent of the DNA was acid soluble with staphylococcal nuclease (50 μg/ml) for 15 minutes at 37°C in RSB supplemented with $10^{-4}M$ CaCl$_2$. More than 80 percent of the acid insoluble DNA from this preparation consisted of 20 to 145 base pairs, as analyzed on 6 percent acrylamide gels. The isolation of resistant DNA from the digested nuclear preparations and the conditions and analysis of hybridization, including DNA concentrations and times of incubation, were as described (Fig. 2 legend). (○——○) Fibroblast DNA digested with deoxyribonuclease I and hybridized to globin cDNA; (□——□) brain DNA digested with deoxyribonuclease I and hybridized to globin cDNA; (△——△) 18-day red blood cell DNA digested with staphylococcal nuclease and hybridized to globin cDNA. Fig. 4 (right). Hybridization kinetics of globin cDNA and mixtures of 5-day erythroid RNA and DNA from 18-day or 5-day RBC nuclei digested with pancreatic deoxyribonuclease. Embryonic erythroid RNA was prepared from the cytoplasm of RBC's from 5-day chick embryos. The 5-day RBC's were obtained by vein puncture, washed repeatedly with autoclaved PBS, and lysed in autoclaved RSB supplemented with 0.5 percent NP-40 and mouse liver ribonuclease inhibitor (17). Nuclei were sedimented in a table-top centrifuge, and the supernatant was extracted several times with equal volumes of a solvent mixture containing phenol-chloroform (1 : 1) and then several times with a mixture of chloroform and isoamyl alcohol (24 : 1). The aqueous phase was made 0.1M with respect to NaCl, and the nucleic acid was precipitated with two volumes of 95 percent ethanol (overnight at −20°C) and recovered by centrifugation for 60 minutes at 11,000 rev/min (HB-4 head of a Sorvall RC-5 centrifuge); it was suspended in a solution of 10 mM NaCl, 10 mM tris-HCl (pH 7.4), and 5 mM MnCl$_2$, incubated with ribonuclease-free deoxyribonuclease (10 μg/ml) (Worthington) for 30 minutes at 37 °C, and again extracted with the above solvent mixtures. The RNA was precipitated with ethanol and suspended in a solution of 10 mM NaCl, 10 mM tris-HCl (pH 7.4), and the amount of RNA was determined at A_{260} in a Zeiss spectrophotometer (RNA at 1 mg/ml + 24 A_{260}). The nuclei were digested with pancreatic deoxyribonuclease (10 μg/ml) (Worthington) for 30 minutes at 37°C, and again extracted with the conditions, and analysis of hybrid formation are described in Fig. 2. The ratio of DNA to RNA in the hybridization mixture was 10 : 1. The nucleic acid concentration ranged from (per milliliter) 19 mg of DNA and 1.9 mg of RNA to 1.9 mg of DNA anf 0.19 mg of RNA. Because of the minor contribution of RNA to the total nucleic acid concentration, the RNA concentration was not considered in the calculation of C_0t'. (●——●) DNA from pancreatic deoxyribonuclease-treated 18-day red blood cells plus RNA from 5-day embryonic red blood cells hybridized to globin cDNA. (○——○) DNA from pancreatic deoxyribonuclease-treated 5-day red blood cells plus RNA from 5-day embryonic red blood cells hybridized to globin cDNA.

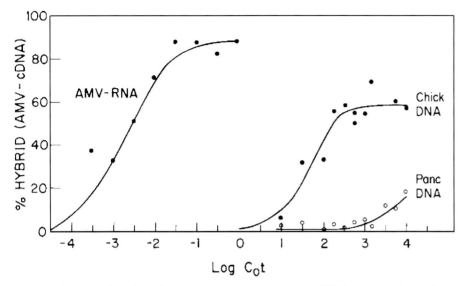

Fig. 5. Preferential digestion of endogenous avian tumor virus (ATV) genes by deoxyribonuclease I, in the embryonic line of chick erythroblasts. Preparations of DNA and procedures for digestion and hybridization were as described (Figs. 1 and 2). The AMV cDNA was prepared by the method described (17) for globin cDNA. The template for the reaction was the 35S RNA prepared from purified virus (a gift from J. Beard). The specific activity of the probe was about 4×10^7 to 6×10^7 count/min per microgram, and the probe hybridized to about 90 percent with its template, with a log $C_0t_{1/2}$ of about -2.75.

treatment of embryonic red cell nuclei. Figure 6 shows that 78 percent of the total ^{3}H-labeled DNA is protected at saturation when DNA from nuclei treated with deoxyribonuclease I is used to drive the reaction. In contrast, when the driver DNA is DNA obtained from the 11S monomers produced by staphylococcal nuclease treatment from nuclei, 94 percent of the labeled DNA is protected from S1 nuclease. Addition of total DNA (but not deoxyribonuclease I–treated DNA) to the reaction after 78 percent saturation is achieved increases the extent of hybridization to 95 percent (Fig. 6 and Table 2A). This suggests that the smallest deoxyribonuclease-digested fragments do not inhibit the hybridization of longer fragments and that the failure to reach full saturation is due to the absence of a particular subset of sequences in the deoxyribonuclease-treated DNA. The saturation value decreases to 65 percent after 52 percent digestion (Table 2).

Whether the specific nuclear DNA sequences that are preferentially digested by deoxyribonuclease I are related to the sequences that are actively being transcribed was tested as follows. The DNA from nuclei digested to 20 percent acid solubility was hybridized to total [^{3}H]thymidine-labeled tracer DNA. After saturation had been reached (78 percent), nuclear RNA was added to the hybridization reaction. Under these conditions, the saturation value increases to 89 percent (Fig. 6 inset). The difference in saturation between the reactions that occur in the presence and absence of added nuclear RNA (78 versus 89 percent) is very reproducible and is probably due to RNA-DNA hybrid formation

nous RNA sequences were present per cell. Similar very low but detectable levels of accumulation of RNA sequences from endogenous chicken viruses have been reported by Hanafusa *et al.* (20). Despite the fact that this gene has a low activity in erythroblasts, it is nevertheless sensitive to digestion by pancreatic deoxyribonuclease in isolated nuclei (Fig. 5). From the $C_0t_{1/2}$ of the reactions, it is possible to calculate that these endogenous viral sequences are about 100 times more sensitive to deoxyribonuclease I than the bulk of the nuclear DNA. Thus, two specific classes of active genes are preferentially digested by deoxyribonuclease I the

very active globin genes and the less active endogenous RNA tumor virus genes.

Digestion of a Specific Class of
Nuclear DNA Sequences

The next question is whether after 20 percent digestion of nuclei by deoxyribonuclease I a specific 20 percent of the DNA sequences is absent from the remaining DNA. Total [^{3}H]thymidine-labeled chick DNA was hybridized with a 10,000-fold excess of driver DNA obtained after mild deoxyribonuclease

Table 1. Observed and predicted saturation of globin cDNA by various nucleic acid preparations. Globin gene terminology is based on the tryptic peptide analysis of Brown and Ingram (18), as is the representatin of globin chains in the adult (18-day) and embryonic (5-day) red blood cell (RBC) populations; (+) indicates either the presence of the DNA sequences coding for particular globin chains or the presence of the globin polypeptide chains in the respective embryonic or adult populations; (−) indicates the absence of the globin genes or the absence of the globin polypeptide chains. In the nucleic acid mixtures containing both DNA and RNA, the contribution of each nucleic acid to the protection of cDNA from S1 nuclease digestion is represented by the subscripts D and R, respectively. Saturation values are taken from Figs. 2 and 3.

	Globin genes						Saturation of adult cDNA (%)	
Item	π (α-like)	α_A	α_D	ϵ (β-like)	ρ (β-like)	β	Observed	Predicted
cDNA from adult RBC*		35%	15%			50%		
Active genes in adult RBC	−	+	+	−	−	+		
Active genes in embryonic RBC	+	+	+	+	+	−		
Postulated sequences remaining after treatment of:								
(A) Embryonic RBC (Fig. 2)	−	−	−	−	−	+	50	50
(B) Adult RBC (Fig. 2)	+	−	−	+	+	−	28	25†
(C) Embryonic RBC plus embryonic RBC RNA (Fig. 3)	+$_R$	+$_R$	+$_R$	+$_R$	+$_R$	+$_D$	94	100
(D) Adult RBC plus embryonic RBC RNA	+$_{D,R}$	+$_R$	+$_R$	+$_{D,R}$	+$_{D,R}$		72	75

*Complementary DNA was prepared from adult globin mRNA (17). The percentage representation of specific globin genes in the cDNA population is based on the analysis of Brown and Ingram (18), and on the assumption that a stoichiometric relation exists between the template mRNA and cDNA product.   †The predicted 25 percent saturation of cDNA by DNA of deoxyribonuclease I-digested adult RBC nuclei is based on the shared tryptic peptides of the adult β-chain and the embryonic ϵ- and ρ-chains.

Fig. 6. Kinetics of reassociation of chick total [3H]-labeled DNA and DNA from nuclease-treated nuclei of 18-day RBC's. Total [3H]-labeled DNA was prepared by incubation of RBC's from 4-day chick embryos with [3H]deoxythymidine. Cells were obtained by vein puncture, washed several times in medium F-12 (Gibco), and incubated for 5 hours in F-12 with [3H]deoxythymidine (50 μc/ml; 16 c/mmole; New England Nuclear). The preparations of total DNA and of DNA from nuclei of 18-day RBC's digested with pancreatic deoxyribonuclease I were as described (Fig. 2); and the conditions of staphylococcal nuclease digestion were as described in Fig. 3 except that digestion was for 10 minutes to produce a population of fragments with a weight average molecular weight of 150 bases. Nuclear RNA was prepared from 5-day RBC's by lysis of these cells in the presence of mouse liver ribonuclease inhibitor (18), with an autoclaved solution of 0.5 percent NP-40 in RSB. The nuclei were washed several times in autoclaved RSB and lysed by gentle homogenization in 20 volumes of 0.15M NaCl, 0.05M sodium acetate (pH 5.1) and 0.3 percent SDS. The nuclear lysate was extracted three times with equal volumes of a mixture of phenol and chloroform (1 : 1) and numerous times with a mixture of chloroform and isoamyl alcohol (24 : 1). Ethanol precipitation, elimination of DNA, and determination of RNA concentration were as described (Fig. 4). Hybridization and analysis of

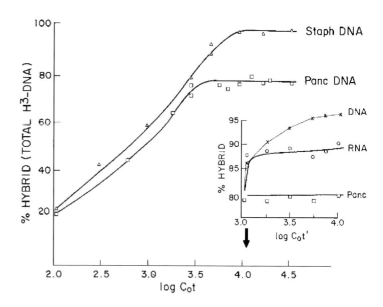

hybrid formation were as described (Fig. 2). The concentration of labeled DNA was 10^{-4} that of driver DNA. Total [3H]-labeled DNA (5000 count/min per 5μl of hybridization mixture) was annealed with DNA (20mg/ml) from deoxyribonuclease I digestion of 18-day RBC nuclei (□——□) or DNA (20mg/ml) from 18-day RBC's (△——△). All points are plotted as DNA C_0t. (Inset) After the reaction with the digested DNA from 18-day RBC's reached saturation (log C_0t = 4.0), and equal volume of the digested DNA was added to the reaction at 20 mg/ml (□——□), or an equal volume of total DNA at 20 mg/ml (×——×), or an equal volume of nuclear/RNA at 10mg/ml (○——○). The reaction was allowed to proceed until a second saturation was achieved. The arrow shows when the additional nucleic acids were added to the original hybridization reaction.

between nuclear RNA and its [3H]-labeled template DNA since treatment of the hybrids with ribonuclease H (which specifically degrades RNA in DNA-RNA hybrids) makes an additional 15 percent of the hybrids sensitive to S1 nuclease (Table 2A).

The experiments described in Fig. 6 suggest that pancreatic deoxyribonuclease preferentially digests nuclear DNA sequences active in transcription of nuclear RNA. To verify this, [3H]thymidine-labeled nuclei were partially digested with staphylococcal nuclease to produce a random population of small, resistant DNA fragments of predominantly 180 and 360 base pairs. This labeled DNA was hybridized to saturation with DNA obtained from red cells treated with pancreatic deoxyribonuclease, and the unhybridized labeled DNA (about 20 percent of the total) was isolated by passing the mixture over hydroxyapatite (HAP). According to the data in Fig. 6, the single-stranded flow-through DNA from HAP should be enriched in those sequences coding for active genes. To test this, excess total nuclear RNA was hybridized with the HAP flow-through DNA. About 48 percent of the labeled DNA is protected from S1 nuclease at saturation, and in a control experiment, nuclear RNA saturated only 9.8 percent of the total [3H]-labeled DNA (Table 2B). Thus the HAP flow-through DNA is enriched in DNA sequences complementary to nuclear RNA.

These conclusions can be tested in another way. Nuclear RNA, which has a kinetic complexity five to ten times greater than cytoplasmic RNA (21), was labeled for a short period and then hybridized to an excess of total red cell DNA or to an excess of embryo red cell DNA that had been treated with deoxyribonuclease I (Table 2C). Whereas 91 percent of the labeled RNA was protected from ribonuclease A digestion after being saturated by total DNA and 86 percent was protected by the staphylococcal nuclease DNA fragments, only 16 percent of the labeled RNA was protected at saturation by the pancreatic deoxyribonuclease I–treated DNA. Since 70 to 90 percent of the nuclear RNA is predominantly one major kinetic class of sequences (21), these experiments also suggest that pancreatic deoxyribonuclease preferentially digests much of the DNA coding for nuclear transcripts.

Altered Histone Conformation

Associated with the Globin Genes

It is possible that the preferential digestion of the globin genes by deoxyribonuclease I is related to the way the 11S monomers are packaged into higher order structures within the cell nucleus. In order to test this possibility, the purified 11S monomers were mildly digested with pancreatic deoxyribonuclease in the presence of 3mM MgCl$_2$ until 15 percent of the DNA was acid soluble. When the

isolated DNA was hybridized to the globin cDNA, the adult-specific globin genes were absent (Table 2D). Thus, although the 11S monomers contain all the globin sequences, in the appropriate ionic conditions the proteins protecting the globin genes adopt a configuration that renders the associated globin DNA sensitive to digestion by pancreatic deoxyribonuclease. If these proteins are histones, then perhaps they are modified, either by direct chemical modification or by association with nonhistone proteins.

The sensitivity of monomers to pancreatic deoxyribonuclease apparently rules out a class of explanations for the digestion of active genes by deoxyribonuclease I based on the higher order packaging of monomers within the nucleus. The following experiment also supports the view that monomer packaging is not the basis for the digestion of active genes. The 11S monomers isolated from 18-day red cell nuclei were treated with trypsin (22) to remove 20 to 30 residues from the NH$_2$-terminus of each of the histones (7). The trypsin was then inactivated by addition of soybean trypsin inhibitor, and the particles were redigested with staphylococcal nuclease. The resistant DNA (about 20 percent of the total) was isolated and hybridized to tracer amounts of total [3H]-labeled DNA and to globin [3H]-labeled cDNA. Whereas more than 80 percent of the total [3H]-labeled DNA hybridizes to the resistant DNA, only 25 percent of the globin cDNA forms stable hybrids. (We have made no attempt to show that the 25 percent hybridization represents cross reaction with inactive embryo genes as we describe in Fig. 4). Thus, active globin 11S monomers become sensitive to staphylococcal nuclease after but not before treatment with trypsin, suggesting again that the conformation of transcriptionally "active" monomers is different from that of the inactive ones. The increased accessibility of actively transcribing genes to staphylococcal nuclease after treatment with trypsin is in good agreement with the in vivo experiments of Roberts and Kroeger (23), who showed

Table 2. Observed saturation of trace amounts of ^{3}H-labeled probes by vast excesses of driver nucleic acids. (A) The treatment of nuclei, isolation of DNA, and reannealing were as described (Fig. 6). Percentage digestion was determined by the absorbancy at 260 nm of the perchloric acid soluble fraction. The 11S monomers were prepared as described (D) below. The weight average molecular weight of the DNA fragments decreased from 200 bases at 10 percent digestion to about 80 bases at 52 percent digestion. The date for the 19 percent digestion are taken from Fig. 6. In the remaining experiments, DNA was at a concentration of 15 mg/ml and hybridization was assayed as described (Fig. 6). For ribonuclease H digestions, the hybridized mixture of total ^{3}H-labelled DNA, the DNA from nuclei treated with deoxyribonuclease to 19 percent acid solubility, and the nuclear RNA was desalted by passage over Sephadex G25 equilibrated with a solution of 40 mM tris-HCl (pH 7.7), 4 mM MgCl$_2$ and 1 mM dithiothreitol. The sample was treated with ribonuclease H (Miles; 7 units) in the presence of bovine serum albumin (30 μg/ml) and glycerol (4 percent) for 30 minutes at 37°C. This procedure was followed by digestion with S1 nuclease, as described. (B) The 4-day RBC's were incubated with [^{3}H]thymidine (Fig. 6). Monomers were prepared from these labeled cells (D, below). The DNA from ^{3}H-labeled monomers were annealed with excess DNA from RBC nuclei treated with deoxyribonuclease I (Fig. 1) as described (Fig. 2). At log C_0t of 4.25, approximately 80 percent of the ^{3}H-labeled DNA was hybridized as assayed by S1 digestion. Single-stranded DNA was isolated by HAP chromatography. Portions of the reaction mixture (100,000 count/min) were pipetted into 0.15M sodium phosphate (PB) (pH 7.0) and placed on a water-jacketed column containing 10 g of HAP (Bio-Rad-DNA grade) at 60 °C. Repeated washings with 0.15 MPB resulted in the elution of single-stranded DNA (approximately 20,000 count/min). This fraction is referred to as "HAP flow-through DNA." It is 98 percent sensitive to S1 nuclease and contains more than 90 percent single-copy sequences. The remaining double-stranded DNA (80,000 count/min) was eluted with 0.48M PB. The HAP flow-through was desalted by passage over Sephadex G25 (equilibrated in 0.1M NaCl, 0.01M tris-HCl, PH 7.4) and precipitated with 100 μg of (carrier) yeast transfer RNA (tRNA) (Sigma) in two volumes of 95 percent ethanol. The HAP flow-through DNA was recovered by centrifugation for 30 minutes (11,000 rev/min, HB-4 head of a Sorvall RC5 centrifuge) and suspended in a solution of 10 mM NaCl, 10 mM tris-HCl) (pH 7.4). Nuclear RNA (12 mg/ml) was prepared from 10-day old chick embryo reticulocytes (Fig. 6 legend) and used to drive the reaction against total ^{3}H-labeled DNA or the HAP flow-through fraction. The hybridization conditions and analysis of hybrid formation were as described (Fig. 2). Seven percent of the total DNA and 1 percent of the HAP flow-through DNA behaved as foldback DNA, as assayed by S1 nuclease, and was subtracted from the observed saturation values to yield the figures of 9.8 percent and 48 percent, respectively. Saturation was achieved by log $Cr_0t = 3$, although points were taken to log $Cr_0t = 4$ (Cr_0t, moles of ribonucleotide per liter × seconds). (C) ^{3}H-labeled nuclear RNA was prepared by incubation of 4-day chick embryo RBC's for 30 minutes in medium F-12 (Gibco) with [^{3}H]uridine (100 μc/ml) (New England Nuclear; 16 c/mmole). The isolation of nuclear RNA and the preparation of the staphylococcal nuclease limit digest were as described. Total DNA and DNA from nuclei treated with deoxyribonuclease I (isolated as described above) were annealed with tracer ^{3}H-labeled nuclear RNA (5000 count/min per 5 μl of the hybridization mixture). At intervals, a sample (10 μl) of the hybridization mixture was pipetted into 400 μl of double-strength SSC (SSC consists of 0.14M NaCl; 0.014M sodium citrate); half of this sample was immediately precipitated with trichloroacetic acid, and the other half was incubated for 30 minutes at 37°C with previously boiled ribonuclease A (20 μg/ml) (Worthington). The percentage hybridization was then determined (Fig. 2 legend). Of the labeled RNA 6 percent was resistant to ribonuclease in the absence of DNA and this was subtracted from the observed saturation values obtained at log $Cr_0t = 4.25$. (D) Conditions and analysis of hybridization for reactions with trace amounts of ^{3}H-labeled globin cDNA and the isolation of total DNA and the DNA from digested RBC and fibroblast were as described (Fig. 2 legend). The posterior one-third of de-embryonated blastoderms (area vasculosa) were dissected from approximately 500 24-hour chick embryos, washed several times in PBS (Gibco), and gently lysed by homogenization in RSB with 0.5 percent NP-40. The resultant nuclei were then treated with pancreatic deoxyribonuclease to 20 percent acid solubility and the DNA was isolated as described (Fig. 1 legend). Monomers were prepared by incubation of 18-day RBC nuclei with staphylococcal nuclease (50 μg/ml) (Worthington) for 5 minutes at 37°C in RSB supplemented with 10^{-4}M CaCl$_2$. The nuclei were then centrifuged and washed in 0.075M NaCl, 0.02M EDTA, and 0.01M tris-HCl (pH 7.4). The resultant pellet was suspended in 5 mM sodium phosphate (pH 6.8), and insoluble material was removed by low speed centrifugation. Of the soluble material, 90 percent sedimented at 11S and 10 percent at 15S. Removal of contaminating 15S material had no effect on our results. Monomers were treated with pancreatic deoxyribonuclease in either 5 mM sodium phosphate (pH 6.8) or in RSB. The isolation of DNA was as described (Fig. 2 legend). (E) The 18-day RBC monomers were prepared as described above. Trypsinized monomers were prepared by the incubation of RBC monomers with trypsin (100 μg/ml) in RSB for 20 minutes at 37°C. After the addition of soybean trypsin inhibitor (200 μg/ml) (Worthington), the suspension was made 10^{-4} M with respect to CaCl$_2$ and digested with staphylococcal nuclease (50 μg/ml) for 30 minutes at 37°C. The isolation of the remaining DNA (50 to 80 bases in length) and the conditions and analysis of hybridization were as described (Fig. 2 legend). About 20 percent of the DNA was resistant after these procedures: yet, this resistant DNA saturates more than 80 percent of the labeled total DNA, an indication that the inactive regions of the chromosome are randomly covered by the trypsin-resistant histone cores.

Item	Saturation*(%)
(A) Tracer ^{3}H-labeled DNA (total) annealed with driver DNA from	
11S monomers	94 ± 3†
18-day RBC DNA—10% digestion with deoxyribonuclease I	85 ± 4
18-day RBC DNA—19% digestion with deoxyribonuclease I	78 ± 3
18-day RBC DNA—52% digestion with deoxyribonuclease I	65 ± 5
18-day RBC DNA—19% digestion with deoxyribonuclease I	78 ± 3
plus DNA—19% digestion with deoxyribonuclease I	79 ± 4
plus total DNA	95 ± 4
plus nuclear RNA	89 ± 3
plus nuclear RNA and then ribonuclease H	74 ± 4
(B) Driver nuclear RNA annealed with tracer	
^{3}H-DNA (HAP flow-through)	48 ± 8
^{3}H-DNA (total)	9.8 ± 1
(C) Tracer ^{3}H-nuclear RNA annealed with driver	
DNA (total)	91 ± 6
18-day RBC DNA (staphylococcal nuclease, limit digest)	85 ± 5
18-day RBC DNA digestion with deoxyribonuclease I	16 ± 5
(D) Tracer ^{3}H-labeled globin cDNA annealed with driver	
DNA (total)	93 ± 3
18-day RBC DNA digestion with deoxyribonuclease I	25 ± 3
Fibroblast DNA digestion with deoxyribonuclease I	94 ± 3
24-hour chick blastoderm DNA digestion with deoxyribonuclease I	94 ± 4
18-day RBC monomers (deoxyribonuclease I–5 mM sodium phosphate; 0.5 mM MgCl$_2$)	91 ± 4
18-day RBC monomers (deoxyribonuclease I–3 mM MgCl$_2$; 10 mM NaCl)	25 ± 4
(E) Driver DNA from 18-day RBC monomers treated with	
trypsin and then staphylococcal nuclease and annealed with tracer	
^{3}H-globin-labeled cDNA	25 ± 5
^{3}H-DNA (total)	83 ± 4

*Saturation refers to the plateau in percent hybridization of tracer ^{3}H-labeled probe by driver DNA or RNA. All experiments were taken out to log C_0t or log $Cr_0t = 4.25$. The plateau was defined in all cases by 4 to 10 points.   †Mean ± standard deviation of the mean.

that, after injection of trypsin in to salivary glands, the puffed regions were morphologically more sensitive than the unpuffed regions and were consequently even more accessible to transcription by the endogenous RNA polymerase.

Implication for Gene Regulation

Experiments with staphylococcal nuclease and deoxyribonuclease II have led to the conclusion that active genes are probably associated with histones (*10, 13, 19*), while our experiments are best understood if the conformation of these histones is different from that of the bulk of the histones—a conformation that renders the associated DNA particularly sensitive to digestion by pancreatic deoxyribonuclease I in the proper ionic environment. While our experiments with the globin cDNA show that the nontranscribed strand of the DNA double helix is digested by deoxyribonuclease I, the experiments with nuclear RNA (Table 2) and those previously described with ribosomal RNA (*14*) show that the transcribed strand of the double helix is also digested by deoxyribonuclease I.

Much of our data involves hybridization with very small DNA fragments 20 to 250 bases in length (the weight average molecular weight is about 150 to 200 bases). Almost certainly, the smallest fragments do not participate in the hybridization reaciton; consequently, our estimate of the effective DNA concentration during renaturation is likely to be slightly high. The hybrids formed between self-annealed fragments from the pancreatic deoxyribonuclease digestion of nuclei have a melting temperature (T_m) (measured by hyperchromicity and by thermal elution from HAP) of $77°C$ in $0.15M$ sodium phosphate buffer at pH 7.0 (data not shown). This is significantly lower than the T_m of $84°C$ that we observe for self-annealed fragments that are much longer (500 base pairs). Although we cannot offer a conclusive explanation for these results, it is likely that the difference in T_m is a consequence of the small size of the DNA fragments from digestion. Nevertheless, the difference in T_m suggests the interesting possibility that active genes may yield a subclass of deoxyribonuclease I–resistant fragments that are protected by histones, but smaller in size than that needed for stable hybrid formation at $65°C$. This is probably not the case since the preferential digestion of the globin genes can still be demonstrated even after the hybridization reaction is performed under less

stringent conditions at $50°C$ (data not shown). Given these reservations, we think our interpretation of the deoxyribonuclease digestions is probabley correct sinct the biological controls are so striking. Thus, in adult erythrocytes the staphylococcal nuclease fragments retain globin sequences, while the DNA fragments from pancreatic deoxyribonuclease digestion, which are about the same size, do not contain hybridizable adult globin DNA. Similarly, the same-sized deoxyribonuclease I fragments from fibroblasts or brain retain the globin and ovalbumin sequences, whereas those from embryonic and adult red cells retain only the ovalbumin sequences and not their respective activated globin sequences.

It is unlikely that the described preferential digestion of active genes is a consequence of the transcription process per se since the adult globin genes are sensitive to the nuclease in mature adult erythrocytes that have stopped synthesizing RNA. This observation also demonstrates that the inactivation of the avian red cell during erythroid development is not necessarily a consequence of an altered chromosome structure imposed at this primary level of DNA folding.

Since our cDNA is complementary to the globin structural gene and since this region of the transcription unit is clearly in an altered conformation (as revealed by deoxyribonuclease I digestion) in red blood cells, but not in brain or fibroblasts, a simple model of gene activation involving only the activation of some promoter sequence at the $5'$ end of the transcription unit can probably be excluded. We therefore propose that gene activation requires the assembly of an altered subunit structure throughout the entire transcription unit. The mechanism by which an altered subunit structure is propagated across the entire length of a transcription unit represents the primary conceptual problem raised by these results [see also (*24*)].

The deoxyribonuclease-resistant DNA protects less of the tracer ^{3}H-labeled total DNA as digestion of nuclei is increased (Table 2A). Similar effects are not observed with DNA fragments obtained after digestion of nuclei with staphylococcal nuclease. Interpretation of these data is complicated by the fact that the DNA fragments become progressively smaller as the digestion proceeds; nevertheless, $C_0 t_{1/2}$ for the reaction of those sequences that do hybridize is not significantly different from control values, and the saturation values increase by no more than 4 to 8 percent when the

reaction mixture is shifted to less stringent conditions ($50°C$) after reaching saturation at $65°C$. These observations suggest the possibility that there may actually be a spectrum of transcribing (or potentially transcribable) chromosomal structures which can be defined by their sensitivity to deoxyribonuclease I.

The findings from these experiments with pancreatic deoxyribonuclease raise the possibility that gene activation during the development of the red cell lineage may involve the sequential assembly of a different type of chromosome structure (one that is deoxyribonuclease I sensitive) about the specific genes to be activated. To study this question, we have isolated a population of precursor erythroid cells from the developing yolk sac of 25-hour chick embryos. This population has been reported to contain more than 80 percent precursor red blood cells (*25*). When nuclei from these cells were digested with deoxyribonuclease I, the globin genes were not preferentially digested (Table 2D), and the kinetics of hybridization were essentially the same as those obtained from total DNA (Fig. 2). Thus, between 25 and 35 hours of development (when hemoglobin first appears in the chick embryo) there appears to be a new type of chromosome structure imposed on the globin genes in cells within the erythroid lineage. Since the chromosome is assembled at the time of DNA replication (*26*), it is not unreasonable that this new type of structure is actually dictated as the globin genes are replicating (*14, 27*). These observations are consistent with previous findings (*17*) that globin RNA sequences are not detectable in embryos at 25 hours of development, but begin to appear at 35 hours in coordination with the appearance of detectable hemoglobin. Thus, even though posttranscriptional controls are important in gene regulation (*21*), a major component of regulation is a transcriptional one mediated through chromosome structure.

Summary

Ten percent digestion of isolated nuclei by pancreatic deoxyribonuclease I preferentially removes globin DNA sequences from nuclei obtained from chick red blood cells but not from nuclei obtained from fibroblasts, from brain, or from a population of red blood cell precursors. Moreover, the nontranscribed ovalbumin sequences in nuclei isolated from red blood cells and fibroblasts are retained after mild deoxyribonuclease I digestion. This suggests that active genes are preferentially digested by deoxyribonuclease I. In contrast, treatment of red cell nuclei with staphylococcal nuclease results in no preferential digestion of active globin genes. When the 11S monomers obtained after staphylococcal nuclease digestion of nuclei are then digested with deoxyribonuclease I, the active globin genes are again preferentially digested. The results indicate that active genes are probably associated with histones in a subunit conformation in which the associated DNA is particularly sensitive to digestion by deoxyribonuclease I.

References and Notes

1. J. D. Watson and F. H. C. Crick, *Nature (London)* 171, 964 (1953).
2. M. F. Lyon, *ibid.* **190**, 372 (1961); E. B. Lewis, *Adv. Genet.* **3**, 73 (1950); W. K. Baker, *ibid.* **14**, 133 (1968); H. D. Berendes, *Int. Rev. Cytol.* **35**, 61 (1973); M. Ashburner, *Cold Spring Harbor Symp. Quant. Biol.* **37**, 655 (1973); J. H . Frenster, *Nature (London)* **206**, 680 (1965); B. J. McCarthy, J. T. Nishiura, D. Doenecke, D. S. Nasser, C. B. Johnson, *Cold Spring Harbor Symp. Quant. Biol.* **38**, 763 (1973); R. T. Simpson, *Proc. Natl. Acad. Sci U.S.A.* **71**, 2740 (1974); K. Marushige and J. Bonner, *ibid.* **68**, 2941 (1971).
3. J. Gottesfeld, R. F. Murphy, J. Bonner, *Proc. Natl. Acad. Sci. U.S.A.* **72**, 4404 (1975).
4. A. L. Olins and D. E. Olins, *Science* **183**, 330 (1974); C. L. F. Woodcock, *J. Cell. Biol.* **59**, 3689 (1973); J. P. Baldwin, P. G. Boseley, M. Bradbury, K. Ibel, *Nature (London)* **253**, 245 (1975); R. D. Kornberg and J. O. Thomas, *Science* **184**, 865 (1974); C. G. Sahasrabuddhe and K. E. Van Holde, *J. Biol Chem.* **249**, 152 (1974); J. A. D'Anna and I. Isenberg, *Biochemistry* **13**, 2098 (1974); S. C. R. Elgin and H. Weintraub, *Annu. Rev. Biochem.* **44**, 725 (1975); D. Hewisch and L. Burgoyne, *Biochem. Biophys. Res. Commun.* **52**, 504 (1973); M. Noll, *Nature (London)* **251**, 249 (1974); R. Axel, W. Melchior, B. Sollner-Webb, G. Felsenfeld, *Proc. Natl. Acad. Sci. U.S.A.* **71**, 4101 (1974); B. M. Honda, D. L. Baillie, E. P. M. Candido, *FEBS Lett.* **48**, 156 (1974); D. R. Oosterhof, J. C. Hozier, R. L. Rill, *Proc. Natl. Acad. Sci. U.S.A.* **72**, 633 (1975); P. Oudet, M. Gross-Bellard, P. Chambon, *Cell* **4**, 281 (1975); H. J. Li, *Nucleic Acid Res.* **2**, 1275 (1975); V. V. Bakayev, A. A. Melnickov, V. D. Osicka, A. J. Varshavsky, *ibid.*, p. 1401; H. G. Martinson and B. J. McCarthy, *Biochemistry* **14**, 1073 (1975); J. D. Griffith, *Science* **187**, 1202 (1975).
5. J. L. Germond, B. Hirt, P. Oudet, M. Gross-Bellard, P. Chambon, *Proc. Natl. Acad. Sci. U.S.A.* **72**, 1843 (1975); R. Clark and G. Felsenfeld, *Nature (London)* **229**, 101 (1971); J. P. Langmore and J. C. Wooley, *Proc. Natl. Acad. Sci U.S.A.* **72**, 2691 (1975).
6. H. Weintraub, K. Palter, F. Van Lente, *Cell* **6**, 85 (1975).
7. H. Weintraub and F. Van Lente, *Proc. Natl. Acad. Sci. U.S.A.* **71**, 4249 (1974).
8. F. H. C. Crick and A. Klug, *Nature (London)* **255**, 530 (1975).
9. In the strictest sense all nu bodies cannot be homogeneous since the histones themselves are not homogeneous. Thus, histones are extensively modified [A. Ruiz-Carrillo, L. Wangh, V. Allfrey, *Science* **190**, 117 (1975)] and are also genetically polymorphic (L. H. Cohen, K. M. Newrock, A. Zweidler, *ibid.*, p. 994).
10. Recent papers by Gottesfeld *et al.* (*3*) and Lacy and Axel *et al.* (*19*) as well as older experiments of Axel *et al.* (*13*) make it very likely that histones are associated with actively transcribed regions of DNA.
11. R. Axel, H. Cedar, G. Felsenfeld, *Proc. Natl. Acad. Sci. U.S.A.* **70**, 2029 (1973); R. S. Gilmour and J. Paul, *ibid.*, p. 3440; A. W. Steggles, G. N. Wilson, J. A. Kantor, *ibid.* **71**, 1219 (1974); T. Barrett, P. Maryanka, P. Hamlyn, H. Gould, *ibid.*, p.5057.
12. R.F. Itzhaki, *Biochem. J.* **125**, 221 (1971); A. E. Mirsky, *Proc. Natl. Acad. Sci. U.S.A.* **68**, 2945 (1971); M. Noll, *Nucleic Acid Res.* **1**, 1573 (1974); D. Oliver and R. Chalkley, *Biochemistry* **13**, 5093 (1974); T. Pederson, *Proc. Natl. Acad. Sci. U.S.A.* **69**, 2224 (1972).
13. R. Axel, H . Cedar, G. Felsenfeld, *Cold Spring Harbor Symp. Quant. Biol.* **38**, 773 (1973).
14. H. Weintraub, in *Results and Problems in Cell Differentiation*, J. Reinert and H. Holtzer, Eds. (Springer-Verlag, Berlin, 1975), vol. 7, p. 27.
15. R. J. Billing and J. Bonner, *Biochem. Biophys. Acta* **281**, 453 (1972).
16. C. Berkowitz and P. Doty, *Proc Natl. Acad. Sci. U.S.A.* **72**, 3328 (1975).
17. M. Groudine, H. Holtzer, K. Scherrer, A. Therwath, *Cell* **3**, 243 (1974); M. Groudine and H. Weintraub, *Proc. Natl. Acad. Sci. U.S.A.* **72**, 4464 (1975).
18. J. Brown and V. Ingram, *J. Biol. Chem.* **249**, 3960 (1974).
19. E. Lacy and R. Axel, *Proc. Natl. Acad. Sci. U.S.A.* **72**, 3978 (1975).
20. H. Hanafusa, W. S. Hayward, J. H. Chen, T. Hanafusa, *Cold Spring Harbor Symp. Quant. Biol.* **39**, 1139 (1974).
21. K. Scherrer and L. Marcaud, *J. Cell. Physiol.* **72** (Suppl. 1), 181 (1968); L. Grouse, M. D. Chilton, B. J. McCarthy, *Biochemistry* **11**, 798 (1972); I. R. Brown and R. B. Church, *Dev. Biol.* **29**, 73 (1972); E. H. Davidson and R. J. Britten, *Q. Rev. Biol.* **48**, 565 (1973); M. J. Getz, G. D. Birnie, B. D. Young, E. MacPhail, J. Paul, *Cell* **4**, 121 (1975); B. R. Hough, M. J. Smith, R. J. Britten, E. H. Davidson, *ibid*, **5**, 291 (1975).
22. After treatment with trypsin, the repeating pattern of monomer, dimer, trimer (and so on), DNA fragments generated by partial digestion of nuclei with staphylococcal nuclease is largely preserved. In addition, more than 90 percent of the 11S monomers, after treatment with trypsin, retain a trypsin-resistant core composed of interacting histone COOH-terminal cleavage fragments (H. Weintraub, in preparation).
23. M. Roberts and H. Kroeger, *Experientia* **20**, 326 (1957).
24. K. Yammamoto and B. Alberts, *Ann. Rev. Biochem.*, in press.
25. M. Wenk, *Anat. Rec.* **169**, 453 (1971).
26. H. Weintraub, *Cold Spring Harbor Symp. Quant. Biol.* **38**, 247 (1973); R. L. Searle and R. T. Simpson, *J. Mol. Biol.* **94**, 479 (1975).
27. H. Holtzer, H. Weintraub, R. Mayne, B. Mochan, *Curr. Top. Dev. Biol.* **9**, 299 (1973).
28. R. J. Britten and D. Kohne, *Science* **161**, 529 (1968).
29. We thank N. Powe and R. Blumental for technical assistance, the National Science Foundation and American Cancer Society for support, R. Axel for the ovalbumin cDNA, V. Vogt for the S1 nuclease, and A. J. Levine for critically reading the manuscript. M. G. thanks the Medical Scientific Training Program of the University of Pennsylvania and the National Institutes of Health.

Stalder et al. (1980) Cell **20**, 451—460

Tissue-specific DNA Cleavages in the Globin Chromatin Domain Introduced by DNAase I

Jurg Stalder*†, Alf Larsen†, James D. Engel‡,
Maureen Dolan‡, Mark Grodine† and
Harold Weintraub†

† Genetics Division
Hutchinson Cancer Center
1124 Columbia Street
Seattle, Washington 98104
‡ Departments of Biochemistry
and Molecular Biology
Northwestern University
Evanston, Illinois 60201

Summary

Using recombinant chicken DNA clones as probes we have investigated the DNAase I sensitivity of chromosomal DNA regions bordering the α– and β–globin structural genes. By both a blot hybridization assay and solution hybridization we find that regions around these globin genes are preferentially sensitive (relative to the ovalbumin gene) to DNAase I after mild digestion of isolated red cell nuclei. These regions are resistant in cells that do not express globin. The preferential DNAase I sensitivity extends to at least 8 kb on the 3′ side of the β–globin gene cluster and to 6 or 7 kb on the 5′ side, where relatively resistant DNA fragments have been identified. Using low levels of DNAase I to titrate the sensitivities of coding and adjacent non-coding regions, it was observed that coding regions are organized into a very sensitive structure while adjacent non-coding regions are organized into a moderately sensitive structure. The blot hybridization assay has also revealed that DNAase I introduces specific, double-stranded cuts into both the α– and β–globin gene clusters. Many of these cuts are tissue-specific. Several α gene-specific sites occur toward the 3′ side of the α–coding sequences. The β sites are different in embryonic and adult red cells. In embryonic cells the cut occurs near the 5′ end of an embryonic β gene, while In adult cells there are two cuts, one at approximately 2 kb and the other at approximately 6 kb from the 5′ side of an adult gene. Based on the observation that the general region around the origin for replication and promotors for transcription in the SV40 minichromosome is also very sensitive to specific, double-stranded scissions by DNAase I, we speculate that the specific cuts in the globin domain may be structural modifications of the chromatin that are associated with origins for DNA replication or promotors for transcription.

*Present address: Universität Bern. Institut für Hygience und Medizinische Mikrobiologie, 3000 Bern, den Friedbuhlstrasse 51, Switzerland.

Introduction

During development of the chick embryo, Hb and Hb mRNA are first detected at approximately 35 hr of incubation in cells from the primitive (or embryonic) lineage of red cells. Steady-state globin mRNA is not detectable in precursor hematocytoblasts (Groudine et al., 1974) present in the area vasculosa before 20–25 hr. As assayed by solution hybridization, the globin chromatin is resistant to DNAase I before 25 hr, but becomes sensitive to DNAase I after overt erythroid differentiation at 35 hr (Weintraub and Groudine, 1976).

The primitive erythroblasts present in the embryonic circulation between 2 and 5 days produce only embryonic β–globin chains, yet both embryonic and adult β–globin genes are sensitive to DNAase I (Stalder et al., 1980) as assayed by blot hybridization. The sensitivity of the adult β–globin gene in embryonic red cells has been interpreted to reflect a "preactivation" chromosome structure associated with this nonexpressed adult globin gene. At day 6 of development a new red cell line, the definitive line, appears in the embryonic circulation. These cells are morphologically very distinct from primitive cells and they produce adult β–globin chains. The Hb "switching" observed in the chick erythropoietic system at this time is reflected in an apparent conversion of the embryonic β–globin gene from a very DNAase I-sensitive state in primitive cells to a more resistant state in definitive cells. We presume that these events are taking place at the level of a common precursor cell (see Discussion).

The sensitivity of an actively transcribed gene to DNAase I reflects among other characteristics its association with HMG 14 and 17 (Weisbrod and Weintraub, 1978; Weisbrod, Groudine and Weintraub, 1980). When these proteins are gently eluted from individual nucleosomes, the active nucleosomes lose their sensitivity to DNAase I. Upon reconstitution with pure HMG 14 or 17 these nucleosomes regain their sensitivity to DNAase I. Nucleosomes associated with nonexpressed genes do not have a high affinity binding site for these proteins so the interaction of HMG 14 and 17 with active nucleosomes is very specific.

Here we show that the actively transcribed globin chromatin is very sensitive to DNAase I; however, adjacent noncoding regions are also structurally distinct in that they display a moderate level of sensitivity; thus these regions are cut more readily by DNAase I than, for example, the ovalbumin gene. We define the contiguous chromosomal structure exhibiting high and moderate DNAase sensitivity as a "chromosomal domain". We have also found that DNAase I introduces specific double-strand breaks into both the α–globin domain and the β–globin domain. Many of these cuts are cell type-specific. The specific β–related cuts were found to be different in the primitive and definitive red cell lines.

Results

Preferential DNAase Sensitivity of Chromosomal Regions surrounding Globin Genes

Figure 1 shows restriction maps for four recombinant DNA clones containing various chicken globin genes: λCβG1 (or β1) carries both an embryonic β gene (e) and the adult β–globin gene (a) (Dodgson, Stommer and Engel, 1979); λCβG2 (or β2) carries both a second embryonic β gene (e) and an adult β-like gene, probably the hatching β gene; λCβG3 (or β3) is the 5′ extension of λCβG2 and carries part of the second embryonic β gene (M. Dolan, J. B. Dodgson and J. D. Engel, manuscript in preparation); finally, λCαG2 (or α2) carries the two major α gene loci coding for the αa and αd polypeptide chains. Both of these α chains are expressed in adult and embryonic red cells (Brown and Ingram, 1974).

To examine the DNAase I sensitivity of regions surrounding the various globin genes, nuclei from mature definitive red cells (containing adult and hatching Hb and obtained from 14 day old chick embryos) were digested with various levels of DNAase I, and the DNA was purified and then restricted with Bam HI. The restricted DNA was separated on 1% agarose gels, blotted to nitrocellulose filters (Southern, 1975) and hybridized to the various nick-translated λ–globin clones or to ovalbumin cDNA.

Figure 2a shows a very low level digest from red cell nuclei treated with 1.5 μg/ml of DNAase. Before

restriction, this DNA had an average size of approximately 15 kb. As predicted from the restriction map (Figure 1), DNA which has not been DNAase-treated yields five major bands at roughly molar intensities when digested with Bam HI and then hybridized to λCαG2. A nonstoichiometric band at approximately 7.2 kb probably represents a junction fragment. The second junction fragment (at 0.8 kb) is very faint, probably because it contains a very low percentage of the probe.

Surprisingly, all the fragments hybridizing to the α clone are preferentially digested by DNAase I. As controls, there is no preferential digestion of the ovalbumin gene (6.5 kb) in the same DNA sample. Several tubulin-related bands and the RAV-O ev-1 locus (Astrin, 1978) are also not digested. In addition, comparable levels of digestion of MSB nuclei (a lymphoid line of chicken cells that does not synthesize globin RNA) or brain nuclei (not shown) result in no preferential digestion of the α–globin-related DNA, nor of ovalbumin.

Figure 2b shows that a similar level of sensitivity is displayed by the Bam fragments that hybridize with β1 and β2 probes using the same DNA preparation. As in the case of the α genes, all the fragments hybridizing to these β clones are preferentially digested by DNAase I, whereas the ovalbumin structural gene displays no preferential sensitivity. The same results are obtained with a variety of restriction enzymes: Hind III, Eco RI, MSP and Kpn. We conclude that for α2, β1 and β2, both coding regions and adjacent regions are preferentially sensitive to DNAase I, using the blot hybridization assay.

A Region of Relative Resistance at the 5′ Side of the β–Globin Domain

With β3 a slightly different story emerges (Figure 2b). The DNA hybridizing to the 8.5 and 4 kb fragments is digested rapidly; however, the DNA hybridizing to the 5 kb and 1.5 kb fragments is relatively resistant. These fragments are located approximately 7 kb from the 5′ end of the embryonic β–globin gene cluster (Figure 1).

One technical problem we have encountered using genomic clones as hybridization probes (particularly β2 and β3 and to a lesser extent α2) is that there appears to be a rather high "apparent" background within the lanes. This "apparent" background is very resistant to low salt and high temperature washes and on many occasions resolves into rather discrete bands (see, for example, Figure 3, below). We presume that the globin domains also contain many repetitive sequences which are scattered throughout the genome (Shen and Maniatis, 1980), and that this is the reason for much of the apparent background. Additional evidence for this interpretation comes from the fact that most of the noncoding regions from β2 and β3 clones hybridize to a great number (50–300 per genome) of

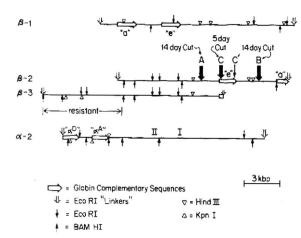

λ CHARON 4A / GLOBIN RECOMBINANTS

Figure 1. Restriction Maps for Four λCharon 4a Recombinant Chicken Clones Carrying Globin Genes

"a" and "e" denote adult and embryonic β–globin genes extending over the distance indicated by the thick arrows. The direction of transcription (5′→3′) was determined by Dodgson et al. (1978) and M. Dolan, J. B. Dodgson and J. D. Engel (unpublished results). The thick vertical arrows point to sites of preferred cutting by DNAase I as referred to in the text.

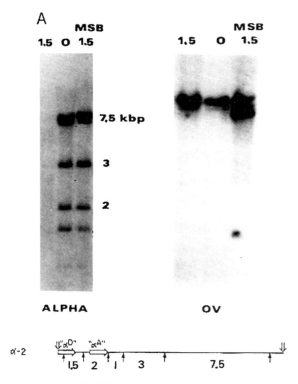

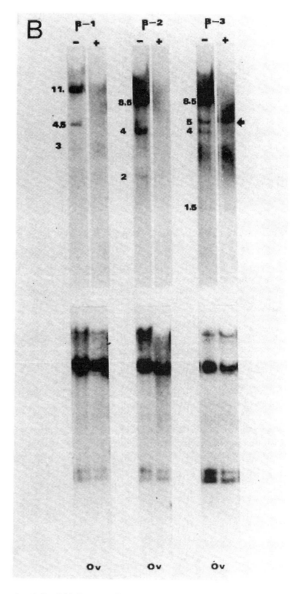

Figure 2. Preferential DNAase I Sensitivity of Noncoding Regions Adjacent to Coding Globin Genes

(A) 14 day red cell or MSB nuclei were digested with 1.5 μg/ml DNAase I to a size of 10–15 kb. The purified DNA from these digests was restricted with Bam HI and together with control chicken DNA (○) electrophoresed in duplicate on a 1% agarose gel and blotted onto nitrocellulose (Southern, 1975). One set of digests was hybridized to λCαG2 (left) and the other set to ovalbumin cDNA (right). Only fragments at 2 and 1.5 kb contain α–globin coding sequences. The additional MSB ovalbumin band probably represents an ovalbumin polymorphism present in these cells. [As discussed previously (Stalder et al., 1980) several restriction enzymes yield ovalbumin gene fragments that are different from those reported by other laboratories. We assume that this represents a populational polymorphism].

(B) The same red cell samples as shown in (A) were hybridized to λCβG1, λCβG2, λCβG3 or ovalbumin cDNA (Ov.). (−) and (+) indicate without and with (1.5 μg/ml) DNAase I respectively. The 4 kb fragment in β2 the 3 kb fragment in β1 and the 5, 4 and 1.6 kb fragments in β3 are all noncoding. The arrow in the β3 blot points to a relatively DNAase-resistant fragment located at the 5′ end of the β domain (see Figures 1 and 3).

recombinant λ clones when the chicken library is probed with these sequences (H. Weintraub, unpublished observations).

To reduce the background so that the kinetics of DNAase I digestion could be examined and the relative resistance of the 5′ region of β3 could be established, blots containing DNA digested to rather high levels were hybridized to labeled β3 in the presence of an excess of a cold, competing DNA fragment from β2 (we used the 8.5 kb Bam fragment in β2). Under these hybridization conditions, we detect only the 1.6, 5 and 4 kb Bam fragments present in β3 toward the 5′ end of the β–globin gene domain (Figure 3A). Many cross-hybridizing fragments are also observed. Figure 3B shows that there is a very gradual decrease in hybrid-

ization to the 6.5 kb ovalbumin fragment at these high DNAase concentrations. A similar decrease is observed with the 5 kb Bam fragment associated with β3; however, the 4 kb Bam fragment in β3 is significantly more sensitive [separate experiments show it to have an intermediate level (see below) of sensitivity] and the 1.6 kb fragment slightly more resistant. The resistance of the 1.6 kb band might be related in part to its small target size; however, for β2 and α2, correspondingly small Bam fragments are clearly sensitive (Figure 2) to DNAase I. Thus while most of the β gene cluster is sensitive to DNAase I, a region that begins approximately 7 kb from the 5′ end of the embryonic β gene in λCβG3 is relatively resistant to DNAase I digestion. Figure 3C shows the results from

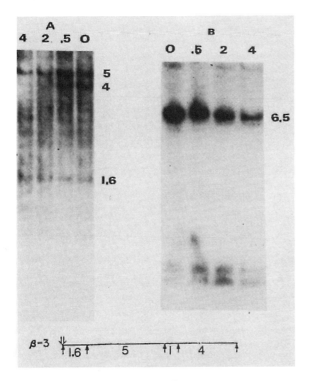

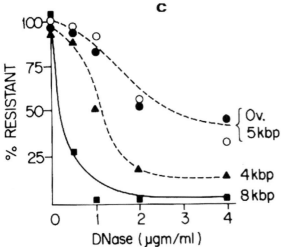

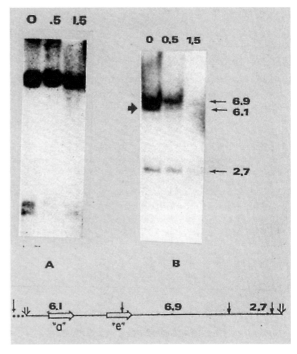

densitometer tracings of related experiments hybridizing ovalbumin and whole β3 to the same samples of DNAase-treated DNA from embryonic cells. While interpretation of the data is complicated by difficulties associated with differential transfer and binding of the DNA (and also by the background due to repetitive sequences), these effects probably do not change the rather qualitative conclusion that the region 7 kb from the 5' end of the β gene cluster is relatively resistant to DNAase I.

A Moderate DNAase I-Sensitive State for Noncoding Adjacent Fragments

Using very low levels of DNAase I, we asked whether the entire domain defined by these probes had the same sensitivity to DNAase I. Figure 4B shows an RI digest (see Figure 1) hybridized to β1. At low levels of digestion, the 6.1 kb fragment coding for the adult globin gene (arrow) is much more sensitive in adult red cell nuclei than are the adjacent fragments (see also Figure 3). At slightly higher levels of digestion, all these fragments become digested faster than the ovalbumin standard (Figure 4A). In analogous experiments, we have also noted that, using α2 and β2 as probes, coding regions are very sensitive to DNAase I, while adjacent noncoding regions are usually of intermediate sensitivity (data not shown). For conven-

Figure 3. Resistance of the 5' Region of β3 to DNAase I

(A) Red cell nuclei from 5 day red cells were digested with increasing doses of DNAase I and the digest was analyzed, as described in Figure 2. The samples were again digested with Bam HI, but in this case hybridization was to λCβG3. An excess (20 μg) of the 8.5 kb Bam fragment from β2 was included in the hybridization to decrease the number of CPM hybridizing to repetitive sequences in the chick genome so that the "effective" hybridization probe is that part of β3 shown at the bottom of the figure.

(B) The same samples as shown in (A) were hybridized to ovalbumin cDNA.

(C) Blots similar to the one shown in (A) and (B), except that whole β3 was used, were scanned with a soft laser densitometer (Zeineh) and the area under the peaks (above the repetitive sequence hybridization) was determined at each DNAase concentration. The data represent the average from two separate blots. In the top curve, the open circles are from ovalbumin. The closed symbols refer to the Bam fragments present in β3.

Figure 4. Moderate Sensitivity of Noncoding Adjacent Sequences in the Globin Domain

(A) The same DNA samples as in (B) were digested with Bam HI and hybridized to ovalbumin cDNA. (B) DNAase I-treated samples were digested with RI and hybridized to β1. The 6.1 kb fragment contains the coding region for both the adult and embryonic genes present in β1 (Ginder, Wood and Felsenfeld, 1979).

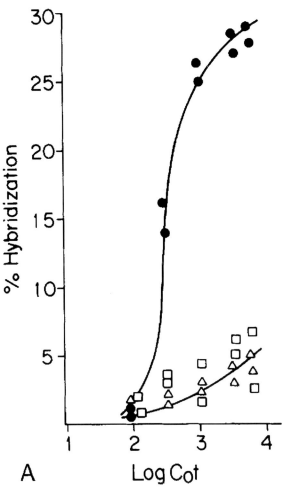

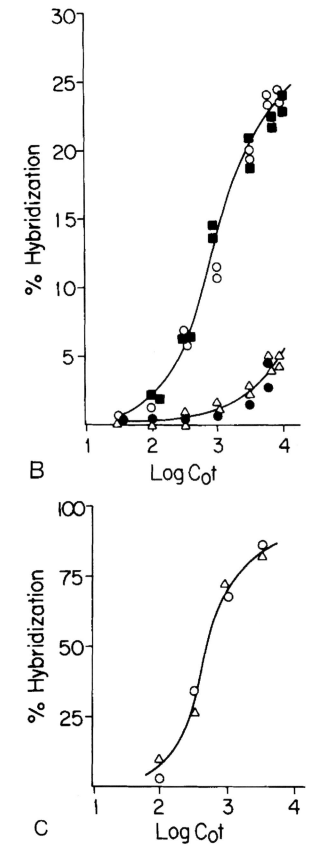

Figure 5. Preferential Sensitivity to DNAase I of Noncoding Regions in the α2 and β1 Domains As Assayed by Solution Hybridization

14 day nuclei were digested to 15% acid solubility with DNAase I or to 40% acid solubility with staph nuclease. The resistant DNA was then used tor solution hybridization.

(A) Hybridization of λCαG2 (4×10^8 cpm/μg, 50,000 CPM input, 12 mg/ml DNA; 20 μl volume). (● —— ●) Hybridization to total DNA; (△——△) hybridization to DNAase I-treated DNA; (□——□) self-hybridization of probe alone. Approximately 33% of λCαG2 Is represented by chicken sequences.
(B) Hybridization of λCβG2 (conditions as in Figure 5A). (○ —— ○) Hybridization to total DNA; (■——■) hybridization to staph nuclease-treated DNA; (△——△) hybridization to DNAase I-digested DNA; (● —— ●) self-hybridization of probe alone.
(C) Hybridization to ovalbumin cDNA (6×10^8 cpm/μg; 20,000 cpm input; DNA at 12 mg/ml; 20 μl volume). (○ —— ○) Hybridization to staph nuclease-digested DNA; (△——△) hybridization to DNAase I-digested DNA. cDNA background was 5–10% and was subtracted from each point as determined from a parallel hybridization.

ience, we refer to the adjacent regions of moderate and high sensitivity as a "chromosomal domain." For the β genes, we have identified the apparent junction of this domain at the 5′ side, but not at the 3′ side (Figure 3). Although we have not yet found recombinant clones to "hook-up" λCβG1 with λCβG2 and λCβG3, evidence from cellular DNA blots indicates that only 1.8 kb separates β1 from β2 and that the direction of transcription of each β gene is in the same orientation; thus it is probable that all the β-related genes are in the same chromosomal domain, as defined by DNAase sensitivity.

Digestion of Chromatin Domains by DNAase I As Assayed by Solution Hybridization

To determine whether adjacent, noncoding sequences were also sensitive to DNAase I as assayed by solution hybridization, red cell nuclei were digested to 15% acid solubility with DNAase I. The purified DNA was hybridized in solution to tracer amounts of $\lambda C\alpha G2$ or $\lambda C\beta G1$. Figures 5a and 5b show that this preparation fails to hybridize to these probes; however, hybridization to ovalbumin cDNA was normal (Figure 5c). Since coding regions comprise only 10–20% of the hybridizing component of these probes, we conclude that at these levels of digestion, the entire chromosomal domain defined by these probes is digested to pieces too small to hybridize when the nuclei are digested to 15% acid solubility. Similar results have also been obtained for $\lambda C\beta G2$. We do not yet know whether lower levels of digestion also discriminate between coding and noncoding sequences in the globin chromatin domains, as we have observed by blot hybridization.

Specific Cutting Sites for DNAase I

At very low levels of DNAase I, specific, discrete subbands appear when the Bam-digested DNA is hybridized to $\alpha 2$ (Figure 6). The same sub-fragments are

observed in adult and embryonic red cells, both of which produce αA and αD globin chains. These types of discrete bands are not observed when the DNAase I-treated DNA is not restricted. (We do, however, observe very broad $\alpha 2$-related bands without restriction.) Thus we presume that one end of these bands comes from a Bam cleavage, while the other arises from a specific DNAase I cut. Low-level digests of pure DNA with DNAase I fail to generate these bands after Bam restriction. Similarly, digestion of MSB or brain nuclei with DNAase I also does not lead to subbands. While we have not yet mapped the exact location of the α-related cleavages, it is clear that they are from regions of the α domain that do not code for α–globin and, by virtue of their size, must come from the 7.5 or 3 kb Bam fragments, both of which are 3' to the structural genes. More recently, we have also identified a specific cleavage that occurs right at the 5' side of αD. Treatment of nuclei with increasing concentrations of DNAase I (Figure 6) shows that these sub-fragments appear very early during the course of digestion. They also tend to persist throughout the digestion, and they comprise a very significant fraction of the total radioactivity hybridizing to the 7.5 kb fragment, presumably the fragment from which the larger bands are derived.

Specific DNAase I Cutting Sites in the β–Globin Domain

Figure 7A shows a control and a low level DNAase I digest of adult (14 day) and embryonic red cell nuclei after restriction with Bam and hybridization to pHb1001, an adult β–globin cDNA clone (a gift from W. Salser). The clone detects two adult β–globin

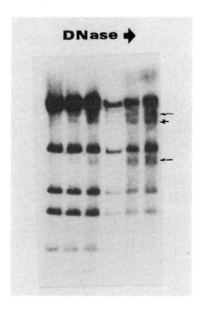

Figure 6. Generation of Specific Subfragments after Low-Level Digestion of Adult (14 Day) Nuclei with DNAase I

14 day red cell nuclei were digested with increasing concentrations of DNAase I (left to right: 0; 0.2; 0.4; 0.6; 0.8; 1.0 μg/ml DNAase I) and the digested DNA was purified and restricted with Bam HI and analyzed as described in the legend to Figure 2. Hybridization was to $\lambda C\alpha G2$.

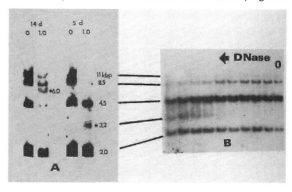

Figure 7. Generation of Specific Subfragments after Low-Level Digestion of Adult (14 Day) and Embryonic (5 Day) Red Cell Nuclei with DNAase I (1 μg/ml)

(A) The experimental details were as described in the legend to Figure 6, except that hybridization was to pHb1001. The 11 and 8.5 kb Bam genes are embryonic; the 4.5 and 2.0 kb genes are adult. Most of these genes are detected by cross-hybridization.

(B) 5 day nuclei were digested with increasing concentrations of DNAase I and hybridized to pHb1001 after Bam digestion. In this experiment, hybridization was in 50% formamide at 45°C. Under these conditions pHb1001 hybridizes most intensely with the 4.5 kb Bam gene rather than the 2 kb gene, as observed in (A).

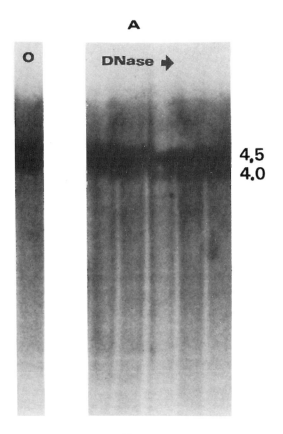

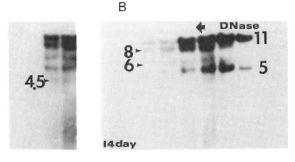

Figure 8. Mapping the Specific Cuts in the Adult β Domains

(A) Adult (14 day) nuclei digested wlth DNAase I (0; 0.2; 0.4; 0.6; 0.8; 1.0 μg/ml). All samples were digested with Hind III and hybridized to pB2H2, a subclone of λCβG2 (in pBR322) that contains the embryonic gene flanked by the 4.5 kb Hind III sites (see Figure 1).
(B) Control and DNAase-treated DNA from 14 day nuclei were treated with Eco Rl and, after blotting, hybridized to λCβG2. The two lanes on the left are longer exposures of the corresponding lanes on the right.

genes at 2 and 4.5 kb, and two embryonic β genes at 8.5 and 11 kb. Low level DNAase I digestion generates a 6 kb sub-band in adult nuclei and a 3.2 kb sub-band in embryonic nuclei. Thus these β sub-bands are not only tissue-specific (we do not see them in brain or MSB cells), but they are different in adult and embryonic red cell lines. They are also not seen when free DNA is digested with low levels of DNAase I. Because they are detected with a cDNA clone, these sub-bands

must include β–globin coding regions. Figure 7B is a dose response to DNAase that shows that in embryonic red cells, the 3.2 kb band appears very early during digestion, and also that no detectable 6 kb sub-band is observed in embryonic cells. Similarly, the 3.2 kb sub-band is also not detected in adult nuclei over a wide range of DNAase concentrations.

To map the sites of DNAase I cutting, blots were hybridized with λCβG1, λCβG2 and λCβG3. λCβG1 gave no detectable sub-bands; λCβG3 gave weak sub-bands; λCβG2 gave very prominent sub-bands (data not shown). The failure to see sub-bands from λCβG1 (which contains the 11 and 4.5 kb Bam fragments) indicates that neither the 6 kb adult sub-bands nor the 3.2 kb embryonic sub-band arise from the 11 or 4.5 kb β genes. This means that they must arise from regions of the genome defined by λCβG2 and λCβG3, presumably from the region around the 8.5 kb Bam embryonic β–globin gene since, by elimination, this gene is the only one remaining that is large enough to give rise to discrete fragments of 6 and 3.2 kb detectable by hybridization to pHb1001.

In 14 day red cells, Bam digested DNA yields a subfragment of 6 kb, presumably derived from the 8.5 kb gene in λCβG2. Thus the adult cut is 2.3 kb from either or both Bam sites (Figure 1). Digestion of the DNA with Hind III and hybridization with labelled pB2H2 [a subclone of λCβG2 in pBR322 that contains the embryonic β-coding region flanked by the two Hind III (Figure 1) sites 4.5 kb apart] yields the parent fragment at 4.5 kb and a subfragment at 4 kb (Figure 8A). The results with Hind III, together with the Bam digestion data, clearly establish one cut in adult nuclei at the 5′ side of the embryonic gene in β2 (see arrow A in Figure 1). Digestion with RI and hybridization wilh λCβG2, however, yields parent molecules of 11 and 5 kb and subfragments of 8 and 2.8 kb (the 2.8 kb fragment is barely visible at this exposure). These fragment sizes are consistent with the previous identification from the Bam and Hind III digests. Unexpectedly, subfragments of 6 and 4.5 kb (Figure 8B) are also generated. This second pair of fragments would arise if there were actually a second site for DNAase I-specific cleavage (see arrow B in Figure 1) approximately 2.3 kb from the other Bam site in λCβG2. Thus we believe that in 14 day red cells there are actually two sites for preferential DNAase cutting, one approximately 2 kb from the 5′ end of the adult gene and the second approximately 6 kb from the 5′ end of this same gene in λCβG2.

Figure 1 also shows our estimation of the approximate position for the specific DNAase I cut observed in 5 day embryonic red cell nuclei (arrow C). The subfragment obtained after Bam digestion and hybridization to pHb1001 is approximately 3.2 kb. Since we do not see a 3.2 kb subfragment after digestion with DNAase I alone, it follows that one end of the subfragment is generated by DNAase I and the second end by Bam. This places the DNAase I-sensitive site 3.2 kb from either the left or right Bam sites (arrow C or C'). The two possible cutting sites must therefore be just within either the 3' or 5' side of the coding region for the gene, since the subfragment is detected with the cDNA clone, pHb1001. To determine whether the specific cut occurred at arrow C or C' in Figure 1, DNAase-treated chromosomal DNA was digested with Hind III and the blot was hybridized to pB2H2 (Figure 9). A cut at C' would predict the generation of two subfragments of approximately 3.5 and 1 kb from the parent 4.5 kb fragment; a cut at C would predict the generation of two co-migrating fragments, each about 2.2 kb. When this experiment was performed, a single broad subfragment was obtained (which occasionally resolves into several discrete fragments) at about 2.2 kb, indicating that the specific cut in 5 day primitive erythroblast nuclei occurred near the 5' side (arrow C and not arrow C') of the embryonic β–globin gene in λCβG2. This site was further confirmed in separate experiments, where Bam-digested DNA hybridized to pB2H2 yielded two subfragments, one at approximately 3.2 kb and a second at approximately 4.8 kb. More recently, we have also found that there is a second embryo-specific DNAase I cut at the 5' side of the embryonic gene present in β1. As with the embryonic gene in β2, this cleavage is observed only in embryonic red cells and not in adult red cells.

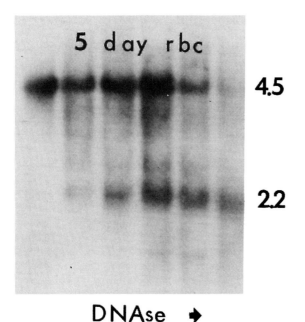

Figure 9. Mapping the Specific DNAase Cleavage in the Embryonic Red Cell β Domain

5 day red cell nuclei were digested with DNAase I (0; 0.1; 0.2; 0.4; 0.5; 0.8 μg/ml). All samples were restricted with Hind III and hybridized to pB2H2 (see Figure 8A). Unlabeled bands probably represent hybridization to repetitive DNA sequences.

Discussion

Structural Domains for Active Genes

Using the blot hybridization assay, we have shown that the chromosome region directly associated with actively transcribed DNA sequences is extremely sensitive to DNAase I. In addition, chromatin regions on either side of these sequences are also preferentially digested; however, these regions display a moderate degree of sensitivity. This type of structure (a chromosomal domain) extends to at least 8 kb on the 3' side of the β and ex domains and to approximately, 7 kb to the 5' side of the β–globin gene domain. In the case of the β chromosomal domain the junction with resistant chromatin can be observed, since several restriction fragments beyond this point are relatively insensitive to DNAase I under our conditions.

It has been shown (Flint and Weintraub, 1977; Friova et al., 1978) that for hamster cells transformed by adenovirus, only those integrated adenovirus DNA sequences that were transcribed were sensitive to DNAase I (after 15% digestion) as assayed by solution hybridization; adjacent adenovirus sequences that were not transcribed were not preferentially digested using this assay. More recently, we have confirmed this observation by blot hybridization; however, because we do not yet have good probes for inactive rat genes, it is difficult to say with certainty whether the nontranscribed adeno sequences fall into the insensitive or moderately sensitive class and, consequently, the exact nature of the integrated adenovirus domain is not certain.

The structural basis for the globin domains is not known. While it is clear that HMG 14 and 17 are responsible, in part, for establishing the DNAase I sensitivity of a large percentage of the very sensitive coding regions (Weisbrod et al., 1980), preliminary results suggest that HMG 14 and 17 are not responsible for the moderately sensitive state, and that treatments that destroy higher-order nuclear structure also eliminate the intermediate level of sensitivity. It is also possible that the adjacent regions actually code for some as yet unidentified red cell specific RNA; however, preliminary evidence suggests that this is not the case, at least for the great majority of the globin domain.

The results presented here can be considered in terms of one very popular model for chromosome organization (Figure 10). The model is based on the organization of lampbrush loops and, to some extent, polytene chromosomes. Most recently, it has its biochemical basis in the experiments first done by Benyajati and Worcel (1976) and Laemmli et al. (1977). [See also Cook and Brazell (1975) and Igo-Kemenes and Zachau (1977)]. When the loop is relaxed, a moderately sensitive DNAase I domain results. This may be the state of the globin domain, but not of the ovalbumin domain in the red cell stem cell (hemato-

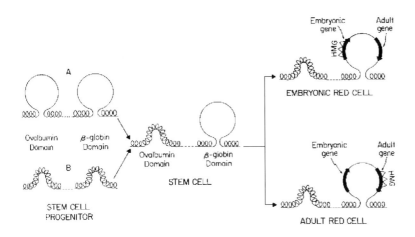

Figure 10 Interpretation of Moderate Sensitivity to DNAase I in Terms of "Lampbrush" Chromosome-Like Loops or Domains

Chromosome domains are either condensed or relaxed. Transcription units become activated in relaxed domains by virtue of their association with HMG 14 and 17 (open triangles). The red cell stem cells (hematocytoblasts) are unique in that they are predicted to be defined by a relaxed globin chromatin domain not yet fully activated by HMGs. The hematocytoblast progenitors can have either relaxed (A) or constrained (B) domains; however, whichever proves to be the case, the state of the domains is seen to be the same for ovalbumin and globin. Finally, we should point out that the association of HMGs with the adult gene in embryonic red cells is not indicated, even though this gene has a structure displaying a marked sensitivity to DNAase I. Until proven otherwise, we postulate that DNAase I sensitivity per se is not necessarily always indicative of HMG association; consequently, in order to prove an association with HMGs, the appropriate reconstitution experiments must be performed (Weisbrod et al, 1980).

cytoblast). With terminal differentiation to an embryonic red cell, HMG 14 and 17 (and possibly other changes as well; see McGhee and Ginder, 1979) become associated with the newly activated embryonic gene. As the embryo ages, the hematocytoblast (or its descendants) switch to adult red cell production and the adult β gene is now assembled with HMG 14 and 17. Besides focusing one level of control to the base of the loops, the figure also illustrates two extreme states for the progenitor to the hematocytoblast. Ovalbumin and globin domains are both seen to be "open" initially and then differentially "repressed," or repressed initially and then differentially activated. Using very young chick blastoderms, we are currently trying to distinguish between these two possibilities.

Specific Cutting by DNAase I

Specific cutting by DNAase I was first observed by Wu et al. (1979) in several Drosophila genes. We have shown here that DNAase I introduces specific double-stranded cuts into regions associated with the $\alpha-$ and β-globin domains. We have also observed specific cutting in regions near the 5' side of the active, RAV-O, ev-3 locus (but not the inactive RAV-O, ev-1 locus) and regions at the 5' side of the integrated adenovirus genome. We suspect that the specific DNAase cleavages will reflect a number of different types of structural modifications, and that eventually roles for each will have to be defined more precisely.

The most significant observation from our work is that many of the specific cleavages introduced by DNAase I are tissue-specific. This is most dramatically seen in the case of the β-globin genes where the DNAase I sites differ, even between adult and embryonic red cells.

At present, we think that the best clue to the function of these modified chromatin regions comes from analogous observations with the SV40 minichromosome. Here the general region around the origin for replica-

tion and the probable promotors for transcription is preferentially digested by nucleases (Scott and Wigmore, 1978; Waldeck et al., 1978; Varshavsky, Sundin and Bohn, 1979). While many of the specific cellular cutting sites seem to be located at the 5' side of transcribing genes, we would guess that some of these sites are not promotor structures, since several are present on the 3' side of the α-globin genes where no transcription has yet been detected. If the same type of structure as seen in SV40 is also responsible for some of the specific cutting sites associated with the globin genes, and if some of these structures also act as cellular origins for DNA replication, then the results with the β-globin domain would indicate that the origin is specific (see also Seidman, Levine and Weintraub, 1979), and actually changes during the switch from primitive to definitive (adult) erythropoiesis. [These specific cleavage sites may also contain potential polymerase III transcription units that have homologies to known origins of DNA replication (Jelinek et al., 1980).]

We previously proposed (Weintraub et al., 1977) a mechanism for generating chromosomal diversity (and hence cellular diversity) during the unfolding of a given lineage. The idea was based on the conservative segregation and assembly of nucleosomes (Leffak, Grainger and Weintraub, 1977), and required that cellular origins change in a precise way during development. We are now trying to test whether the sites for DNAase I cutting in the β-globin domain can act as origins in an appropriate recombinant DNA system.

Experimental Procedures

Cells and Nuclear Digestions

Red blood cells from either 5 day or 14 day White Leghorn chicken embryos were isolated and the nuclei prepared as described (Weintraub and Groudine, 1976). MSB cells were grown as described by Weisbrod and Weintraub (1978).

After the nuclei had been washed several times in reticulocyte standard buffer (RSB) [0.01 M Tris–HCl (pH 7 4), 0.01 M NaCl, 3 mM

MgCl$_2$] without NP40, they were digested in RSB with 0.1 –1.5 μg/ml of pancreatic deoxyribonuclease (DNAase I; Sigma) at a DNA concentration of 1 mg/ml at 37°C for 10 min. The digestion was terminated by adding Na$_3$EDTA (pH 7.2) to a final concentration of 2.0 mM, and the DNA was isolated as described by Stalder et al. (1980).

Blotting and Restriction Enzyme Digestions

Purified DNA was digested with restriction enzymes (Biolabs) according to the manufacturer's recommendations, and the DNA fragments were separated on 1% agarose slab gels. DNA was transferred to nitrocellulose filters (Schleicher & Schuell BA 85) according to the method of Southern (1975) and hybridized to nick-translated DNA probes (spec. act. 2–4 × 10^8 cpm/μg) (Weinstock et al., 1978) in 1 M NaCl, 2 mM EDTA, 0.05% SDS, 0.05% Na-Sarcosyl, 0.1% Na-pyrophosphate, 50 mM Tris–HCl (pH 8.0), 5 × Denhardt's solution (Denhardt, 1966), 50 μg/ml each of sheared salmon sperm DNA and poly(A), and 5 × 10^6 cpm of ^{32}P-labeled probed, for 3–5 days at 65°C, after prehybridizing the filters at 65°C overnight in the same solution without the DNA probe. Filters were washed twice at 65°C for 1 hr intervals in hybridization solution, twice in 0.3 M NaCl, and once in 0.15 M NaCl, and then exposed for 1 day to 1 week at −70°C using an intensifying screen (Dupont)

Solution Hybridization

For solution hybridization, nuclei were routinely digested with DNAase I to 15% acid solubility as described by Weintraub and Groudine (1976). The DNA was then purified. treated with alkali (0.3 M NaOH) at 37°C overnight, neutralized, and ethanol-precipitated. Hybridization was performed as described previously (Weintraub and Groudine. 1976).

Ovalbumin cDNA

Ovalbumin mRNA was a gift from P. Thomas. cDNA was synthesized under conditions similar to those described by Friedman and Rosbash (1977). A 100 μl mixture contained 50 mM Tris–HCl (pH 8.1),10 mM MgCl$_2$, 10 μg of actinomycin D, 5 mM dithiothreitol, 0.6 μg of (dT) $_{12–18}$, 0.5 mM dGTP, 0.5 mM TTP, 5 nM dCTP, 5 nM ^{32}P-dCTP (11.1 × 10^{12} becquerels/mmole), 5 μg of RNA, and 40 μl of avian myeloblastosis virus polymerase. Incubation was for 4 hr at 37°C, and the reaction was stopped by the addition of NaDodSO$_4$ to 0.1%. The reaction mixture was centrifuged through a 1 ml column of packed Sephadex G-50 layered over sterile sea sand. The excluded cDNA was adjusted to 0.3 M NaOH and incubated at 65°C for 0.5 hr. The mixture was then neutralized, and the cDNA was precipitated with 2.5 vol of ethanol overnight at −20°C. Specific activity of the probe was approximately 2 × 10^8 cpm/μg.

References

Astrin, S. (1978). Proc. Nat. Acad. Sci. USA 75, 5941–5945.

Benyajati, C. and Worcel, A. (1976). Cell 9, 393–407.

Brown, J. L. and Ingram, U. M. (1974). J. Biol. Chem. 249, 3960–3972.

Cook, I. and Brazell, P. (1975). J. Cell Sci. 19, 261–279.

Denhardt, D. T. (1966). Biochem. Biophys. Res. Commun. 23, 641–646.

Dodgson, J. B., Stommer, J. and Engel, J. D. (1979). Cell 17, 879–887

Flint, S. J. and Weintraub, H. (1977). Cell 12. 783–794.

Friedman, E. and Rosbash, M. (1977). Nucl. Acids Res. 4, 3455–3471.

Frlova, E. I., Zalmanzon, E. S., Lukanidin, E. M. and Georgiev, G. P. (1978). Nucl. Acids Res. 5, 1–11.

Ginder, G. D., Wood, W. and Felsenfeld, G. (1979). J. Biol. Chem. 254, 8099–8102.

Groudine, M., Holtzer, H., Scherrer, K. and Therwath, A. (1974). Cell 3, 243–247.

Igo-Kemenes, T. and Zachau, H. G. (1977). Cold Spring Harbor Symp. Quant. Biol. 42, 109–118.

Jelinek, W. R., Toomey, R. P., Leinwand, L., Duncan, C. H., Biro, P. A., Choudary, P. V., Weissman, S. M., Rubin, C. M., Houck, C. M., Deininger, P. L. and Schmid, C. W. (1980). Proc. Nat. Acad. Sci., in press.

Laemmli, U. K., Cheng. S. M., Adolph, K. W., Paulson, J. R., Brown, 1 J. and Baumbach, W. R. (1977). Cold Spring Harbor Symp. Quant. Biol. 42, 351–360.

Leffak, I. M., Grainger, R. and Weintraub, H. (1977). Cell 12, 837–845.

McGhee, J. D. and Ginder, G. S. (1979). Nature 280, 419–420.

Scott, W. A. and Wlgmore, D. J. (1978). Cell 15, 1511–1518.

Seidman, M. M., Levine, A. J. and Weintraub, H. (1979). Cell 18, 439–449.

Shen, C.-K. J. and Maniatis, T. (1980). Cell 19, 379–391.

Southern, E. (1975). J. Mol. Biol. 93, 503–517.

Stalder, J., Groudine, M., Dodgson, J. B., Engel, J. D. and Weintraub, H. (1980). Cell 19, 973–980.

Varshavsky, A. J., Sundin, O. and Bohn, M. (1979). Cell 16, 453–466.

Waldeck, W., Fohring, B., Chowdhury, K., Gruss, D. and Sauer, G. (1978). Proc. Nat. Acad. Sci. USA 75, 5964–5968.

Weinstock, R., Sweet, R., Weise, M.. Cedar, H. and Axel, R. (1978). Proc. Nat. Acad. Sci. USA 75, 1299–1303.

Weintraub, H. and Groudine, M. (1976). Science 93, 848–853.

Weintraub, H., Flint, S. J., Leffak, M., Groudine, M. and Grainger, R. (1977). Cold Spring Harbor Symp. Quant. Biol. 43, 401–407.

Weisbrod, S. and Weintraub, H. (1978). Proc. Nat. Acad. Sci. USA 76, 328–332.

Weisbrod, S., Groudine, M. and Weintraub, H. (1980). Cell 19, 289–301.

Wu,-C., Bingham, P. M., Livak, K. J., Holmgren, R. and Elgin, S. C. R. (1979). Cell 16, 797–806.

Note Added in Proof

We have found that while the 3.2 kb β–globin sub-band (Figure 7) is not found in digestive brain or MSB cells, it is detectable in digested DNA from cultured fibroblasts. Other sub-bands have thus far only been observed in erythrocytes and not in brain, MSB cells or fibroblasts. While the biological relevance of the 3.2 kb sub-band in fibroblasts is uncertain at the moment, the observation again reintorces the notion that these sub-bands probably reflect a number of different types of chromosomal modifications involved with several different types of chromosomal functions. After this work was submited, Kuo, Mandel and Chambon (Nucl. Acids Res. 7, 2105–2113) described similar sub-bands from the 3′ side of the conalbumin gene and a similar intermediate level of DNAase 1 sensitivity for flanking regions of the ovalbumin gene.

McGhee & Ginder (1979) Nature (London) **280**, 419–420

Specific DNA methylation sites in the vicinity of the chicken β-globin genes

METHYLATION of cytosine is the only post-synthetic modification so far detected in the DNA of higher eukaryotes and thus has been made the basis of several proposed mechanisms of gene activity and cellular differentiation[1-3]. All evidence indicates, however, that the overall content of 5-methylcytosine in DNA does not vary significantly between different tissues of the same organism and does not change throughout at least some steps of differentiation[4-6]. As 5-methylcytosine occurs predominantly in the dinucleotide sequence CpG (ref. 7) and as a number of bacterial restriction endonucleases can distinguish whether this sequence is methylated or unmethylated, it is now possible to investigate DNA methylation patterns in the vicinity of single genes[8]. In this report, we use this technique to provide a correlation between DNA methylation and the activity of the chicken β-globin genes.

The restriction endonuclease *Hpa*II cleaves the DNA sequence CCGG (ref. 9) but does not cleave the methylated sequence C(MeC)GG (ref. 10). A second restriction endonuclease, *Msp*I, is able to cleave this same DNA sequence whether it is methylated or not[11]. Thus DNA from any cell or tissue type can be digested with either *Hpa*II or *Msp*I, the cleaved DNA electrophoresed on an agarose gel, transferred to a nitrocellulose filter[12] and hybridised to a gene-specific probe. Any bands seen in the autoradiogram which differ between the two enzyme digests can reasonably be ascribed to differences in methylation of CCGG sites in the vicinity of the specific gene. This kind of experiment, used here to examine the chicken β-globin genes, has recently been used to investigate a single methylation site in the large intervening sequence of the adult rabbit β-globin gene[13].

Figure 1 shows the autoradiogram from an experiment with chicken DNA, using the plasmid pHb1001 as a ^{32}P-labelled hybridisation probe (pHb1001 is a cDNA clone of the adult chicken β-globin gene[14], provided by Dr W. Salser). Figure 1*a* represents the *Hpa*II digest of adult chicken erythrocyte DNA and shows bands at 4.2 and 1.3 kilobase pairs. These fragment sizes should correspond to distances between unmethylated CCGG sites in the DNA regions containing β-globin genes. Figure 1*b* represents the *Msp*I digest of the same adult erythrocyte DNA and shows bands at 2.0 and 1.3 kilobases, as well as a taint but reproducible band at 0.4 kilobases. These fragment sizes should correspond to distances between both methylated and unmethylated CCGG sites in the β-globin gene DNA. The 1.3-kilobase fragment is of similar intensity in both digests. However, the *Hpa*II digest (*a*) lacks both the 0.4 and 2.0 kilobase fragments of the *Msp*I digest as well as any other bands which could correspond to a combination of the *Msp*I bands. This suggests that the CCGG sites at the ends of the 1.3-kilobase fragment are completely (>90%) unmethylated in adult chicken erythrocyte DNA and furthermore, that at least two other specific sites are completely methylated. DNA isolated from the circulating reticulocytes of phenylhydrazine-induced anaemic adult chickens gives digestion patterns which are essentially identical to those of adult erythrocyte DNA (data not shown). These results are in contrast to those for the rabbit β-globin gene[13], where in both marrow and spleen cells of anaemic rabbits, only partial methylation of a site in the large intervening sequence was observed; this difference could be due to the different positions of the cleavage sites within the gene sequence or, as the authors have pointed out[13], due to a heterogeneous cell population.

Figure 1*c* shows the *Hpa*II digest of oviduct DNA. The observed band pattern is not only different from the *Msp*I digest of the same DNA (Fig. 1*d*) but also different from the *Hpa*II digest of erythrocyte DNA (Fig. 1*a*). There are now prominent bands at 4.2, 3.0, 1.6 and 1.3 kilobases (as well as several weaker

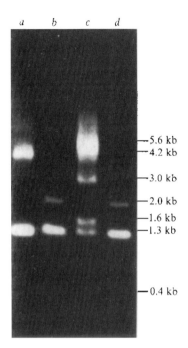

Fig. 1 DNA was isolated from chicken tissues by extensive digestion with proteinase K (Merck) in the presence of 0.2% SDS with or without previous isolation of nuclei. This was followed by at least three extractions with an equal volume of phenol–chloroform–isoamyl alcohol (25:24:1), three extractions with two volumes of chloroform–isoamyl–alcohol (24:1) and three ethanol precipitations. In a typical digestion protocol, 200 units of restriction enzyme were added to 100 μg DNA in a final buffer of 6 mM KCl, 10 mM Tris-HCl (*p*H 7.5), 10 mM MgCl₂, 1 mM dithiothreitol and 100 μg ml⁻¹ autoclaved gelatin. (The enzyme *Msp*I was obtained from New England Biolabs. The enzyme *Hpa*II was obtained from either New England Biolabs, Bethesda Research Laboratories, or Boehringer–Mannheim: all gave identical results. It was verified that, in the digestion conditions used, both *Hpa*II and *Msp*I gave identical cleavage patterns with the plasmid pBR322.) The reaction was incubated at 37°C for 2 h, a further 200 units of enzyme were added and incubation was continued for an additional 2 h at 37°C. Identical results were obtained if this second incubation was continued for 10–15 h The completion of digestion was monitored by adding a small amount of bacteriophage λ DNA to an aliquot of either the first or the second incubation mixture and verifying by electrophoresis that the required cleavage products were present. After digestion, the DNA was repurified by repeating the proteinase K digestion and organic extraction steps described above. and 25–50 μg electrophoresed on each slot of a 1% agarose slab, 6 mm thick, for 2 h at 200 V. DNA was then transferred to Schleicher and Schuell nitrocellulose filters by the method of Southern[12], the filters hybridised to ³²P-labelled nick-translated pHb1001 for 24 h at 70°C in 6 × SSC, washed in 3 × SSC, and finally washed for 30 min in 0.1 × SSC. Filters were exposed to Kodak XR5 X-ray film for 24–96 h at −80°C, using two Dupont Cronex intensifying screns. A typical autoradiogram is shown above and described in more detail in the text. *a*, Adult erythrocyte DNA digested with *Hpa*II; *b*, adult erythrocyte DNA digested with *Msp*I; *c*. oviduct DNA digested with *Hpa*II; *d*, oviduct DNA digested with *Msp*I. Band sizes were calibrated relative to a *Hin*dIII digest of λ and a *Hae*III digest of ΦX174 DNA, run in an adjacent gel slot, and are probably accurate to 5%. Identical *Hpa*II digestion patterns were obtained with DNA from erythrocytes or reticulocytes isolated from several individual adult roosters and chickens, of both White Rock and Leghorn breeds. *Hpa*II digestion of a mixture of erythrocyte and oviduct DNA yielded a pattern which was the expected sum of the patterns of slots *a* and *c* above. A digest of the same DNA with a mixture of *Hpa*II and *Msp*I yielded only the *Msp*I digestion pattern. The *Msp*I digestion pattern of all DNA samples investigated was identical to that in slots *b* and *d* above. Digestion with several restriction endonucleases which do not have CG as part of their recognition site gave the same digestion patterns in different tissues (see also ref. 16).

but poorly resolved bands at higher molecular weight). As Fig. 2*b* and *c* show, all these bands appear as partial digestion products of adult erythrocyte DNA by the enzyme *Msp*I. This suggests that, in the oviduct, CCGG sequences in the vicinity of the β-globin genes are only partially methylated. Nevertheless, the absence of the 2.0-kilobase band suggests that at least one such sequence is completely methylated. As shown in Fig. 2*e*, DNA isolated from the circulating red blood cells of 5-d-old embryos (in which the adult β-globin gene is not being expres-

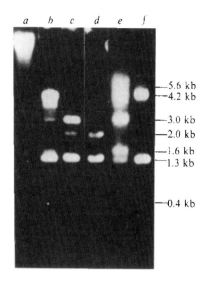

a b c d e f

—5.6 kb
—4.2 kb
—3.0 kb
—2.0 kb
—1.6 kb
—1.3 kb
—0.4 kb

Fig. 2 *a–d*. Partial digestion of adult erythrocyte DNA with the restriction endonuclease *Msp*I; *a*. no enzyme control; *b*, 2 units; *c*, 12 units; *d*, 100 units of enzyme. All incubations were for 2 h at 37 °C with 51 µg DNA and using the buffer conditions described in Fig 1. *e*, Complete *Hpa*II digest of erythrocyte DNA from 5-d-old embryos. For comparison, slot *f* is a complete *Hpa*II digest of adult erythrocyte DNA.

It is undoubtedly premature to relate DNA methylation patterns to a specific biological activity. It does seem clear, however, that the observed tissue differences do not solely reflect different rates of cell division, as both adult and embryonic red cells have ceased division and yet their DNA shows different methylation patterns. Moreover, brain, a non-proliferating tissue, shows the same pattern as the proliferating oviduct. On the other hand, at least some correlation can be made between these methylation patterns and globin gene activity. In cells which are expressing or have expressed the adult β-globin gene (adult reticulocytes and erythrocytes) the CCGG sites near the ends of the gene sequence (that is at the ends of the 1.3-kilobase fragment) seem to be completely unmethylated. In cells which are not expressing this gene (oviduct, brain and embryonic red blood cells) these sites can be at least partially methylated. The determination of how these specific methylation events occur during cellular differentiation must await a more detailed description of their exact positions in the gene sequence. Nevertheless, we believe these results to be the first correlation between site-specific DNA methylation and eukaryotic gene expression.

We thank Dr Gary Felsenfeld for advice and encouragement, and Dr W. Salser for providing an unpublished characterisation of the plasmid pHb1001.

JAMES D. MCGHEE

GORDON D. GINDER

Laboratory of Molecular Biology,
National Institute of Arthritis, Metabolism
  and Digestive Diseases,
National Institutes of Health,
Bethesda, Maryland 20205

Received 12 March; accepted 31 May 1979.

sed[14,15]) or from adult brain (data not shown) gives digestion patterns essentially identical to that of oviduct DNA.

The legend to Fig. 1 summarises control experiments which show that the above results are not due to partial digestions, DNA isolation methods, enzyme inhibitors, DNA sequence rearrangements or individual differences between chickens. Figure 2*a* also provides an important control, showing that undigested DNA is still transferred to the nitrocellulose filter for hybridisation; thus, any high molecular weight bands would have been detected.

One current difficulty in assigning the observed bands to their positions within the gene sequence arises from the fact that the adult chicken β-globin gene in pHb1001 shares considerable sequence homology with at least one of the embryonic β-globin genes[16], and both sequences can show up on the filter hybridisations. However, our preliminary results using a genomic clone (G.D.G., W. I. Wood and G. Felsenfeld, unpublished) indicate that the prominent 1.3-kilobase fragment shown in Figs 1 and 2 is derived from the adult β-globin gene sequence; moreover one end of this fragment lies within about 30 bases of the 5′-end of the coding sequence[14].

1. Scarano, E. *Adv. Cytopharmac.* **1**, 13–24 (1971).
2. Holliday, R. & Pugh, J. E. *Science* **187**, 226–232 (1975).
3. Riggs, A. D. *Cytogenet. Cell Genet.* **14**, 9–25 (1975).
4. Vanyushin, B. F.. Mazin. A. L., Vasilyev, V. IC. & Belozersky, A. N. *Biochim. biophys. Acta* **299**, 397–403 (1973).
5. Razin, A. & Cedar, H. *Proc. natn. Acad Sci. U.S.A.* **74**, 2725–2728 (1977).
6. Pollock, J. M., Swihart, M. & Taylor, J. H. *Nucleic Acids Res.* **5**, 4855–4863 (1978).
7. Grippo, P., Iaccarino, M.. Parisi. E. & Scarano, E. *J. molec. Biol.* **36**, 195–208 (1968).
8. Bird, A. P. & Southern. E. M. *J. molec. Biol.* **118**, 27–47 (1978).
9. Garfin, D. E. & Goodman. H. M. *Biochem. biophys. Res. Commun.* **59**, 108–116 (1974).
10. Mann, M. B. & Smith, H. O. *Nucleic Acids Res.* **4**, 4211–4221 (1977).
11. Waalwijk, C. & Flavell, R. A. *Nucleic Acids Res.* **5**, 3231–3236 (1978).
12. Southern, E. M. *J. molec. Biol.* **98**, 503–517 (1975).
13. Waalwijk, C. & Flavell. R. *Nuclelc Acids Res.* **5**, 4631–4641(1978).
14. Salser, W. A. *et al.* in *Cellular and Molecular Regulation of Hemoglobin Switching* (ed. Nienhuis. A. W. & Stamatoyannopoulos, G.) (Grune and Stratton, New York. 1979).
15. Brown, J. L. & Ingram, V. M. *J. biol. Chem.* **249**, 3960–3972 (1974).
16. Engel, J. D. & Dodgson, J. B. *J. biol. Chem.* **253**, 8239–8246 (1978).

Promoter elements

6

As discussed in the previous section, gene activation or repression in prokaryotes is brought about by the binding of particular regulatory proteins to specific DNA sequences within the target gene. Such binding then results in the specific activation or repression of the gene. By analogy, therefore, it is likely that a similar process takes place in eukaryotes. Thus, once a particular gene has become accessible by an alteration in chromatin structure, specific regulatory factors would bind to it and stimulate its transcription. In this model, genes that are activated in response to a particular stimulus would therefore contain a specific DNA sequence within their regulatory regions which would result in such activation, whereas this sequence would not be present in genes that are not modulated by the particular stimulus.

The paper by Pelham presented here represents the first evidence for the existence of such a regulatory sequence in higher eukaryotes. Pelham studied the gene encoding the 70 kDa heat-shock protein (hsp70) from the fruit fly *Drosophila*. As its name implies, the transcription of this gene is activated in response to exposure of these cells to elevated temperatures (Ashburner and Bonner, 1979). In the experiment illustrated in Figure 2 of the paper, Pelham introduced a plasmid containing the transcribed region of the *Drosophila hsp70* gene together with 1100 bases of 5′ flanking sequence into mammalian cells. He demonstrated that the expression of this *hsp70* gene was indeed increased when the mammalian cells were exposed to elevated temperature. In contrast, a control gene, encoding the herpes simplex virus thymidine kinase enzyme, was not activated upon exposure to elevated temperature. This result therefore establishes that the signals necessary for the heat-induced activation of the *hsp70* gene reside within the gene itself, and that the *Drosophila* signals are recognized in mammalian cells.

Based upon this result, Pelham then linked portions of the 5′ upstream region of the *hsp70* gene to the non-inducible thymidine kinase gene and tested whether such regions could confer heat-inducibility upon the thymidine kinase gene. As illustrated in Figure 3 of the paper, a small region of the *hsp70* gene regulatory region (from positions −10 to −186 relative to the start site of transcription) was indeed capable of conferring heat-inducibility upon the thymidine kinase gene. Further deletion analysis, illustrated in Figure 4 of the paper, showed that this effect could be achieved with an even shorter region located at positions −47 to −66 relative to the transcriptional start site. Hence this 20 bp sequence suffices to render a completely distinct gene heat-inducible (Figure 6.1).

Moreover, having identified this sequence in the *Drosophila hsp70* gene, Pelham was able to show that similar sequences also exist in a number of other heat-inducible genes, as illustrated in Figure 7 of the paper. Hence this so-called heat-shock element (HSE) represents a DNA regulatory sequence that is present in all heat-inducible genes and which can confer this response upon an

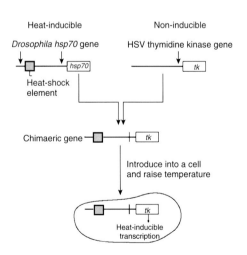

Figure 6.1 Transfer of the HSE from the heat-inducible *hsp70* gene to the thymidine kinase gene renders the latter heat-inducible. Abbreviation: HSV, herpes simplex virus; HSE heat-shock element.

unrelated gene that is not normally inducible. Moreover, in many cases the HSE is located within the 5′ upstream region adjacent to the transcriptional start site, paralleling the similar position of regulatory sequences adjacent to the gene promoter in bacteria.

Once the principle of a specific regulatory sequence that confers a particular effect had been established in the case of the HSE, a number of other distinct DNA sequences that mediate the responses to particular stimuli have been identified. Such sequences include the cyclic AMP (cAMP) response element (CRE), which mediates the induction of a number of genes in response to treatment with cAMP, the glucocorticoid response element, which mediates the response to glucocorticoid hormones, and many others (for a review, see Latchman, 1995a). Hence the paper by Pelham established the principle of short DNA sequences located adjacent to the gene promoter that produce a specific pattern of gene expression in response to a particular stimulus.

The Pelham paper is, however, of significance for an additional reason. Thus, as indicated above, Pelham assessed the heat-inducibility of the *Drosophila hsp70* gene when introduced into mammalian cells. The fact that the *Drosophila* HSE can function in mammalian cells provides an indication of the evolutionarily conserved nature of the response. It is also, however, of particular significance in terms of the mechanism by which the HSE acts. Thus the HSE used by Pelham was taken from the cold-blooded organism *Drosophila*, and would therefore normally be activated by the thermally stressful temperature for this organism of 37 °C. In contrast, the mammalian cells grow normally at 37 °C and only express the heat-shock genes at the higher temperature of 42 °C. Thus if the *Drosophila* HSE were a DNA thermostat, it would be set to activate gene expression at 37 °C regardless of the cell type in which it is present (Figure 6.2).

In fact, however, this is not observed, and the *Drosophila* HSE was activated at 42 °C when introduced into mammalian cells in the experiments described by Pelham. This indicates that the HSE is not a thermostat, but rather must act by being recognized by a cellular protein which is activated in response to elevated temperature. Obviously, in a mammalian cell, this protein would be activated at the appropriate mammalian heat-shock temperature and would then bind to the HSE and activate gene expression.

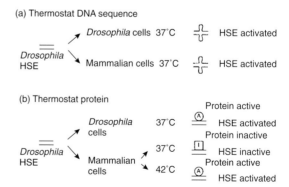

Figure 6.2 Possible models of HSE action. In (a), the HSE acts as a thermostat, detecting increased temperature directly. In (b) it acts indirectly by binding a protein that is activated by increased temperature. Note that only possibility (b) can account for the finding that the *Drosophila* HSE activates transcription in mammalian cells at the mammalian heat-shock temperature of 42 °C and not at the *Drosophila* heat-shock temperature of 37 °C.

This paper, therefore, provided one of the earliest pieces of evidence that the DNA sequence elements in the gene promoter act by binding specific regulatory proteins, which are now referred to as transcription factors. In the case of the HSE this was confirmed in the paper by Parker and Topol presented here, which directly identified a specific protein factor that binds to the HSE, thereby confirming the indirect evidence provided by the work of Pelham.

In a previous paper (Parker and Topol, 1984), these authors had reported the preparation of an extract from non-heat-shocked *Drosophila* cells which was able to transcribe specific genes that were added to the extract. In the subsequent paper presented here, they prepared a similar extract from heat-shocked *Drosophila* cells and showed that it transcribed an added *hsp70* gene more actively than the extract prepared from non-heat-shocked cells. These data are illustrated in Figure 2 of the paper, which also shows that an added actin gene was transcribed more actively by the extract prepared from non-heat-shocked cells compared with the level of transcription in the extract from heat-shocked cells (Figure 6.3).

Based on this difference, Parker and Topol then used a chromatographic procedure to purify a factor from the heat-shocked cell extract which specifically stimulated the transcription of the hsp70 gene while having no effect on the transcription of the non-heat-inducible actin gene. They named this factor heat-shock transcription factor or HSTF, although it is now more commonly known as HSF (heat-shock factor).

As illustrated in Figure 7 of the paper, they then used DNase I footprinting analysis (Galas and Schmitz, 1978) to determine the binding site for HSF within the *hsp70* gene promoter. In a direct link with the Pelham paper, they were able to show that HSF binds to the region of the promoter identical to that shown by Pelham to contain the HSE which is responsible for induction of the gene in response to elevated temperature (Figure 6.3).

Hence HSF represents a transcription factor which appears to be specifically active in extracts prepared from heat-shocked cells and which binds to the HSE responsible for gene activation in response to elevated temperature. Interestingly Parker and Topol were also able to purify HSF from non-heat-shocked cells but, as shown in Figure 6 of their paper, the factor was much less active than that purified from heat-shocked cells. This suggests that HSF exists in unheated cells in an inactive form that can be activated by exposure of the cells to elevated temperature. In agreement with this idea, subsequent studies (Zimarino and Wu, 1987) indicated that the heat-shock genes can be activated by HSF in the presence of inhibitors of *de novo* protein synthesis, exactly as would be expected if pre-existing HSF is activated in cells exposed to elevated temperature.

Although the mechanisms involved in transcription factor activation in response to specific stimuli will be discussed in a later section, the work presented here provides an important framework for understanding the role of DNA binding sequences and the transcription factors that bind to them. Thus a specific sequence, the HSE, confers the response to elevated temperature by binding a specific factor, HSF, which is activated and binds to DNA only in cells exposed to elevated temperature. Such binding of HSF to the HSE then results in enhanced transcription of the gene (for a review, see Morimoto, 1993).

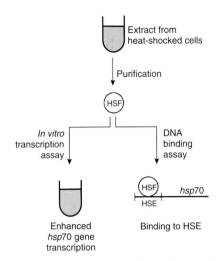

Figure 6.3 Purification of the HSF from a cell extract prepared from heat-shocked cells. Note that the HSF protein is able to stimulate *hsp70* gene transcription when added to an *in vitro* transcription assay. In addition, it was also demonstrated to bind to the HSE in the *hsp70* gene, which had been shown previously to be responsible for its heat-inducibility (see Figure 6.1).

Pelham (1982) Cell **30**, 517—528

A Regulatory Upstream Promoter Element in the Drosophila Hsp 70 Heat-Shock Gene

Hugh R. B. Pelham
M. R. C. Laboratory of Molecular Biology
Hills Road
Cambridge CB2 2QH, England

Summary

Deletion mutants of the Drosophila hsp 70 heat-shock gene have been assayed in COS cells using a vector that contains an SV40 replication origin. COS cells are SV40-transformed monkey cells that support high-level replication of the vector. S1 mapping of transcripts shows that the hsp 70 gene is heat-inducible in these cells, whereas the herpes virus thymidine kinase gene is not. Residues −10 to −66 of the hsp 70 gene are apparently sufficient for heat-inducible promotion, and residues between −47 and −66 are necessary. This region upstream of the TATA box contains sequence features common to other heat-shock genes. Expression of the hsp 70 gene can be forced at low temperature by SV40 sequences that include the 72 bp repeat, but only if these are present on the 5′ side of the gene. It seems that the upstream element of the hsp 70 promoter is analogous to that of other promoters, but is only functional in heat-shocked cells.

Introduction

In response to an increase in temperature, cells from a wide range of organisms rapidly synthesize a small number of proteins (the heat-shock proteins or hsp's) while repressing the synthesis of most other proteins. This appears to be an ancient, evolutionarily conserved response to thermal stress, and it evidently serves a protective function. A similar response can be induced at constant temperature by other stressful agents, including certain metabolic inhibitors, ionophores and recovery from anoxia (reviewed by Ashburner and Bonner, 1979).

In Drosophila, heat shock has been shown to result in greatly increased transcription of seven different protein-coding genes (Ashburner and Bonner,1979). Copies of all seven of the heat-shock genes from D. melanogaster have been cloned and their promoter regions sequenced (Karch et al., 1981; Torok and Karch, 1980; Ingolia and Craig, 1981; Ingolia et al., 1980; Holmgren et al., 1981).

The most highly conserved hsp is a 70 kd polypeptide (hsp 70). Most D. melanogaster strains have five copies of this gene per haploid genome, at two different cytogenetic loci (Ashburner and Bonner, 1979). Several copies of the gene have been sequenced; they are closely homologous to each other, and the homology extends for about 350 bases on the 5′ side of the transcribed portion of the genes (Karch et al., 1981; Ingolia et al., 1980). Recent genetic evidence

indicates that no more than 479 bases of flanking sequence is required for activity of one of the genes in vivo, and all transcriptional and regulatory signals probably are contained in the homologous region (Udvardy et al., 1982). The genes contain no intervening sequences.

Recently it has been shown that expression of cloned copies of the hsp 70 gene introduced into the genome of mouse tissue culture cells can be controlled by heat shock, indicating a considerable conservation of the regulatory signals (Córces et al., 1981). Regulation must involve interaction of specific cellular factors with the gene because the temperature required to induce activity is characteristic of the mouse cells, not of the flies from which the gene was originally obtained.

I describe experiments designed to locate the regulatory signals of the hsp 70 gene and to define the level at which regulation occurs. The assay I have used involves the introduction of plasmids containing an SV40 origin of replication into COS cells. These are monkey cells containing a defective integrated SV40 genome; they produce T antigen constitutively, and thus support high-level replication of the input DNA (Gluzman, 1981; Mellon et al., 1981).

The results can be interpreted in the light of what is known about eucaryotic promoters. These can be divided into at least two functional elements. One contains the TATA box which is involved in defining the position at which RNA polymerase II initiates transcription both in vivo (Grosveld et al., 1982; Benoist and Chambon, 1981; Grosschedl and Birnstiel, 1980a) and in vitro (Corden et al., 1980; Grosveld et al., 1981). The other is an upstream element, somewhat variable in position, that is required in vivo for efficient use of the TATA box (Benoist and Chambon, 1981; Dierks et al., 1981; Grosschedl and Birnstiel, 1980a, 1980b; Grosveld et al., 1982; Mellon et al., 1981; McKnight et al., 1981; Moreau et al., 1981). The evidence suggests that heat-shock genes have an upstream element with specific sequence characteristics that directs efficient use of the TATA box in heat-shocked cells, but is nonfunctional under normal conditions.

Results

Characteristics of the COS Cell System

The plasmids used in this work are diagrammed in Figure 1; further details are given in Figure 8 and Experimental Procedures. The basic vector is a pBR322 derivative that lacks the "poison sequences" that inhibit replication in animal cells (Lusky and Botchan, 1981). To this is joined a 240 bp fragment of SV40 DNA that contains the origin of replication, but has only a partial copy of one of the 72 bp repeats, and thus does not contain an efficient promoter (Benoist and Chambon, 1981). The standard plasmid pHT1

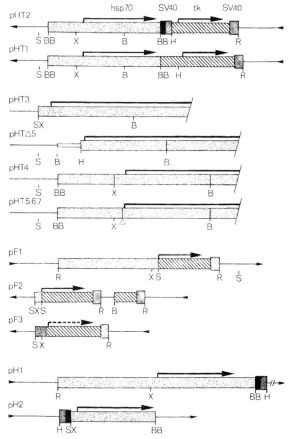

Figure 1. Structure of Plasmids

Thin lines: procaryotic vector sequences; diagonal stripes: tk sequences; stipples: hsp 70 gene and flanking sequence. Darker stippled areas are derived from plasmid 132E3, lighter ones from plasmid 56H8. Wider shaded boxes: SV40 sequences, with the 72 bp repeat region in black. Arrows above genes indicate transcripts. For further details see Figures 5 and 8 and Experimental Procedures. Note that pHT1 is the second plasmid shown. Only the 5′ part of the hsp 70 gene, enlarged twofold, is shown for plasmids pHT3–7 and the pHTΔ5′ series; the rest of the plasmid is in each case identical to pHT1. The thin box in the pHTΔ5′ plasmids represents 5S DNA spacer sequences, or in the case of pHTΔ5′-66* sequences from upstream of the tk gene.

The plasmids are shown linearized arbitrarily in the vector sequences; arrows on these indicate their orientation, pointing in the direction from the Eco RI site toward the Sal I site in the Tet[R] gene. Selected restriction sites are abbreviated as follows: B = Bam HI; BB = Bam HI fused to Bgl II; R = Hind III; R = Eco RI; S = Sal I; SX = Sal I fused to Xho I; X = Xho I. Plasmid pF2 was constructed as part of a separate experiment, and the extra tk sequence shown on the right of the plasmid is irrelevant to the results presented here.

(shown on the second line of Figure 1) contains also a complete hsp 70 gene, including 1100 bases of 5′ flanking sequence, and the herpes virus thymidine kinase (tk) gene as an internal control.

The system was characterized using the S1 mapping technique (Berk and Sharp, 1977; Weaver and Weissman, 1979) to detect and quantitate cytoplasmic tk transcripts. DNA was introduced into the cells in the presence of DEAE-dextran, a method that for these cells is simpler and at least as efficient as

calcium phosphate coprecipitation. The ultimate concentration of transcripts was not affected by varying the amount of input DNA between 2 and 20 μg per 75 cm^2 flask, and was achieved within 30 hr of exposure of the cells to DNA; the transcripts remained at this level for at least another 10–15 hr. Plasmids containing two copies of the SV40 replication origin replicated faster, the transcript level saturating by about 20 hr, but the final abundance of transcripts was not significantly higher than that obtained with plasmids containing a single origin (see below). Plasmids lacking SV40 sequences yielded no detectable transcripts under these conditions. In subsequent experiments, flasks of cells were transfected with 2–5 μg of DNA, and the cells were harvested 35 hr after removal of the DNA.

Because the 72 bp repeat region of SV40 has been implicated as a long-range cis-acting enhancer of the expression of other genes (Banerji et al., 1981; Moreau et al., 1981), the properties of plasmid pHT1 were compared to those of pHT2, which has an extra copy of the SV40 origin including the complete 72 bp repeat region (see Figure 1). (The partial copy of the 72 bp repeat present on all the plasmids is not sufficient for enhancer activity, as shown by Banerji et al., 1981.) Figure 2A shows the results obtained by S1 mapping the 5′ ends of tk transcripts obtained from these plasmids. The probe was a 131 bp Eco RI–Bgl II DNA fragment, ^{32}P-labeled at the Bgl II site with polynucleotide kinase (see Figure 2). Hybridization of authentic tk mRNA to this fragment protects a labeled fragment of about 54–56 bases from S1 nuclease, and with probe added in excess, the amount of label recovered in the protected fragments reflects the abundance of tk transcripts (McKnight et al., 1981).

In addition to the expected tk transcripts, a number of transcripts originating upstream of the normal 5′ end were observed in COS cells, indeed, these were often more abundant than the correct transcripts. Figure 2A also shows that the presence of the second SV40 replication origin and 72 bp repeat apparently increased the overall abundance of tk transcripts (at least in non-heat-shocked cells). This result was not, however, obtained reproducibly. The effect seen in Figure 2A may be due to the second origin of replication; this alone can increase the copy number of the plasmid at early times, and hence the level of transcripts (data not shown). The variability between experiments probably reflects slight differences in the rate of plasmid replication in the COS cells, as well as some inherent variability of the S1 assay. In general, the 72 bp repeat appeared to have little effect on tk gene expression with the plasmid constructions and assay conditions that I have used. This was true whether or not the hsp 70 gene was also present on the plasmid.

The extra 5′ ends of the tk transcripts map upstream from the TATA box, and some are within the essential upstream region defined by deletion mapping of the

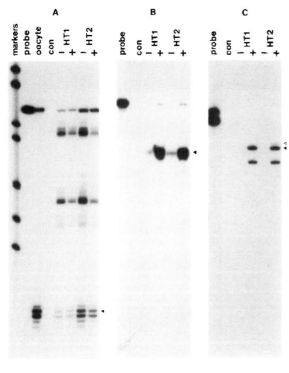

tk 5′

EcoRI ◄—— 131 bp ——► BglII

~55bp

hsp70 5′

HaeIII ◄—— 573bp ——► HaeIII

~395bp

hsp70 3′

BamHI ◄—— >3Kb ——► BamHI

~1090bp

Figure 2. S1-Mapping of tk and hsp 70 Transcripts Synthesized in COS Cells

The same RNA samples were hybridized to the 5′ tk probe (A), the 5′ hsp 70 probe (B) and the 3′ hsp 70 probe (C). The hybrids were digested with nuclease S1 and analyzed on gels as described in Experimental Procedures. Autoradiographs of portions of the gels are shown. Markers visible in (A) are fragments of 162, 149, 124, 112, 92, 78 and 69 bp (pBR322/Hpa II). Also shown for reference is the S1 map of tk transcripts obtained after injection of the gene into Xenopus oocytes. Cells were transfected with pHT1 or pHT2 (see Figure 1) and either heat shocked (+) or not (−). Control RNA (con) was from cells that were mock-transfected and heat-shocked. The line drawing outlines the S1 mapping schematically (not to scale). Thin lines represent mRNA, with arrowheads at the 3′ end, thick lines represent the DNA probes, with asterisks indicating the position of ^{32}P label. The positions of the expected protected fragments on the gels are indicated by solid triangles in A–C. The open triangle in C indicates a band derived from readthrough transcripts which map to the point of sequence divergence between probe and template. The hsp 70 3′ probe contains two hybridizing species and yields two protected fragments as explained in the text. Note that panels A–C are not directly comparable; the probes were not of the same specific activity, the gels were of different acrylamide concentrations, and the autoradiographic exposures differ.

tk promoter (McKnight et al., 1981). They have been found consistently, whether or not there are SV40 or hsp 70 sequences upstream from the tk gene, and they are also synthesized in vitro in HeLa cell extracts (P. Farrell and H. Pelham, unpublished observations). This suggests that the extra 5′ ends correspond to true transcriptional starts, but their significance remains unclear.

Expression of the Hsp 70 Gene Is Heat-Inducible in COS Cells

Hsp 70 transcripts were detected using a 5′ S1 probe consisting of a 573 bp Hae III fragment continuously labeled in the mRNA complementary strand (see Figure 2 and Experimental Procedures). Authentic mRNA from heat-shocked Drosophila cells protects a ~400-base fragment from S1 nuclease (data not shown; see also Torok and Karch, 1980; Ingolia et al., 1980). There was also some weak cross-hybridization with monkey cell heat-shock RNA, but the protected fragments were less than 200 bases long and did not interfere with the assay.

Figure 2B shows that incubating the COS cells at 42°C from 30 to 35 hr after transfection increased the abundance of hsp 70 transcripts about 50-fold, and these transcripts had the expected 5′ ends. In contrast, heat shock slightly reduced the abundance of tk transcripts (Figure 2A). Since most of these were synthesized prior to the heat shock, it is impossible to say whether tk transcription continues at 42°C. However, in many experiments the level of at least the larger tk transcripts was reduced by heat shock; this may indicate a reduction in the rate of transcription and some turnover of preexisting RNA during heat shock.

The 3′ ends of the hsp 70 transcripts were also located by S1 mapping (Figure 2C). The probe consisted of the entire plasmid 132E3, which contains two copies of the hsp 70 gene. This was labeled at the Bam HI sites in the centers of the genes (see Figure 8B). Because the gene sequences diverge on the 3′ side of the coding region, two protected fragments were obtained; one ends at the point of divergence, the other at the true 3′ end of the transcript. The latter mapped to the expected position within the hsp 70 sequence, but the large size of the protected fragment precluded accurate mapping. As expected, abundant protected fragments were only obtained with RNA from heat-shocked cells.

These results establish that expression of the Drosophila hsp 70 gene can be controlled in monkey cells, even when the gene is present on a plasmid at many thousands of copies per cell and is adjacent to a gene that is expressed at low temperature. The level of induction of hsp 70 mRNA (typically 20 to 50 fold) is comparable to the induction of monkey heat-shock RNA under the same conditions, as detected by cross-hybridization with the Drosophila probe (data not shown). The presence of the SV40 72 bp repeat at the 3′ end of the hsp 70 gene (in pHT2 but not in pHT1) had no effect on this control or on the final level of transcripts obtained (Figure 2).

Fusion of the Hsp 70 Gene to the tk Gene

The results described above do not rule out the possibility that control of hsp 70 gene expression is at a posttranscriptional level in COS cells; for example, the transcripts might be selectively degraded at low temperature. To address this question, fusions of the hsp 70 promoter region to the tk gene were constructed (see Figure 1) and tested (Figure 3). The first of these fusions, pF1, contains about 2 kb of the 5′ flanking region of a hsp 70 gene up to position −10 joined via a synthetic linker to a Hae III site 8 bases upstream from the tk initiation site, the distance from TATA box to start site being exactly maintained. (Throughout this paper positions are numbered relative to the transcription start point of the gene concerned.) The hsp 70 gene used in this construction was derived from the 87A7 cytogenetic locus; it has a sequence very similar to the 87C1 gene used previously out to about position −350. In subsequent constructions, parts of both genes were used for convenience, but because of small differences in the untranslated portion of the mRNAs it was necessary always to use the appropriate probe for S1 mapping.

The fused gene in plasmid pF1 was clearly heat-inducible, transcription initiating at the normal tk start site and also a few bases upstream (Figure 3). In fact, transcripts synthesized during a 5 hr heat shock were considerably more abundant than those normally produced from the tk promoter. At low temperature the only transcripts detectable were initiated upstream from the fusion point, and thus mapped to the point of divergence of the S1 probe from the template. A further construction (pF2) containing only residues −10 to −186 of the heat-shock gene was equally inducible, although the level of readthrough transcripts was higher (Figure 3). Since tk mRNA is clearly stable at low temperature, this result implies that the principal control of hsp 70 gene expression in COS cells is at the level of transcription initiation, and that the promoter and regulatory sequences are contained within the −10 to −186 5′ flanking region.

A third fusion (pF3, see Figures 1 and 5) contains the entire tk gene including the TATA box and surrounding sequences, out to a Hha I site at position −50, but lacks the important upstream promoter region which lies between residues −52 and −109 (McKnight et al., 1981). This gene is joined to 5′ flanking sequences from the hsp 70 gene, extending from one of the overlapping Hha I sites at positions −64/−66 out to position −400. The resultant hybrid gene was transcribed extremely poorly in COS cells at both high and low temperatures (Figure 3). Prolonged exposure of the autoradiogram revealed very

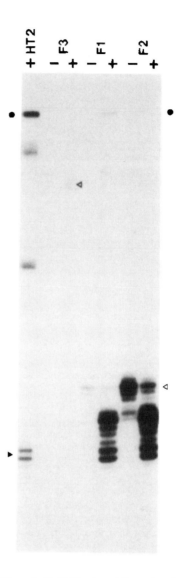

Figure 3. 5′ S1 Mapping of Transcripts from the tk Gene Fused to the hsp 70 Promoter

Solid circles indicate the position of intact probe; open triangles indicate the position corresponding to upstream transcripts mapping to the point of sequence divergence between probe and template. Solid triangles indicate fragments corresponding to initiation at the correct tk sequence. Cells were transfected with pF1, pF2 or pF3, and either heat-shocked (+) or not (−). A sample from Figure 2 (pHT2) is included for reference. The sequence at the junction of hsp 70 and tk DNA in pF1 and pF2 is ...TATAAATAGAGGCGCTTCGTCGACCCCTCGAACA*CCGAG..., where the underlined bases are derived from the tk gene and the asterisked A residue probably corresponds to the first base of authentic tk mRNA. The junction in pF3 is described in the text.

faint bands corresponding to the normal tk transcripts (including the upstream starts between −30 and −50), but these were completely unaffected by heat shock. Thus the sequences upstream of −63 in the hsp 70 gene are not sufficient to replace functionally the upstream region of the tk promoter at either 37°C or 42°C.

Pelham (1982) Cell **30**, 517–528

Deletion Mapping the Heat-Shock Promoter

To locate a DNA sequence essential for the heat-shock response, sequential 5' deletions were prepared with Bal 31 nuclease, synthetic Hind III linkers ligated to the remaining DNA, and the plasmid reconstructed as shown in Figure 1 (pHTΔ5' series). These plasmids were introduced into COS cells, and the hsp 70 transcripts were detected by 5' S1 mapping. Figure 4 shows that deletion to position −186, −108, −97, −68 or −66 did not prevent the normal heat induction. A deletion to position −44 gave a very low level of transcripts after heat shock, comparable to the normal low temperature level, while a deletion to position −28, within the TATA box, gave no detectable correct 5' ends at all.

To confirm a consistent efficiency of transfection and RNA recovery, each RNA sample was also hybridized to the tk probe. As expected, the level of tk transcripts was similar in all cases (examples are shown in Figure 4). A further internal control is provided by transcripts that originate upstream of the hsp 70 promoter. With the 5' S1 probe, these map to the point of sequence divergence between probe and template (Figure 4). The abundance of these transcripts was not greatly affected by heat shock, but a slight induction (about 2-fold) was sometimes observed even though the concentration of tk transcripts in the same samples remained constant. The mechanism of this apparent induction may be posttranscrip-

tional; it is clearly distinct from the true transcriptional induction. Thus even though mutants such as Δ5'-44 show a very weak apparent induction (Figure 4), I have considered them uninducible because the ratio of correct to upstream starts is not increased by heat shock.

The construction of deletion mutants places novel DNA sequences adjacent to the promoter, and it is possible that these might substitute in some way for the deleted sequences. In these constructions I have used a DNA fragment from the spacer region of Xenopus 5S DNA in the hope that this would be completely irrelevant to a polymerase II promoter. As a control, however, the remaining hsp 70 sequences in Δ5'-66 were also joined to another DNA fragment from 200 to 700 bp upstream of the tk gene (see Experimental Procedures). This sequence has no effect on tk transcription in Xenopus oocytes (McKnight et al., 1981). The resultant plasmid, pHT Δ5'-66*, behaved identically to pHT Δ5'-66 (Figure 4). Thus induction of the hsp 70 promoter in COS cells requires DNA sequences between −44 and −66, but sequences farther upstream from this appear completely dispensable.

Further Characterization of the Upstream Promoter Element

Three additional mutants help to define the important sequences between −44 and −66. Their sequences

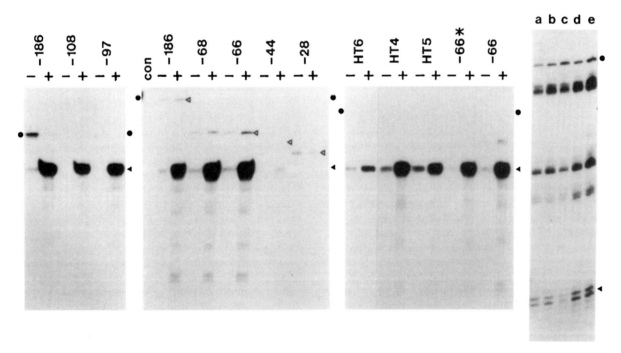

Figure 4. 5' S1 Mapping of Transcripts from Deletion Mutants of the hsp 70 Gene

Symbols have the same meaning as those in Figure 3. Parts of four separate gels are shown. The plasmids tested were members of the pHTΔ5' series. and also pHT4 (a control), pHT5 and pHT6. The critical sequences of these are shown in Figure 5. See text for an explanation of plasmid pHΔ5'-66*. Control RNA was from heat-shocked mock-transfected cells. Tracks a, b, c, d and e show mapping of the tk transcripts in the RNA samples derived from heat-shocked cells that had been transfected with pHΔ5'-186, -68, -66, -44 and -28, respectively.

in this region are shown in Figure 5 (see also Figure 1). One, pHT5, has residues −45 through −50 replaced by a decameric Hind III linker. This introduces four base changes and moves the −47 to −66 region four bases farther from the TATA box. Despite these changes, pHT5 was still clearly heat-inducible (Figure 4), although its activity was somewhat reduced; in duplicate experiments the abundance of induced transcripts varied from 40% to more than 70% of the control, pHT4 (see Figure 1). Comparison with pHTΔ5′-66 suggests that at most 18 bp in the region upstream of −45 are essential (Figure 5), and the distance of this sequence from sequences downstream of −45 is not critical.

Two other deletions were obtained after S1 digestion of Hind III–cut pHT5 and were identified by DNA sequencing. In one, pHT6, residues −45 through −69 were replaced by a single C residue, although the upstream sequences show some homology to those deleted (Figure 5). This plasmid produced very few transcripts after heat shock (Figure 4). Although it showed an apparent low level of induction (about 3-fold), the upstream transcripts also appeared inducible (not visible in Figure 4), and as discussed earlier this may not reflect a true transcriptional induction. The second mutant, pHT7, lacks a further 2 bp (Figure 5), and had equally low activity (data not shown).

These results confirm the conclusion derived from the inactivity of pF3 that hsp 70 sequences upstream of −66 are themselves unable to stimulate normal levels of transcription from an intact TATA box region. They also suggest that at least some of the 9 bases in the −47 to −66 region that differ between pHT5 and pHT6 are essential for efficient promoter function.

Forced Expression of the Hsp 70 Gene at Low Temperature

A reasonable interpretation of the data presented above is that the hsp 70 gene has a perfectly normal TATA box and transcription initiation site, but has an upstream region that is only functional in heat-shocked cells. A prediction from this would be that expression of the gene at low temperature could be achieved by provision of a functional upstream element, such as that contained in the SV40 promoter region.

Figure 2 shows that the presence of SV40 sequences including the 72 bp repeat at the 3′ end of the hsp 70 gene (in pHT2) does not affect its expression; a similar result is shown in Figure 6 with a plasmid (pH1) containing a different hsp 70 gene and lacking the tk gene. However, a completely different result was obtained with plasmid pH2, which has the SV40 origin region on the 5′ side of the gene, with the late side of the 72 bp repeat some 200 bases from the start site (see Figure 1). This plasmid gave a substantial level of hsp 70 transcripts at 37°C (Figure

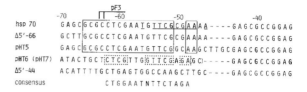

Figure 5. Sequences of the 5′ Flanking Region of the hsp 70 Gene and Some Mutants

The basic structures of the mutant plasmids are shown in Figure 1. The sequence shown is that of the mRNA sense strand from the gene derived from plasmid 56H8 (Karch et al., 1981). The corresponding region of the gene from plasmid 132E3 differs only at positions −3 (A) and −12 (T). Transcription is initiated at position +1; the TATA box and the inverted repeat at position −50 are underlined. The points of fusion with the tk gene in pF1, pF2 and pF3 are indicated above the sequence. Boxes enclose those sequences that are present in all the genes that show good heat-inducibility but are absent or disrupted in those that do not (see Figure 4). Dotted boxes indicate the bases in pHT6 that are homologous to the parent gene in this region. For an explanation of the consensus sequence see Figure 7 and the Discussion.

6), the total abundance of these transcripts being comparable to that obtained from other plasmids after heat shock. However, mapping of the transcripts shows that while many of them were initiated at the normal site, there was also considerable use of upstream start sites (Figure 6). Such upstream starts are detectable at an extremely low level with plasmids that lack the 72 bp repeat, but they are not heat-inducible. The preferential use of these upstream start sites in cells transfected with plasmid pH2 suggests that the SV40 sequences may act as a polymerase "entry site" in this plasmid (see Discussion).

Although plasmid pH2 retains all sequences normally required for heat induction, the presence of the SV40 sequences upstream apparently prevented this response. Superimposition of the normal heat-shock response on the pattern of transcripts obtained at 37°C would have been readily detected; in fact there was no change in either the absolute level of transcripts or in the ratios of normal to upstream starts after heat shock (Figure 6). This suggests that the SV40 sequence can have a repressive effect, perhaps via chromatin structure, at least 150 bases away. Alternatively, the SV40 sequences may simply remain active at high temperature, thus not actually repressing the hsp 70 promoter element, but making it functionally redundant.

From these results, I conclude that hsp 70 mRNA is not exceptionally unstable in cells at 37°C; that the TATA box region is capable of efficiently directing polymerase initiation events at low temperature; and that a promoter function analogous to that which is regulated by heat shock can be provided at low tem-

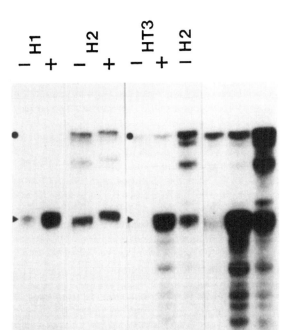

Figure 6. Effect of SV40 sequences on hsp 70 Transcription

Transcripts were detected by 5′ S1 mapping, and symbols have the same meaning as in Figure 3. The SV40 72 bp repeat is on the 3′ side of the gene in plasmid pH1, on the 5′ side in plasmid pH2, and absent in plasmid pHT3 (see Figure 1). Results from two different experiments with plasmid pH2 are shown. The three tracks on the right are a longer exposure of the adjacent three tracks; the bands below the normal ~400 bp band reflect slight degradation of the RNA in this experiment.

perature by a piece of SV40 DNA that includes the 72 bp repeat.

Discussion

Functional Domains of the Hsp 70 Promoter

Sequences upstream of the hsp 70 coding region promote efficient transcription of the gene in monkey cells only when the cells are exposed to heat shock. This regulation is maintained despite the heterologous nature of the assay, with the gene present on a multicopy replicating plasmid, and whether or not it is adjacent to a gene (tk) that is active at low temperature. The results presented place limits on the sequences responsible for this function, and they suggest that the response to heat shock is an intrinsic property of the promoter itself.

The functional analysis of fusions between the hsp 70 and tk genes shows that heat-inducible promotion does not require any of the transcribed hsp 70 sequences, nor the first 10 bases of the 5′ flanking region. Analysis of 5′ deletion mutants indicates that sequences upstream of position −66 have no detectable function in COS cells. Thus the region −10 to −66 appears sufficient for regulated promotion of transcription. This region can be divided into two

functional domains on the basis of this work and by analogy with other systems. The first of these consists of the TATA box and succeeding GC-rich sequence. The second is an element upstream of the TATA box that is required for efficient promotion in vivo.

The TATA Box Region

There is nothing particularly unusual about the TATA box region of the hsp 70 and other heat-shock genes. All the genes show homology to the sequence (A/G)C(C/A)GGCGC immediately following the TATA box, but many polymerase II promoters have a similar region. For example, equally strong homology to this sequence is found in the Drosophila histone H4 and alcohol dehydrogenase genes (Goldberg, 1979; H. Haymerle, personal communication).

Evidence from a number of systems indicates that the function of the TATA box region in vivo is to direct the polymerase to initiate transcription at a discrete site approximately 30 bases downstream from the TATA box itself (Grosschedl and Birnstiel, 1980a; Benoist and Chambon, 1981; Grosveld et al., 1982). It seems that the hsp 70 TATA box can perform this function equally well in normal and heat-shocked cells. First, the residual accurate transcription from the deletion mutant pHTΔ5′-44 is significantly higher than that from pHTΔ5′-28, which lacks the TATA box, but this transcription is not greatly affected by heat shock. The TATA box also directs a significant fraction of polymerases to the correct start site when transcription is forced at low temperature by SV40 sequences (Figure 6). Furthermore, this region directs initiation in whole-cell extracts prepared from non-heat-shocked HeLa cells with an efficiency comparable to that of other promoters, whether or not the upstream region is present (P. Farrell and H. Pelham, unpublished observations). These results suggest that the TATA box region plays a passive, nonregulatory role in the transcription of the hsp 70 gene.

The Upstream Promoter Element

The upstream element appears to lie between residues −47 and −66 (see Figure 5). Mutations that affect this region have a profound effect on promoter activity in heat-shocked cells. The element can be separated from the TATA box by a further 4 base pairs without abolishing promoter function.

Since the heat shock genes probably share a common induction mechanism, I have looked for similarities between the sequence of the hsp 70 upstream element and the corresponding position in other Drosophila heat-shock genes. An obvious feature of the hsp 70 sequence is the inverted repeat centered at position −51/−50. Larger inverted repeats are present in approximately the same position in 5 of the 6 other heat-shock genes, and when these repeats are aligned, significant sequence homology between the genes is apparent (Figure 7). This homology is largely restricted to positions −48 through −62 on the hsp 70 sequence, within the region containing the functional element, and for these bases a consensus sequence CTGGAAT(N)TTCTAGA can be derived. The hsp 70 gene fits this consensus at 11 of the 14 positions; of these, three are altered in pHT6, which is at best only weakly inducible (see Results).

Ten of the 14 bases in the consensus sequence themselves form an inverted repeat (CT-GAA--TTC-AG), a property characteristic of many protein recognition sites on DNA. This inverted repeat is, however, offset from the center of the large inverted repeats; many of the bases in the latter do not match the consensus sequence or are outside it (Figure 7). It seems that there has been selection for two independent features: a sequence similar to the consensus, and a larger inverted repeat at the appropriate position nearby, regardless of the exact sequence of the latter. One can speculate that the larger inverted repeat might be a general feature that aids recognition of the functional sequence by a protein (for example, by influencing the position of a nucleosome). Alternatively, it might facilitate the function of such a protein, for example, by forming loops that would make a local unwinding of the DNA helix more energetically favorable. However, the activity of pHT5, albeit lower than that of the parental plasmid, indicates that perturbation of the inverted repeat in the hsp 70 gene can at least be tolerated.

Not all the heat-shock genes fit the general pattern closely. The hsp 68 gene fits the consensus sequence poorly, although a better homology exists on the 3′ side of the inverted repeat (Figure 7). The hsp 23 gene also does not fit the consensus in this region, although the homologous sequence CGA-GAAGTTTCGTG is present 100 bases upstream from the TATA box. It will be interesting to locate the functional sequences in this gene.

Despite these exceptions, the general correlation between the positions of the homologous sequences, the inverted repeats, and the hsp 70 functional region suggests that the consensus sequence and the inverted repeat together may constitute a functional domain. The variable distance between this region and the TATA box in the various genes is consistent with the flexibility of this distance in the hsp 70 gene (as exemplified by pHT5), and suggests that the upstream region is a physically independent functional unit. This model can be tested by mutational analysis of other heat-shock promoters.

Function of the Upstream Element

The upstream element of the hsp 70 promoter seems to be the site at which regulation of transcription occurs. The regulation is positive, in the sense that deletions in this region greatly reduce transcription during heat shock rather than causing constitutive expression of the gene. This can explain how control can be maintained despite the presence of the gene in thousands of copies per cell: there may be no specific repressor to be titrated out.

The role of the upstream region in heat-shocked cells appears to be analogous to that of the upstream regions of other polymerase II promoters, namely, to stimulate use of the TATA box region. This is also suggested by the observation that its function can be mimicked at low temperature by SV40 sequences containing the 72 bp repeat. When these are on the

```
Consensus              CTGGAATNTTCTAGA
Function               ========================
hsp 70    CGAGAGACCGCGCCTCGAATGTTCGCGAAAAGAGCGCCGGAGTATAAA
                                   :
hsp 83    CATCCAGAAGCCTCTAGAAGTTTCTAGAGACTTCCAGTTCGGTCGGGTTTTTCTATAAA
                                   :
hsp 22    ATTCGAGAGAGTGCCGGTATTTTCTAGATTATATGGATTTCCTCTCTGTCAAGAGTATAAA
                                   :
hsp 26    TTTCTGTCACTTTCCGGACTCTTCTAGAAAAGCTCCAGCGGGTATAAA
                                   :
hsp 27    GTTCCGTCCTTGGTTGCCATGCACTAGTGTGTGTGAGCCCCAGCGTCAGTATAAA
                                   :
hsp 68    TGACCCTTTCTCGCAGGGAAATCTCGAATTTTTCCCCTCCCGGCGACAGAGTATAAA

hsp 68    CTCGCAGGGAAATCTCGAATTTTCCCCTCCCGGCGACAGAGTATAAA

hsp 23    CGCCGACGGGCGCACGCACACTACGATAGCCGAGCGGTTGTATAAA
```

Figure 7. Homologies between the Drosophila Heat-Shock Genes in the Upstream Region

Sequences are from Karch et al. (1981), Holmgren et al. (1981) and Ingolia and Craig (1981). The sequences of the genes for hsp 70, hsp 83, hsp 22, hsp 26, hsp 27 and hsp 68 were aligned at the center of the major inverted repeats (underlined), and the consensus sequence was derived for the homologous region. The inverted repeat in the consensus sequence is also underlined. Dots indicate the dyad axis of the major inverted repeats. Lines above the hsp 70 sequence indicate the region that is essential for heat induction of this gene (see Figure 5). The bottom two sequences show an alignment of the hsp 68 gene that matches the consensus sequence more closely, and the best match that can be obtained with the hsp 23 gene in this region. There is no corresponding inverted repeat in the hsp 23 gene, and the significance of the weak homology is doubtful. The second base shown in the hsp 70 sequence is a G, not a C as reported previously (Karch et al., 1981).

5′ side of the gene, polymerase molecules tend to initiate transcription at preferred sites close to the SV40 sequence, but many of them initiate some 200 bp farther along the DNA, downstream from the TATA box. The ability of the SV40 72 bp repeat to induce initiation in such a manner supports the suggestion of Moreau et al. (1981) that it acts as an entry site in the chromatin from which polymerase II can search for a TATA box or preferred start site. It is not clear whether this activity is the same as the enhancer activity associated with the SV40 72 bp repeat. In assays that do not involve high-level replication of the DNA the enhancer sequences greatly stimulate (200-fold) transcription from normal, intact promoters in a way that is cis-acting but apparently independent of the distance of the enhancer from the promoter, and of its orientation (Banerji et al., 1981). Such an effect of the 72 bp repeat was not observed in the COS cell system with either the hsp 70 gene or the tk gene, presumably because the high copy number of the genes in these cells somehow makes enhancer sequences redundant.

The results presented here are compatible with a model in which the hsp 70 upstream element itself functions as a regulated entry site. it is not clear what such a site might be in physical terms, but one of its properties could be to organize the local chromatin structure such that an appropriate region of the DNA is exposed. The SV40 origin and 72 bp repeat region is evidently capable of organizing nucleosomes such that the 72 bp repeat is exposed (see Jakobovits et al., 1980; Saragosti et al., 1980, and references therein) and such forced organization might also affect the function of the hsp 70 upstream region when this is nearby. The hsp 70 genes in Drosophila cells also have an organized chromatin structure: there is a DNAase I-hypersensitive site that has been mapped to a diffuse region covering about 200 bp upstream from the transcription initiation site (Wu, 1980). This site is present before heat shock, and it has been suggested that it is a necessary but not sufficient prerequisite for transcription (Elgin, 1981). If this is true in the COS cell system, then any DNA sequences required for chromatin organization must also be present in the −10 to −66 region, and the upstream element becomes a candidate for a chromatin organizer at low temperature as well as a promoter element at high temperature. Analysis of the chromatin structure of mutant plasmids in COS cells should clarify this point.

COS cells are a heterologous environment for the Drosophila hsp 70 gene, and they respond to heat shock at a different temperature from flies (which are fully heat-shocked at 37°C). Also, the replicating plasmid system is a somewhat abnormal state for a cellular gene. DNA sequences (such as enhancers) that facilitate gene expression in the homologous chromosome may appear functionless when assayed in this way. It

is thus possible that DNA sequences other than those described in this paper also contribute to the regulation of hsp 70 gene expression in Drosophila cells. However, it is not necessary to postulate the existence of such sequences to account for the heat-inducibility of the gene.

Finally, it may be useful to compare the upstream region of the hsp 70 promoter with the so-called CCAAT box. This is a conserved sequence found about 80 bp upstream from the start site of all globin genes and also some other genes (Efstratiadis et al., 1980). There is good evidence in the globin system that the CCAAT box is a functional upstream element (Dierks et al., 1981; Grosveld et al., 1982; Mellon et al., 1981). Despite this functional analogy, the CCAAT box differs from the hsp 70 upstream region in several respects: it is farther away from the TATA box, is not generally associated with an inverted repeat, and its activity is constitutive in COS cells. It remains to be seen whether sequences with properties similar to those of the hsp 70 upstream element are a common feature of inducible genes.

Experimental Procedures

Plasmid Constructions

All plasmids were grown in Escherichia coli HB101. The basic structures of the plasmids are shown in Figure I and details of some of them are shown in Figure 5. Figure 8C outlines the construction of pHT1. The starting plasmid was pMLRIIG (Lusky and Botchan, 1981). This is a pBR322 derivative (pML) in which bases 1092–2485 are deleted, with the RIIG fragment of SV40 (see Figure 8A) inserted into the Eco RI site via synthetic Eco RI linkers, the late side of the SV40 insert being close to the ampicillin resistance gene. This was digested with Bam HI and Hind III, and the herpes tk gene was inserted to form pXTK1. The tk gene was derived from the plasmid ptk/Δ3′-1.13 sequenced by McKnight (1980). The Hind III site in pXTK1 was removed by filling in and religation, and then a decameric Hind III linker was inserted at the Pvu II site present 200 bp upstream of the transcription initiation site in the tk gene, to form pXTK10 (Figure 8C). This plasmid still retains the origin fragment from SV40 (Figure 8A). The hsp 70 genes were obtained from the plasmids 56H8 and 132E3 (Figure 8B), large portions of which have been sequenced by Torok and Karch (1980) and Karch et al. (1981). The Bgl II fragment containing the left-hand gene of 132E3 was inserted into the Bam HI site of pXTK10 to form pHT1 (Figure 8C). This fragment contains about 1100 bases of 5′ flanking sequence and about 70 bases of 3′ flanking sequence (Karch et al., 1981).

Other plasmids are diagrammed in Figure 1. pHT2 was constructed by first replacing the Hind III–Bam HI fragment of pXTK10 with the origin + 72 bp rpt fragment of SV40 (see Figure 8A; the SV40 Pvu II site was converted to a Bam HI site with a CCGGATCCGG linker) and then inserting the hsp 70 Bgl II fragment. pHT3 was derived from pHT1 by deletion of the Sal I–Xho I fragment on the 5′ side of the hsp 70 gene (Figure 1); it retains 194 bp of 5′ flanking sequence. pHT4 was formed by substitution of the Xho I–Bam HI fragment of pHT1 with the corresponding fragment from 56H8 (Figure 8B).

The pHTΔ5′ series was prepared as follows: 10 μg of Xho I–cut pHT4 DNA was incubated at 30°C with 3 units of Bal 31 nuclease (BRL) in the recommended buffer, and samples were taken after 15, 40, 65 and 90 seconds. Digestion was terminated by phenol extraction. DNA from the samples was pooled, ligated to Hind III linkers (GCAAGCTTGC) and digested with Hind III and Bgl II (which cuts within the tk gene). The Hind III–Bgl II gene-containing fragment was isolated on an agarose gel and recloned between the Hind III and Bgl II sites of plasmid pXTK12. This plasmid is similar to pXTK10 (Figure

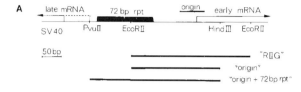

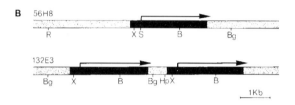

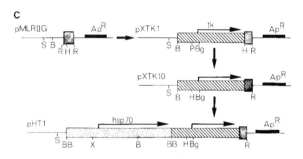

Figure 8. Origins of DNA Fragments Used in Plasmid Construction

(A) The SV40 origin region, showing the 72 bp repeats and the early and late promoters. Late transcripts have very heterogeneous 5′ ends. The fragments used are indicated and have the following coordinates (SV numbering system: see Tooze, 1981): RIIG, 160–5092; origin, 160–5171; origin + 72 bp rpt, 270–5171. (B) Relevant segments of Drosophila DNA from plasmids 56H8 and 132E3. Black boxes indicate the sequences that are strongly conserved between all three hsp 70 gene copies shown. Arrows above these indicate transcripts. (C) Steps in the construction of pHT1. Plasmids are shown arbitrarily linearized, and symbols have the same meaning as in Figure 1. For details see Experimental Procedures. Selected restriction sites are indicated in B and C:, and are abbreviated as follows: B = Bam HI; Bg = Bgl II; 8B = Bam HI joined to Bgl II; H = Hind III; Hp = Hpa I; P = Pvu II; R = Eco RI; S = Sal I; X = Xho I.

8C), except that the Hind III–Bam HI fragment has been replaced by a 360 bp Hind III–Hpa II fragment containing spacer sequences from a Xenopus 5S DNA clone (pXls 11; for sequence see Peterson et al., 1980), modified by ligation of a 10 base Bam HI linker (BRL) to the filled-in Hpa II site. The final construction (Figure 1) is thus the same as pHT1, but with the sequences upstream of the hsp 70 gene deleted and replaced by the 5S spacer DNA. pHTΔ5′-66* has this DNA replaced by the Bam HI–Hind III fragment from upstream of the tk gene in pXTK10 (Figure 8C). Deletion mutants were mapped by accurate sizing of small (<50 bp) restriction fragments on sequencing gels, and the end points of pHTΔ5′-68, -66, -44 and -28 were confirmed by partial DNA sequencing (Maxam and Gilbert, 1977). pHTΔ5′-186 was prepared by ligating linkers directly to the filled-in Xho I site. The nomenclature of these mutants indicates the last base of the gene sequence remaining, even if this is in fact derived from the Hind III linker.

pHT5 was derived from pHT1 by replacing the Xho I–Bam HI fragment of the hsp 70 gene with two fragments: the Xho I–Nru I fragment from positions −190 to −50 of the hsp 70 gene, with a Hind III linker added to the Nru I end, and the Hind III–Bam HI fragment from pHTΔ5′-44. The junction sequence is shown in Figure 5. pHT6 and pHT7 were derived from pHT5 by digestion with Hind III, treatment with S1 nuclease and transformation of the linear molecules into E. coli HB101. Sequencing of the resultant clones showed two classes of deletion; the most common was that in pHT7, and is effectively the result of recombination between the GAGCGC sequences at −44 and −70 (see Figure 5).

pF1 contains the Eco RI–Sal I fragment from the 5′ side of the hsp 70 gene in 56H8 (Figure 8B) joined via an 8 base Sal I linker to the Hae III site 8 bp upstream from the transcription start site of the tk gene. The Sal I site is at position −12 of the hsp 70 gene; the junction sequence is given in the legend to Figure 3. The tk gene is joined to the SV40 origin fragment (Figure 8A) as in pHT1, and this tripartite piece of DNA is inserted at the Eco RI site of pML (see Figure 1). pF2 contains a similar insert, but the hsp 70 sequences extend only from the Sal I site to the Xho I site at −186, which is fused to the Sal I site of pML. The plasmid also contains a second SV40 origin fragment and a piece of the tk gene that extends from the Bam site upstream of the gene to a Hind III linker positioned at the start site of transcription (Figure 1). This fragment was derived from a deletion mutant constructed by S. McKnight, and it lacks all the transcribed sequences of the tk gene. pF2 was originally constructed for a different purpose, and the presence of this extra tk sequence is in fact irrelevant to the results presented in this paper.

The hsp 70 sequences in pF3 are derived from the right-hand gene in 132E3 (Figure 8B). They extend from the Hpa I site (converted to a Sal I site via an 8-base linker) about 400 bases upstream of the gene to one of the two overlapping Hha I sites at position −65 (see Figure 5). This sequence is very similar to that of the left-hand hsp 70 gene of 132E3, and is identical to it downstream of the Xho I site. The Sal I (Hpa I) site is joined to the Sal I site in pML, and the Hha I site is joined to the Hha I site at position −50 of the tk gene. The rest of the plasmid is identical to pXTK1 (Figure 8C).

pH1 is derived from pML by insertion of Drosophila and SV40 sequences between the Eco RI and Hind III sites (Figure 1). The Drosophila DNA consists of the Eco RI–Bgl II fragment from 56H8 (Figure 8B), which is joined to the late side of the origin + 72 bp rpt fragment of SV40 (Figure 8A), the SV40 Pvu II site having been converted to a Bam HI site by ligation of a linker (CCGGATCCGG). pH2 is a Bam HI–Hind III insert in pML (Figure 1). The Pvu II site on the SV40 origin + 72 bp rpt fragment was converted to a Sal I site by ligation of a linker (GGTCGACC). This was joined to the Xho I–Bgl II fragment containing the left-hand hsp 70 gene from 132E3 (Figure 8B). The Xho I site is at position −195 of this hsp 70 gene.

Transfection of COS Cells

Plasmid DNA was prepared by the rapid alkaline lysis method described by Ish-Horowicz and Burke (1981). High molecular weight RNA was removed by precipitation with 2.5 M LiCl, and the DNA was purified further by phenol extraction. Such DNA was as efficient at transfection as CsCl-gradient-purified DNA.

COS 7 cells (Gluzman, 1981) were maintained in Dulbecco's MEM with 10% calf serum, and transfected when slightly subconfluent. Each 75 cm^2 flask was exposed to 2–5 μg of DNA in 3 ml of medium (without serum) containing 200 μg/ml DEAE–dextran for 8 hr as described by Sompayrac and Danna (1981). Similar results were obtained with 30 min exposure to DNA in 2 ml containing 500 μg/ml DEAE–dextran. The DNA solution was then replaced with fresh medium (with serum) and the cells maintained at 37°C. For heat shock, the flasks were placed in a water bath at 42–43°C from 30 to 35 hr after transfection. Cells were then trypsinized, spun down and lysed on ice in 140 mM NaCl, 5 mM KCl, 3 mM MgCl$_2$, 25 mM Tris–HCl (pH 7.5), 1% NP40. Nuclei were removed by centrifugation, and RNA was prepared from the supernatant by phenol extraction in the presence of 1% SDS, followed by isopropanol precipitation.

S1 Mapping

For tk transcripts, the probe was the 131 bp Eco RI–Bgl II fragment that spans the transcription start (McKnight, 1980), labeled at the Bgl II site with polynucleotide kinase before Eco RI digestion, and purified by agarose gel electrophoresis. For the 5' end of the hsp 70 transcripts, probe was prepared from 132E3 or 56H8. One microgram of plasmid DNA was digested with Xho I in 33 mM Tris–acetate (pH 8), 66 mM KOAc, 10 mM MgCl$_2$, 0.5 mM DTT, 0.1 mg/ml gelatin. Five units of T4 DNA polymerase (BRL) were added and incubation continued for 30 min at 37°C Then 40 μCi α-^{32}P-dATP (400 Ci/mmole) and 100 μM each of dCTP, dTTP and dGTP were added, and 30 min later 250 μM dATP was added. After a 30 min chase, Hae III was added. Finally, the labeled 568 bp (56H8) or 573 bp (132E3) fragment was isolated by agarose gel electrophoresis. Digestion of this with other restriction enzymes showed that it was uniformly labeled by this procedure.

A similar method was used to prepare 3' probe from 132E3. In this case the initial digestion was with Bam HI, the Hae III digestion was omitted and the entire plasmid was used as probe after phenol extraction.

One tenth of the total cytoplasmic RNA from a flask of cells (about 10 μg) was mixed with the appropriate probe in 20 μl of 80% formamide, 10 mM PIPES (pH 6.5), 1 mM EDTA, 0.4 M NaCl, heated to 85°C for 5 min and hybridized overnight at 50°C. The hybrids were diluted with 200 μl of S1 buffer (225 mM NaCl, 75 mM Na$_{0.5}$H$_{0.5}$OAc (pH 4.5), 0.7 mM ZnSO$_4$), digested with 100–200 units of S1 nuclease (BRL) for 10 min at 37°C or 30 min at 22°C, mixed with 50 μl 5× stop buffer (100 mM EDTA, 0.4 M Tris, pH 9.5), ethanol-precipitated and analyzed on sequencing gels.

Acknowledgments

I thank S. McKnight and M. Botchall for providing plasmids and J. Gurdorn, A. Travers, R. Laskey and M. Bienz for helpful comments on the manuscript

Received May 7, 1982; revised June 17, 1982

References

Ashburner, M. and Bonner, J. J. (1973). The induction of gene activity in Drosophila by heat shock. Cell *17*, 241–254.

Banerji, J., Rusconi, S. and Schaffner, W. (1981). Expression of a β-globin gene is enhanced by remote SV40 DNA sequences. Cell *27*, 299–308.

Benoist, C. and Chambon, P (1981). In vivo sequence requirements of the SV40 early promoter region. Nature *290*, 304–310.

Berk, A. J. and Sharp, P. A. (1977). Sizing and mapping of early adenovirus mRNAs by gel electrophoresis of S1 endonuclease-digested hybrids. Cell *12*, 721–732.

Corces, V., Pellicer, A., Axel, R. and Meselson, M (1981). Integration, transcriptions and control of a Drosophila heat shock gene in mouse cells. Proc. Nat. Acad. Sci. USA *78*, 7038–7042.

Corden, J. Wasylyk, B., Buchwalder, A., Sassone-Corsi, P. and Uedinger, L. (1980). Promoter sequences of eucaryotic protein-coding genes. Science *209*, 1406–1414.

Dierks, P., van Ooyen, A., Mantei, N. and Weissman, C. (1981). DNA sequences preceding the rabbit β-globin gene are required for formation in mouse L cells of β-globin mRNA with the correct 5' terminus. Proc. Nat. Acad. Sci. USA *78*, 1411–1415.

Efstratiadis, A., Posakony, J. W., Maniatis, T., Lawn, R. M., O'Connell, C., Spritz, R. A., DeRiel, J. K., Forget, B. G., Weissman, S. M., Slightom, J. L., Blechl, A. E., Smithies, O., Baralle, F. E., Shoulders, C. C. and Proudfoot, N. J. (1980). The structure and evolution of the human β-globin gene family. Cell *21*, 653–668.

Elgin, S. C. R. (1981) DNAase I-hypersensitive sites of chromatin. Cell *27*, 413–415.

Gluzman, Y. (1981). SV40-transformed simian cells support the rep-

lication of early SV40 mutants. Cell *23*, 175–182.

Goldberg, M . (1979). Sequence analysis of Drosophila histone genes. Ph.D. thesis, Stanford University, Stanford, California.

Grosschedl, R. and Birnstiel, M. L. (1980a). Identification of regulatory sequences in the prelude sequences of an H2A histone gene by the study of specific deletion mutants in vivo. Proc. Nat. Acad. Sci. USA *77*, 1432–1436.

Grosschedl, R. and Birnstiel, M. L. (1980b). Spacer DNA sequences upstream of the T-A-T-A-A-A-T-A sequences are essential for promotion of H2A histone gene transcription in vivo. Proc. Nat. Acad. Sci. USA *77*, 7102–7106.

Grosveld, G. C., de Boer, E., Shewmaker, C. K. and Flavell, R. A. (1982). DNA sequences necessary for transcription of the rabbit β-globin gene in vivo. Nature *295*, 120–126.

Grosveld, G. C., Shewmaker, C. K., Jat, P. and Flavell, R. A. (1981) Localization of DNA sequences necessary for transcription of the rabbit β-globin gene in vitro. Cell *25*, 215–226.

Holmgren, R., Corces, V., Morimoto, R., Blackman, R. and Meselson, M. (1981). Sequence homologies in the 5' regions of four Drosophila heat-shock genes. Proc. Nat. Acad. Sci. USA *78*, 3775–3778.

Ingolia, T. D. and Craig, E. A. (1981). Primary sequence of the 5' flanking region of the Drosophila heat shock genes in chromosome subdivision 67B. Nucl. Acids Res. *9*, 1627–1642.

Ingolia, T. D., Craig, E. A. and McCarthy, B. J. (1980). Sequence of three copies of the gene for the major Drosophila heat shock induced protein and their flanking regions. Cell *21*, 669–679.

Ish-Horowicz, D. and Burke, J. F. (1981). Rapid and efficient cosmid cloning. Nucl. Acids Res. *9*, 2989–2998.

Jakobovits, E. B., Bratosin, S. and Aloni, Y. (1980). A nucleosome-free region in SV40 minichromosomes. Nature *285*, 263–265.

Karch, F., Torok, I. and Tissières, A. (1981). Extensive regions of homology in front of the two hsp 70 heat shock variant genes in Drosophila melanogaster. J. Mol. Biol. *148*, 219–230.

Lusky, M. and Botchan, M. (1981). Inhibition of SV40 replication in simian cells by specific pBR322 DNA sequences. Nature *293*, 79–81.

Maxam, A. M. and Gilbert, W. (1977). A new method for sequencing DNA. Proc. Nat. Acad. Sci. USA *74*, 560–564.

McKnight, S. L. (1980). The nucleotide sequence and transcript map of the herpes simplex virus thymidine kinase gene. Nucl. Acids Res. *8*, 5949–5964.

McKnight, S. L., Gavis, E. R., Kingsbury, R. and Axel, R. (1981). Analysis of transcriptional regulatory signals of the HSV thymidine kinase gene: identification of an upstream control region. Cell *25*, 385–398.

Mellon, P., Parker, V., Gluzman, Y. and Maniatis, T. (1981). Identification of DNA sequences required for transcription of the human α1-globin gene in a new SV40 host–vector system. Cell *27*, 279–288.

Moreau P., Hen, R., Wasylyk, B., Everett, R., Gaub, M. P. and Chambon, P., (1981). The SV40 72 base pair repeat has a striking effect on gene expression both in SV40 and other chimeric recombinants. Nucl. Acids Res. *9*, 6047–6068.

Peterson, R. C., Doering, J. L. and Brown, D. D. (1980). Characterization of two Xenopus somatic 5S DNAs and one minor oocyte-specific 5S DNA. Cell *20*, 131–141.

Saragosti, S., Moyne, G. and Yaniv, M. (1980). Absence of nucleosomes in a fraction of SV40 chromatin between the origin of replication and the region coding for the late leader RNA. Cell *20*, 65–73.

Sompayrac, L. M. and Danna, K. J. (1981). Efficient infection of monkey cells with DNA of simian virus 40. Proc. Nat. Acad. Sci. USA *78*, 7575–7578.

Tooze, J., ed. (1981). DNA tumour viruses. Cold Spring Harbor, N.Y.: Cold Spring Harbor Laboratory.

Torok, I. and Karch, F. (1980). Nucleotide sequences of heat shock activated genes in Drosophila melanogaster. I. Sequences in the regions of the 5' and 3' ends of the hsp 70 gene in the hybrid plasmid

56H8. Nucl. Acids Res. *8*, 3105–3123.

Udvardy, A., Sumegi, J., Toth, E. C., Gausz, J., Gyurkovics, H., Schedl, P. and Ish-Horowicz, D. (1982). Genomic organization and functional analysis of a deletion variant of the 87A7 heat shock locus of Drosophila melanogaster. J. Mol. Biol. *155*, 267–280.

Weaver, R. F. and Weissman, C. (1979). Mapping of RNA by a modification of the Berk–Sharp procedure: the 5′ termini of 15S β-globin mRNA precursor and mature 10S β-globin mRNA have identical map coordinates. Nucl. Acids Res. *7*, 1175–1193.

Wu, C. (1980). The 5′ ends of Drosophila heat shock genes in chromatin are hypersensitive to DNase I. Nature *286*, 854–860.

Parker & Topol (1984) Cell **37**, 273—283

A Drosophila RNA Polymerase II Transcription Factor Binds to the Regulatory Site of an hsp 70 Gene

Carl S. Parker and Joanne Topol
Division of Chemistry 147-75
California Institute of Technology
Pasadena, California 91125

Summary

A Drosophila RNA polymerase II transcription factor that is specific for at least one of the heat-shock genes has been isolated (designated HSTF for heat-shock transcription factor). This factor is required for active transcription of an hsp 70 gene in addition to RNA polymerase II and another general transcription factor, the A factor. Footprint analysis of the HSTF on the hsp 70 gene reveals that it binds specifically to a 55 bp region upstream from the TATA box. Both coding and noncoding DNA strands are completely protected from DNAase I cleavage by the HSTF. HSTF binding occurs in the apparent absence of RNA polymerase II. The HSTF is present in both heat-shocked and nonshocked cells, although it is more transcriptionally active when isolated from heat-shocked cells. The previously described B factor (an RNA polymerase II transcription factor that binds to the TATA box), isolated from nonshocked cells, is significantly reduced in both binding and transcriptional activity in heat-shocked cells. The potential role of the HSTF and the B factor in the activation of heat-shock gene transcription is discussed.

Introduction

The heat-shock response in Drosophila (as well as in a wide range of organisms) offers a potentially useful system for biochemical studies of transcription in eucaryotes. When Drosophila cells are subjected to elevated temperatures the cells rapidly activate the transcription and translation of seven different protein-coding genes, known as heat-shock genes. As the seven heat-shock genes are induced, the transcription and translation of other genes in the cell is greatly reduced (for reviews see Ashburner and Bonner, 1979; Peterson and Mitchell, 1983).

Several laboratories have shown that one of the Drosophila heat-shock genes, the hsp 70 gene (hsp for heat-shock protein), can be activated by heat shock in heterologous in vivo expression systems. For example, an hsp 70 gene is heat-inducible when integrated into the genome of mouse tissue culture cells (Corces et al., 1981), as is the hsp 70 gene in transient-expression assays in monkey COS cells (Pelham, 1982; Mirault et al., 1982). These experiments have indicated that a relatively small element of DNA is necessary and sufficient for transcriptional activation of the hsp 70 gene. This regulatory site consists of approximately 20 bp of DNA positioned just 5'-ward from the TATA box (Pelham, 1982).

Beginning with an active nuclear-extract preparation

derived from heat-shocked Drosophila K_c cells, we have identified a transcription factor specific for the heat-shock genes. The heat-shock-gene-specific transcription factor (abbreviated HSTF) is chromatographically distinct from the previously described B factor (Parker and Topol, 1984). The B factor is a Drosophila RNA polymerase II transcription factor that binds to a region of DNA surrounding the start point of transcription including the TATA box. The B factor is required, in addition to another chromatographically distinct transcription factor, the A factor, for accurate and efficient initiation of transcription on the Drosophila histone H3, H4, and actin (5C) genes. In this report we show that specific initiation of transcription in vitro on the hsp 70 gene by highly purified RNA polymerase II is dependent upon the HSTF and the A factor. Although addition of the B factor to the in vitro transcription reaction is not required for efficient transcription of the hsp 70 gene, we cannot rule out the presence of a low level of B factor in the transcription reactions.

We show by footprint analysis that the HSTF binds to a 55 bp region of DNA 5'-ward from the TATA box of the Drosophila hsp 70 gene. The binding site includes the 20 bp activator-promoter element described above. Footprint analysis also reveals a significant reduction in binding activity of the B factor in heat-shocked cells. We present a hypothesis, based on this and other information to be described, that suggests a possible role for the HSTF and B factor in the transcriptional activation of the hsp 70 genes.

Results

Effects of Heat Shock on In Vitro Transcription with K_c Cell Nuclear Extracts

To determine whether differences between the transcriptional activity of a heat-shock gene and a gene that is not transcribed during heat shock could be duplicated in vitro, we compared the level of transcription of these genes in a heat-shocked and a nonshocked K_c cell nuclear extract. The hsp 70 gene derived from the 87A7 cytological locus of Drosophila melanogaster (clone 56H8 of Artavantis-Tsakonas et al., 1979) was chosen as our standard heat-shock gene for all of the experiments described in this report. When the hsp 70 gene (designated 56H8 in Figure 1) is truncated by digestion at the Bgl I site and transcribed in a heat-shocked nuclear extract, a transcript of approximately 390 bases is synthesized (Figure 2A). This is the size of the in vitro transcription product predicted if initiation occurs at the same site as it does in vivo (Torok and Karch, 1980). Fingerprint analysis of this in vitro transcription product has demonstrated that initiation of transcription in vitro occurs precisely at the known in vivo start point (Scott and Parker, unpublished observations). The transcriptional activity observed for the hsp 70 gene in a heat-shocked extract is approximately 0.5 transcripts per gene in a 30 min reaction (Figure 2A). This level of synthesis is similar to that observed for the hsp 70 gene in a nonshocked nuclear extract (Fig. 2, compare A with B).

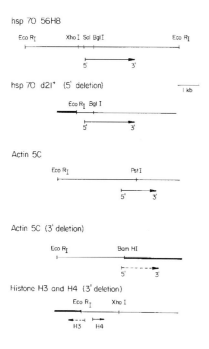

Figure 1. Templates and DNA Fragments Used in the Transcription and Footprinting Assays

The bold line indicates pBR322 sequences.

Similarly, an actin gene derived from the 5C cytological locus (Fyrberg et al., 1983) was chosen as a gene representative of those genes not normally transcribed during heat shock. This actin gene is a particularly good one for this purpose because Findly and Pederson (1981) have previously shown that this gene is not efficiently transcribed during heat shock in K_c cells in vivo. The actin 5C gene (designated Actin 5C in Figure 1), truncated at the Pst I site, is efficiently transcribed in a nonshocked nuclear extract (Figure 2B). The actin gene, however, is transcribed very inefficiently in a heat-shocked nuclear extract (Figure 2A). The low level of actin (5C) transcription observed in vivo is also found in vitro.

These preliminary observations suggested that a transcription factor specific for the heat-shock genes might be active in the heat-shocked K_c cell nuclear extract. They further suggested that one or more factors required for actin (5C) gene transcription in vitro are reduced in activity or abundance in the heat-shocked nuclear extract. We therefore attempted to isolate the factor required for hsp 70 gene transcription in vitro from heat-shocked K_c cells using standard biochemical procedures.

Isolation of a Transcription Factor Specific for the Heat-Shock Gene

The chromatographic procedures previously used to fractionate a nonshocked K_c cell nuclear extract were used to fractionate the heat-shocked nuclear extract (Parker and Topol, 1984). A diagrammatic summary of the chromato-

graphic steps used is shown in Figure 3. The results of chromatography of a heat-shocked K_c cell nuclear extract on a DEAE cellulose (DE 52) column is shown in Figure 4A. Gradient elution of the proteins adsorbed to the resin was performed from 0.1 to 0.6 M KCl in our standard chromatography buffer (see Experimental Procedures). Specific transcription of the hsp 70 gene requires the addition of the DE 52 flow-through (fraction A) to an aliquot from the appropriate DE 52 column fraction (all components are derived from heat-shocked cells). Transcription-factor activity is more clearly observed when 4 U Drosophila RNA polymerase II (also derived from heat-shocked cells; see Experimental Procedures) is added to the reconstitution reaction. When this was done, the transcription activity was found to elute from the DE 52 column just ahead of the peak of RNA polymerase II activity (fractions 20–23 in Figure 4A). Analysis of the transcriptional activity of the DE 52 column fractions on the actin 5C gene reveals that the actin gene is very inefficiently transcribed (Figure 4A).

The indicated fractions (labeled HSTF) were chosen so as to minimize the levels of contaminating RNA polymerase II activity. Subsequent chromatography of these pooled fractions on a DEAE Sephadex (A25) column resulted in the separation of the HSTF from RNA polymerase II (Figure 4B). Transcription of the hsp 70 gene is now dependent upon the addition of RNA polymerase II to the A factor and the HSTF (Figure 4B, compare lanes 1 and 2).

The A 25-derived HSTF was then chromatographed on a Biorex 70 column (Figure 5A; see Experimental Procedures). The HSTF eluted from the column between 75 and 250 mM KCl, a position distinct from our previously reported B factor (B factor elutes between 300 and 450 mM KCl; Parker and Topol, 1984). The most active Biorex 70 fractions were pooled and chromatographed on a double-stranded DNA cellulose column (Figure 5B; see Experimental Procedures for details). Fractions that possessed transcriptional activity in the presence of the DE 52 A fraction and 4 U RNA polymerase II were frozen separately in small aliquots.

Efficient transcription of the hsp 70 gene in vitro requires addition of RNA polymerase II to the HSTF and A fractions. However, because of the chromatographic similarities of the HSTF and RNA polymerase II on the resins we have used for isolation, we cannot rule out the presence of either a small amount of active RNA polymerase II or the presence of certain subunits of RNA polymerase II (perhaps noncatalytic). These putative RNA polymerase II components may be important for the HSTF transcriptional activity we observe. The net purification and yield of HSTF activity is summarized in Table 1. Specific activity measurements at the Biorex 70 and DNA cellulose steps listed in Table 1 are based on the titration assays shown in Figures 6A and 6B, respectively.

It is interesting to note that once the HSTF has been purified through several chromatographic steps, the in vitro transcripts synthesized from the hsp 70 gene often show

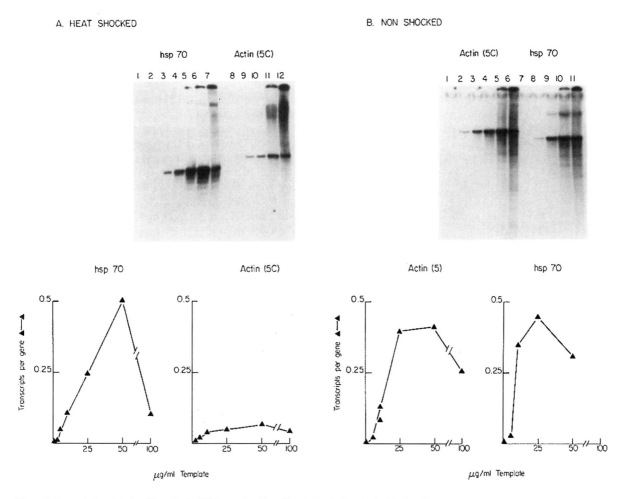

Figure 2. Transcription of the hsp 70 and Actin (5C) Genes in a Heat-Shocked and a Nonshocked Nuclear Extract

(A) Heat-shocked nuclear extract. The indicated amount of DNA template was transcribed with 15 μl of nuclear extract. In lanes 1–7, the template is the hsp 70 gene contained in the 56H8 plasmid linearized with Bgl I. In lanes 8–12, the template is the actin (5C) gene contained in the plasmid DmA2 linearized with Pst I. Lane 1: no DNA; lane 2: 20 ng; lane 3: 0.1 μg; lane 4: 0.2 μg; lane 5: 0.5 μg; lane 6: 1 μg; lane 7: 2 μg; lane 8: 0.1 μg; lane 9: 0.2 μg; lane 10: 0.5 μg; lane 11: 1 μg; lane 12: 2 μg.

(B) Nonshocked nuclear extract. The indicated amount of DNA template was transcribed with 15 μl of nuclear extract. In lanes 1–6, the template is the actin 5C gene contained in the plasmid DmA2 linearized with Pst I. In lanes 7–11, the template is the hsp 70 gene contained in the 56H8 plasmid linearized with Bgl I. Lane 1: 20 ng; lane 2: 0.1 μg; lane 3: 0.2 μg; lane 4: 0.5 μg; lane 5: 1 μg; lane 6: 2 μg; lane 7: 20 ng; lane 8: 0.1 μg; lane 9: 0.2 μg; lane 10: 0.5 μg; lane 11: 1 μg.

length heterogeneity (see Figure 6, A, B. and C). The length heterogeneity is always smaller than the normal 390 base transcript observed when the unfractionated nuclear extracts are used (compare the results shown in Figure 4 to those in Figure 6). We have not determined the source of the heterogeneity in any detail; however, one possibility is that the heterogeneity is due to the absence of the B factor (the HSTF and B factor are separated chromatographically on Biorex 70). In the absence of the B factor RNA polymerase II may not be positioned precisely at the start site of transcription. Reconstitution-transcription assays in which highly purified B factor and HSTF are combined to test this possibility have not yet been performed. Other explanations, including premature termination or degradation of the transcript, are possible and cannot be ruled out at this time.

The most highly purified HSTF DNA cellulose fraction was assessed for its ability to reconstitute specific initiation on the Drosophila actin 5C gene in combination with RNA polymerase II and the A factor. As shown in Figure 6D, no specific initiation was observed on the actin gene. The lack of specific initiation on the actin gene was observed even when saturating levels of the HSTF for the hsp 70 gene were used (compare lanes 11–14 in 6D to lanes 4–7 in 6B).

Footprint Analysis of the HSTF

One possible mechanism by which the HSTF could effect specific initiation on the hsp 70 gene and not on the Drosophila actin 5C gene would involve sequence-specific DNA binding by the transcription factor To address this possibility, footprint analysis (Galas and Schmitz, 1978)

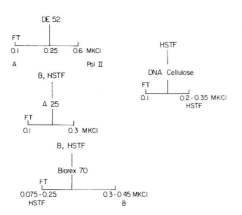

Figure 3. Fractionation Scheme of the Heat-Shocked Nuclear Extract

was performed on the hsp 70 gene with the HSTF at the Biorex 70 and DNA cellulose purification steps. Specific binding to a 55 bp region 5'-ward (upstream) from the TATA box was observed on the hsp 70 gene with those Biorex 70 fractions that reconstituted specific initiation (as indicated by the bracket labeled "HSTF-binding" in Figure 6A). The same region of the hsp 70 gene was also protected from DNAase I cleavage with the DNA cellulose fractions that reconstituted specific transcription (indicated by the bracket labeled "HSTF-binding" in Figure 5B).

A detailed analysis of the HSTF footprint is shown in Figure 7. The coding strand is completely protected from position −37 to −92. The noncoding strand is also completely protected in a slightly more 5'-ward position from −41 to −99. The sequences present in the binding site include the region shown by Pelham (1982) to be critical for the transcriptional activation of a Drosophila hsp 70 gene in COS cells (labeled "consensus sequence" in Figure 7). (A diagrammatic summary of the binding studies is shown in Figure 10A.)

Titration of the HSTF using the footprint assays allows us to estimate the purity of the HSTF preparations. We estimate that the final preparations are between 10%–20% HSTF, assuming two proteins of 50,000 daltons (for the sake of argument) are required to protect fully the 55 bp binding site. This value assumes that all binding proteins in the preparation are active and is likely to be an underestimate of the actual purity. For these reasons we argue that the components that bind specifically to DNA are likely to be an important component of the transcription factor preparation. Transcription and binding studies with deletion mutants in the binding site also support this claim (Topol and Parker, unpublished data). Only when the transcription factor is completely purified will we know with certainty whether the component that binds specifically to DNA is the transcription factor.

When we compare the transcriptional activity concentration of the HSTF fractions at the Biorex 70 step to the DNA cellulose step (Table 1) we observe no significant increase. The concentration of binding protein, however, has increased by at least a factor of three (see Figure 5 and the

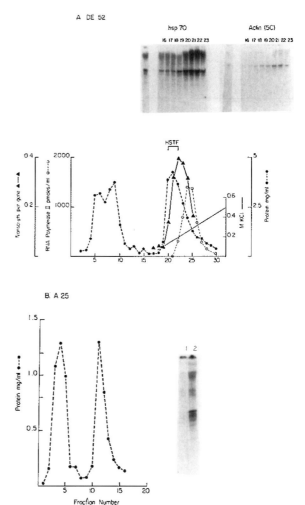

Figure 4. DEAE Cellulose (DE 52) and DEAE Sephadex (A25) Chromatography of the HSTF

(A) DEAE cellulose gradient chromatography. To 10 μl of each of the indicated DE 52 fractions, 10 μl of the DE 52 flow-through (fraction A), 4 U highly purified RNA polymerase II, and 250 ng of a Bgl I-cleaved hsp 70 (56H8) template were combined in the reconstitution-transcription assays. Note that the peak of transcriptional activity (fractions 22–23) precedes the RNA polymerase II peak (fractions 24–25) The HSTF activity present in fractions 22–24 was pooled and chromatographed on the DEAE Sephadex column. Alongside of the results obtained with the hsp 70 gene are shown the results obtained with an actin 5C gene The same amounts of RNA polymerase II, A factor, and DE 52 column fractions were combined with 250 ng of Pst I-cleaved actin (SC) template. Lane i is the input to the column, in which 15 μl of the heat-shocked nuclear extract was used to transcribe 250 ng of the hsp 70 gene (56H8) linearized with Bgl I, or 250 ng of the actin (SC) gene linearized with Pst I.
(B) DEAE Sephadex step-elution chromatography The template used in these assays was 200 ng of Bgl I-cleaved hsp 70 gene (56H8). Lane 1: 10 μl DE 52 fraction A plus 10 μl of the 0.3 M KCl step-eluted A25 column fraction. Lane 2: 10 μl of DE 52 fraction A plus 10 μl of the 0.3 M KCl step-eluted A25 column fraction plus 4 U partially purified RNA polymerase II.

legend to Figure 5 for details). Thus it seems likely that the HSTF may be more labile as a transcription factor than a binding protein. We know that further addition of either the DE 52 A factor or RNA polymerase II to the DNA cellulose-

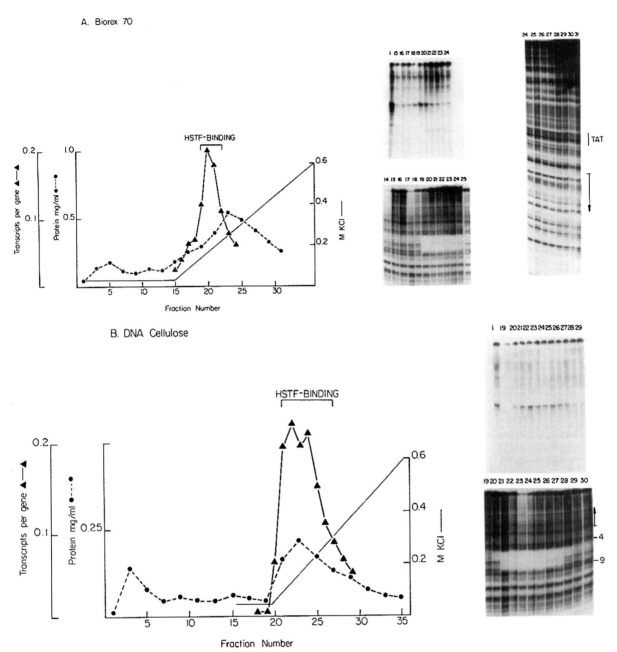

Figure 5. Biorex 70 and Double-Stranded DNA Cellulose Chromatography of the HSTF

(A) Biorex 70 gradient elution of the step-eluted A25 material. The upper portion of the panel shows the reconstitution-transcription assays obtained by assaying 10 μl of each of the indicated Biorex 70 fractions with 10 μl of the DE 52 A fraction and 4 U partially purified Drosophila RNA polymerase II. The template used in these assays was 250 ng of the hsp 70 gene (56H8) cut with Bgl I. Directly below the transcription assays are the results of the footprint assays. The footprint assays employed 15 μl of the indicated Biorex 70 fractions, and 5 ng of the hsp 70 DNA fragment designed d21+ diagrammed in Figure 1. The fragment was labeled by DNA polymerase I (Klenow fragment) filling at a unique Eco RI site approximately 250 bp upstream (5′-ward) from the start point of transcription. The fragment was then gel-purified after secondary cleavage with Bgl I. The footprint assays shown in the rightmost portion of the panel are those in which the histone H3 and H4 genes were used to identify B-factor binding activity. Fifteen microliters of each of the indicated Biorex 70 fractions was also used in the histone-gene footprinting assays. The histone gene containing Eco RI to Xho I fragment used is diagrammed in Figure 1 (designated histone H3 and H4 3′ deletion). The fragment was labeled by DNA polymerase I (Klenow fragment) filling at the Eco RI site, which is approximately 55 bp 3′-ward from the start point of transcription of the histone H3 gene. The TATA box for the histone H3 gene is indicated by the line alongside of the footprint and the start point of transcription is indicated by the arrow. The bracketed fractions (labeled HSTF-binding) were pooled and subsequently chromatographed on a double-stranded DNA cellulose column.

(B) DNA cellulose gradient-elution chromatography. The reconstitution-transcription assays shown were performed with 10 μl of each indicated column fraction and 10 μl of DE 52 fraction A plus 4 U partially purified RNA polymerase II. The template used was 250 ng of the hsp 70 gene (56H8) cut at the Bgl I site. Directly below the transcription assays are the results of footprint analysis using 5 μl of the indicated DNA cellulose fractions. The DNA fragment used in these assays was the hsp 70 DNA fragment described in (A) of this figure. Fractions containing the maximal reconstitution activity were stored separately in small aliquots at −70°C after dialysis.

Table 1. Purification Table of the HSTF

Fraction	Protein Concentration	Total Volume	Total Protein	Transcription[a] Units/ml	Specific Activity[b]	Total Transcription Units	Yield (%)
K_c nuclear extract (heat-shocked)	8.5 μg/ml	30 ml	255 mg	33		990	100
DE52(pooled)	4.6μg/ml	24ml	110mg	27		648	65
A 25 (pooled)	686 μg/ml	12 ml	8.2 mg	13		156	15.7
Biorex 70 (pooled)	350 μg/ml	4.4 ml	1.54 mg	40	0.1	176	17.7
DNA cellulose	150 μg/ml	0.9 ml	135μg	53	0.35	47	3.6

[a]Transcriptional Units/ml = number of transcripts per gene in a 30 min reaction per ml of material assayed This was determined where a linear relationship existed between the volume of material assayed and the number of transcripts synthesized.

[b]Specific activity = number of transcriptional units per microgram of protein assayed. This was also measured in the linear range of the material being assayed. Only in the cases of the Biorex 70 and DNA cellulose fractions can this value be directly applied to the HSTF transcriptional activity. DNA cellulose column fractions were not pooled, the value shown is a calculated average based on the total transcription units and the total amount of protein present in the active fractions

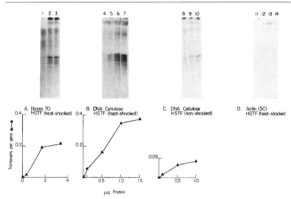

Figure 6. HSTF Concentration and Specific Activity Determinations

All reactions in (A), (B), and (C) contained 250 ng of a Bgl I-cut hsp 70 gene (56H8) and 10 μl of the DE 52 A fraction plus 4 U partially purified RNA polymerase II. (A) Biorex 70 pooled fractions from heat-shocked cells. The indicated amounts (in mass of protein) of the pooled Biorex 70 fractions were added as follows: Lane 1: 330 ng; lane 2: 1.6 μg; lane 3: 3.3 μg. (B) DNA cellulose HSTF from heat-shocked cells. The following amounts (in mass of protein) of a DNA cellulose fraction were added to the transcription assays. Lane 4: 100 ng; lane 5: 0.5 μg: lane 6: 1 μg; lane 7: 1.5 μg. (C) DNA cellulose HSTF from nonshocked cells. The following amounts (in mass of protein) of a DNA cellulose fraction of the HSTF from nonshocked cells were added to the transcription assays. Lane 8: 100 ng; lane 9: 0.5 μg; lane 10: 1 μg. (D) DNA cellulose HSTF from heat-shocked cells The following amounts of a DNA cellulose fraction of the HSTF from heat-shocked cells were added to transcription assays that contained 250 ng of the actin (5C) gene linearized with Pst I. The reactions contained the same amounts of the A factor and RNA polymerase II as described for (A), (B), and (C). Lane 11: 100 ng; lane 12: 0.5 μg; lane 13: 1 μg; lane 14: 1.5 μg.

derived HSTF does not result in a significant stimulation of specific hsp 70 gene transcription. Similarly, addition of the DNA cellulose flow-through fraction to the reconstitution reactions employing the DNA cellulose-derived HSTF does not stimulate hsp 70 gene transcription. We have not yet tested the effects of B-factor addition to transcription reactions employing Biorex 70 or DNA cellulose-derived HSTF.

The HSTF is also found in nonshocked K_c cell nuclear extracts. As shown in Figure 6C, a highly purified HSTF preparation derived from nonshocked cells (purified by the

same four chromatographic steps that were used for the HSTF derived from heat-shocked cells) can reconstitute specific synthesis on the hsp 70 gene. The HSTF derived from nonshocked cells is significantly less active, possessing only 0.06 U/μg of protein, than the equivalent HSTF derived from heat-shocked cells, which has a specific activity of approximately 0.35 U/μg of protein (compare Figure 6C with 6B). The pattern of protection from DNAase I cleavage on the hsp 70 gene with the highly purified HSTF derived from nonshocked cells is, however, identical with that obtained from heat-shocked cells (data not shown). These observations suggest that the HSTF isolated from heat-shocked cells may be more active as a transcription factor than the equivalent factor derived from nonshocked cells.

The B-Factor Activity Is Reduced in Heat-Shocked Cells

As described previously, a Drosophila RNA polymerase II transcription factor, designated the B factor (Parker and Topol, 1984), is present in nonshocked K_c cells and is required for efficient transcription of the actin 5C and histone H3 and H4 genes in vitro. Highly purified preparations of this factor were shown to bind to a 65 bp region of DNA including the TATA box and leader region of the Drosophila actin 5C and histone H3 and H4 genes (Parker and Topol, 1984). When footprint analyses with the histone H3 and H4 genes were carried out on the Biorex 70 column fractions derived from heat-shocked cells, no specific binding was observed in fractions that would contain the B activity in nonshocked cells (Figure 5A, fractions 24–31). The sensitivity of the footprint assays are such that we estimate a minimal reduction of 5 to 10 fold in B-factor binding activity. The reduction in binding activity is similar to the decline in transcriptional activity observed with heat-shocked nuclear extracts when either the actin 5C gene (Figure 2A) or the histone H3 and H4 genes (data not shown) are used as templates. We suggest that the apparent absence of B-factor activity is responsible, at least in part, for the reduction in transcription of the actin 5C

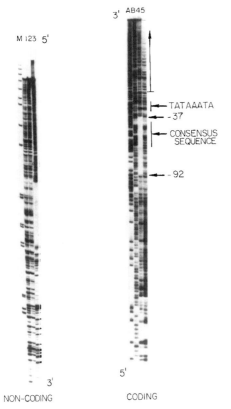

Figure 7. Footprint Analysis of the HSTF on the hsp 70 Gene

The noncoding strand was labeled at the Xho I site of the plasmid 56H8 by DNA polymerase I (Klenow fragment) filling (see Figure 1). This results in a fragment labeled approximately 200 bp 5′-ward from the start point of transcription. The coding strand was labeled by T4 polynucleotide kinase at the same Xho I site of 56H8. The resulting end-labeled DNA was cut with Bgl I and the fragment isolated. Approximately 5 ng of DNA fragment was used in each experiment. Lane A: chemical cleavage at purine residues; lane B: chemical cleavage of pyrimidine residues. Lanes 1, 2, and 5: no HSTF additions; lanes 3 and 4: 100 ng of DNA cellulose HSTF (from heat-shocked cells).

and histone H3 and H4 genes in vitro with nuclear extracts derived from heat-shocked cells.

This suggestion is strengthened by the observations that addition of the B factor to the heat-shocked nuclear extracts can stimulate specific transcription of the actin 5C gene approximately 8-fold (Figure 8). Similarly, addition of B factor to the heat-shocked nuclear extract can also stimulate transcription of the hsp 70 gene (3-fold in the example shown in Figure 8). It is apparent, however, that the B factor need not be present in normal levels for the hsp 70 gene to be transcribed efficiently in vitro.

Both the HSTF and the B factor can bind to the hsp 70 gene simultaneously as revealed by the footprint analysis shown in Figure 9. We observe the characteristic hypersensitive sites (to DNAase I cleavage) near the start point of transcription on the coding strand of the hsp 70 gene that were previously observed for the histone H3 gene upon B-factor binding (Parker and Topol, 1984). The 3′ border of protection by the B factor on the hsp 70 gene

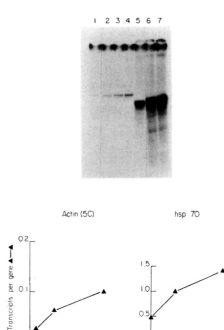

Figure 8. Effects of the Addition of Highly Purified B Factor to Heat-Shocked Nuclear Extracts

In lanes 1–4, the template is the actin (5C) gene truncated at the Pst I site. In lanes 5–7, the template is the hsp 70 gene truncated at the Bgl I site. All reactions contained 15 μl of a heat-shocked nuclear extract The indicated amounts of B factor (DNA cellulose derived, see text) were added to the reactions 5 min prior to the addition of the nuclear extract. Lane 1: no additions; lane 2: 0.25 μg; lane 3: 1 μg; lane 4: 3 μg; lane 5: no additions; lane 6: 1 μg; lane 7: 3 μg

(position +29) is similar to that observed on the histone H3 gene (position +30). A diagrammatic summary of these results is shown in Figure 10B. We do not know at the present time whether B factor will bind efficiently to the hsp 70 gene in the absence of the HSTF. This question is currently being investigated.

Discussion

The heat-shock response in Drosophila offers an interesting system to approach the molecular mechanism of specific activation and general repression of eucaryotic gene transcription. Beginning with a very active nuclear extract obtained from heat-shocked Drosophila cultured cells, we have isolated a transcription factor (abbreviated HSTF) specific for the heat-shock gene. This factor, in combination with the A factor and a highly purified preparation of RNA polymerase II, will reconstitute specific initiation of transcription on an hsp 70 gene but not on an actin 5C gene. Footprint analysis of the HSTF reveals that it binds to a 55 bp region upstream from the TATA box of an hsp 70 gene. This region includes a consensus sequence that, when placed in an analogous position upstream from the herpes simplex virus thymidine kinase (tk)

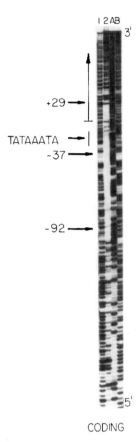

1 2 A B

3'

+29 →

TATAAATA →

-37 →

-92 →

5'

CODING

Figure 9. Simultaneous Binding of the B Factor and the HSTF to an hsp 70 Gene

The DNA fragment was labeled by T4 polynucleotide kinase at the Xho I site of 56H8 and isolated after cleavage with Bgl I. Approximately 5 ng of fragment was used in the footprint reaction shown, along with 1 μg of DNA cellulose-derived B factor (Parker and Topol, 1984) and 0.2 μg of DNA cellulose-derived HSTF. Lane A: chemical cleavage of the fragment at purine residues. Lane B: chemical cleavage of the fragment at pyrimidine residues. Lane 1: no protein additions. Lane 2: the footprint reaction.

gene, confers heat inducibility to the tk gene (Pelham, 1982). The role of DNA sequences in similar positions (50 to 120 bp) upstream from the TATA homology has been determined for a variety of other eucaryotic genes. Studies with the SV40 early, metallothionein, globin, and herpes virus thymidine kinase genes demonstrate that in the absence of these "upstream" sequences, there is a marked reduction in transcription (Fromm and Berg, 1982; Brinster et al., 1982; Dierks et al., 1981; Mellon et al., 1981; Grosveld et al., 1982; McKnight et al., 1981; McKnight and Kingsbury, 1982). Thus, for a variety of different eucaryotic genes, upstream sequences play a critical role for transcription in vivo. Dynan and Tjian (1983) have recently demonstrated that an SV40-specific transcription factor (Sp1) can bind to upstream sequences. The Sp1 transcription factor may thus represent an activity analogous to the HSTF we describe.

The footprint assays allow us to calculate the number of templates that could potentially be completely protected

from DNAase I cleavage by the HSTF in the in vitro transcription assays. When a saturating amount of DNA cellulose-derived factor (determined by footprinting assays) is present in an in vitro transcription reaction, only 0.25 transcripts per gene are synthesized. This observation suggests that not all of the genes that bind factor are transcribed. It is possible that a certain fraction of proteins that bind DNA are inactive as transcription factors, or some other component is limiting in the reactions. Another possibility that cannot be ruled out at this time is that nonspecific inhibitors of transcription and HSTF binding are present in the DE 52 A factor or in the RNA polymerase II preparation.

We observe that the HSTF (as a binding protein) is present in nonshocked cells at approximately the same levels as in heat-shocked cells. This is consistent with the finding that protein synthesis is apparently not required for the activation of heat-shock-gene transcription in vivo (Ashburner and Bonner, 1979). The HSTF derived from heat-shocked cells is, however, generally more active in transcription assays than the equivalent factor derived from nonshocked cells (approximately 6-fold for the experiment shown in Figure 6). It is interesting to speculate that, upon heat shock, the preexisting HSTF is activated and thus can positively affect heat-shock-gene transcription. However, the differences in specific activity that we observe between a heat-shocked and nonshocked HSTF probably do not account, in and of themselves, for the greatly enhanced level of transcription observed in vivo for the hsp 70 gene. From these observations it is apparent that binding of the HSTF to DNA alone is not sufficient to bring about positive modulation of RNA polymerase II initiation rates. We know, however, that binding is necessary for positive effects of the HSTF on RNA polymerase II initiation rates when an HSTF derived from heat-shocked cells is used in the reconstitution assays (Topol and Parker, unpublished data). The possibility exists that when we prepare nuclear extracts from the nonshocked cells, the cells are subjected to a certain amount of stress (anoxia) and for this reason may be somewhat heat-shocked. The low level of transcriptional activity that we observe with the HSTF isolated from nonshocked cells may be due to heat shock induced by the initial extraction procedure, and may not be an inherent feature of the HSTF found in nonshocked cells. It is essential that the active form of the HSTF be identified to learn precisely what, if any, modifications occur to the HSTF upon heat shock.

The observation that the B factor is very difficult to distinguish in the appropriate Biorex 70 column fractions from heat-shocked material leads us to suggest that the reduced activity or abundance of the B factor may be responsible, at least in part, for the reduced transcriptional activity observed on the actin (5C) and histone H3 and H4 genes in vitro. The reduced effective activity of the B factor may also be responsible, to a certain extent, for the reduction in transcription observed for many genes upon heat shock in vivo. It is likely that more than just the B

A

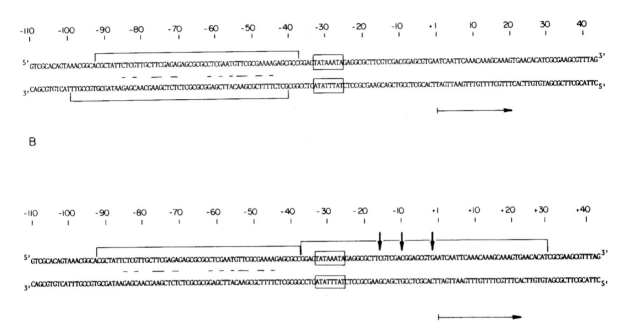

B

Figure 10. Diagrammatic Summary of the Protection Experiments

(A) Protection from DNAase I cleavage by DNA cellulose-derived HSTF on both the coding and noncoding strands. The protected region is indicated by the brackets.

(B) Protection from DNAase I cleavage by DNA cellulose-derived HSTF and B factor. Only the coding strand was analyzed. The contiguous region protected by addition of the B factor is indicated by the raised bracket. The three hypersensitive sites are indicated by the arrows.

factor is affected by heat shock, because addition of B factor alone does not fully restore nonshocked levels of actin (5C) transcription in a heat-shocked nuclear extract. Other explanations for the loss of B-factor activity upon heat shock include nonspecific proteolysis of the B factor, its rapid export from the nucleus, or its tighter association with the chromatin (it is therefore less readily extracted from the nuclei). Nonspecific proteolysis is less likely because neither the HSTF nor RNA polymerase II activity decreases upon heat shock, but we cannot exclude the latter two possibilities at this time.

Both the HSTF and B factor can bind to the hsp 70 gene simultaneously. Therefore, in a nonshocked cell, the B factor and the HSTF may both be bound to the hsp 70 gene and allow a certain level of transcription to occur from the hsp 70 gene. Once the cell has been heat-shocked, however, the B-factor activities (with regard to both binding and transcription) are reduced (we estimate at least 5 to 10 fold). However, the HSTF is active after heat shock and genes to which the HSTF can bind have the potential of being transcribed during heat shock. Genes that require the B factor to be transcribed but can no longer bind B factor (because of its reduced abundance or activity) may not be transcribed efficiently in vivo.

Although it is tempting to speculate that upon heat shock the HSTF replaces the B factor, it is possible that some B

factor remains active after heat shock and is essential for efficient hsp 70 gene transcription in vivo. This may explain why the actin (5C) gene is transcribed, albeit at a low level, in a heat-shocked nuclear extract. We cannot rule out the possibility that a certain low level of B factor is present in our reconstitution reactions, even with the most highly purified material. Therefore, it will be useful to know if mutations in the TATA box region of the hsp 70 gene that eliminate B factor binding will also prevent transcription in reconstitution assays with the highly purified HSTF. Should future investigations reveal that the B factor is required for the optimal level of hsp 70 gene transcription in vitro and in vivo, we would want to know what effects the HSTF has on B-factor binding to the hsp 70 promoter. It might be possible that HSTF binding aids in the binding of the B factor to the hsp 70 promoter. Thus, even at reduced levels of B-factor activity, B factor might be present at sufficient levels to bind stably to the hsp 70 gene but not to other genes.

Positive control of transcription in procaryotes is often brought about by the binding of activator proteins to specific sequences of DNA (Rosenberg and Court, 1979; Johnson et al., 1981; Hochschild et al., 1983; Hawley and McClure, 1983). These DNA sequences are generally located 5′-ward (upstream) from the bacterial promoter consensus sequence or Pribnow box. Although direct analogies to procaryotic positive control mechanisms at this point are rather premature, one cannot overlook the similarities between the spacial [*sic*] distribution of the positive effector molecules of procaryotes and the HSTF of Drosophila in the promoter region. It is interesting to speculate that protein–protein contacts between the HSTF and RNA polymerase II may be required for initiation of transcription at the hsp 70 promoter.

Experimental Procedures

Cloned Drosophila DNA Segments
The templates used in the in vitro transcription assays were an hsp 70 gene, 56H8 (Artavantis-Tsakonas et al., 1979), derived from the 87A7 cytological locus and an actin gene, DmA2 (Fyrberg et al., 1983), derived from the 5C cytological locus The 56H8 plasmid contains a complete copy of one hsp 70 gene. DmA2 is an Eco RI to Sal I subclone, and is known to be actively transcribed in the K_c cells used in this study (Fyrberg et al., 1983). These plasmids are shown in Figure 1.

Nuclear-Extract Preparation
The procedures used have been described in detail (Parker and Topol, 1984). The K_c cells were heat-shocked by immersion in a 42°C water bath until the media temperature rose to 36°C. The cells were then transferred to a 36°C water bath for 10 min. and subsequently harvested by centrifugation The cell densities were between 2–6×10^6 cells/ml.

In Vitro Transcription Assays
The standard reaction consists of the following components in a final reaction volume of 50 μl: 25 mM Hepes (pH 7.6), between 12 to 70 mM KCl (60 mM optimum), 5 mM MgCl$_2$, 0.6 mM each of the three unlabeled nucleoside triphosphates, 35 μM of the labeled nucleoside triphosphate (generally CTP with a specific activity in the reaction of approximately 2,600 cpm/pmole), 3% glycerol, 0.3 mM DTT, 0.03 mM EDTA, and DNA as indicated. Approximately 33% of the reaction volume consists of the nuclear extract (15 μl of 50 μl) or 20%–70% of the reaction volume was obtained from the various column fractions (10–35 μl of 50 μl) The appropriate figure legends contain the details of the amount of each component added. Reactions were initiated by the addition of extract to an otherwise complete reaction. The reactions were performed at room temperature (20–25 °C) for 30 min. Reactions were terminated by the addition of 100 μl of termination buffer (1% Sarkosyl, 100 mM NaCl, 100 mM Tris [pH 8.0], 10 mM EDTA) and extracted with 100 μl of redistilled and neutralized phenol. After separation of the organic and aqueous phases, nucleic acids were precipitated from the aqueous phase by the addition of 500 μl of 90% ethanol, 0.1 M NaOAc.

The template concentration used in the reconstitution assays was 10 μg/ml The hsp 70 gene was truncated at the Bgl I site, producing a 390 base in vitro transcript (see Figure 1). The actin (5C) gene (DmA2 of Fyrberg et al., 1983) was truncated at the Pst I site, producing a runoff transcript of approximately 550 bases (see Figure 1). Only runoff transcripts were synthesized from both templates.

The in vitro-synthesized RNA was analyzed by electrophoresis in denaturing 5% polyacrylamide 6 M urea gels as described (Parker and Topol, 1984). Standard RNA polymerase assays were performed essentially as described by Schwartz et al. (1974). Protein determinations were made according to the procedure of Bradford (1976) with bovine serum albumin as the standard

HSTF Chromatography
Buffers and Resins
The standard chromatography buffer was HGED, which consists of 25 mM Hepes (pH 7.6), 10% (v/v) glycerol, 0.1 mM EDTA, 1 mM dithiothreitol. KCl was added to the standard buffer and its molarity is indicated as follows: 0.1 HGKED is 0.1 M KCl. DEAE cellulose (Whatman DE 52), DEAE Sephadex (Pharmacia) A 25, phosphocellulose (P11), and Biorex 70 (Biorad) resins were all prepared according to the manufacturers' recommendations. Double-stranded DNA cellulose was prepared following the method of Alberts and Herrick (1971).

Chromatographic Procedures
Thirty milliliters of the heat-shocked Drosophila nuclear extract (obtained from two 30 l preparations) were chromatographed on a 30 ml DE 52 column. A 150 ml 0.1 to 0.6 M KCl gradient elution was carried out. The transcriptional activity of each fraction was independently determined after each fraction was dialyzed against 0.04 HGKED for 1–2 hr.

Fractions containing HSTF activity were pooled (approximately 110 mg of protein) and chromatographed on a 15 ml DEAE Sephadex (A 25) column. The protein that bound to the column was step-eluted with 0.3 HGKED, pooled, and dialyzed against 0.05 HGKED for 1 to 2 hr. The pooled material (8.2 mg of protein) was next chromatographed on a 5 ml Biorex 70 column. The protein bound to the column was eluted with a 25 ml 0.1 to 0.6 M KCl gradient Each fraction was dialyzed separately against 0.04 HGKED and assayed for HSTF transcriptional and DNA binding activity.

The active fractions were pooled (approximately 1.5 mg of protein) and applied to a 0.4 ml double-stranded DNA cellulose column. The proteins adsorbed were eluted with a 2 ml 0.1 to 0.6 M KCl gradient. Individual fractions (0.15 ml) were dialyzed for 1 to 2 hr prior to transcription and footprint analysis. Active fractions were divided into three 50 μl aliquots and stored at -70°C.

RNA Polymerase II Isolation
RNA polymerase II fractions from the above-described large-scale DEAE cellulose column that did not contain any assayable HSTF activity were pooled (approximately 50 mg of protein). The enzyme was applied to a 10 ml A 25 column and eluted with a 5-column volume gradient. RNA polymerase II activity was determined essentially as described by Schwartz et al (1974). The RNA polymerase II-containing fractions were pooled (approximately 75 mg of protein), with the exception of the leading edge of the peak This was done to avoid including any residual HSTF activity. After a 4 hr dialysis, the enzyme was chromatographed on a 0.5 ml phosphocellulose column. The enzyme was eluted with a 5-column volume KCl gradient from 0.1 to 0.4 HGKED. This step, in addition to concentrating the enzyme further, removed any remaining HSTF activity from the enzyme because the HSTF does not bind to phosphocellulose (data not shown). A final RNA polymerase II concentration of approximately 150 μg per ml was obtained in a volume of 1.2 ml (starting with 50 ml of the nuclear extract). The yield of enzyme was generally 10% of the starting material. The yield was particularly low because of the amount of enzyme that was discarded to avoid HSTF contamination. The purity of the enzyme was estimated by polyacrylamide gel electrophoresis (Laemmli, 1970) to be approximately 50%.

Footprint Analysis
Fragment Preparation
The fragments used in the footprint analyses of the hsp 70 and histone genes are as indicated in the respective figures. Briefly, two fragments were used for the footprint experiments on the hsp 70 gene. The footprints shown in Figure 7 employed a Xho I to Bgl I fragment from the 56H8 plasmid labeled either by T4 polynucleotide kinase at the Xho I site (coding strand) or by DNA polymerase I (Klenow fragment) filling at the Xho I site (noncoding). The footprints shown in Figure 5 employed a Bal 31 mutant, d21$^+$, of the hsp 70 gene, which will be described in detail elsewhere (Topol and Parker, unpublished data). The d21$^+$ fragment was labeled by Pol I filling (Klenow fragment) at an Eco RI site approximately 260 bp upstream (5'-ward) from the start point of transcription (see Figure 1) The histone gene containing fragment shown in Figure 5A was also selected from a set of Bal 31 deletion mutants (Parker and Hogness, unpublished data). This fragment contains a unique Eco RI site approximately 50 bp downstream from the start point of transcription of the H3 gene, and was labeled at this site by DNA polymerase I (Klenow fragment) filling (see Figure 1) For further details on these methods see Maniatis et al. (1982).

Footprint Reactions
All footprint reactions were carried out as previously described (Parker and Topol, 1984)

Acknowledgments
The helpful comments of Stewart Scherer and Barbara Wold on the manuscript were greatly appreciated We also appreciated the advice and

critical comments of Stewart Scherer, Herschel Mitchell, and Nancy Peterson during the course of these experiments. This research is supported by the National Institutes of General Medical Sciences, and by a young investigator award to C. S. P. from the Camille and Henry Dreyfus Foundation. Continued financial support from the Division of Chemistry is also acknowledged.

The costs of publication of this article were defrayed in part by the payment of page charges. This article must therefore be hereby marked "*advertisement*" in accordance with 18 U.S.C. Section 1734 solely to indicate this fact.

Received January 27, 1984; revised March 6, 1984

References

Alberts, B., and Herrick, G. (1971). DNA cellulose chromatography. Meth. Enzymol. *21*, 198–213.

Artavantis-Tsakonas, S., Schedl, P., Marault, M.-E., Moran, L., and Lis, J. (1979). Genes for the 70,000 dalton heat shock protein in two cloned D. melanogaster DNA segments. Cell *17*, 9–18.

Ashburner, M., and Bonner, J. J. (1979). The induction of gene activity in Drosophila by heat shock. Cell *17*, 241–254.

Bradford, M. M. (1976). A rapid and sensitive method for the quantitation of microgram quantities of protein utilizing the principle of protein-dye binding. Anal. Biochem. *72,* 248–254.

Brinster, R. L., Chen, H. Y., Warren, R., Sarthy, A., and Palmiter, R. D. (1982). Regulation of metallothionein–thymidine kinase fusion plasmids injected into mouse eggs. Nature *296*, 39–42.

Corces, V., Pellicer, A., Axel, R., and Meselson, M. (1981). Integration, transcription and control of a Drosophila heat shock gene in mouse cells. Proc. Nat. Acad. Sci. USA *78*, 7038–7042.

Dierks, P., van Ooyen, A., Mantei, N., and Weissman, C. (1981). DNA sequences preceding the rabbit β-globin gene are required for formation in mouse L cells of βglobin RNA with the correct 5′ terminus. Proc. Nat. Acad. Sci. USA *78*, 1411–1415.

Dynan, W. S., and Tjian, R. (1983). The promoter-specific transcription factor Sp1 binds to upstream sequences in the SV40 early promoter. Cell *35*, 79–87.

Findly, R. C., and Pederson, T. (1981). Regulated transcription of the genes for actin and heat shock proteins in cultured Drosophila cells. J. Cell Biol. *88*, 323–328.

Fromm, M., and Berg, P. (1982). Deletion mapping of DNA regions required for SV40 early region promoter function in vivo. J. Mol. App. Genet. *1*, 457–481.

Fyrberg, E. A., Mahaffey, J. W., Bond, B. J., and Davidson, N. (1983). Transcripts of the six Drosophila actin genes accumulate in a stage- and tissue-specific manner. Cell *33*, 115–123.

Galas, D., and Schmitz, A. (1978). DNAse footprinting: a simple method for the detection of protein–DNA binding specificity. Nucl. Acids Res. 5, 3157–3170.

Grosveld, G. C., de Boer, E., Shewmaker, C. K., and Flavell, R. A. (1982). DNA sequences necessary for transcription of the rabbit β-globin gene in vivo. Nature *295*, 120–126.

Hawley, D. K., and McClure, W. R. (1983). The effect of a lambda repressor mutation on the activation of transcription initiation from the lambda P$_{RM}$ promoter. Cell *32*, 327–333.

Hochschild, A., Irwin, N., and Ptashne, M. (1983). Repressor structure and the mechanism of positive control. Cell *32*, 319–325.

Johnson, A. D., Poteete, A. R., Lauer, G., Sauer, R. T., Ackers, G. K., and Ptashne, M. (1981). λ repressor and *cro*-components of an efficient molecular switch. Nature *294*, 217–223.

Laemmli, U. (1970). Cleavage of structural proteins during the assembly of the head of bacteriophage T4. Nature *227*, 680–685.

Maniatis, T., Fritsch, E. F., and Sambrook, J. (1982). Molecular Cloning, A Laboratory Manual. (Cold Spring Harbor, New York: Cold Spring Harbor Laboratory).

McKnight, S. L., and Kingsbury, R. (1982). Transcriptional control signals of a eukaryotic protein-coding gene. Science *217*, 316–324.

McKnight, S. L., Gavis, E. R., Kingsbury, R., and Axel, R. (1981). Analysis of transcriptional regulatory signals of the HSV thymidine kinase gene: identification of an upstream control region. Cell *25*, 385–398.

Mellon, P., Parker, V., Gluzman, Y., and Maniatis, T. (1981). Identitication of DNA sequences required for transcription of the human α1-globin gene in a new SV40 host-vector system. Cell *27*, 279–288.

Mirault, M.-E., Southgate, R., and Delwart, E. (1982). Regulabon of heat shock genes: a DNA sequence upstream of Drosophila hsp 70 genes is essential for their induction in monkey cells. EMBO J. *1*, 1279–1285.

Parker, C. S., and Topol, J. (1984). A Drosophila RNA polymerase II transcription factor contains a promoter-region-specific DNA-binding activity. Cell *36*, 357–369.

Pelham, H. R. B. (1982). A regulatory upstream promoter element in the Drosophila hsp 70 heat-shock gene. Cell *30*, 517–528.

Peterson, N. S., and Mitchell, H. K. (1984). Heat shock proteins. In Comprehensive Insect Physiology Biochemistry and Pharmacology, G. A. Kerut and L. Gilbert, eds. (New York: Pergamon Press), in press.

Rosenberg, M., and Court, D. (1979). Regulatory sequences involved in the promotion and termination of transcription. Ann. Rev. Genet. *13*, 319–353.

Schwartz, L. B., Sklar, V. E. F., Jachning, J. A., Weinmann, R., and Roeder, R. G. (1974). Isolation and partial characterization of the multiple forms of deoxyribonucleic acid-dependent ribonucleic acid polymerase in the mouse myeloma, MOPC 315. J. Biol. Chem. *249*, 5889–5897.

Torok, I., and Karch, F. (1980). Nucleotide sequences of heat shock activated genes in Drosophila melanogaster. I. Sequences in the regions of the 5′ and 3′ ends of the hsp 70 gene in the hybnd plasmid 56H8. Nucl. Acids Res. *8*, 3105–3123.

Long-distance regulatory elements

7

As described in the previous section, in eukaryotes many regulatory DNA sequences are located adjacent to the gene promoter, as is the case in prokaryotes. In addition, however, many eukaryotic genes also contain regulatory elements that are located at a distance from the gene promoter. This was first shown by the demonstration that sequences over 100 bp upstream of the transcriptional start site of the histone H2A gene were essential for its high-level transcription (Grosschedl and Birnsteil, 1980).

More detailed studies of such distant regulatory sequences were carried out in a number of viruses that infect eukaryotic cells. These viruses thus contain gene regulatory sequences typical of those in eukaryotic genes, but are relatively easy to study because of their simplicity. Such studies led to the concept of an enhancer element that can activate an unrelated promoter from a distance (for reviews of enhancer elements, see Hatzopoulos et al., 1988; Muller et al., 1988; Thompson and McKnight, 1992). As defined in this way the basic characteristics of an enhancer are, firstly, that it can activate a promoter at a distance of up to several thousand bases from the promoter; secondly, that it can activate a promoter when placed in either orientation relative to the promoter; and thirdly, that it can activate the promoter when placed either upstream or downstream of the promoter or within the transcribed region (Figure 7.1).

Despite this progress in the characterization of viral gene enhancers, at the time that the paper by Banerji et al. presented here was published, no such identification of an element with the characteristics of an enhancer had been performed for a cellular gene. Similarly, since most viral enhancers are active in all cell types, no enhancer had been identified that was specifically active in a particular cell type and could therefore play a role in producing tissue-specific gene expression. The paper by Banerji et al. presented here identified such a tissue-specific enhancer element in the immunoglobulin heavy chain gene, a finding that was also reported by Gillies et al. (1983) in the same issue of *Cell*.

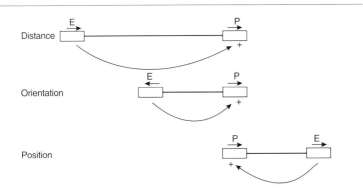

Figure 7.1 An enhancer element (E) can activate a promoter (P) independently of its distance from the promoter, its orientation relative to the promoter, or its position upstream or downstream of the promoter

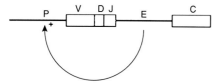

Figure 7.2 Position of the enhancer (E) within an intron of the immunoglobulin heavy chain gene located between the exons that encode the J (joining) and C (constant) regions of the heavy chain. Note the position of the enhancer downstream of the promoter (P). Abbreviations: V, variable region; D, diversity region.

Banerji et al. focused on an intron in the immunoglobulin heavy chain gene which is located between the exons that encode the J (joining) and C (constant) regions of the immunoglobulin heavy chain (Figure 7.2). This region of the immunoglobulin heavy chain gene was then introduced into a plasmid that contained two other genes under the control of their corresponding gene promoters. These were the gene encoding the large-T antigen of the eukaryotic virus simian virus 40 (SV40) and the cellular β-globin gene. As indicated in Figure 2 and Table 1 of the paper, the intron of the immunoglobulin gene was able to enhance the activity of the SV40 promoter, resulting in increased production of large-T antigen. Similarly, as indicated in Figure 3 of the paper, it was able to enhance the activity of the β-globin promoter. This effect occurred regardless of the distance, orientation or position of the intron relative to the two promoters. Hence the sequence within the immunoglobulin intron fulfils the definition of an enhancer element that was based upon the elements contained within eukaryotic viruses. It thus represented the first example of an enhancer element within a cellular gene and, as illustrated in Figure 4 of their paper, Banerji et al. were able, by deletion analysis, to define a minimal 300 bp region which was essential for the enhancer effect.

Most importantly, however, Banerji et al. also demonstrated that, unlike the viral enhancers, the enhancer in the immunoglobulin heavy chain gene was active in a tissue-specific manner. Thus, as illustrated in Figure 2 and Table 2 of the paper, the enhancer was able to activate the T-antigen or β-globin promoters only in B-lymphocytes and not in a variety of other cell types. This enhancer thus also represented the first example of an enhancer with cell-type-specific activity. The B-cell-specific activity of the immunoglobulin heavy chain gene enhancer plays a key role in the cell-type-specific expression pattern of this gene, which is transcribed only in B-lymphocytes and not in other cell types. Thus the enhancer would act to stimulate the immunoglobulin heavy chain gene promoter only in B-lymphocytes. Interestingly, the immunoglobulin heavy chain gene promoter also exhibits B-cell-specific activity even in the absence of the enhancer (Garcia et al., 1986). Hence the combination of a B-cell-specific promoter with a B-cell-specific enhancer produces the high-level transcription of the immunoglobulin gene in B-lymphocytes (Figure 7.2).

As illustrated in Figure 7.2, the immunglobulin heavy chain gene promoter is located upstream of the region of the gene that encodes the V (variable) region of the protein, whereas the enhancer is located between the J and C regions. In non-B-lymphocytes the DNA encoding the V region is separated from the J and C regions by over 100 kb of DNA. During the maturation of B-lymphocytes, however, a DNA rearrangement takes place which brings the V region DNA adjacent to the J and C regions to produce the arrangement illustrated in Figure 7.2 (for a review, see Gellert, 1992). This rearrangement has the effect of bringing the enhancer adjacent to the promoter, allowing it to exert a maximal effect and hence producing high-level gene expression in B-lymphocytes. However, its primary purpose lies elsewhere, in producing the diversity of immunoglobulins required to defend the body against large numbers of foreign proteins. The rearrangement process ensures that in each case a different one of the many V regions is brought adjacent to the J and C regions, so creating many immunoglobulin molecules with different variable regions. Thus this example represents a special case of DNA rearrangement and does not controvert our previous general conclusion in Section 1 that in most cases the DNA in all cell types is essentially identical.

Although the immunoglobulin heavy chain gene represents a special case in terms of its use of DNA rearrangement, cell-type-specific enhancers, as identified originally in this gene, are found in a very wide variety of other genes that exhibit tissue-specific gene expresssion without any DNA rearrangement (for a review, see Muller et al., 1988).

One such enhancer that confers specific expression in cells of pancreatic origin was identified within the 5′ upstream region of the insulin gene by similar methods to those used by Banerji et al., involving the introduction of gene constructs into various cell lines (Walker et al., 1983). In the paper by Hanahan presented here, the analysis of this enhancer was taken one stage further to analyse its effect within the intact animal *in vivo*. In this work, a region of the insulin gene including 520 bp immediately 5′ to the transcriptional start site was linked to the gene encoding the SV40 large-T antigen. This linkage was carried out with the insulin gene sequence in both orientations relative to the SV40 large-T antigen gene so as to test for bidirectional enhancer activity. The two constructs were then used to create transgenic mice containing this gene in every cell of the body. The expression pattern of this gene was then studied in two ways. Firstly, since large-T antigen is an oncoprotein capable of transforming cells to a cancerous phenotype, it was possible to monitor its expression by studying which cells in the transgenic mouse became cancerous. In addition, expression was also monitored directly at the protein level using an antibody capable of recognizing the large-T antigen.

These two methods of analysis gave the same results. Thus, by both criteria, the insulin enhancer was able to target gene expression specifically to the pancreas. As illustrated in Fig. 2 of the paper, the transgenic mice contained tumours of pancreatic origin, whereas no tumours were seen in other tissues. Similarly, as illustrated in Fig. 5, the large-T antigen could be detected at the protein level only in extracts of pancreatic tissue and not in a variety of other tissues. Most interestingly, this tissue-specific expression directed by the insulin regulatory elements was detected in both sets of transgenic mice produced with constructs containing the insulin gene regulatory region in either orientation relative to the large-T antigen gene. This indicates that the regulatory sequence in the insulin gene is acting as an enhancer element and so works independently of orientation to produce pancreatic-specific gene expression.

Moreover, the enhancer was able to direct gene expression specifically to the cells within the pancreas that would normally express the insulin gene. Thus Fig. 4 of the paper shows that the tumours produced within the pancreas in the transgenic mice were composed of insulin-producing cells and did not represent proliferation of other cell types in the pancreas that do not normally express insulin. Similarly, as illustrated in Fig. 6 of the paper, the cells that expressed the large-T antigen, as detected by antibody staining, were cells that normally produce insulin, and not the other cells in the pancreas that produce other hormones such as glucagon or somatostatin.

Hence the insulin enhancer is not only capable of directing the expression of an unrelated gene to a specific organ, but is also capable of directing such expression to specific cells within the organ. Thus this enhancer can confer the normal cell-type-specific pattern of insulin gene expression upon a heterologous gene. This work extended the previous *in vitro* studies such as that of Banerji et al. by showing that an enhancer can act *in vivo* to produce the correct cell-type-specific pattern of gene expression.

The two papers described above provide a clear indication of the critical role played by enhancers in producing cell-type-specific patterns of gene activity both *in vitro* and *in vivo*. However, while for some genes the correct promoter and enhancer combination can produce the appropriate pattern of cell-type-specific gene expression *in vivo*, for other genes this does not appear to be the case. Thus in some situations where a particular gene (with its appropriate promoter and enhancer elements) is introduced into transgenic mice, it does not exhibit the correct tissue-specific pattern of gene expression. Rather, its expression appears to be influenced by regulatory elements within the region of the genome into which it has been inserted. It thus exhibits position-dependent gene expression, which varies depending on the site into which it has been inserted, rather than the position-independent expression that it would exhibit if it contained all the appropriate regulatory elements.

This phenomenon was studied in detail by Grosveld et al. (1987) using the β-globin gene locus, which contains the adult β-globin gene and a number of other related globin genes that are expressed in embryonic development (Figure 7.3). Although each of these globin genes contains its own promoter and enhancer elements, Grosveld et al. (1987) found that, in order to direct expression of the β-globin genes to the appropriate erythroid cells in transgenic mice, it was necessary to include an additional short region present at the 5′ end of the gene cluster. Thus in the presence of this region a β-globin gene would show position-independent, erythroid-specific gene expression in transgenic mice, whereas in its absence the expression would vary depending on the position in the genome into which the construct had been integrated. This region was originally referred to as the dominant control region, but has now been renamed the locus control region (LCR), and is indicated in Figure 7.3.

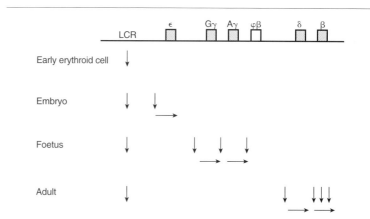

Figure 7.3 Globin gene locus containing five active genes encoding β-globin-type proteins (hatched boxes) and one inactive pseudogene (open box). Note the presence of multiple DNase I-hypersensitive sites within the LCR (shown as a single arrow for simplicity). These sites appear in early erythroid cells before the appearance of other hypersensitive sites adjacent to each individual gene which appear later in development, as these genes are expressed sequentially in the embryo, foetus and adult.

Although the original paper by Grosveld et al. (1987) identified the existence of the LCR, it did not characterize its properties in detail. This was achieved in a subsequent paper from the Grosveld group by Blom van Assendelft et al. which is presented here. In particular, they investigated both the cell-type-specific activity of the LCR and its ability to activate a heterologous gene as well as the β-globin gene. Thus in Figures 3 and 4 of the paper it is shown that the LCR can enhance the expression of the thymidine kinase and Thy-1 genes in cells of erythroid origin. This effect is only observed in constructs containing the LCR and not in constructs from which it has been specifically deleted, as illustrated in Figure 4(B). Hence the LCR is indeed capable of activating the expression of other genes apart from its normal role in activating the β-globin gene. It thus represents a regulatory element capable of activating heterologous gene expression, paralleling the ability of promoters and enhancers to achieve this effect.

Moreover, as illustrated in Figure 7 of the paper by Blom van Assendelft et al., this effect is cell-type-specific. Thus the LCR is capable of activating β-globin gene expression when constructs are introduced into the K562 cell line, which is of erythroid origin, but cannot achieve this effect in L cells, which are not of erythroid origin. Hence the LCR acts in a cell-type-specific manner, being functional only in erythroid cells and not in other cell types.

The paper by Blom van Assendelft et al. established the LCR as an element which, like promoter and enhancer elements, can activate gene expression in a cell-type-specific manner. It is now believed that within the β-globin gene locus the LCR acts at a very early point in embryonic development to open up the chromatin structure of the β-globin gene region. Subsequently individual genes encoding the β-globin-like molecules produced in the embryo and subsequently in the foetus and then in the adult are activated by their individual promoter and enhancer elements. In agreement with this idea, DNase I-hypersensitive sites (see Section 5) appear in the LCR very early in erythroid development, and subsequently appear at the 5′ end of each gene just prior to its expression (Figure 7.3).

Subsequent studies have suggested that LCR elements exist in a variety of other genes including, for example, the CD2 gene, where the LCR directs a T-lymphocyte cell-specific pattern of gene expression (for review of LCR elements, see Townes and Behringer, 1990; Dillon and Grosveld, 1993). It is likely, therefore, that LCR elements play a critical role in producing tissue-specific gene expression acting in concert with cell-type-specific promoter and enhancer elements. Thus a wide variety of different regulatory elements located adjacent to the gene promoter and at various distances from it play a critical role in producing the appropriate cell-type-specific pattern of gene expression.

Banerji et al. (1983) Cell **33**, 729–740

A Lymphocyte-Specific Cellular Enhancer Is Located Downstream of the Joining Region in Immunoglobulin Heavy Chain Genes

Julian Banerji, Laurel Olson, and Walter Schaffner
Institut für Molekularbiologie II
der Universität Zürich, Hönggerberg
CH-8093 Zürich, Switzerland

Summary

Transcriptional enhancers, originally discovered in viral genomes, are short, *cis*-acting, regulatory sequences that strongly stimulate transcription from promoters of nearby genes. We demonstrate the existence of an enhancer within a mouse immunoglobulin heavy chain gene. A DNA fragment located between the joining region and the switch recombination region in the intron upstream of the immunoglobulin μ constant region has been linked, in both orientations, to genes coding for rabbit β-globin or SV40 T antigen. This, element enhances the number of correct β-globin gene transcripts by at least two orders of magnitude and also stimulates production of T antigen. It acts from several hundred to several thousand base pairs up or downstream of a promoter without amplifying template copy number. Of the various cell lines tested, the immunoglobulin gene enhancer functions only in lymphocyte-derived (myeloma) cells. We propose that this tissue-specific enhancer contributes to the activation of somatically rearranged immunoglobulin variable region genes and possibly to abnormal expression of other genes (e.g. *c-myc*) that become translocated to its domain of influence.

Introduction

The enormous repertoire of genes coding for immunoglobulin (Ig) molecules is achieved, in part, by rearranging pieces of DNA that are noncontiguous in chromosomes of germ-line cells (Hozumi and Tonegawa,1976; reviewed by Tonegawa, 1983). To create a functional Ig heavy chain gene, one of a large number of variable (V) region DNA segments (Cory and Adams,1980; Rabbitts, Matthyssens, and Hamlyn, 1980) is joined to one or more diversification (D) elements (Schilling et al., 1980; Sakano et al., 1981), which is joined to one of four joining (J) regions (Early et al., 1980; Sakano et al., 1980) upstream of the constant region exons (Gough et al., 1980; Rogers et al., 1980) of the complete gene.

In myeloma cells the number of transcripts derived from a functionally rearranged kappa light chain V region gene is about four orders of magnitude higher than those derived from unrearranged V_K sequences (Mather and Perry, 1981). In the similarly organized Ig heavy chain genes, the DNA upstream from each variable region remains essentially unchanged by the rearrangements that create a transcriptionally active Ig heavy chain gene (Davis et al., 1980; Sakano et al., 1980; Clarke et al., 1982). Therefore,

transcription of the functionally rearranged gene might be a consequence of signals encoded in the newly juxtaposed downstream sequences.

A DNA sequence that has been directly shown to influence transcription in such a fashion is the enhancer element in the genome of simian virus 40 (SV40). In a manner independent of distance and orientation, this DNA element can serve, in cis, to enhance the transcription of a nearby β-globin gene by more than two orders of magnitude even when it is located several thousand base pairs downstream from the promoter (Banerji, Rusconi and Schaffner, 1981; see also Moreau et al., 1981). Enhancers have also been identified in the genomes of polyoma virus (de Villiers and Schaffner, 1981), bovine papilloma virus (BPV) (Lusky, Berg, and Botchan, 1982), and Moloney sarcoma virus (MSV) (Laimins et al., 1982). Viral enhancers function in a wide variety of cell types but perform optimally in their natural host cells (de Villiers and Schaffner, 1981; Laimins et al., 1982; de Villiers et al., 1982b). They have proved to be useful for recombinant DNA studies of gene expression (for a review see Banerji and Schaffner,1983; de Villiers and Schaffner, 1983). In particular, the SV40 enhancer has allowed efficient expression of an Ig lambda gene in HeLa cells, where it is not otherwise expressed (Picard and Schaffner, 1983). However, a nonviral enhancer with the above described properties has not yet been described in association with a cellular transcription unit, though it remains to be seen whether the "upstream sequences" identified in certain cellular promoters (e.g., see Grosschedl and Birnstiel, 1980) are functionally related to enhancers.

We tested the hypothesis that an enhancer element is encoded in the DNA of the mouse immunoglobulin constant region locus. Using a transient expression assay, we found such a DNA element in the intron between the J_H and the μ constant region exons. This enhancer element behaves similarly to the SV40 viral enhancer except for a striking tissue specificity—it functions in B-lymphocyte-derived (myeloma) cells, but not in other cell types we have tested.

Results

We began our studies with cloned DNA consisting of two Eco RI fragments encoding the J_H cluster and the μ constant region of the unrearranged (germ-line) mouse immunoglobulin heavy chain gene (clone M 23; J. Banerji, Master's thesis, State University of New York at Stony Brook, 1980; Marcu et al., 1980; a gift from K. Marcu). This DNA was digested with the restriction endonuclease Xba I, which released three fragments of interest encompassing almost the entire intron between the J and constant region exons. These three fragments, labeled B, E, and C (Figure 1C), were subcloned in either orientation into the plasmid p24β (Figure 1D). This plasmid contained both a rabbit β-globin gene (Figure 1) and SV40 sequences coding for T antigen, including the early promoter but

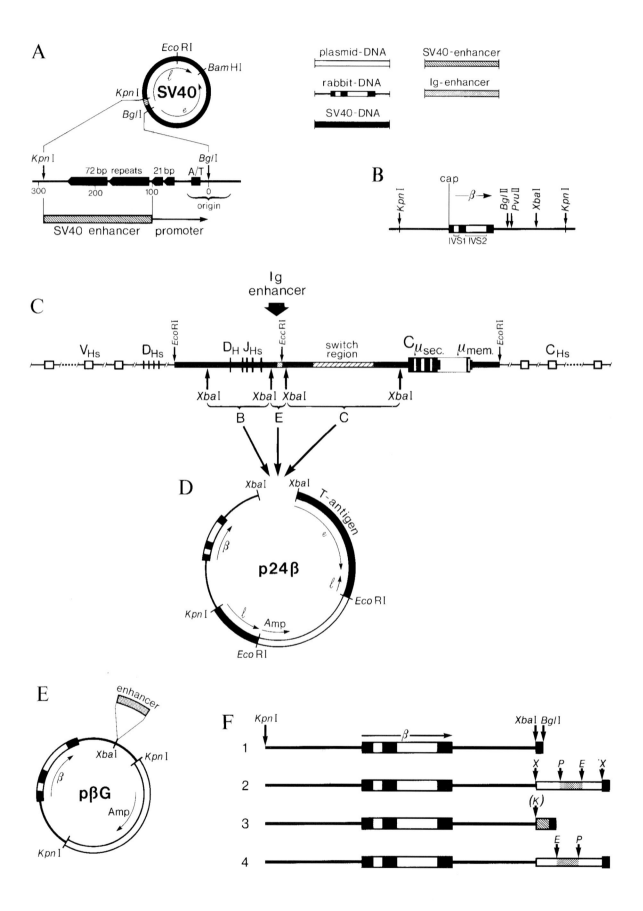

excluding the SV40 enhancer (Figure 1A). The six recombinants were named p24βB+, p24βB−, p24βE+, p24βE−, p24βC+, and p24βC−. As a negative control we used the plasmid p24β (the parental recombinant without an insert), and as a positive control we used pDept (containing the early region of SV40, including the viral enhancer shown in Figure 1A). These eight recombinants were transfected into HeLa cells which, after 2 days, were fixed and stained by indirect immunofluorescence for T antigen. No fluorescent cells were found (i.e., no enhancer activity was detected) when any of the three Ig segments were tested in HeLa cells (Figure 2).

The situation was different when we transfected the recombinants into a cultured mouse myeloma cell line, X63-Ag8, derived from B-lymphocytes in which immunoglobulin genes are normally heavily transcribed. Using a modified DEAE-dextran transfection protocol and the transient assay described above, a bright nuclear T-antigen fluorescence was seen in a fraction of the cells transfected with the recombinants p24βE+ or p24βE− (containing the Ig Xba I E fragment), or with pDept (the positive control containing the SV40 enhancer), but not with any of the other recombinants (Figure 2 and data not shown).

The Ig Enhancer Acts Only in *Cis*

To determine whether the enhanced expression of T antigen observed with the immunoglobulin Xba I E fragment was a *cis*- or a *trans*-acting phenomenon, we cotransfected two DNA clones into X63-Ag8 cells: one, pTAR7, contained the Xba I E fragment devoid of SV40 sequences (see Experimental Procedures for a complete description of this and other clones); the other, pET3, contained the SV40 early region devoid of an enhancer. No T-antigen-positive cells were detected in the transient expression assay, indicating that the immunoglobulin Xba I E fragment cannot act in *trans* but only in *cis* (Table 1).

Long-Range Effect

To determine whether this fragment could also enhance the correct expression of another gene, such as the rabbit β-globin gene, seven of the eight recombinants used above were again transfected into X63-Ag8 cells. However, as a positive control we used pβGXsv512, consisting of a cloned β-globin gene pβG (see Figure 1E) with a small

fragment of SV40 containing the enhancer inserted 1.5 kb behind the gene. Thirty-six hours after transfection the cells were harvested, their RNA was extracted, and the number of correctly initiated β-globin transcripts was measured by the S1 nuclease hybridization assay (Figure 3A).

The results (Figure 3B) show that in myeloma cells the number of correctly initiated β-globin gene transcripts is greatly increased in clones containing the Ig Xba I E fragment in either orientation but not in clones containing the B and C fragments. Thus this Ig gene segment is similar to the SV40 enhancer in that it can serve, in a transient expression assay, to enhance transcription of a gene when it is placed in either orientation, up to 500 bp upstream, or 2500 bp downstream, from the promoter of a nearby gene.

Enhancement Does Not Require Template Amplification

The clones p24βE+ and p24βE− contain more viral sequences than pβGXsv512, the positive control. To determine whether part of the Ig enhancer effect was due to template amplification via the SV40 origin mediated by enhanced levels of T antigen, appropriate recombinants were restricted with the endonucleases Kpn I and Bgl I. A fragment of DNA containing the β-globin gene, linked in some cases to an SV40 or an Ig enhancer but devoid of a functional SV40 replication origin and the T-antigen structural gene, was purified by gel electrophoresis (see Figure 1F). These linear DNAs were transfected into X63-Ag8 cells. S1 mapping of the resultant transcripts again showed that the Ig enhancer is more effective in myeloma cells than the SV40 enhancer, and that the observed enhancement of at least two orders of magnitude was not due to T-antigen-mediated replication of the transfected templates (Figure 3C).

We also conducted a transient expression assay in the presence of hydroxyurea, which inhibits DNA replication (Manteuil and Girard, 1974). We used three clones containing the T-antigen coding region and the SV40 early promoter, pET3, pDept, and p24βE+. Clone pET3 contained no enhancer; pDept included an SV40 viral enhancer; and p24dE+ included an Ig enhancer. Each recombinant was transfected onto two plates of X63-Ag8 cells. In addition, pDept was transfected onto two plates of HeLa cells. One plate of each pair was treated normally; the

---

Figure I. Schematic Representation of Recombinant DNAs Used

(A) Map of the 5.2 kb SV40 genome (Tooze, 1981) with the Kpn I to Bgl I region expanded to show SV40 enhancer (fine hatching) and SV40 promoter.

(B) Map of the rabbit hemoglobin β1 gene within a 4.7 kb Kpn I fragment showing relevant restriction sites.

(C) Schematic diagram of the immunoglobulin heavy chain gene family in its germ-line configuration. Thin lines: regions not drawn to scale. Thick lines: two Eco RI fragments containing the μ constant region (Marcu et al., 1980). Coarse hatching in switch recombination region indicates sequences that are deleted in this phage clone. Stippled area under large arrow shows location of the Ig enhancer. Arrows pointing upwards mark the restriction sites used to generate Ig Xba I fragments B, E, and C.

(D) Map of p24β, a clone containing a unique Xba I site into which DNA fragments were cloned and assayed for their ability to enhance expression of β-globin or T-antigen (the synthetic Xba I linker in SV40 is located 99 bp from the Bgl I site between the 72 bp and 21 bp tandem repeats; a gift from Y. Gluzman; see also Figure 1A). Clones thus generated were designated p24βB+ and p24βB− (Xba I B fragment in either orientation), p24βE+ and p24βE− (Xba E enhancer fragment in either orientation)

(E) Map of PβG, a clone containing only the 4.7 kb Kpn I fragment with the rabbit β-globin gene. This clone was used to test enhancer DNA fragments as above—e.g, clone pβGxba32 contained the SV40 enhancer.

(F) Schematic diagram of Kpn I–Bgl I fragments used for the experiment shown in Figure 3C. (1) derived from p24β (without enhancer); (2) and (4) derived from p24βE+ and p24βE−, respectively (with Ig enhancer); (3) derived from pβGxba32 (with SV40 enhancer). (K) refers to a destroyed Kpn I site.

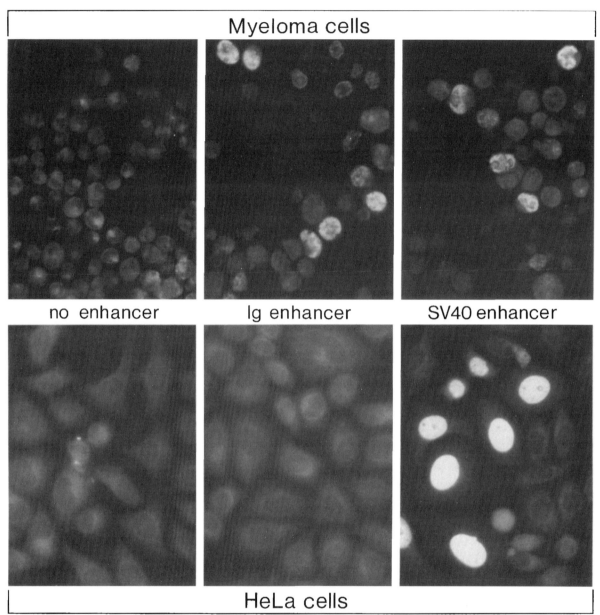

Figure 2. Myeloma-Cell-Specific Activity of the Ig Enhancer

SV40 T-antigen clones were transfected in parallel into both myeloma X63-Ag8 cells and HeLa cells. Thirty-six hours later, cells were fixed and stained for nuclear T antigen by indirect immunofluorescence (see Experimental Procedures). Left: negative control, clone p24β with a deletion of the entire 72 bp repeat enhancer region (for details see Figure 1 and Experimental Procedures). Center: clone p24βE+ with the Ig enhancer replacing the SV40 enhancer. Nuclear T-antigen fluorescence is visible in transfected myeloma cells but not in HeLa cells (see also Table 1). Right: positive control, clone pDept, which contains the SV40 enhancer in its wild-type configuration.

other was treated with hydroxyurea (see Experimental Procedures). When the levels of T-antigen expression directed by an SV40 and an Ig enhancer were compared, the percentages of T-antigen-positive cells were about the same, either with hydroxyurea treatment or without (see Table 1). The percentage of T-antigen-positive cells was always smaller under conditions of hydroxyurea treatment. However, this appears to be a general phenomenon since it is also seen in hydroxyurea-treated HeLa cells transfected with pDept (containing the SV40 enhancer). Essen-

tially the same results were obtained when DNA replication was inhibited by cytosine arabinoside (araC) (data not shown). We conclude that the enhancing ability of the Ig Xba I E fragment parallels that of the SV40 enhancer, which does not act via template amplification (Banerji et al., 1981).

Cell-Type Specificity

We wanted to know if the Ig enhancer would function in mouse cells other than the X63-Ag8 cell line derived from

Table 1. Enhancer-Directed T-Antigen Synthesis in Transfected Cells

				Duplicate Plate Treated with Hydroxyurea	
T Antigen Clone (Enhancer)	Cells (Type)	Total Cells[a] (per 8 × 8 mm)	Positive for T Antigen	Total Cells (per 8 × 8 mm)	Positive for T Antigen
pET3 (no enhancer)	X63 Ag8 (myeloma)	50,000	0	14,000	0
p24βE+ (Ig enhancer)	X63-Ag8	54,000	2130 (4%)	11,500	158 (1.4%)
pDept (SV40 enhancer)	X63-Ag8	51,000	2860 (5%)[b]	13,000	187 (14%)
pDept	HeLa (human carcinoma)	49,000	8400 (17%)	21,000	1160 (6%)
pET3 (no enhancer) and pTAR7 (Ig enhancer but no T antigen gene) }	X63-Ag8	46,000	0		

All transfections were done without chloroquine (see Experimental Procedures).

[a] The total number of cells in the 8 × 8 mm area was extrapolated from counting three areas of 0.145 mm² each (400-fold magnification); that is, a total of 150–300 cells was counted. The number of T-antigen-positive cells was counted either in the entire 8 × 8 mm area or in a defined section thereof such that at least 100 positive cells were scored.

[b] pDept, with the wild-type arrangement of the SV40 enhancer-promoter, gives higher values of T-antigen-positive cells than p24βE+ containing the Ig enhancer and the truncated T-antigen gene. However, when both enhancers are explanted and placed 2500 bp downstream of the β-globin gene, the Ig enhancer is stronger than the SV40 enhancer.

mouse lymphocytes. In other words, is its inability to function in human HeLa cells a species restriction, or would the enhancer also be nonfunctional in mouse 3T6 cells, implying that it is cell-type specific? Transfections into several different cell lines are compared in Table 2. The Ig enhancer does not function detectably in the cell line 3T6, derived from mouse fibroblast, nor does it function in mink lung cells or human HeLa cells (however, we cannot exclude a weak but significant activity below the threshold of the immunofluorescence assay). In contrast, the Ig enhancer functions well in all mouse myeloma cells tested: X63-Ag8; and two related cell lines, F0 and Sp6Bu6Bu.

Delimitation of Important Sequences

To narrow the region responsible for the enhancing effect, we constructed the deletion mutants shown in Figure 4. The Xba I E fragment can be cleaved once by each of the endonucleases Pst I or Eco RI. We tested both the large and the small Pst I to Xba I fragments in clones named pPTA-L and pPTA-S, and both the large and the small Eco RI to Xba I fragments in clones named pET1 and pET3. After assaying these clones for enhanced T-antigen expression (see Figure 4 for results), we concluded that most, if not all, of the enhancing activity is in the central Eco RI to Pst I fragment. This fragment was sequenced in its entirety from both ends (Figure 5A).

A 171 bp Alu I restriction fragment (Figure 4) from the central portion of the Eco RI–Pst I region was subcloned into a Pvu II site behind the β-globin gene (Figure 1B) in the clone pβGX (Figure 1E). This clone, called pβGXalu, was transfected into X63-Ag8 cells along with five other clones, and the β-globin gene transcripts were analyzed as before. The other clones were all derivatives of the parental clone pβG. The first, pβG, contained no insert; the second, pβGXsv512, contained the SV40 enhancer; the third, pβGXpy, the polyoma enhancer (de Villiers et al., 1982b; a gift from J. de Villiers) and the last two, pTAR1 and pTAR7, contained the Xba I E fragment cloned in

either orientation. The SV40 and polyoma virus enhancers and the Ig enhancer Alu I subfragment were about equally efficient, whereas the Ig Xba I E fragment was two to three times more efficient than the viral enhancers or the Alu I subfragment (data not shown).

To delineate further the sequences responsible for the enhancing effect, exonuclease Bal 31 was used to generate a series of deletion mutants. The large Eco RI to Xba I fragment was subjected to resection from the Eco RI site, and synthetic Xho I linkers were used to mark the mutant breakpoints. The fragments were then inserted into a plasmid vector pBAG, containing SV40 T-antigen coding sequences including the early promoter, but excluding the SV40 enhancer. These recombinants were then transfected into X63-Ag8 cells, which were assayed for enhanced T-antigen expression. A fragment retaining considerable enhancer activity, pBAG-A, was then subjected to a Bal 31 resection from the other (Xba I) end, and synthetic Xba I linkers were used to mark these deletion endpoints. The new deletion fragments were replaced into the vector pCAB and assayed for T-antigen expression. Figure 4 depicts the entire collection of DNA fragments that were generated and tested for enhancer activity, along with an indication of their relative enhancing abilities. The endpoint of each deletion mutant was determined by sequencing.

The results of the Bal 31 resection indicate that sequences important for the enhancing effect are distributed over most of the 300 bp Eco RI to Pst I fragment. Similar results have been obtained by testing deletion mutants of the SV40 and polyoma enhancers (Benoist and Chambon, 1981; S. Lupton and R. I. Kamen, personal communication), suggesting that an enhancer does not consist of one short sequence that is necessary and sufficient for full activity. Rather, a number of short but significant homologies have been found by computer-aided comparison of this sequence with those of viral enhancers from SV40, BKV, polyoma virus, adenovirus (E1a), MSV, and BPV (see Gluzman and Shenk, 1983). However, when any arbitrary

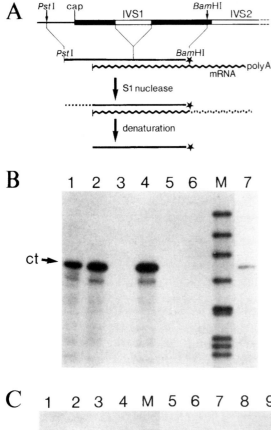

Table 2 Cell-Type Specificity of the Ig Enhancer

DNA Clone[a] (Amount)	Cells (Species)	Total Cells (per 8 × 8 mm)	Positive for T Antigen
p24βE+ (0.25 μg)	X63-Ag8 (mouse)	31,000	1,320 (4%)
pDept (0.25 μg)	X63-Ag8	41,000	2,920 (7%)
pDept (0.025 μg)	X63-Ag8	32,000	502 (1.6%)
pDept (0.0025 μg)	X63-Ag8	38,000	59 (0.2%)
p24βE+ (0.25 μg)	3T6 (mouse)	27,000	0
pDept (0.25 μg)	3T6	35,000	3,700 (11%)
pDept (0.025 μg)	3T6	30,000	790 (3%)
pDept (0.0025 μg)	3T6	34,000	62 (0.2%)
p24βE+ (0.25 μg)	Lung (mink)	30,000	0
pDept (0.25 μg)	Lung	32,000	2,820 (9%)
pDept (0.025 μg)	Lung	33,000	380 (1.2%)
pDept (0.0025 μg)	Lung	36,000	12 (0.03%)
p24βE+ (0.25 μg)	HeLa (human)	47,000	0
pDept (0.25 μg)	HeLa	43,000	13,000 (30%)
pDept (0.025 μg)	HeLa	45,000	2,240 (5%)
pDept (0.0025 μg)	HeLa	40,000	240 (0.6%)

[a]All transfections were done using 35 mm plates where 0.25 μg DNA are normally used (see Experimental Procedures).

criterion is used to compare these enhancers, no most compelling "core" sequence can be extracted. The polyoma enhancer shares many homologies with the Ig enhancer. However, except for homology blocks K, N and C, these homologies cannot be assembled into a distinct spatial framework (Figure 5B). Some sequences seem to be absolutely required for enhancer activity: T-antigen expression is no longer detectable in mutants deleted extensively from either end. This central region contains a duplication of the sequence 5'-GTGGTTT-3', which also appears on the opposite strand. This sequence, or its derivatives, appears in other enhancers (see Weiher et al., 1983) and is present 150 bp upstream, and on the other strand 150 bp downstream, of the DNAase I hypersensitivity site (Chung et al., 1983) of the Ig kappa light chain intron (Max et al., 1981).

Figure 3. β-Globin Gene Transcription Directed by the Ig Enhancer

(A) S1 nuclease mapping scheme (Weaver and Weissmann, 1979). A β-globin gene clone lacking the first intervening sequence (IVS1; see also Figure 1B; Weber et al, 1981; a gift from H Weber) was used as a radioactive probe (for further details see Rusconi and Schaffner, 1981). DNA end-labeled at the Bam HI site was hybridized to unlabeled RNA from 10[6] transfected X63-Ag8 myeloma cells, treated with S1 nuclease, denatured, fractionated by gel electrophoresis, and autoradiographed.

(B) Gel autoradiography. Hybridization to RNA from cells transfected with recombinant pβGXsv512, which contains the SV40 enhancer downstream of the β-globin gene (lane 1; positive control; Figure 1E); with the recombinant p24βE+ containing the Ig Xba I E enhancer fragment in the same orientation with respect to the β-globin gene promoter as it is to the Ig gene promoter in vivo (lane 2); with the parental globin recombinant p24β not containing an enhancer (lane 3, negative control); with the recombinant p24βE− containing the Ig enhancer fragment in the inverted orientation (lane 4); with recombinants p24βB+ and p24βB−, a 1:1 mixture with both

orientations of the Xba I B fragment (lane 5); with p24βC+ and p24βC−, a 1:1 mixture with both orientations of the Xba I C fragment (lane 6; see also Figure 1D). (Lane 7) Control hybridization to rabbit reticulocyte RNA. (Lane M) End-labeled marker DNA fragments (pBR322 digested with Hpa II). ct: fragment with correct terminus, mapping 354 nucleotides upstream from the Bam HI site.

(C) Gel autoradiography after transfection of purified linear template DNA devoid of the SV40 replication origin and T-antigen coding sequences (see Figure 1F). (Lane 1) Hybridization to RNA from cells transfected with the large Kpn I–Bgl I fragment of p24β (not containing an enhancer). (Lane 2) Kpn I–Bgl I fragment of p24βE+ (enhancer) (Lane 3) Kpn I–Bgl I fragment of pβGxba32 (SV40 enhancer; see also Figure 1E). (Lane 4) Kpn I–Bgl I fragment of p24βE− (inverted Ig enhancer). (Lanes 5–9) A dilution series with the full-length probe. The dilution factor between neighboring lanes is 3.16—i.e., lanes 7 and 9 are one order of magnitude apart. (Lane M) Marker DNA fragments. fl: full-length probe (453 nucleotides). ct: correct terminus fragment (354 nucleotides).

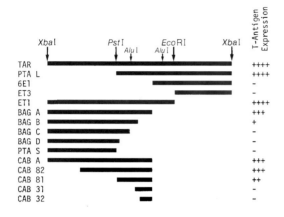

Figure 4. Localization of the Ig Enhancer Activity

Top line: the Xba I E fragment. Black stars indicate DNA still present in the deletion mutants. Immunofluorescence quantitation of the fraction of T-antigen-positive cells was done as indicated in Table 1. − no positive cells, + less than 5%, ++ between 5% and 30%, +++ between 30% and 75%, ++++ more than 75% of the fraction of T-antigen-positive cells seen with the standard recombinant p24βE+. Endpoints of the Bal 31 resection series are as follows: p6E1 extends up to and includes base pair 198; pBAG-A, base pair 198; pBAG-B, base pair 120; pBAG-C, base pair 77; pBAG-D, base pair 21; pCAB-A, base pair 198, pCAB-81, base pair 24; pCAB-31, base pair 138; pCAB32, base pair 164 (see Figure 5 for numeration)

Discussion

We have shown that the mouse genome contains a sequence that, in *cis*, enhances the expression of a nearby gene. Simple deletion of this region from the Ig transcription unit and measurement of a reduction in the number of Ig mRNAs would not allow one to distinguish whether this region acts in *cis* or in *trans* to enhance transcription or to ensure the stability (or processing) of the primary transcript. Since DNA segments that act in *trans* to stimulate transcription have recently been identified in the genomes of pseudorabies virus and adenovirus (see Gluzman and Shenk, 1983), we have explicitly demonstrated here that the Ig enhancer does not act in *trans*. We also show that it is not required in *cis* for RNA processing by explanting the enhancer outside of a transcription unit and showing that the transcripts are correctly initiated. We considered it necessary to S1 map the transcripts since it has been shown (Picard and Schaffner, 1983) that abundant production of Ig transcripts that appear properly sized on Northern blots can result from aberrant splicing of longer transcripts that initiate at sites farther upstream on the same molecule (e.g., at SV40 control signals). Such aberrant transcripts have been found to direct the synthesis of Ig protein. (M. Neuberger, personal communication).

Placement of the enhancer element in either orientation from several hundred to several thousand base pairs up or downstream from a promoter can increase the number of correctly initiated transcripts of the β-globin gene by at least two orders of magnitude. It has also been shown that this Ig μ Xba I E fragment enhances correct transcription

from the promoter of an immunoglobulin lambda light chain gene (D. Picard and W. Schaffner, unpublished results). The Ig enhancer element functions in a manner similar to the SV40 viral enhancer in that it does not act by raising the copy number of the transfected template. It differs from the SV40 enhancer in that it functions only in myeloma cells, derived from B-lymphocytes, but not in cells that stem from mouse fibroblasts or from human or mink epithelial cells. Though studies comparing the efficiencies of mouse viral enhancers in mouse and primate cells have noted a species preference for these enhancers (de Villiers et al., 1982b; Laimins et al., 1982), the Ig enhancer is the first example of an enhancer that functions in mouse cells of one specific lineage, but not another.

We propose that this cell-type-specific enhancer element serves to enhance correct transcription from the promoter of the most proximal upstream immunoglobulin variable region which becomes joined to the Ig constant region by the somatic recombination events that create a complete Ig heavy chain gene. Furthermore, because of its location in front of the switch recombination region, this DNA element would continue to be associated with, and would promote transcription from the original variable half of that gene throughout subsequent class switching events. Thus, whether the class switch involves actual DNA rearrangements (Kataoka et al., 1980; Davis et al., 1980; Sakano et al., 1980) or novel splicing of longer transcripts (Maki et al., 1981; Yaoita et al., 1982), normal B-cells probably use the enhancer described here to express Ig heavy chains of all classes (Figure 6). In addition, the most proximal upstream V region promoter probably acts as a sink or "road block" for information traveling along the DNA from the enhancer (see below), preventing enhanced transcription from more distal V region promoters.

The model that this region of the genome contains sequences that are important for the expression of Ig genes is compatible with observations of Alt et al. (1982a). Continued growth of transformed lymphoid cells in culture leads to deletions in the tandemly repeated switch sequences, similar to those which have been previously described (Marcu et al., 1980). One of these deletions extends into the Xba I E fragment; and the loss of the ability to express Ig heavy chains has been correlated, in one instance, with the loss of sequences that we have now shown to contain the Ig enhancer.

There are reports that the unrearranged μ constant region is transcribed to some extent (Kemp et al., 1980; Alt et al., 1982b). The 5′ ends of these "sterile" μ transcripts have been mapped to a position close to the location of the Ig enhancer described here (Nelson et al. 1983) We previously found that information is transmitted in both directions from an enhancer to proximal promoters (a property that seems intrinsic to all enhancers), even from a position 3' to the promoter (Banerji et al. 1981). We also noted that functional promoters and plasmid sequences attenuate the propagation of this information along the DNAs, though long stretches of DNA from be-

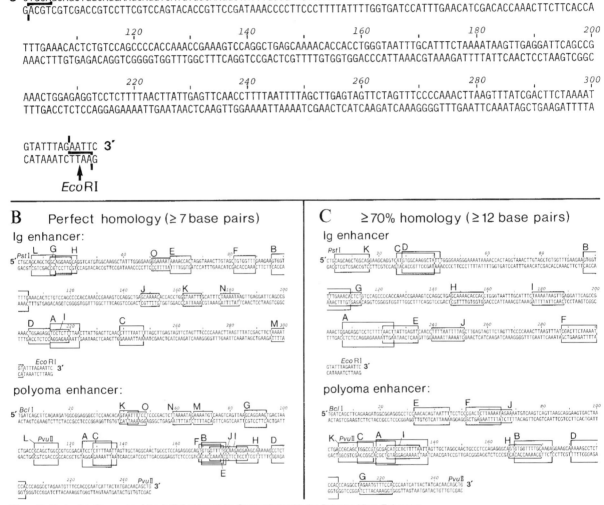

Figure 5. Nucleotide Sequence of the Ig Enhancer and a Comparison with the Polyoma Virus Enhancer

(A) Nucleotide sequence of the 313 bp Pst I to Eco RI Ig enhancer segment from the unrearranged Ig μ constant region. The sequence of part of this region, from a rearranged Ig gene in the IgM expressing plasmacytoma HPC76, has been presented before (Gough and Bernard, 1981). Among several differences between our sequence and theirs is a 6 bp deletion (our coordinates 166–171) of unknown physiological relevance.

(B and C) Homologies between the Ig and polyoma enhancers according to different criteria. Homologous sequences are boxed; any pair of short homologies is labeled identically (not all homologies are on the same DNA strand). The polyoma sequence is from Tyndall et al. (1981).

tween the mouse major and minor β-globin genes have no attenuating effect in these experiments (Banerji and Schaffner in de Villiers et al., 1982a). Enhancer-activated pseudopromoters have indeed been found in plasmid DNA (Wasylyk et al., 1983; see also Moreau et al., 1981). It seems likely therefore that the sterile μ transcripts are due to transcription from pseudopromoters located near the Ig enhancer (see also Figure 6). Sterile kappa transcripts that begin at a pseudopromoter upstream of the J segments of the unrearranged kappa light chain constant region (Van Ness et al., 1981) may arise by a similar mechanism.

The chromatin of actively expressed Ig kappa genes contains a general DNAase I sensitivity (Storb et al., 1981), and specific DNAase I hypersensitivity sites exist in the intron between the J region and constant region of Ig genes (Parslow and Granner, 1982; Chung et al., 1983). Such DNAase I hypersensitivity sites are also present in polyoma or SV40 chromatin extracted from infected cells (Herbomel et al., 1981; Cremisi, 1981). These sites coincide with the locations of the papovavirus enhancers. In other genes (putative) enhancers might also be situated near such hypersensitivity sites (Muskavitch and Hogness, 1982).

If enhancers activate genes within their domain of influ-

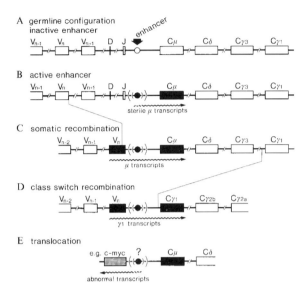

Figure 6. Model of the Ig Enhancer Effect on Gene Transcription

(A) Germ-line configuration of the Ig heavy chain gene family. This region is not transcribed when the Ice enhancer is inactive.

(B) Cells in which the enhancer is active give rise to sterile μ transcripts that start at pseudopromoters in the unrearranged locus.

(C) Creation of a functional Ig gene by somatic recombination (VDJ joining) gives rise to correct Ig μ heavy chain gene mRNAs. Alternate processing of longer transcripts can give rise to mRNAs coding for Ig heavy chains of other classes.

(D) Class switching by acetic: deletion gives rise to mRNAs coding for Ig heavy chains of invariant idiotype but different isotype.

(E) Aberrant translocation of the Ig enhancer to the vicinity of other, nonimmunoglobulin genes gives rise to transcripts beginning at nearby promoters and pseudopromoters. (The last line of this figure describes a hypothetical arrangement. In fact, it has been shown, in the case of a mouse plasmacytoma, that the 5' end of *c-myc* and the 5' end of a functionally rearranged Ig gene (including the Ig enhancer) have been juxtaposed by a reciprocal crossover occurring within both genes (Cory et al., 1983: see also Stanton et al., 1983; Shen-Ong et al., 1982). Taub et al., 1982, have described lymphomas in which the human Ig μ constant region is juxtaposed to a *c-myc* gene; however, the location of the putative human Ig enhancer is unknown.

ence, it follows that when they are translocated they may also cause alterations in the transcription pattern of the newly juxtaposed DNA. Oncogenically transformed animal cells have novel DNA sequences inserted near real (or presumed) cellular oncogenes in chicken bursal lymphomas (Hayward et al., 1981; Neel et al., 1981; Payne et al., 1982), in mouse mammary tumors, thymic lymphomas, and plasmacytomas (Nusse and Varmus,1982; Tsichlis et al., 1983; Shen-Ong et al., 1982; Cory et al., 1983; Stanton et al., 1983), and in human Burkitt's lymphomas (Taub et al., 1982; see also Figure 6). Transcriptional enhancers may be responsible for altering gene expression in many of the above examples, and we consider it likely that enhancers are integrally involved in the normal (and abnormal) development of eucaryotic cells.

Experimental Procedures

Construction of Recombinant DNAs

All clones were constructed according to standard recombinant DNA techniques (Maniatis et al, 1982). Clones not described in detail in the text

were made as follows. For pDept, the small Eco RI to Bam HI fragment of pML (Lusky and Botchan, 1981) was replaced by the large Kpn I to Bam HI fragment of SV40 (see Figure 1A) using Eco RI linkers at the Kpn I site. For pET1, the Eco RI fragment of p24βE+ containing part of the Ig Xba I E fragment and all of the SV40 early region was cloned into pUC8 (Viera and Messing, 1982) with the enhancer being proximal to the polylinker. pET3 is a derivative of p24βE−, analogous to pET1. pTA-S is a derivative of p24βE+, lacking the Pst I fragment which encompasses part of the Ig Xba I E fragment and part of the β-globin gene. pTA-L is a derivative of p24βE−, analogous to pTA-S. pβG is the 4.7 kb Kpn I fragment containing the rabbit hemoglobin β1 gene (a gift from T. Maniatis) cloned into the Kpn I site of pJC-1 (a gift from J. Jenkins; see Banerji et al., 1981; see Figure 1 E). pβG− is a derivative of pβG containing the Kpn I insert in the opposite orientation. pβGX is a derivative of pβG containing an Xho linker at the Bgl II site (shown in Figure 1B) behind the β-globin gene. pBAG is a derivative of p24β with its β-globin gene replaced by that from pβGX. This clone was used as a recipient vector for the Xba I to Xho I fragments of the Bal 31 resection. pCAB is a derivative of pET1 with the Bgl II to Xba I fragment of pBAG-A replacing the enhancer-containing Xba I to Bam HI (polylinker) fragment of pET1. It was used as a recipient vector for Xba I to Xho I fragments of the second Bal 31 resection. pTAR1 is the Ig Xba I E fragment cloned into the Xba I site of pβG−. The orientation of the insert with respect to the β-globin gene is the same as that of p24βE+ (Figure 1F, second line). pTAR7 is identical with pTAR1 except for the inversion of the Xba I insert. pβGXalu is the Alu fragment depicted in Figure 4 inserted into the Pvu II site (shown in Figure 1B) behind the globin gene of pβGX. pβGXpy is the 246 bp Bcl I to Pvu II fragment containing the polyorna enhancer (de Villiers and Schaffner, 1981; see also Figure 5) cloned with Xho I linkers into pβGX. pβGXsv512 is a 201 bp fragment extending from the Kpn I site of SV40 and containing the SV40 enhancer, cloned with Xho I linkers into pβGX. pβGXba32 is a 366 bp Kpn I to Hind III fragment containing the SV40 enhancer cloned with synthetic linkers into the Xba I site of pβG (see Figure 1E). p6E1 is a derivative of p24βE+ digested with Bgl II, then digested with Bal 31, and finally reclosed in the presence of Xho I linkers.

Cell Growth and Transfection with DEAE-Dextran
Cell Growth

Myeloma cells X63-Ag8 (Kohler and Milstein, 1975; a gift from S. Y. Chung), F0 (Fazekas de St. Growth and Scheidegger, 1980; a gift from H. Hengartner), and Sp6Bu6Bu (a gift from G. Köhler) were grown in Dulbecco's modified Eagle's medium with 4.5 g glucose/liter (Gibco), containing 2.5% calf serum and 12.5% fetal calf serum, and buffered with 3.7 g sodium bicarbonate/liter. Mouse 3T6 (a gift from G. Magnusson), mink lung (a gift from H. Diggelmann), and HeLa cells (a gift from U. Pettersson) were grown in the same medium except that the fetal calf serum concentration was 2.5%. Cells were seeded the day before transfection. At the time of transfection, they were less than 20% confluent (for HeLa and mink cells, 50% confluent). Myeloma cells were by then moderately adhered to the petri dish such that they were not washed off the plate upon careful changes of the medium.

DNA

For the transfection of one large (90 mm diameter) petri dish, 1 μg of the appropriate DNA clone was diluted with Tris-EDTA buffer (10 mM Tris-HCl, 1 mM EDTA [pH 7.8]) to 50 μl. Tris-buffered saline, 250 μl (TBS; 25 mM Tris-HCl, 137 mM NaCl, 5 mM KCl, 0.7 mM CaCl$_2$, 0.5 mM MgCl$_2$, 0.6 mM Na$_2$HPO^4 [pH 7.4], modified from Kimura and Dulbecco, 1982), was then added, and the sample was mixed well at room temperature. A solution of DEAE-dextran, 300 μl (Pharmacia; M$_r$ 2 $\times$ 10^6; aliquots of 1 mg/ml in TBS kept frozen), was added, and the sample was mixed well at room temperature.

Transfection

The medium from the plate was carefully replaced by 10 ml TBS (90 mm plate). TBS was aspirated, and another ml TBS was added and aspirated. The DNA sample (0.6 ml, 1 μg DNA in TBS with 0.5 mg/ml DEAE-dextran) was added to the center of the plate. The plate was kept at room temperature for 30 min with occasional tilting. The DNA was then aspirated and carefully replaced by 10 ml TBS. TBS was aspirated and replaced by 10 ml tissue culture medium containing 0.1 mM chloroquine diphosphate (Luthman and Magnusson, 1983; Sigma; mN stock solution in TBS, aliquots kept frozen in the dark; chloroquine treatment increased the number of T-antigen-positive cells 2 to 5 fold under our assay conditions). After 5 hr of incubation at 37°C, the chloroquine medium was replaced by standard tissue culture medium and incubation at 37°C was continued for another 30–35 hr. During the incubation time, cells were not allowed to become overcrowded but were split into two plates. RNA analysis was performed as described (Weaver and Weissmann, 1979; de Villers and Schaffner, 1983). For immunofluorescence analysis, transfectons were scaled down to 35 mm plates, and only 0.25 μg DNA was used.

Fixation of Cells for the Immunofluorescence Assay

The usual fixation procedure, in which the fixer is added to the wet cell monolayer after washing once with TBS (Banerji et al., 1981) gave best results with fibroblast and epithelial cells. However, for myeloma cells the procedure had to be modified as follows. Medium was replaced by TBS, which was also aspirated. The plate, without a lid, was put in an upright position and allowed to dry completely in the airstream of the laminar flow hood. Fixer (70% v/v methanol, 30% acetone, kept at −20°C) was added, and the plate was incubated at −20°C for 15 min. The fixer was aspirated, and the plate was completely dried and processed for immunofluorescence as described in Banerji et al (1981).

Inhibition of DNA Replication

After DNA transfection with DEAE-dextran, chloroquinine treatment was omitted because cells are killed by treatment with both chloroquine and hydroxyurea. Instead, cells were incubated in medium containing 10 mM hydroxyurea (Sigma; stock aliquots of 1 M in H_2O kept frozen) for 35 hr and then fixed for immunofluorescence. For some experiments DNA replication was also inhibited with cytosine arabinoside (Sigma) at a final concentration of 0.1 mM, with the other conditions as described above. Mitotic cells are no longer visible in the cell population upon treatment with either of the two inhibitors.

Acknowledgments

We are grateful to our colleagues mentioned in the text for gifts of materials; to Lisbeth Cerny, Verena Kurer, and Dr Hans Eppenberger for help with fluorescence microscopy; to Drs. Peter Dierks, Su-Yun Chung, and Jean de Villiers for technical advice; and to Drs. Edgar Serfling, Peter Dierks, Dani Schümperli, Max Birnstiel, and Tom Sneider for critical reading of the manuscript. Philipp Bucher and Andreas Rutishauser developed the computer program we used to find homologies, Fritz Ochsenbein prepared the excellent graphics, and Silvia Oberholzel interfaced with the word processor, for which we thank them sincerely This work was supported by the Swiss National Research Foundation and the Kanton of Zürich.

Received April 28, 1983

References

Alt, F. W. Rosenberg, N., Casanova, R. J., Thomas, E., and Baltimore, D. (1982a). Immunoglobulin heavy chain expression and class switching in a murine leukemia cell line. Nature 296, 325–331.

Alt, F. W., Rosenberg, N., Enea, V., Siden E., and Baltimore, D. (1982b). Multiple immunoglobulin heavy chain gene transcripts in Abelson murine leukemia virus-transformed lymphoid cell lines. Mol. Cell. Biol. 2, 386–400.

Banerji, J., and Schaffner, W. (1983). Transient expression of cloned genes in mammalian cells. In Genetic Engineering—Principles and Methods, 5, J K. Setlow and A. Hollaender, eds. (New York: Plenum Press), in press.

Banerji, J., Rusconi, S., and Schaffner, W. (1981). Expression of a β-globin gene is enhanced by remote SV40 DNA sequences. Cell 27, 299–308.

Benoist, C., and Chambon, P. (1981). In vivo sequence requirements of the SV40 early promoter region. Nature 290, 304–310.

Berk, A. J., and Sharp, P. A. (1977) Sizing and mapping of early adenovirus mRNAs by gel electrophoresis of S1 endonuclease-digested hybrids. Cell 12, 721–732

Chung, S. Y., Folsom, V., and Wooley, J. (1983). DNase I hypersensitive sites in the chromatin of immunoglobulin kappa light chain genes. Proc.

Natl. Acad. Sci. USA 80, 2427–2431.

Clarke, C., Berenson, J., Goverman, J., Boyer, P. D., Crews, S., Siu, G., and Calame, K. (1982). An immunoglobulin promoter region is unaltered by DNA rearrangement and somatic mutation during B-cell development. Nucl. Acids Res 10, 7731–7749.

Cory, S., and Adams, J. M. (1980). Deletions are associated with somatic rearrangement of immunoglobulin heavy chain genes. Cell 19, 37–51.

Cory, S., Gerondakis, S., and Adams, J. (1983). Interchromosomal recombination of the cellular oncogene *c-myc* with the immunoglobulin heavy chain locus in marine plasmacytomas is a reciprocal exchange. EMBO J. 2, 697–703.

Cremisi, C. (1981). The appearance of DNase I hypersensitive sites at the 5′ end of the late SV40 genes is correlated with the transcriptional switch. Nucl. Acids Res. 9, 5949–5964.

Davis, M. M., Calame, K., Early, P. W., Livant, D. L, Joho, R. Weissmann, I. L., and Hood, L. (1980). An immunoglobulin heavy-chain gene is formed by at least two recombinational events. Nature 283, 733–739.

de Villiers, J. and Schaffner, W. (1981). A small segment of polyoma virus DNA enhances the expression of a cloned rabbit β-globin gene over a distance of at least 1400 base pairs. Nucl. Acids Res. 9, 6251–6264.

de Villiers, J. and Schaffner, W. (1983). Transcriptional "enhancers" from papovaviruses as components of eukaryotic expression vectors. In Methods in Nucleic Acid Biochemistry, R. A. Flavell, ed. (Amsterdam: Elsevier Press), 911–919.

de Villiers, J., Olson, L., Banerji, J., and Schaffner, W. (1982a). Analysis of the transcriptional "enhancer" effect. Cold Spring Harbor Symp. Quant. Biol. 47, in press.

de Villiers, J., Olson, L., Tyndall, C., and Schaffner, W. (1982b) Transcriptional "enhancers" from SV40 and polyoma virus show a cell type preference. Nucl. Acids Res. 10, 7965–7976.

Early, P,. Huang, H., Davis, M., Calame, K., and Hood, L. (1980). An immunoglobulin heavy chain variable region gene is generated from three segments of DNA: V_H D and J_H Cell 19, 981–992.

Fazekas de St.Groth, S., and Scheidegger, D. (1980). Production of monoclonal antibodies: strategy and tactics. J. Immunol. Meth. 35, 1–21.

Gluzman, Y., and Shenk, T., eds. (1983). Enhancers and Controlling Elements, Banbury Meeting 1983 (Cold Spring Harbor, New York: Cold Spring Harbor Laboratory), in press.

Gough, N. M, and Bernard, O. (1981). Sequences of the joining region for immunoglobulin heavy chains and their role in generation of antibody diversity. Prod Nat. Acad. Sci. USA 78, 509–513.

Sough, N. M., Kemp, D. J., Tyler, B. M., Adams, J. M., and Cory, S. (1980). Intervening sequences divide the gene for the constant region of the mouse immunorglobulin μ chains into segments, each encoding a domain. Proc. Nat. Acad. Sci. USA 77, 554–558.

Grosschedl, R. and Birnstiel, M. L. (1980). Spacer DNA sequences upstream of the T-A-T-A-A-A-T-A sequence are essential for promotion of H2A histone gene transcription in vivo. Proc. Nat. Acad. Sci. USA 77, 7102–7106.

Hayward, W. S., Neel, G. G., and Astrin, S. M. (1981). Activation of a cellular onc gene by promoter insertion in the ALV-induced lymphoid leukosis. Nature 290, 475–480.

Herbomel, P., Saragosti, S., Blangy, D., and Yaniv, M. (1981). Fine structure of the origin-proximal DNaseI-hypersensitive region in wild-type and EC mutant polyoma. Cell 25, 651–658.

Honjo, T., and Kataoka, T. (1978). Organization of the immunoglobulin heavy chain genes and allelic deletion model. Proc. Nat. Acad. Sci. USA 75, 2140–2144.

Hozumi, N., and Tonegawa, S. (1976). Evidence for somatic rearrangement of immunoglobulin genes coding for variable and constant regions. Proc. Nat. Acad. Sci. USA 73, 3628–3632.

Kataoka, T., Kawakami, T., Takahashi, N., and Honjo, T. (1980). Rearrangement of immunoglobulin γ1-chain gene and mechanism for heavy-chain class switch. Proc. Nat. Acad. Sci. USA 77, 919–923.

Kemp, D. J., Cory, S., and Adams, J. M. (1979). Cloned pairs of variable region genes for immunoglobulin heavy chains isolated from a clone library of the entire mouse genome. Proc. Nat. Acad. Sci. USA 76, 4627–4631.

Kemp, D. J., Harris, A. W., Cory, S., and Adams, J. M. (1980). Expression of the immunoglobulin C μ gene in mouse T and B lymphoid and myeloid cell lines Proc. Nat. Acad. Sci. USA 77, 2876–2880.

Kimura, G. and Dulbecco, R. (1972). Isolation and characterization of temperature sensitive mutants of SV40. Virology 49, 394–403.

Kohler, G., and Milstein, C. (1975). Continuous cultures of fused cells secreting antibody of predefined specificity. Nature 256, 495–497.

Laimins, L. A., Khoury, G. Gorman, C., Howard, B., and Gruss, P. (1982). Host-specific activation of transcription by tandem repeats from simian virus 40 and Mooney murine sarcoma virus. Proc. Nat. Acad. Sci. USA 79, 6453–6457.

Lusky, M., and Botchan, M. (1981). Inhibitory effect of specific pBR322 DNA sequences upon SV40 replication in simian cells. Nature 293, 79–81.

Lusky, M., Berg, L., and Botchan, M. (1982). Enhancement of tk transformation by sequences of bovine papilloma virus. In Eukaryotic Viral Vectors, Y. Gluzman, ed. (Cold Spring Harbor, New York: Cold Spring Harbor Laboratory), pp. 99–107.

Luthman, H., and Magnusson, G. (1983). High efficiency polyoma DNA transfection of chloroquine treated cells. Nucl. Acids Res. 11, 1295–1308.

Maki, R., Roeder, W., Traunecker, A., Sidman, C., Wabl, M., Raschke, W., and Toncegawa, S. (1981). The role of DNA rearrangement and alternative RNA processing in the expression of immunoglobulin delta genes. Cell 24, 353–365.

Maniatis, T., Fritsch, E. F., and Sambrook, J. (1982) Molecular Cloning: A Laboratory Manual. Cold Spring Harbor. New York: Cold Spring Harbor Laboratory.

Manteuil, S., and Girard, N. (1974) Inhibitors of DNA synthesis: their influence on replication and transcription of simian virus 40 DNA. Virology 60, 438–454.

Marcu, K. B., Banerji, J., Penncavage, N. A., Lang, R., and Arnheim, N. (1980). 5' Flanking region of irnmunoglobulin heavy chain constant region genes displays length heterogeneity in germ lines of inbred mouse strains. Cell 22, 187–196.

Mather, E. L., and Perry, R. P. (1981). Transcriptional regulation of immunoglobulin V genes. Nucl. Acids Res. 9, 6855–6866.

Max, E. E., Maizel, J. V., and Leder, P. (1981) The nucleotide sequence of a 5.5 kg DNA segment containing the mouse κ immunoglobulin J and C region genes. J. Biol. Chem. 256, 5116–5120.

Moreau, P., Hen, R. Wasylyk, B., Everett, R., Gaub, M. P., and Chambon, P (1981) The SV40 72 base pair repeat has a striking effect on gene expression both in SV40 and other chimera recombinants Nucl. Acids Res. 9, 6047–6068.

Muskavitch, M. A. T., and Hogness, D. S. (1982) An expandable gene that encodes a Drosophila glue protein is not expressed in variants lacking remote upstream sequences. Cell 29, 1041–1051.

Neel, B. G., Hawyard, W. S., Robinson, H. L., Fang, J. and Astrin, S. M. (1981). Avian leukosis virus induced tumors have common proviral integration sites and synthesize discrete new RNAs: oncogenesis by promoter insertion. Cell 23, 323–334.

Nelson, K J., Haimovich, J., and Perry, R. (1983). Characterization of productive and sterile transcripts from the immunoglobulin heavy chain locus: processing of Mu_m and Mu_s mRNA. Mol. Cell. Biol., in press.

Nusse, R., and Varmus, H. E. (1982). Many tumors induced by the mouse mammary tumor virus contain a provirus integrated in the same region of the host genome. Cell 31, 99–109.

Parslow, T. G., and Granner, D. K. (1982). Chromatin changes accompany immunoglobulin kappa gene activation: a potential control region within the gene. Nature 299, 449–451.

Payne, G. S., Bishop, J. M., and Varmus, H. E. (1982) Multiple arrangements of viral DNA and an activated host oncogene in bursal lymphomas. Nature 295, 209–214.

Picard, D., and Schaffner, W. (1983) Correct transcription of a cloned mouse irrimunoglobulin gene in vivo. Proc. Nat. Acad. Sci. USA 80, 417–421

Rabbitts, T. H., Matthyssens, G., and Hamlyn, P. H. (1980). Contribution of immunogobulin heavy-chain variable-region genes to antibody diversity. Nature 284, 238–243.

Rogers, J., Early, P., Carter, C., Calame, K., Bond, M., Hood, L., and Wall, R. (1980). Two mRNAs with different 3' ends encode membrane-bound and secreted forms of immunoglobulin μ chain. Cell 20, 303–312.

Rusconi, S., and Schaffner, W. (1981). Transformation of frog embryos with a rabbit β-globin gene. Proc. Nat. Acad. Sci. USA 78, 5051–5055.

Sakano, H., Maki, R., Kurosawa, Y., Roeder, W., and Tonegawa, S. (1980). Two types of somatic recombination are necessary for the generation of complete immunoglobulin heavy-chain genes. Nature 286, 676–683.

Sakano, H., Kurosawa, Y., Weigert, M., and Tonegawa, S. (1981). Identification and nucleotide sequence of a diversity DNA segment (D) of immunoglobulin heavy-chain genes. Nature 290, 562–565.

Schilling, J., Clevinger, B., Davie, J. M., and Hood, L. (1980). Amino acid sequence of homogeneous antibodies to dextran and DNA rearrangements in heavy chain V-region gene segments. Nature 283, 35–40.

Shen-Ong, G. L. C., Keath, E. J., Piccoli, S. P., and Cole, M. D. (1982). Novel myc oncogene RNA from abortive irnmunoglobulin-gene recombination in mouse plasmacytomas. Cell 31, 443–452.

Stanton, L. W., Watt, R., and Marcu, K. B. (1983). Translocation, breakage and truncated transcripts of c-myc oncogene in murine plasmacytomas. Nature, in press.

Storb, U., Wilson, R., Selsing, E., and Walfield, A. (1981). Rearranged and germline immunoglobulin K genes: different states of DNase I sensitivity of constant K genes in immunocompetent and nonimmune cells. Biochemistry 20, 990–996.

Taub, R., Kirsch, I., Morton, C., Lenoir, G., Swan, D., Tronick, S., Aaronson, S. and Leder, P. (1982). Translocation of the c-myc gene into the immunoglobulin heavy chain locus in human Burkitt lymphoma and murine plasmacytoma cells. Proc. Nat. Acad. Sci. USA 79, 7837–7841.

Tonegawa, S. (1983). Somatic generation of antibody diversity. Nature 302, 575–581.

Tooze, J., ed. (1981). DNA Tumor Viruses. (Cold Spring Harbor, New York: Cold Spring Harbor Laboratory).

Tsichlis, P. N., Strauss, P. G., and Hu, L. F. (1983). A common region for proviral DNA integration in MoNuLV-induced rat thymic lymphomas. Nature 302, 445–449.

Tyndall, C., La Mantia, G., Thacker, C. M., Favaloro, J., and Kamen, R. (1981). A region of the polyoma virus genome between the replication origin and late protein coding sequences is required in cis for both early gene expression and viral DNA replication. Nucl. Acids Res. 9, 6231–6250.

Van Ness, B. G., Weigert, M., Coleclough, C., Mather, E. L., Kelley, D. E., and Perry, R. P. (1981). Transcription of the unrearranged mouse $C_κ$ locus: sequence of the initiation region and comparison of activity with a rearranged $V_κ$–$C_κ$ gene. Cell 27, 593–602.

Vieira, J., and Messing, J. (1982). The pUC plasmids, an M13mp7-derived system for insertion mutagenesis and sequencing with synthetic universal primers. Gene 19, 259–268.

Wasylyk, B., Wasylyk, C., Augereau, P., and Chambon, P. (1983). The SV40 72 bp repeat preferentially potentates transcription starting from proximal natural or substitute promoter elements. Cell 32, 503–514.

Weaver, R. F., and Weissmann, C. (1979) Mapping of a RNA by a modification of the Berk–Sharp procedure: the 5' termini of 15S β-globin mRNA precursor and mature 10S β-globin mRNA have identical map coordinates. Nucl. Acids Res., 7, 1175–1193.

Weber, H., Dierks, P., Meyer, F., van Ooyen, A., Dodkin, C., Abrescia, P., Kappeler, M., Meyhack, B., Zeltner, A., Mullen, E. E., and Weissmann, C. (1981) Modification of the rabbit chromosomal β-globin gene by restructuring and site-directed mutagenesis. In ICN–UCLA Symposium on Molecular and Cellular Biology, 33, D. D. Brown and C. F. Fox, eds. (New York: Academic Press), pp. 367–385.

Weiher, H., König, M., and Gruss, P. (1983). Multiple point mutations affecting the simian virus 40 enhancer. Science 219, 626–631.

Yaoita, Y., and Honjo, T. (1980) Deletion of immunoglobulin heavy chain from expressed allelic chromosome. Nature *286*, 850–853.

Yaoita, Y., Kumaglli, Y., Okumura, K., and Honjo, T. (1982). Expression of lymphocyte surface IgE does not require switch recombination. Nature *297*, 697–699.

Note Added in Proof

After completion of our experiments we learned that S. Gillies et al. (this issue) and M. Neuberger (EMBO J., in press) have reached a similar conclusion about the presence of a myeloma specific enhancer element in the intron of the Ig heavy chain constant region.

Hanahan (1985) Nature (London) **315**, 115–122

Heritable formation of pancreatic β-cell tumours in transgenic mice expressing recombinant insulin/simian virus 40 oncogenes

Douglas Hanahan

Cold Spring Harbor Laboratory, Cold Spring Harbor, New York 11724, USA

Following the transfer into fertilized mouse eggs of recombinant genes composed of the upstream region of the rat insulin II gene linked to sequences coding for the large-T antigen of simian virus 40, large-T antigen is detected exclusively in the β-cells of the endocrine pancreas of transgenic mice. The α- and δ-cells normally found in the islets of Langerhans are rare and disordered. Well-vascularized β-cell tumours arise in mice harbouring and inheriting these hybrid oncogenes.

INSULIN, a polypeptide hormone involved in the control of carbohydrate metabolism, is synthesized and stored in the β-cells of the endocrine pancreas, from which it is released into the bloodstream in response to various signals. The endocrine pancreas is comprised of isolated islands or islets, which are dispersed throughout the larger mass of the exocrine pancreas; the islets consist of four major cell types, α, β, δ and PP, which synthesize the hormones glucagon, insulin, somatostatin and pancreatic polypeptide, respectively[1,2]. Rat and human insulin genes have been cloned and analysed[3–13]. There are two non-allelic rat insulin genes that differ in nucleotide sequence and in the number of introns; the rat insulin I and II genes have one and two introns, respectively[5,6]. The two genes share a region of considerable sequence homology in the 5' flanking region, extending ~300 base pairs (bp) upstream of the point of transcriptional initiation[12]; the human gene is 75% homologous over the first 240 bp of this region[8]. Expression of hybrid genes using the 5' flanking regions of both the human insulin and rat insulin I genes occurs when they are introduced transiently into cultured cells derived from an insulinoma[11].

Here I seek to address aspects of insulin gene expression by the introduction of recombinant insulin genes into the mouse germ line, using microinjection of DNA into fertilized mouse embryos. This technique has been used previously to examine the control and consequences of expression of a number of genes[14–23], including two oncogenes. The complete simian virus 40 (SV40) early region, including the viral transcriptional enhancer and promoter, is not expressed at detectable levels in normal tissues of transgenic mice harbouring it, but choroid plexus tumours heritably arise, and these tumours produce high levels of large-T antigen[19]. The cellular proto-oncogene c-*myc*, when linked to the long terminal repeat of mouse mammary tumour virus, is expressed in a number of tissues, including mammary cells, and two lines of mice heritably develop breast tumours[20].

The recombinant oncogenes used here consist of regulatory information associated with the rat insulin II gene linked to protein coding information for the oncogene SV40 large-T antigen (Tag). The rationale for using such a hybrid is threefold: (1) to examine the ability of the insulin sequences to mediate correct tissue- and cell-type specific expression in the β-cells of the endocrine pancreas, using a readily identified and distinguishable marker protein (a viral antigen); (2) to examine the consequences to the organism of tissue-specific expression of

Fig. 1 Structure of recombinant insulin/large-T antigen genes. The plasmids pRIP1–Tag (*a*) and pRIR–Tag (*b*) consist of the coding information of the SV40 early region fused to sequences derived from the 5' flanking region of the rat insulin II gene. Boxed regions denote insulin gene flanking DNA and the thick solid line refers to the SV40 early region. The cap site where insulin gene transcription initiates is indicated by an L-shaped arrow, and the associated transcriptional enhancer element[11] (EH) is shown (the insulin gene promoter, which lies between them, is not indicated). The points of SV40 large-T antigen translation initiation and transcription termination are indicated by boxed ATG and A_n respectively. The *Bgl*I site in SV40 was converted to an *Xba*I site using an oligonucleotide linker, and the *Xba*I/*Bgl*I to *Bam* fragment comprising the SV40 early region[24] was linked to two different orientations of the insulin gene flanking region. This fragment of the SV40 early region includes the early mRNA cap sites, but lacks all known components of the SV40 early promoter. For *RIP1–Tag*, an *Xba* linker was inserted into the *Dde*I site at +8 bp relative to the point of initiation of insulin gene transcription, and the *Bam* to *Dde*I/*Xba* fragment from −660 to +8 was combined with the *Xba*/*Bam* fragment of the SV40 early region and inserted as a *Bam* linear in a derivative of pBR322 that lacks the R1 site, and includes a *lacUV5* promoter in place of the R1 to *Bam* fragment of pBR. For the plasmid pRIR–Tag, which carries the insulin flanking region in the opposite orientation, a *Dde* site at −520 was converted to an *Xba* site, and the *Bam*/*Xba* fragment extending from +180 nucleotides past the cap to −520bp upstream was combined with the structural gene for large-T antigen and inserted into the same plasmid vector. The two hybrid genes were similarly prepared for microinjection, except that for *RIR–Tag*, the *Bam* insert was first purified from the plasmid vector by gel electrophoresis and electroelution, followed by passage over a DEAE-Sepacel column and pRIP1–Tag was linearized using *Sal*I. The DNAs were extracted with phenol, chloroform, precipitated in 70% ethanol, resuspended in 10 mM NaCl, 5 mM Tris *p*H 7.4, 0.1 mM EDTA and passed through a 1-ml spin column of Sepharose CL6B, after which the DNA was diluted to concentrations of ~75 copies per pl for pRIP1–Tag/*Sal* or ~20 copies per pl for the *RIR–Tag/Bam* insert. The DNAs were injected into fertilized one-cell embryos essentially as described[14,16], except that the injected embryos were cultured overnight to the two-cell stage before being re-implanted into the oviducts of pseudopregnant females. The F_2 embryos were derived from matings of B6D2F1 (C57BL/6J × DBA/2J) males and females, obtained from the Jackson laboratory.

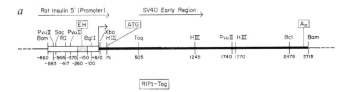

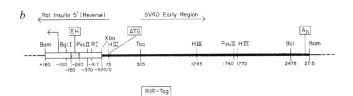

an oncogene, in particular the possibility that SV40 large-T antigen might induce β-cell proliferation and consequent oncogenesis; and (3) to assess the prospects that oncogenes can be used to facilitate the establishment of stable cell lines from rare cell types specifically expressing such fusion genes. This report describes analyses pertaining to the first two of these questions.

Hybrid gene constructs

Two hybrid genes were constructed to combine DNA 5′ to the coding sequences of rat insulin II gene with the large-T antigen structural information. This 5′ flanking region includes the insulin gene promoter as well as a transcriptional enhancer element that stimulates initiation of transcription in cultured insulinoma cells but not in fibroblasts[11]. The first configuration denoted *RIP1* (rat insulin promoter 1) carries the insulin control region aligned to promote transcription of the large-T antigen gene (Fig. 1). The two genes were linked together in regions corresponding to the 5′ untranslated portion of each messenger RNA. The resulting hybrid gene, *RIP1-Tag*, thus comprises a complete transcription unit with 660 bp of sequence located 5′ to the point of transcriptional initiation of the insulin gene linked to protein coding and termination information from the early region of SV40.

The second configuration, denoted *RIR-Tag* (rat insulin reverse), includes 520 bp of DNA 5′ to the insulin gene and the first 180 bp of the transcribed portion of the gene. In this hybrid, the promoter element for insulin gene transcription is inverted with respect to the coding information for SV40 large-T antigen, and, therefore, does not comprise a fusion in the 5′ untranslated regions of the two genes, but rather includes an inverted promoter and enhancer region linked to the *Tag* structural gene, as shown in Fig. 1. Both hybrid genes lack the SV40 early promoter and transcriptional enhancer element, and each includes protein coding information for SV40 small-T antigen, as well as for large-T, because the two proteins are derived from the primary transcript of the SV40 early region by differential splicing of overlapping introns[24].

The two fusion genes were injected into fertilized one-cell mouse embryos, which were then inserted into the oviducts of pseudopregnant female mice and allowed to develop. Four transgenic mice were produced from the *RIP1-Tag* injections and one from the *RIR-Tag* injections. The apparent heterozygous copy number (per diploid genome) of the acquired DNA ranged from ~1 for mouse 87, which carries *RIR-Tag*, to ~5 for mouse 163, which carries and is genotyped as *RIP1-Tag1*, and 5 for mouse 164 (*RIP1-Tag2*). Two of the founder mice (165 and 168) were mosaic for the acquired transgene, using the criteria of infrequent transmission to progeny and lower apparent copy number in the founder than in its progeny. The non-mosaic progeny of M165 (*RIP1-Tag3*) have copy numbers of ~25, whereas those of M168 (*RIP1-Tag4*) have ~12 copies per diploid genome. In each case of multiple copies, genetic and biochemical analysis indicates that the genes are heterozygous and integrated in a single location (data not shown).

Phenotype of transgenic mice

During the initial breeding analysis of the insulin large-T antigen transgenic mice, a phenotype of premature death emerged. The three original non-mosaic mice died at 9–12 weeks of age. Two of these transmitted the transgene to offspring before dying, whereas the third (M163) did not. Surprisingly, this phenotype appeared with the *RIR-Tag* gene, which has the insulin promoter inverted with respect to the large-T antigen coding region, as well as with the *RIP1-Tag* gene, in which the insulin promoter is aligned to transcribe the large-T antigen gene.

The phenotype of premature death proved to be heritable and co-segregated with the acquired hybrid gene; the partial pedigree in Fig. 2 illustrates this observation. The penetrance is virtually complete in the early generations. The analysis that follows deals primarily with the progeny of the two original non-mosaic

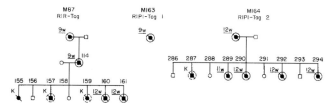

Fig. 2 An early pedigree of insulin/large-T antigen transgenic mice. A partial pedigree analysis is presented, to examine the heritability of the phenotype of premature death and islet hyperplasia. Females are indicated by open circles, males by open squares and transgenic mice by solid circles and squares. The circled slash denotes premature death, with the age at death shown in weeks (w). A slash alone with a superscript K indicates the animal was killed and the dotted circle around a killed mouse denotes evidence of islet hyperplasia following histopathology. (M155 was killed but not examined for hyperplasia.) Each original transgenic mouse is assigned a genotype consisting of the name of the gene that was injected and acquired followed by a number which distinguishes it from the other independent mice carrying the same gene. Therefore, progeny of a particular mouse can be readily identified as such (for example, all mice of genotype *RIP1-Tag2* are progeny of the original transgenic mouse 164).

mice that transmitted—M164 (*RIP1-Tag2*) and M87 (*RIR-Tag*). Progeny of each were killed and subjected to pathological and biochemical analysis. The only obvious abnormal pathology consisted of hyperplasia of the islets of Langerhans, determined following standard histochemical staining of tissue sections. Solid tumours were observed in a few of these mice. Figure 3a shows representative pancreas sections of a normal mouse and of a transgenic mouse (M384, *RIP1-Tag2*) that carried visible tumours; both were ~12 weeks old. Sections were stained with cresol violet, which distinguishes the endocrine islets from the surrounding exocrine tissue. Sections from M384 showed normal-sized islets, enlarged (hyperplastic) islets and solid tumours. All stained with similar characteristics, and were identifiable as islet-like regions by comparison with normal islets. Note,

Fig. 3 (Opposite) Histology and pathology of normal and insulin/T-antigen transgenic mice pancreas. *A*, Cresol violet-stained sections of a normal pancreas and a pancreas derived from a transgenic mouse bearing solid tumours. *a*, Normal (non-transgenic) pancreas with a single islet in the centre and an adjacent blood vessel (×50). *b*, Part of a solid tumour (one of those pictured below) near two normal-sized islets (at the top) in M384 of the *RIP1-Tag2* line (×50). *c*, A normal-sized islet from a non-transgenic mouse (×200). *d*, A normal-sized transgenic islet (from *b*) (×200). In all cases, exocrine cells show purple cytoplasmic staining with somewhat deeper-coloured nuclei, whereas endocrine cell nuclei stain dark violet with minimal cytoplasmic staining. Note the denser packing of nuclei in all sizes of islets derived from the transgenic mouse relative to the normal mouse islet. Compare the normal islet in *a* with the two small islets near the top of *b* and the higher magnifications in *c* and *d*. The extent of cytoplasmic staining observed in the exocrine tissue varies with the length of time the sections were stored before treatment with cresol violet. *B*, Comparison of a whole pancreas from a normal mouse (*a*) with that of M384 (*b*), showing several vascularized solid tumours in the transgenic pancreas (×25).

Methods. Sections (14 μm) were taken on a cryostat from pancreas flash frozen in liquid nitrogen, stored at −70°C and then mounted in tissue Tek OCT embedding medium and sectioned at −20°C on to gelatinized microscope slides. The sections were fixed for 10 min in 4% paraformaldehyde in phosphate-buffered saline (PBS), washed in PBS and air dried. Sections were then rehydrated in PBS, treated in filtered cresol violet for 2 min, washed for 5 min in two changes of 70% ethanol, then washed twice for 2 min each in 90% ethanol, 100% ethanol and xylene and then mounted under coverslips in Gurr Fluromount. The cresol violet solution consists of 0.25% (w/v) thianin that had been dissolved by heating in 200 mM hydroxyacetate plus 26 mM NaOH and stored at 50°C.

Hanahan (1985) Nature (London) **315**, 115–122

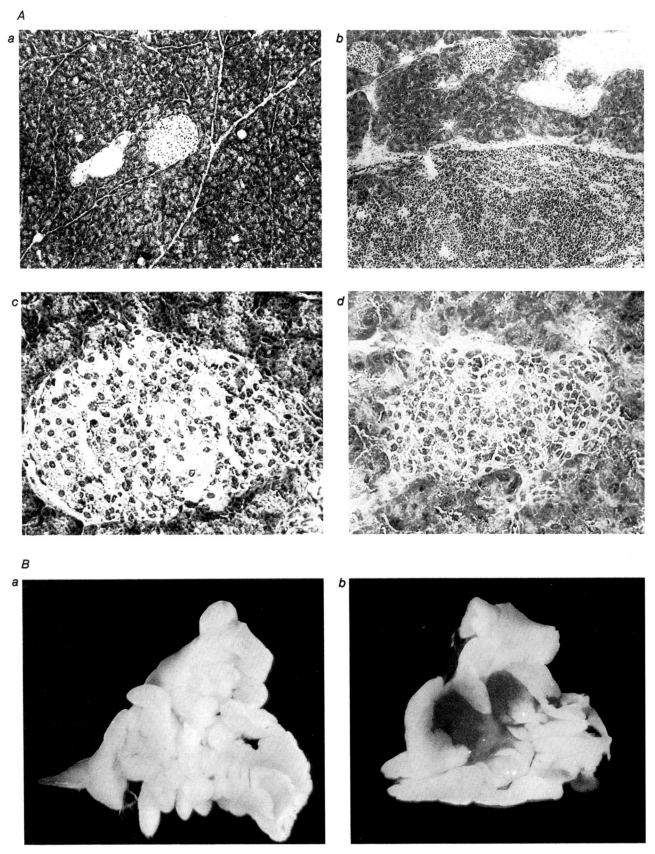

Fig. 3

however, that all islet-like areas in the transgenic mouse appear more densely packed with cell than do the normal mouse islets, demonstrated by comparison of Fig. 3A c and d.

Analysis of the mice listed in Fig. 2 shows that all transgenic mice in these two lineages have the phenotype of islet hyperplasia (if killed) and/or premature death. Before their sudden death, the mice seem normal in appearance and activity. The pedigree stops at the generation during which the mice were placed on a high sugar diet to counteract indications that they were hypoglycaemic which is a probable consequence of islet hyperplasia, and a possible cause of sudden premature death. Subsequent generations, maintained on this diet, now live somewhat longer and heritably develop solid tumours of the pancreas between 10 and 20 weeks of age. An example of a pancreas containing several tumours is shown in Fig. 3B. These tumours eventually comprise a significant fraction of the mass of the pancreas and are highly more vascularized. Individual tumours can reach a size of 5 mm in diameter, with up to five or six separate tumours visible in a pancreas. There is no indication that the tumours metastasize other organs.

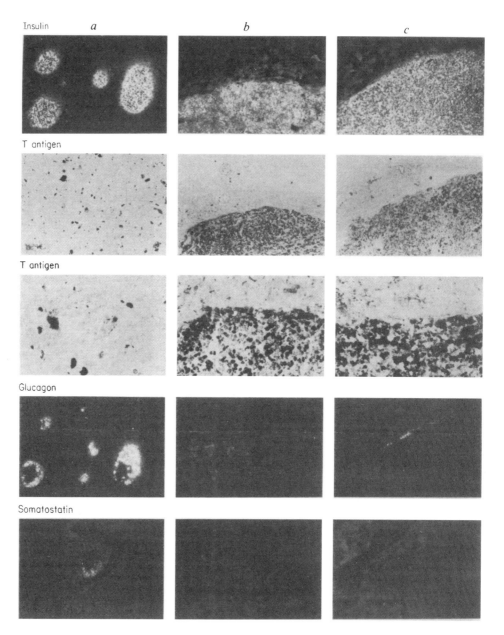

Fig. 4 Analysis of islet cell hormone and large-T antigen expression in a normal mouse and two tumour-bearing transgenic mice. Each column shows an analysis of one mouse: a, normal (non-transgenic); b, 384 from the *RIP1–Tag2* line; c 306 from the *RIR–Tag* line. In each case, near (but not necessarily adjacent) thin sections were analysed for the presence of the specified antigen by using an antiserum to that antigen followed by a conjugated second antibody to visualize the first. The first row examines insulin, the second and third rows large-T antigen, the fourth row glucagon and the fifth row somatostatin. Magnification is ×50 except for the third row, which is ×200. The use of second antibodies alone gives no specific staining. In a, insulin and glucagon plates are of the same cluster of islets, whereas the somatostatin and large-T antigen are of other islets, because that cluster was not present in those sections. The immunostains for the hormones were visualized with rhodamine-conjugated second antibody and epifluroscent [*sic*] illumination, whereas the immunostain for large-T antigen used peroxidase-conjugated second antibody, and was viewed under bright-field illumination. b, c, Orientation of all the plates is the same. The islet tumour is on the bottom, with adjacent exocrine tissue above it. Column b is aligned so the tumour is a centred arc, whereas the plates in c have the tumour ascending to the right. The glucagon and somatostatin plates of b and c are of sections near those used for insulin and large-T antigen; the outline of the exocrine/tumour boundary can be seen in each. The ×200 plates of large-T antigen also show an exocrine/tumour boundary, in which peroxidase staining occurs in nuclei in the tumour tissue but not in the exocrine tissue.

Methods. 14-μm sections were taken as described in Fig. 3 legend, except that for the large-T antigen immunostaining the sections were not fixed, but rather air dried on gelatinized slides. The hormone immunostains were performed on paraformaldehyde-fixed sections as in Fig. 3. Sections were rehydrated in Net-Gel (150 mM NaCl, 5 mM EDTA, 50 mM Tris pH 7.4, 0.25% gelatin, 0.02% NaN$_3$, 0.05% NP40), then treated for 30 min in blocking solution (45% Net-Gel, 45% Dulbecco's modified Eagle's medium, 10% fetal bovine serum, 1% bovine serum albumin (BSA), 0.5% NP40). The sections were then incubated overnight at room temperature in the first antibody, in a binding solution (blocking solution without BSA), washed several times for 30 min in Net-Gel, incubated with second antibody in binding solution for 3 h at room temperature, and again washed in several changes Net-Cel. The sections treated with rhodamine-conjugated second antibody were taken through graded alcohol into xylene and mounted under coverslips in Gurr Fluromount+2% diazabicyclooctane. The large-T antigen sections were visualized with horseradish peroxidase-conjugated second antibody, which was developed by treatment for 10 min in 0.25 mg ml^{-1} diaminobenzidine, 3 mg ml^{-1} nickel sulphate, 0.003% H$_2$O$_2$. The slides were then washed several times in Net-Gel and mounted. The antisera, sources and dilutions are as follows: guinea pig anti-porcine insulin (gift of R. Santerre) 1:5,000; rabbit anti-large-T antigen (a gift of D. Lane) 1:2,000; rabbit anti-human glucagon, Dako A565, 1:1,000; rabbit anti-human somatostatin, Dako A566, 1:1,000; horseradish peroxidase-conjugated swine anti-rabbit IgG Dako P217, 1:200; rhodamine-conjugated goat anti-guinea pig IgG, Capel 2207-0081, 1:200; rhodamine-conjugated sheep anti-rabbit IgG Capel 2212-0084, 1:200.

Cell-type specificity

To evaluate the observed hyperplasia of the islets of Langerhans, I examined normal and transgenic mouse pancreas for expression of both SV40 large-T antigen and polypeptide hormones characteristic for the major cell types. Thin sections were prepared from fresh frozen pancreas obtained from two transgenic mice (M384, *RIP1–Tag*2, and M306, *RIR–Tag*) carrying visible pancreatic tumours as well as from a normal mouse. The sections were analysed by immunostaining using antisera to glucagon, insulin and somatostatin and each visualized with rhodamine-conjugated second antibody to identify the α-, β- and δ-cell types of the islets. Large-T antigen was identified using both a polyclonal antiserum and monoclonal antibodies, each visualized by horseradish peroxidase-conjugated second antibody. Figure 4*a* shows representative immunostains of normal (non-transgenic) islets for the presence of insulin, large-T antigen, glucagon and somatostatin. The islets are comprised predominantly of α- and β-cells with fewer δ-cells; no cells react with antibodies to large-T antigen. Controls using second antibodies alone show no specific staining (not shown).

The islet tumours (Fig. 4*b, c*) in contrast, seem to consist of solely of [*sic*] insulin-producing cells, identifiable as β-cells by immuno- and histochemical staining. Glucagon- and somatostatin-producing cells are rare if not completely absent from the tumours. Characteristic localization of large-T antigen to the nucleus is observed in the islet tumour cells; it is clearly absent from adjacent exocrine cells. Evidence from both polyclonal serum (shown) and monoclonal antibodies (not shown) indicates the same pattern of large-T antigen expression. By these criteria, the islet tumours are essentially pure populations of proliferating β-cells. Occasional clusters of α-cells, perhaps remnants of an islet, can be detected on the edges of a tumour. Thus, both hybrid oncogenes specifically induce proliferation of the β-cells of the islets of Langerhans, eventually resulting in the appearance of pure β-cell tumours. In comparison, naturally occurring insulinomas are characteristically a mixed population of both β- and δ-cells, producing insulin and somatostatin, respectively[25,26].

Tissue-specific expression

The ability of the insulin control region to mediate tissue-specific expression of the SV40 large-T antigen following integration of the hybrid genes into the mouse germ line was assessed by protein blotting analyses, using monoclonal antibodies that specifically recognize large-T antigen. Figure 5*a* presents a comparison of pancreas from several mice, including one normal (non-transgenic) mouse, several insulin/large-T antigen transgenic mice, and one carrying a collagen/large-T antigen fusion gene (M410) which therefore lacks the insulin gene sequences (D.H., unpublished results). M287 and M289 are siblings in the *RIP1-Tag*2 lineage (see Fig. 2); M287 had large solid tumours whereas M289 did not, which illustrates the difference in levels of expression observed in mice of the same age and genotype with and without solid tumours. M277 contains immunoprecipitable large-T antigen, demonstrating that the *RIP1–Tag*3 lineage derived from the founder M165 also expresses the fusion gene. Large-T antigen is not detectable in a normal (F₁) pancreas or in the transgenic mouse (M410) that lacks the insulin gene regulatory region. The *RIR–Tag* transgenic mouse 155 does not show expression in the blot demonstrated here, but other analyses show a low level of large-T antigen. The *RIP1–Tag*4 lineage, exemplified here by M274, does not show detectable expression of large-T antigen by protein blotting. However, pancreatic tumours occasionally arise in this line, which indicates that the *RIP1–Tag*4 insertion can express.

The tissue specificity of fusion gene expression was evaluated further by comparing a number of different tissues from two mice (Fig. 5*b, c*), progeny of the two founders M164 and M165. Both show tissue-specific expression of SV40 large-T antigen, as only pancreas is producing immunoprecipitable protein that co-migrates with authentic large-T antigen. Some degradation of the antigen in pancreatic tissue is observed routinely; it is not clear whether this represents a normal situation in the β-cells or is simply a consequence of the protein isolation procedure. Analysis of a mouse from the *RIR–Tag* lineage also shows expression only in the pancreas, as does a similar examination of the tissues of mouse 163 (*RIP1–Tag1*) (not shown). In a normal pancreas, insulin-producing β-cells comprise only ~1-2% of the total cell mass[1,2]. Therefore, a lower limit can be established for inappropriate expression: either <~1% of the cells in another tissue could be expressing at levels comparable with those in β-cells, or all the cells in an inappropriate tissue could be expressing at levels of <1% of those observed in the islet cells. Within these limits, the expression of both insulin/large-T antigen hybrid genes is tissue specific to the pancreas. Taken together with the immunohistochemical analysis, the transferred insulin sequences are specifying correct expression in the β-cells of the endocrine pancreas.

Penetrance and the onset of oncogenesis

The relationship between large-T antigen expression and the development of β-cell tumours bears on possible mechanisms for oncogenesis. The two previous examples of oncogene expression in transgenic mice give contrasting results[27]. An apparently dormant SV40 early region is activated to express at high levels, and frequently amplified, in the development of choroid plexus tumours[19]. In contrast, a murine mammary tumour virus *myc* fusion gene is expressed in all mammary glands and in several other tissues before oncogenesis, which occurred in only 1 of the 10 mammary glands, implying that a secondary event is required to elicit tumour development[20]. In the insulin/large-T antigen transgenic mice examined to date, a similar pattern emerges. No more than 4 or 5 of the ~100 islets in a pancreas develop into solid β-cell tumours, which are highly vascularized and readily distinguishable from normal pancreatic tissue (Fig. 3*B*). Yet most islets show hyperplasia and all islets examined have increased β-cell density. Furthermore, the β-cells seem to be expressing the hybrid genes well before the development of hyperplasia and tumours, as is demonstrated in an examination of the pancreas from an 8-week-old mouse (M633) in the *RIP1–Tag*2 lineage. This mouse exhibited normal pathology and histology, in particular showing no hyperplasia or visible tumours in the pancreas. Thin sections were prepared from the pancreas and stained with antibodies for insulin and large-T antigen; a representative analysis of a pair of islets is shown in Fig. 6. Large-T antigen is expressing in both of these islets, as well as in all other islets examined in this pancreas. Thus, large-T antigen is expressing in the β-cells of a young mouse before emergence of obvious hyperplasia and solid tumours.

The interpretation of these observations is that the hybrid insulin/large-T antigen genes are expressed in all β-cells, thereby inducing their proliferation, which leads to initial hyperplasia followed by oncogenic transformation. The development of solid tumours occurs in only a few of the islets, which indicates that additional events are probably necessary to mediate the rapid proliferation and extensive vascularization characteristic of the β-cell tumours. It will be of interest to distinguish between two possible classes of transformation: (1) an islet-specific change which affects most of the β-cells in an individual islet; and (2) a rare cell alteration that then produces a clonal tumour arising from that one altered cell.

Cell–cell interaction

Examination of the distribution of cell types in islets of these transgenic mice shows that islets of every size class (small, normal, enlarged, tumour) are densely packed with β-cells. Furthermore, the α- and δ-cells are disordered and rare. Comparison of the glucagon immunostains of normal islets (Fig. 1) and islets of the young transgenic mouse 633 (Fig. 6) shows that the characteristic peripheral distribution of α-cells has been disrupted and that their total number is reduced. Similarly, the

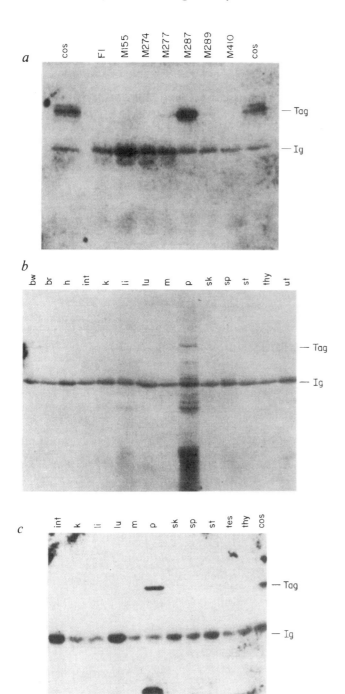

Fig. 5 Tissue specificity of transgene expression. The presence of SV40 large-T antigen in tissues was evaluated by protein blotting[39,40], in which tissue homogenates were immunoprecipitated with a monoclonal antibody (PAb 419) (ref. 41) to T antigen, fractionated on a 10% SDS-polyacrylamide gel, transferred to nitrocellulose by electroblotting and visualized with a second radiolabelled monoclonal antibody (PAb 416), which recognizes a different epitope on the protein[41]. *a*, Pancreas from several mice are compared. F_1 is a non-transgenic mouse, M155 an *RIR–Tag* transgenic, M274 is *RIP1–Tag4*, M277 is *RIP1–Tag3*, M287 and M289 are from siblings from the *RIP1–Tag2* lineage, with M287 having visible tumours and M289 only hyperplastic islets, and M410 is a mouse harbouring a type I collagen/SV40 large-T antigen fusion gene. To provide a control of authentic large-T antigen, total protein from a lysate of the SV40-transformed monkey cell line cos A2 (ref. 42) was immunoprecipitated and similarly treated, except that only 5% as much protein was used compared with the amount of pancreas protein. *b*, *c*, Tissue specificity of large-T antigen expression in the *RIP1–Tag2* (*c*) and *RIP1–Tag3* (*b*) lineages. Equal concentrations of each tissue were examined: bw, body wall; br, brain; h, heart; int, intestine; k, kidney; li, liver; lu, lung; m, muscle; p, pancreas; sk, skin; st, stomach; tes, testes; thy; thymus; ut, uterus. A separate analysis indicates that no large-T antigen can be detected in body wall, brain or heart of M287 (not shown). The radiolabelled antibody PAb416 sticks to the vast excess of immunoglobulin protein used in the immunoprecipitation, thus generating a band of relative molecular mass 55,000 in all samples. This band is further identifiable as IgG using either radiolabelled protein A or rabbit anti-mouse IgG.

Methods. Tissues were flash frozen in liquid nitrogen, stored at $-70\,°C$, and then homogenized in 50 mM Tris *p*H 7.2,5 mM $MgCl_2$, 1 mM $CaCl_2$, 10 mM dithiothreitol, 1% NP40, 100 μg ml^{-1} DNase I, 50 μg ml^{-1} RNase A, 75 μg ml^{-1} phenylmethylsulphonyl fluoride. Protein concentrations were determined using the Bio-Rad protein assay reagent. 500 μg of tissue protein or 25 μg of cos cell protein was incubated overnight at 4 $°C$ with 100 μl of tissue culture supernatant containing the monoclonal antibody PAb419, and then precipitated with 100 μl of 3% protein A–Sepharose (Pharmacia) for several hours. The immunoprecipitate was fractionated on a 10% SDS–polyacrylamide gel, transferred to nitrocellulose by electroblotting overnight, incubated first in a blocking solution (Fig. 4) for 12 h and then in a binding solution (Fig. 4) containing 10^5 c.p.m. ml^{-1} of PAb416, which had been radiolabelled with ^{125}I using the chloramine-T method. The blots were washed in Net-Gel and exposed with intensifying screens onto Kodak XAR film.

normal polar organization of δ-cells is disordered and the δ-cell number severely retarded. This applies to every islet examined in this pancreas, in analyses that included sections taken to include different anatomical regions. These results suggest that during biogenesis of the islets, the α- and δ-cells are either sterically excluded or specifically suppressed by the altered β-cells.

Conclusions

The results presented here demonstrate that DNA within 520 bp upstream of the rat insulin II gene is sufficient to mediate qualitatively correct tissue- and cell-type-specific expression of hybrid insulin/SV40 large-T antigen genes. Protein blotting analysis of four mice harbouring independent insertions of such genes demonstrates expression only in the pancreas among the major tissues examined. The *in situ* analyses of thin sections from the pancreas demonstrates that this expression is cell-type specific, as large-T antigen is only detected in the insulin-producing β-cells of the islets of Langerhans.

The specification of tissue- and cell-type expression by the insulin gene 5′ flanking region occurs in both possible orientations when it is located upstream of the large-T antigen structural gene. Both tissue specificity and bidirectionality are qualities previously associated with transcriptional enhancer elements using transfection of cultured cells[28,29]. It is not clear whether this flanking region carries an additional real or cryptic promoter element oriented away from the insulin gene, or whether non-specific initiation of transcription is occurring in the *RIR–Tag* hybrid gene. RNA analyses will be necessary to address these possibilities.

There are three phases to the phenotype elicited by these hybrid oncogenes: the development of disordered islets, the subsequent enlargement of those islets into obvious hyperplasia and the formation of solid tumours. In young mice, the islets are of normal size, but densely packed with β-cells. The organization of both α- and δ-cells is disrupted and numbers of each severely reduced, δ-cells being particularly rare. These results suggest that islet cells interact to maintain a proper balance, and the transgenic β-cells effectively reduce α- and δ-cell number and organization by retarding their normal cell development. This possibility motivates a similar examination of the distribution of PP cells in the islets and islet tumours of these mice, because PP cells, like β-cells, are often found at significant levels in naturally occurring insulinomas.

The prospect of disruption in islet-cell interactions is intriguing in view of observations that each of the three hormones influences secretion of the other two by their respective cell types. Somatostatin inhibits the release of both insulin and

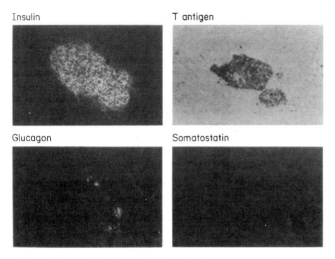

Fig. 6 Immunohistology of islets in the pancreas of a young transgenic mouse. Mouse 633, from the *RIP1–Tag2* lineage, was killed at 8 weeks of age and its pancreas flash frozen in liquid nitrogen, sectioned in a cryostat and analysed for the expression of islet cell hormones and large-T antigen as described in Fig. 4. legend. The two islets shown are representative of all islets in these and other thin sections from this pancreas. The sections used for insulin on large-T antigen are near but not adjacent and the islet sizes have changed somewhat across that distance (×50).

glucagon, insulin inhibits the release of glucagon and somatostatin, whereas glucagon stimulates secretion of insulin and somatostatin (see, for example, ref. 30). The organization of the endocrine pancreas into separate islands has led to suggestions that the islet cells arise from a common stem cell[31,32]. The character of the alterations in islet composition produced by the insulin/large-T fusion gene should allow more detailed examination of islet biogenesis. If islet cell types are interacting through growth factors, the results presented here suggest that they are probably suppressors of cell growth rather than inducers. Insulin itself is one candidate for such a suppressive growth factor.

The dense packing and disorganization of islets is followed by their expansion (hyperplasia), which seems to be a direct consequence of the expression of large-T antigen. Preliminary analysis indicates that serum insulin levels are elevated (two- to ten-fold), but not in proportion to the increased number of insulin-producing cells (>100-fold), which implies that either normal regulation of insulin secretion remains in effect, or that the β-cell structure, organization and/or vascularization is inappropriate for effective secretion. Isolation of β-cell lines from these transgenic mice should be facilitated by the ability of large-T antigen to establish (or immortalize) primary cells to growth in culture, with large-T antigen synthesis here maintained by the co-expression of the insulin and insulin/large-T antigen genes in β-cells. The availability of such β-cell lines will allow more detailed examination of the characteristics of altered β-cells. We may now examine how islet disorganization is correlated with changes in glucagon, somatostatin and insulin serum levels over a detailed timescale. It seems likely that the mice described in this study suffer from serious alterations in regulation of carbohydrate metabolism and that the explanation of their premature death does not simply result from excessive insulin in their serum, but also involves other factors.

Solid β-cell tumours eventually develop from a few of the hyperplastic islets. Tumour development occurs between 10 and 20 weeks of age, is heritable and has occurred in all progeny of the two lines (*RIP1–Tag2* and *RIR–Tag*) in which mice have lived to this age, and is occurring now in the *RIP1–Tag3* line. The solid tumours are frequent in the sense they arise in every mouse harbouring these genes, although they occur only in a fraction of the islets. The characteristics of tumour formation suggest that synthesis of large-T antigen is necessary but insufficient to produce that condition in every islet. Large-T antigen is of a class of oncogenes that can both immortalize primary cells and release (or transform) cultured cells from their normal growth controls, such as contact inhibition and dependence on serum growth factors[24]. Yet, *in vivo* SV40 virus will produce tumours in mice only after a long latent period; oncogenesis seems to be related to the efficacy of the immune response and the presence of large-T antigen on the cell surface as well as in the nucleus[33–35]. There is no evidence of an immune response to hyperplastic islets, nor of large-T antigen on the surface of β-cells; it remains to be established whether the frequency of solid tumour formation is associated with an escape from immune surveillance. A similar hybrid gene, composed of the 5′ flanking region of the rat elastase gene linked to the coding

synthesis is the primary event, there are other potential secondary alterations that may be necessary to induce the development of solid β-cell tumours. These include activation or inactivation of growth factors or receptors, or the cooperation of another oncogene[36,37]. It is well established that nascent tumours cannot exceed certain size constraints without becoming vascularized; thus the induction of angiogenesis[38] is probably one (if not the only) secondary event necessary for solid tumour development. Biochemical and further histochemical comparisons of the islets before and after the predictable development of tumours may provide insight into these possibilities.

The ability to examine the pre- and postnatal development of the endocrine pancreas as well as heritable oncogenesis, using as markers the hormone gene products of these cell types and the SV40 tumour antigen, will provide a means to examine the consequences of cell-specific oncogene expression and the ontogeny of the effects reported here, as well as to address further aspects of the control of insulin-gene expression. It will be of interest to compare different recombinant insulin/oncogenes (such as *myc* and *ras*) for their effects on islet-cell biogenesis and cell–cell interaction, and on secondary events necessary for the development of β-cell tumours.

I thank David Lane, Sue Hockfield, Beth Friedman and Ed Harlow for advice on immunohistology and protein blotting; Jim Watson, Winship Herr, Terri Grodzicker, Pam Mellon and Joe Sambrook for comments on the manuscript; Liz Lacy and Frank Costantini for instruction in microinjection; Stanley Minkowitz and Bartol Matanic for several pathology examinations; David Lane and Robert Santerre for antisera; Rich Cate and Wally Gilbert for the rat insulin gene; Debbie Lukralle for animal care; Mark Lacy for technical assistance; Mike Ockler and Dave Green for artwork; Marilyn Goodwin for preparation of the manuscript; and Jim Watson and Joe Sambrook for support and encouragement. D.H. was initially a Junior Fellow of the Society of Fellows, Harvard University and is currently a Special Fellow of the Leukemia Society of America. This work was funded by the Robertson Research Fund of the Cold Spring Harbor Laboratory and by a grant from Monsanto Company.

Received 18 January; accepted 20 March 1985.

1. Steiner, D. F. & Freinkel, N. (eds) *Endocrine Pancreas. Handbook of Physiology* Vol. 1 (Williams and Wilkins, Baltimore, 1972).
2. Cooperstein , S. J. & Watkins, D. (eds) *The Islets of Langerhans Biochemistry, Physiology, and Pathology* (Academic, New York, 1981).
3. Ullrich, A. *et al. Science* **196**, 1313–1319 (1977).
4. Villa-Komaroff, L. *et al. Proc. natn. Acad. Sci. U.S.A.* **75**, 3727–3731 (1978).
5. Lomedico, P. *et al. Cell* **18**, 545–558 (1979).
6. Cordell, B. *et al. Cell* **18**, 533–543 (1979).
7. Ullrich, A. *et al. Science* **204**, 6121–615 (1980).
8. Bell, G. I. et al. *Nature* **284**, (26–32 (1980).
9. Wu. C. & Gilbert, W. *Proc. natn. Acad. Sci. U.S.A.* **78**, 1577–1580 (1981).
10. Kakita, K., Goddings, S. & Permutt, M. *Proc. nat. Acad. Sci. U.S.A.* **79**, 2803–2807 (1982).
11. Walker, M., Edlund, T., Boulet, A. M. & Rutter, W. J. *Nature* **306**, 557 (1983).
12. Cate, R., Chick, W. & Gilbert, W. *J. biol. Chem.* **258**, 6645–6652 (1983).
13. Episkopou, V., Murphy, A. & Efstratiadis, A. *Proc. natn. Acad. Sci. U.S.A.* **81**, 4657–4661 (1984).
14. Palmiter, R., Chen, H. & Brinster, R. *Cell* **29**, 701–710 (1982).

15. Palmiter, R. *et al. Nature* **300**, 615 (1982).
16. Lacy, E., Roberts, S., Evans, E. P., Burtenshaw, M. D. & Constantini, F. D. *Cell* **34**, 343–358 (1983).
17. McKnight, G., Hammer, R. E., Kuenzel, E. A. & Brinster, R. L. *Cell* **34**, 335–341 (1983).
18. Brinster, R. *et al. Nature* **306**, 332 (1983).
19. Brinster, R. *et al. Cell* **37**, 367–379 (1984).
20. Stewart, T., Pattengale, P. & Leder, P. *Cell* **38**, 627–637 (1984).
21. Grosschedl, R., Weaver, D., Baltimore, D. & Constantini, F. *Cell* **38**, 647–658 (1984).
22. Swift, G., Hammer, R. E., MacDonald, R. J. & Brinster, R. L. *Cell* **38**, 639–646 (1984).
23. Chada, K. *et al. Nature* **314**, 377–380 (1985).
24. Tooze, J., ed. *Molecular Biology of Tumor Viruses* Pt 2 (Cold Spring Harbor Laboratory, New York, 1981).
25. Chick, W. L. *et al. Proc. natn. Acad. Sci. U.S.A.* **74**, 628–632 (1977).
26. Gazdar, A. *et al. Proc. natn. Acad. Sci. U.S.A.* **77**, 3519–3523 (1980).
27. Hanahan, D. *Nature* **312**, 503–504 (1984).
28. Banerji, J., Rusconi, S. & Schaffner, W. *Cell* **27**, 299–308 (1981).
29. Moreau, P. *et al. Nucleic Acids Res.* **9**, 6047–6068 (1981).
30. Raskin, P. in *The islets of Langerhans* (eds Cooperstein, J. J. & Watkins, D.) 467–481 (Academic, New York, 1982).
31. Pictet, R. & Rutter, W.J. in *Endocrine Pancreas Handbook of Physiology* Vol. 1 (eds Steiner, D. F. & Frienkel, N.) (pp. 25–66) (Williams & Wilkins, Baltimore, 1972).
32. Pearse, A. C. E. *Nature* **295**, 96–97 (1982).
33. Hargis, B. J. & Malkiel, S. *J. nat. Cancer Inst.* **63**, 965–967 (1979).
34. Abramczuk, J., Pan, S., Maul, G & Knowles, B. B. *J. Virol* **49**, 540–548 (1984).
35. Lewis, A. M. & Look, J. L. *Science* **227**, 15–20 (1985).
36. Land, H., Parada, L. F. & Weinberg, R. A. *Nature* **304**, 596–609 (1983).
37. Ruley, H. E. *Nature* **304**, 602–606 (1983).
38. Folkman, J., Merler, E., Abernathy, C. & Williams, G. *J. exp. Med.* **133**, 275–288 (1971).
39. Towbin, H., Staehelin, T. & Gordon, J. *Proc. natl. Acad. Sci. U.S.A.* **76**, 4350–4354 (1979).
40. Burnette, W. N. *Analyt. Biochem.* **112**, 195–203 (1981).
41. Harlow, E., Crawford, L. V., Pim, D. C. & Williamson, N. M. *J. Virol.* **39**, 861–869 (1981).
42. Gluzman, Y. *Cell* **23**, 175–182 (1981).

Blom van Assendelft et al. (1989) Cell **56**, 969–977

The β-globin Dominant Control Region Activates Homologous and Heterologous Promoters in a Tissue-Specific Manner

Greet Blom van Assendelft, Olivia Hanscombe, Frank Grosveld, and David R. Greaves

Laboratory of Gene Structure and Expression
National Institute for Medical Research
The Ridgeway, Mill Hill
London NW7 1AA
England

Summary

We have introduced a human β-globin minilocus, containing the recently described dominant control region (DCR), the β-globin or Thy-1 gene, and a thymidine kinase (tk)-neo^R gene into erythroid and nonerythroid cells. Analysis of the transcription levels of the genes shows that the DCR directs high levels of human β-globin, Thy-1 and tk-neo expression independent of integration sites in an erythroid-specific manner. The presence of the DNAase I hypersensitive sites at the 5' end of the locus is required for this effect on the homologous and heterologous gene. An analysis of the DCR chromatin in transfected mouse erythroleukemic cells suggests that the formation of the hypersensitive sites in this region precedes β-globin gene expression.

Introduction

The human β-globin gene is part of a multigene family that is expressed in a tissue- and developmental-specific manner (for review, see Collins and Weissman, 1984). The β-globin gene is first transcribed in the fetal liver, and becomes fully active when the site of erythropoiesis switches to the (adult stage) bone marrow. The activity of the γ-globin genes shows a complementary pattern of expression; i.e., they are fully active in the fetal liver, but are decreased to very low activity in adult bone marrow. The activation of the human β-globin gene during erythroid differentiation can be mimicked in transgenic mice and mouse erythroleukemia (MEL) cells (Magram et al., 1985; Wright et al., 1983; Grosveld er al., 1987). It is clear that these processes are regulated by *trans*-acting factors (Baron and Maniatis, 1986; Wrighton and Grosveld, 1988) that bind to a number of regulatory regions throughout the entire β-globin gene cluster. In particular, the β-globin gene and its immediate flanking regions contain a promoter element and two enhancers, which alone, or in combination, give rise to regulated, erythroid-specific expression in transgenic mice (Behringer et al., 1987; Kollias et al., 1987a; Trudel et al., 1987) and MEL cells (Antoniou et al., 1988). Each of these regions is able to bind a number of ubiquitous and erythroid-specific protein factors (Wall et al., 1988; deBoer et al., 1988) but it is, as yet, not clear which of these plays an important role in the tissue- or developmental-specific regulation of the gene. In addition to the immediate flanking regions of the gene, the entire

gene cluster is regulated by a region at the 5' end (and possibly 3' end) of the gene cluster (Grosveld et al., 1987). Deletion of this region in vivo leads to a classical position effect and the silencing of the β-globin gene in γδβ-thalassemia (van der Ploeg et al., 1980; Kioussis et al., 1983). The deletion is 100 kb (Taramelli et al., 1986) and contains a 20 kb region characterized by a set of tissue-specific DNAaseI hypersensitive sites (Tuan et al., 1985; Forrester et al., 1987; Grosveld et al., 1987). When this region is added to a human β-globin gene construct, it results in very high levels of human β-globin gene expression in transgenic mice, which is related to the copy number, and is independent of the integration site of the transgene (Grosveld et al., 1987). However, it was not clear from these data whether this effect was developmentally specific, nor whether the same effect would be observed if the minilocus were introduced directly into the cells of an erythroid lineage. It was also not clear whether the same would be observed for a heterologous, nonerythroid-specific gene. To answer these questions, we introduced the β-globin minilocus, including a β-globin or Thy-1 gene and a thymidine kinase promoter–driven AGPT hybrid gene (tk-neo) into three types of cells: MEL cells that express β-globin, K562 cells that express γ- but not β-globin, and mouse L cells that do not express any globin genes, but are permissive for the transcription of transfected globin genes. The results show that the DCR sequence is active and regulates both the β-globin, Thy-1 and tk-neo gene in the erythroid cells. No effect is observed in L cells.

Results

Expression in MEL Cells

The β-globin minilocus (Figure 1) was constructed by cloning the upstream and downstream sequences of the β-globin locus with the β-globin gene into the cosmid pTCF (Grosveld et al., 1982) from which the ClaI site had been removed (Grosveld et al., 1987). This cosmid vector contains the aminoglycoside phosphotransferase (AGPT or neo) gene coupled to the Herpes Simplex Virus thymidine kinase (tk) gene promoter and poly(A) addition sequences (Grosveld et al., 1982). This 45 kb β-globin locus recombinant cosmid was linearized with PvuI in the cosmid, leaving the tk-neo at the 5' end of the molecules and subsequently introduced into C88 MEL cells (Deisseroth and Hendrick, 1978) by electroporation (see Experimental Procedures). Stably transfected populations or individual clones were grown by selection in medium containing the aminoglycoside G418 and analyzed before and after erythroid differentiation by the addition of DMSO to one-half of each culture (Antoniou et al., 1988). RNA was prepared from both stages to determine the level of β-globin expression, and DNA was prepared to determine the structural integrity and copy number of the integrated locus.

Southern blots were carried out with EcoRI digested DNA from eight populations, and Figure 2A shows an ex-

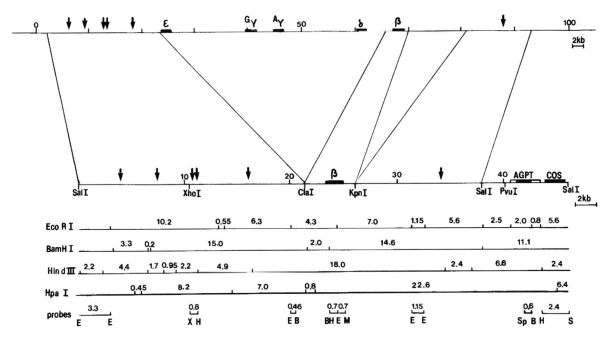

Figure 1. Human β-Globin Minilocus

The top line shows the human β-globin locus; the bottom part the constructed human β-globin minilocus. A number of unique restriction sites is indicated directly on the constructs, while the sizes of the EcoRI, BamHI, HindIII, and HpaI restriction fragments in the minilocus are shown below in kb. On the cosmid pTCF part of the construct, the tk-neo^R is indicated as the AGPT-box.

    At the bottom, the hybridization probes for the 5′-flanking region (3.3 kb E-E, 0.6 kb X-H and 0.46 kb E-B), the β-globin gene (0.7 kb B-E and 0.7 kb E-M), the 3′-flanking region (1.15 kb E-E), and the neo (0.6 kb Sp-B) and cos (2.4 kb H-S) regions of the construct are shown.

E = EcoRI, X = XhoI, H = HindIII, B = BglII, M = MspI, Sp = SphI, S = SalI. Digestion with PvuI results in a cleavage of the amp^R region of the cosmid, and produces a linear molecule that carries the tk-neo gene at the 5′ end of the minilocus.

ample of three of the populations using a human β-globin BamHI–EcoRI second intron probe (Figure 1) and a mouse Thy-1 probe as a control. After densitometer scanning of the bands, an estimate of the copy number of the transfected β-globin gene can be made (Table 1). This is compared with the expression levels of the β-globin gene after induction of the MEL cells (Figure 3A, left panel, and Table 1), resulting in very high RNA levels per copy of the transfected human β-globin gene. In most cases, this level is similar to that obtained in transgenic mice (lane 12, Figure 3; Table 1), and is at least 100-fold higher than the levels obtained without the DCR (Antoniou et al., 1988; data not shown). Additional Southern blot data on the flanking regions of the human β-globin gene in the populations showed that some of these contained a limited number of end fragments, and thus consisted of a limited number of individual clones (4–5); moreover, some of these carried deletions (e.g., population 3, Figure 2B). We therefore also analyzed nine individual clones, which were derived from the first three populations. A similar analysis was carried out to estimate the copy numbers and expression levels of the transfected gene (Figure 1A, left panel; Figure 3; Table 1). Similar results were obtained, although the overall expression levels in the clones were lower, probably due to the fact that the ratio of mouse α- to mouse β-globin mRNA has increased, resulting in relatively lower expression of mouse β-globin mRNA (Table 1). We have at present no explanation why such a selection has taken place.

However, there are a number of exceptions; in this case, clone k, which does not express, and clones t, i, and e, which are exceptionally high. Additional Southern blots were therefore carried out on all of the samples to determine the structure of the entire transfected locus. The analysis (Figure 2B) of a ClaI–SalI digest probed with an XhoI–HindIII fragment (Figure 1) shows that clone k contains the upstream region (Figure 2B) but not the β-globin gene itself (Figure 2A). Additional blots show that the tk-neo gene is present but not linked to the upstream region in this clone (data not shown). An EcoRI digest of clone i probed with a 3′ β-globin probe (Eco–Msp, Figure 1) shows that one of the two copies of the β-globin locus carries a 3′ end deletion. Further blots (EcoRI–XbaI and EcoRI-PstI; data not shown) show that this deletion includes the enhancer that has been mapped in the 3′ flanking regions of the β-globin gene (Kollias et al., 1987a; Behringer et al., 1987; Trudel et al., 1987; Antoniou et al., 1988). Analysis of the other overexpressing clones shows that all the copies in clone t contain a 3′ end deletion, leaving the enhancer in place (data not shown), but deleting the hypersensitive site in the 3′ region of the locus (Figure 2C). This suggests that the absence of the 3′ flanking region may result in an increased expression of the β-globin gene, although it should be stressed that the other overexpressing clone (e) has no apparent deletions. In fact, it appears to contain the same end fragment as clone f (Figure 2D), and we have presently no reasonable explanation for

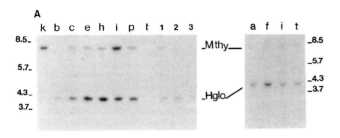

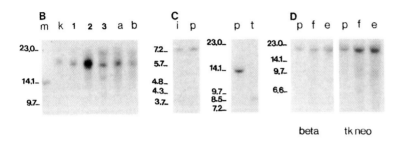

beta   tk neo

Figure 2. Structural Analysis of the Human β-Globin MEL Clones and Populations

(A) DNA (approximately 5 μg) was digested with EcoRI, gel-electrophoresed, Southern blotted, and hybridized to the human β-globin BamHI–EcoRI IVSII probe (see Figure 1) and a mouse Thy-1 probe.

(B) A SalI–ClaI double digest was Southern blotted and hybridized to the 5′-flanking probe XhoI-HindIII (Figure 1).

(C) The Southern blot of the EcoRI digest (see A) was washed (0.2 M NaOH) and reprobed (left two lanes) with the 3′ human β-globin probe EcoRI–MspI (Figure 1). Right two lanes: DNA was digested with BamHI, Southern blotted, and hybridized with the human β-globin BamHI–EcoRI IVSII probe (Figure 1).

(D) An HpaI digest was, after Southern blotting, hybridized to either the human BamHI–EcoRI IVSII probe (Figure 1) or the neo SphI–BglII probe (Figure 1).

On all blots, marker DNA (λ-DNA digested with BstEII or HindIII) was run alongside the DNA samples; indicated sizes are in kb. The numbers or letters indicating the MEL populations and clones (respectively) are above the lanes.

the expression levels of clone e. Nevertheless, we tested directly whether a deletion of the 3′ region would cause an increase in human β-globin expression. Three MEL cell populations transfected with a PvuI–KpnI fragment were generated (hss-5 1, 2, and 3), and their expression levels

determined by comparison with the nondeletion population 3 (Figure 3A, right panel). All three populations showed an increase of expression per copy of the human β-globin gene (Figure 3C; Table 1), and we conclude that the deletion causes an increase of expression. It is pres-

Table 1. Human β-Globin Expression Levels in MEL Cell Populations and Clones

Population Number	Hβ-mRNA/Mα-mRNA (cpm ± 15%)	Copy Number (± 25%)	Hβ/Mα RNA per Hβ DNA	Hβ/Mβ RNA per Hβ DNA	Mα/Mβ (± 30%)
1	2800/1370	2.3	0.9	0.3	0.4
2	19400/2100	6.7	1.4	0.5	0.5
3	6550/990	5.9	1.1	0.4	0.4
4	3090/1700	1.8	1.0	0.5	0.5
5	5850/1100	4.0	1.3	0.3	0.2
6	3060/770	3.1	1.3	0.3	0.2
7	3410/1100	2.7	1.2	0.3	0.2
8	860/1370	0.5	1.2	0.4	0.3
mouse 12	3080/1450	2.0	0.9	0.4	0.4
Clone					
a	450/220	5	0.4	0.1	0.3
b	250/160	5	0.3	0.2	0.7
c	130/120	4	0.3	0.2	0.7
e	3280/220	6	2.5	1.6	0.7
f	380/200	7	0.3	0.2	0.9
h	300/180	7	0.2	0.2	0.9
i	240/200	2	0.6	0.4	0.6
p	320/190	5	0.3	0.2	0.8
t	1880/320	3	2.0	0.9	0.5
Population					
hss-5,1	3770/1360	0.2	13.9	4.6	0.3
hss-5,2	9740/1650	0.6	9.8	2.8	0.3
hss-5,3	15940/1360	1.0	11.7	3.1	0.3

Hβ-mRNA/Mα-mRNA represents the average amount of human β-globin and mouse α-globin mRNA in two separate experiments by excising S1 nuclease-protected fragments from the gel and counting in a liquid scintillation counter. The copy number of the human β-globin DNA per cell is obtained by densitometry of the β-globin DNA signal on Southern blots and normalization to that obtained for Thy-1. Hβ/Mα per Hβ DNA represents the ratio of human β- and mouse α-globin DNA (in cpm) divided by the copy number.

Hβ/Mβ RNA per Hβ DNA is the same as the previous column but using mouse β-globin RNA as the standard. Mα/Mβ represents the ratio of the mouse α- to mouse β-globin mRNA levels.

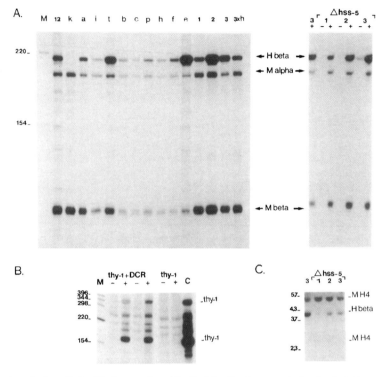

Figure 3. Expression Analysis of the Human β-Globin and Thy-1 MEL Clones and Populations by Nuclease S1 Mapping

(A) Left: ten micrograms of RNA from induced MEL clones and populations (numbered above the lanes) transfected with the β-globin construct was analyzed by S1 nuclease protection (Kollias et al., 1986), using a mixture of three different probes: a 3′ human β-globin probe (760 bp EcoRI–MspI fragment, Figure 1) resulting in a 210 nucleotide protected fragment (H beta); a mouse α-globin probe (260 bp BamHI second exon probe), resulting in a 170 nucelotide protected fragment (M alpha) and a mouse βmaj probe (230 bp HindIII–NcoI second exon probe) to give a 100 nucleotide protected fragment (M beta).

The positive control is a transgenic mouse 12, made with the 38 kb SalI fragment of the human β-globin minilocus (Grosveld et al., 1987). In lane 3xh 30 μg of RNA from clone h was used as a control of excess probe.

Right: S1 nuclease protection analysis on 10 μg of RNA from uninduced (−) and induced (+) MEL populations transfected with the β-globin minilocus after deletion of the 3′ region. The first lane (labeled 3) is the same sample 3 as in the left panel.

(B) S1 nuclease protection analysis (Kollias et al., 1987b) of 20 μg of RNA from uninduced (−) and induced (+) MEL cell populations after transfection with the Thy-1 gene alone (Thy-1) of the Thy-1 gene replacing the β-globin gene in the minolocus cosmid (Figure 1). Positive control C contains 1 μg of brain RNA.

Lane M in (A) and (B) is a marker (pBr322 × Hinfl) indicated in nucleotides.

(C) Southern blot analyses of the DNA from MEL cell populations transfected with the β-globin minilocus without the 3′ hypersensitive region (i.e., minus the KpnI–PvuI fragment in Figure 1).

Five micrograms of DNA was digested with EcoRI, Southern blotted, and hybridized to the mixed human β-globin IVSII probe (BamHI–EcoRI) and the mouse histone H4 (MH4) probe. The control lane contains EcoRI-digested DNA from β-globin minilocus MEL population 3.

Marker DNA (λ-DNA digested with BstEII) was loaded alongside the DNA samples; marker band sizes are indicated in kb.

---

ently not clear whether this is caused by a "distance" effect or the absence of an active hypersensitive site 5 region.

Further S1 protection analysis showed that all of the clones have very low levels of human β-globin expression before induction, with the exception of clone e (not shown) and population hss-5 no. 3 (Figure 3A, right panel). The same expression pattern is observed when the levels of human β-globin and tk-neo RNA were analyzed on Northern blots after reprobing of the blot for the presence of mouse α-globin and histone H4 mRNA as the induction of differentiation and RNA loading controls, respectively (Figure 4). Not surprisingly, the tk-neo gene is expressed at low levels before induction to provide G418 resistance, but it is dramatically increased after the induction of MEL cell differentiation following the pattern of human β-globin gene expression. The tk-neo is at least 50–100-fold higher than that observed in the absence of the flanking regions (clone k and Antoniou et al., 1988), and shows that these regions exert a dramatic effect on this heterologous promoter.

To test whether the induction of the tk-neo gene requires the presence of the human β-globin gene and its enhancers, we replaced the globin gene with a human-murine hybrid Thy-1 gene (Kollias et al., 1987b). The resulting transfected populations were compared with populations containing the Thy-1 and tk-neo genes without the DCR.

Northern blot (Figure 4, right panel) and S1 nuclease protection analysis (Figure 3B) show that the tk-neo gene is still induced at high levels without the β-globin gene and that the transfected Thy-1 gene is also induced to express. Very low levels (at least 50-fold lower for tk-neo) or undetectable levels (for Thy-1) are observed without the DCR, demonstrating that both inductions are completely dependent on the presence of this region.

We subsequently tested whether the DNAaseI hypersensitive sites in the 5′-flanking region had been reformed after transfection into MEL cells (Figure 5). Surprisingly, all of the hypersensitive sites are already present before the induction of the MEL cells, including the hypersensitive site in the 3′ flanking region, which was not reformed in transgenic mice (Grosveld et al., 1987). It should be pointed out, however, that this site does not increase in intensity on the blots with increasing DNAaseI digestion, and therefore behaves differently to the others. In addition to the human globin-specific hypersensitive sites, a very strong hypersensitive site is observed in the tk-neo promoter region.

Expression in K562 and L Cells

We then asked the question whether the minilocus construct would be active in erythroid cells representing a different stage of development or in a non-erythroid cell

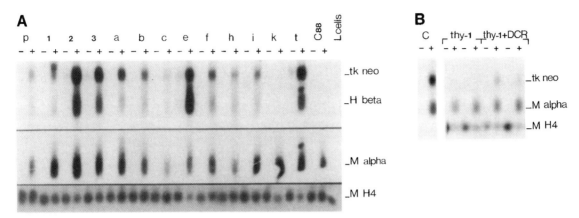

Figure 4. Expression Analysis of the Human β-Globin and Thy-1 MEL Clones and Populations by Northern Blotting

Fifteen micrograms of RNA from uninduced (−) as well as induced (+) clones and populations was Northern blotted after gel electrophoresis, and hybridized to either a mixture of the SphI–BglII neo probe (tk-neo, Figure 1), together with the EcoRI–MspI human β-globin probe (H beta, Figure 2), or the mouse α BamHI second exon probe (M alpha) or mouse H 4 probe (MH₄) in a separate rehybridization. RNA from untransfected MEL-C88 and L cells was used as a negative control.

line. The human cell line K562 (Lozzio and Lozzio, 1975), which expresses the human γ-, but not the β-globin gene, was chosen for the first purpose, since it had been shown previously that transfected γ-, but not β-globin genes (without flanking regions) can be expressed in these cells (Kioussis et al., 1985; Antoniou et al., 1987; Rutherford and Nienhuis, 1987). Mouse L cells were chosen as a non-erythroid cell, since it has been shown that this cell is capable of expressing transfected β-globin genes (Dierks et al., 1981; Busslinger et al., 1983; Murray and Grosveld, 1987).

Southern blots using mouse Thy-1, human β-globin, and tk-neo as probes were used again to establish the copy

number and the integrity of the transfected locus. Figure 6 shows that each of the L cell populations contains a fairly low number of β-globin and tk-neo genes (3, 2, 2, and 5, respectively), while the K562 cell populations contain one high and two low copy numbers (10, 2, and 1, respectively). Surprisingly, the transfected K562 cells express the human β-globin gene, and the copy number of the transfected gene is reflected in the expression levels of the β-globin gene by S1 nuclease protection analysis (Figure 7, middle panel) and Northern blots (Figure 7, lower panel). Comparison of the levels of γ-and β- globin RNA shows that the level of expression per β-globin gene is approximately 5- to 10-fold lower than that of the γ-globin

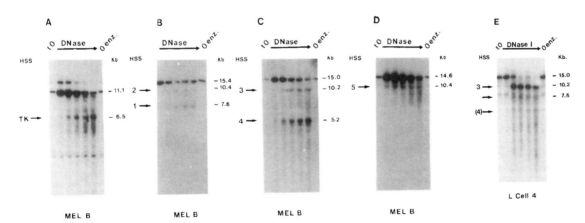

Figure 5. Analysis of the DCR DNAase Hypersensitive Sites

(A)–(D) Nuclei were prepared from approximately 10⁸ uninduced cells of MEL clone B, and digested with 10 μg of DNAaseI per ml for increasing times at 37°C. DNA was prepared and digested to completion with restriction endonucleases BamHI (A), (C), (D) or BglII (B). After electrophoresis and transfer to nitrocellulose, DNA samples were probed with a 3.3 kb EcoRI probe (A) and (B), a 0.46 kb EcoRI–BamHI fragment (C), or a 1.15 kb EcoRI fragment (D). Samples labeled tO are DNA prepared from nuclei before the addition of DNAaseI; samples labeled Oenz. are from nuclei incubated in the absence of DNAaseI for 20 min at 37°C.

(E) Nuclei were prepared from approximately 2 × 10⁷ L cells of population 4, and digested with 10 μg of DNAaseI per ml for increasing times at 37°C. DNA was prepared and digested with BamHI. After electrophoresis and transfer to nitrocellulose, DNA samples were hybridized as (C). Positions where erythroid-specific DNAaseI sub-bands would be expected are indicated by arrows 3 and (4). One of the myeloid-specific sub-bands at 7.5 kb is indicated with a plain arrow. The blots to detect sub-bands corresponding to sites 1 and 2 used a BglII digest and probe as in (B). No sub-bands were detected (data not shown).

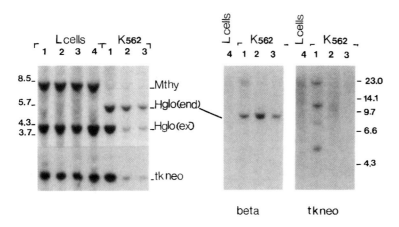

Figure 6. Structural Analysis of the Human β-Globin K562 and L Cell Populations

Left: DNA of the different populations was digested with EcoRI, gel electrophoresed, and Southern blotted. The blot was hybridized to the human β-globin BamHI–EcoRI IVSII probe (H glo, Figure 1), a mouse Thy-1 probe (M thy), and the SphI–BglII neo probe (tk-neo, Figure 1). In the K562 populations, the endogenous human β-globin genes (3 copies) show up as a 6.2 kb EcoRI fragment (H glo [end]).

Right: HpaI digests of the DNA samples were run out on an agarose gel and Southern blotted. The blot was hybridized either with the human β-globin BamHI–EcoRI IVSII probe (Figure 1), or the SphI–BglII neo probe (Figure 1).

The different L cell and K562 populations are indicated by their number above the lanes.

gene. Analysis of the β-globin expression in the L cell populations shows that the gene is expressed in all cases (Figure 7, L cells, lanes 1-4) at levels that are comparable to those obtained without the flanking regions (lane 5). Analysis of the tk-neo expression levels shows one curious anomaly (Figure 7, lower panel). In tile K562 populations the tk-neo mRNA levels are high, but in the case of population 1, it does not follow the expression levels of the β-globin gene. Southern blot analyses (Figure 6 and data not shown) indicate that the majority of the tk-neo genes are intact in this population. Apart from the suggestion that some limiting factor would prevent higher expression levels and that the expression levels of tk-neo would already be maximum in the lower copy number populations, we have at present no explanation for this result. Analyses of the populations for the presence of the DNAase I hypersensitive sites show that the sites return in the K562 cells as expected (data not shown), but that they cannot be detected in the L cell population, with the exception of hypersensitive site 3 (Figure 5E and data not shown). This site is reformed very efficiently in these non-erythroid cells together with two other sites (at 7.5 kb and 5.5 kb in Figure 5E), which were originally mapped by Forrester et al. (1987) to be present in the chromatin of non-erythroid hematopoietic cells.

Discussion

Erythroid-Specific Expression

It is clear from these data that the stable transfection of a β-globin gene minilocus, including the flanking regions, results in a copy number–dependent and integration position–independent expression of the gene in differentiated MEL cell populations and clones. The expression level per gene copy is at a similar level to that of the endogenous globin genes. The effect of the flanking regions is erythroid-specific, since copy number–dependent expression is also obtained in K562 cells. This effect is not observed in the non-erythroid L cells, where low levels of expression are obtained, similar to those obtained without the flanking regions (Dierks et al., 1981; Busslintger et al., 1983; Murray and Grosveld, 1987). These results are therefore very similar to those obtained in transgenic mice (Gros-

veld et al., 1987), and show that the β-globin locus does not need to undergo a complete developmental differentiation program to be expressed at high levels. Neverthe-

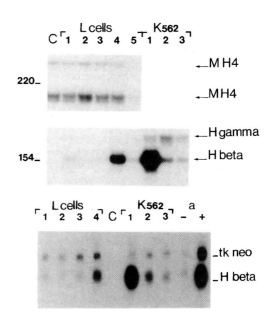

Figure 7. Expression Analysis of the Human β-Globin L Cell and K562 Populations

Top: five microgram aliquots of RNA from the L cell and K562 populations were analyzed by S1 nuclease protection (Kollias et al., 1986) using a mixture of a 5′ human β-globin (AccI) probe (protected fragment: H beta) and a β human γ-globin probe of 10-fold lower specific activity (protected fragment: H gamma). Only 0.5 μg of L cell RNA was used for the endogenous histone H4 control in the S1 nuclease protection experiments (protected fragments MH4). Untransfected L cell RNA (c) was included as negative control (lane c) and RNA from an L cell population containing just the human β-globin gene, but not the DCR sequences, was included as positive control (lane 6; Murray and Grosveld, 1987). Labeled pBr322 DNA digested with Hinfl was run alongside the experiments; sizes are indicated in nucleotides.

Bottom: fifteen micrograms of RNA of each population was Northern blotted after gel electrophoresis and probed with the human β-globin EcoRI–MspI probe (Figure 1; H beta) as well as the SphI–BglII neo probe (Figure 1; tk-neo). As positive control, RNA from both the uninduced (−) and the induced (+) MEL clone was included.

The different populations are indicated by their number above the lanes.

less, some differences are apparent. In the cell cultures, a selection is applied (G418) for the expression of the linked tk-neo gene, which will therefore ensure that the transfected DNA is integrated in a transcriptionally active domain. This is reflected in the variations of the low levels of transcription of the tk-neo gene (and the β-globin gene) before the induction of the MEL cells (Figure 3), and implies that the minilocus does not isolate (silence) the gene from the neighboring chromatin. Once the cells have been induced by DMSO, the levels of expression increase by at least 100-fold and become independent of the position of integration. Such effects could not be observed in transgenic mice, because there is no selection for integration into active chromatin, nor does an "uninduced" stage of expression exist in isolation. As a consequence, it is only possible in the cell culture system to detect the presence of the hypersensitive sites before the induction of β-globin transcription, and their presence at this stage suggests that the erythroid-specific (open) chromatin structure is not coupled β-globin transcription. Interestingly, hypersensitive site 3 and the myeloid hypersensitive sites (Forrester et al., 1987) also return in the non-erythroid-specific L cells (Figure 5). The formation of this site can therefore be mediated by non-erythroid-specific factors, although these apparently do not exert any transcriptional effect in the (non-erythroid) L cells. The significance of this is presently unclear, but it suggests that the formation of site 3 may be dependent on it being present in "active" chromatin. The latter is assured in these experiments by virtue of the G418 selection after transformation.

The deletion of the 3' hypersensitive site appears to have a stimulatory effect on the transcription of the β-globin gene. At present, this could also be explained by a position effect resulting from the large deletion, but it is interesting to note that the expression of the γ-globin gene in the deletion forms of HPFH, which removes this whole region, is higher than the level observed in patients who carry a deletion of only (parts of) the β-globin gene (for review, see Poncz et al., 1989).

Expression of the Thy-1 and tk-neo Gene
The presence of the flanking regions also results in a dramatic stimulation of transcription of the non-erythroid-specific promoter of the HSV thymidine kinase (tk) gene after differentiation of the MEL cells. Interestingly, when an SV40–neo hybrid gene is introduced into the human β-globin locus on chromosome 11 of a MEL–human hybrid cell line (Hu11) by homologous recombination, a similar stimulation of the SV40–neo gene is seen (Nandi et al., 1988). Such erythroid-specific transcription of the SV40–neo construct is not seen when the hybrid gene is integrated elsewhere in the genome outside the β-globin domain. By contrast, when the dominant control region is linked to the tk-neo gene, the promoter shows the high erythroid-specific expression, regardless of where it is integrated in the genome. This effect on the tk-neo gene is unrelated to the presence of the two β-globin gene enhancers, since their presence in large constructs without the flanking DCR does not result in a dramatic stimulation of the tk promoter (Wright et al., 1983). This implies that the DCR itself con-

tains an inducible element and, indeed, when the β-globin gene is replaced with the Thy-1 gene, both tk-neo and Thy-1 are induced. Since all the hypersensitive sites are already formed before the induction of the MEL cells and the start of globin transcription, we therefore suggest that the DCR contains elements of more than one function whose action is required or at least effected at different times during the differentiation of erythroid cells. The changes in chromatin structure would take place early (Figure 5) and clearly precede high levels of transcription, while a second inducible function (not detected in the chromatin structure of the DCR in these experiments) is linked to the actual transcription of the gene(s). The latter function would resemble the function of the β-globin gene enhancers (Behringer et al., 1987; Kollias et al., 1987a; Trudel et al., 1987; Antoniou et al., 1988), but be developmental stage-independent. Indeed, preliminary experiments suggest that one of the hypersensitive sites is associated with an erythroid-specific enhancer (Collis et al., unpublished data). In this respect, it is interesting to note that the β-globin gene as part of the minilocus is expressed in K562 cells that represent a different developmental stage.

Expression in K562 Cells and the Regulation of Stage-Specific Expression
K562 cells do not express the endogenous β-globin gene or transfected β-globin genes without the upstream region (Kioussis et al., 1985; Antoniou et al., 1987; Rutherford and Nienhuis, 1987). This could simply mean that the proximity of the upstream DCR or the absence of some (unknown) repressive element in our minilocus construct is responsible for the low (10% of γ-globin levels), but substantial transcription of the β-globin gene. Interestingly, a similar phenomenon is observed in a number of human globin disorders, suggesting another possibility. Single point mutations have been detected in the γ-globin gene promoter (for review see Poncz et al., 1989), which appear to result in elevated levels of γ-globin gene transcription and a concomitant decrease in β-globin gene transcription. Conversely, deletion of the β-globin gene promoter (Anand et al., 1988; Atweh et al., 1987) correlates with an elevated expression of the γ-globin gene. These in vivo observations suggest that the expression of the γ- and β-globin genes influence each other (for review, see Collins and Weissman, 1984; Poncz et al., 1989). We suggest that this phenomenon could be explained if the genes compete for some element in *cis* (perhaps the DCR or the β-globin gene enhancers) when they are present in the same domain. This competition would not be for any factor in *trans*, because many copies of the exogenous β-globin gene do not compete with the endogenous genes (Grosveld et al., 1987; this paper). Support for such a model in vivo has recently been provided by transient expression experiments with the chicken adult and embryonic globin genes in primitive and definitive chicken red cells (Choi and Engel, 1988). If such a mechanism is true for the human globin genes, the DCR of the β-globin domain would activate both the γ- and the β-globin genes independent of the stage of development. Each of the

genes; i.e., the γ or β-globin genes, would have a high efficiency of transcription at each of their optimal development stages, but have a significant, but lower (10%–20%), efficiency at the inappropriate stages of development when the single genes are present in the domain (as in the minilocus construct). When the other genes are present, this low level would be further decreased by competition with a more efficient gene. We are presently testing this model by using miniloci containing γ-and β-globin genes in isolation or together in *cis*.

Experimental Procedures

Tissue Culture and Cell Transfections

The β-globin minilocus cosmid (Grosveld et al., 1987) was linearized at the PvuI site in the vector Transfection into the diploid MEL cell line C88 (Deisseroth and Hendrick, 1978) and K562 cells (Lozzio and Lozzio, 1975) was performed by electroporation as described (Smithies et al., 1985; Antoniou et al., 1988), into L cells by calcium phosphate precipitation (Wigler et al., 1979).

Directly after electroporation, the cells were split to give rise to three different populations. One percent of each of the C88 populations was plated out separately in small dishes from which individual clones were picked.

Selections for stable populations and clones were performed by adding G418 to the medium 2 days after transfection. The final G418 concentration for K562 and MEL cells was 800 μg/ml; for L cells 300 μg/ml. The obtained MEL populations and clones were induced as described (Antoniou et al., 1988). Copy number and possible deletions of parts of the transfected construct in all populations and clones were analyzed by Southern blotting (Southern, 1975).

RNA Analysis

RNA extraction of the transfected cell populations and clones was done in 3 M lithium chloride, 6 M urea including a 2 min sonication step (Auffray and Rougeon, 1980).

S1 nuclease protection analysis was carried out as described previously (Kollias et al., 1986).

Northern blotting was done as described by Krumlauf et al. (1987).

DNAaseI Sensitivity

DNAase I sensitivity assays were carried out on isolated nuclei of confluent C88, K562, and L cell populations as described previously (Grosveld et al., 1987).

β-Globin Minilocus

The construction of the β-globin minilocus was as previously described by Grosveld et al. (1987). The Thy-1 gene was first cloned as an EcoRI mouse–human hybrid gene (Kollias et al., 1987b) in a polylinker between a ClaI and KpnI site. The gene was then exchanged for the β-globin ClaI–KpnI fragment in minilocus.

Acknowledgments

We are grateful to Mike Antoniou, Khai Siew, Jacky Hurst, and Gloria Charters for their expert technical assistance in tissue culture, to Fiona Watson for the Thy-1 probe, and to Cora O'Carroll for the preparation of the manuscript. G. B. was self-supported. This work is supported by the Medical Research Council (UK).

The costs of publication of this article were defrayed in part by the payment of page charges. This article must therefore be hereby marked "*advertisement*" in accordance with 18 U.S.C. Section 1734 solely to indicate this fact.

Received September 8, 1988; revised December 7, 1988.

References

Anand, R., Brehm, C. D., Kazazian, H. H., and Vanin, E. E (1988). Molecular characterization of a β⁰-thalassaemia resulting from a 1.4kb deletion. Blood *72*, 636–641.

Antoniou, M., deBoer, E., and Grosveld, F. (1987). β-globin gene promoter generates 5' truncated transcripts in the embryonic foetal erythroid environment. Nucl. Acids Res. *15*, 1886.

Antoniou, M., deBoer, E., Habets, G., and Grosveld, F. (1988). The human β-globin gene contains multiple regulatory regions: identification of one promoter and two downstream enhancers. EMBO J. *7*, 377–384.

Atweh, G., Zhu, X., Brickner, H., Dowling, C., Kazazian, H., and Forget, B. (1987). The β-globin gene on the Chinese δβ-thalassaemia chromosome carries a promoter mutation. Blood *70*, 1470–1474.

Auffray, C., and Rougeon, F. (1980). Purification of mouse immunoglobulin heavy-chain messenger RNAs from total myeloma tumour RNA. Eur. J. Biochern. *107*, 303–314.

Baron, M. H., and Maniatis, T. (1986). Rapid reprogramming of globin gene expression in transient heterokaryons. Cell *46*, 591–602.

Behringer, R. R., Hammer, R. E., Brinster, R. L., Palmiter, R. D., and Townes, T. M. (1987). Two 3' sequences direct erythroid specific expression of human β-globin genes in transgenic mice. Proc. Natl. Acad. Sci. USA *84*, 7056–7060.

Busslinger, M., Hurst, J., and Flavell, R. A. (1983). DNA methylation and the regulation of globin gene expression. Cell *34*, 197–206.

Choi, O.-R. B., and Engel, J. D. (1988). Developmental regulation of β-globin switching. Cell *55*, 17–26.

Collins, E S., and Weissman, S. M. (1984). The molecular genetics of human hemoglobin. Proc. Nucl. Acid Res. Mol. Biol. *31*, 315–462.

deBoer, E., Antoniou, M., Mignotte, V., Wall, L., and Grosveld, F. (1988). The human β-globin gene promoter; nuclear protein factors and erythroid specific induction of transcription. EMBO J. *7*, 4203–4212.

Deisseroth, A., and Hendrick, D. (1978). Human α-globin gene expression following chromosomal dependent transfer into mouse Erythroleukemia cells. Cell *15*, 55–63.

Dierks, P. van Ooyen, A., Mantei, N., and Weissman, C. (1981). DNA sequences preceding the rabbit β-globin gene are required for formation of β-globin RNA with the correct 5' terminus in mouse L cells. Proc. Natl. Acad. Sci. USA *78*, 1411–1414.

Forrester, W. C., Takegawa, S., Papayannopoulou, T., Stamatoyannopoulos, G., and Groudine, M. (1987). Evidence for a locus activating region: the formation of developmentally stable hypersensitive sites in globin expressing hybrids. Nucl. Acids Res. *15*, 10159–10177.

Grosveld, F. G., Lund, T., Murray, E. J., Mellor, A. L., Dahl, H. H. M., and Flavell, R. A. (1982). The construction of cosmid libraries which can be used to transform eukaryotic cells. Nucl. Acids Res. *10*, 6715–6732.

Grosveld, E, Blom van Assendelft, G., Greaves, D. R., and Kollias, G. (1987). Position-independent, high-level expression of the human β-globin gene in transgenic mice. Cell *51*, 975–985.

Kioussis, D., Vanin, E., deLange, T., Flavell, R. A., and Grosveld, F. (1983). β-globin gene inactivation by DNA translocation in γβ-thalassaemia. Nature *306*, 662–666.

Kioussis, D., Wilson, F., Khazaie, K., and Grosveld, F. G. (1985). Differential expression of human globin genes introduced into K562 cells. EMBO J. *4*, 927–931.

Kollias, G., Wrighton, N., Hurst, J., and Grosveld, F. (1986). Regulated expression of human ᴬγ-, β-, and hybrid γβ-globin genes in transgenic mice: manipulation of the developmental expression patterns. Cell *46*, 89–94.

Kollias, G., Hurst, J., deBoer, E., and Grosveld, F. (1987a). A tissue and developmental specific enhancer is located downstream from the human β-globin gene. Nucl. Acids Res. *15*, 5739–5747.

Kollias, G., Spanopoulou, E., Grosveld, F., Ritter, M., Beech, J., and Morris, R. (1987b). Differential regulation of a Thy-1 gene in transgenic mice. Proc. Natl. Acad. Sci. USA *84*, 1492–1496.

Krurnlauf, R., Holland, P., McVey, J., and Hogan, B. (1987). Developmental and spatial patterns of expression of the mouse homeobox gene Hox2.1. Development *99*, 603–617.

Lozzio, C. B., and Lozzio, B. B. (1975). Human chronic myelogenous leukemia cell line with positive Philadelphia chromosome. Blood *45*, 321–324.

Magram, J., Chada, K., and Costantini, F. (1985). Developmental regulation of a cloned adult β-globin gene in transgenic mice. Nature *315*, 338–340.

Murray, E., and Grosveld, F. (1987). Site specific demethylation in the promoter of the γ-globin gene does not alleviate methylation mediated suppression. EMBO J. *5*, 2329–2335.

Nandi, A. K., Roginski, R. S., Gregg, R. G., Smithies, O., and Skoultchi, A. I. (1988). Regulated expression of genes inserted at the human chromosomal β-globin locus by homologous recombination. Proc Natl. Acad. Sci. USA *85*, 3845–3849.

Poncz, M., Henthorn, P. Stoeckert, C., and Surrey, S. (1989). Globin gene expression in hereditary persistence of fetal hemoglobin and (δβ)° thalassemia. In Oxford Surveys on Eukaryotic Genes, N. Maclean, ed. (Oxford: Oxford University Press), in press.

Rutherford, T. and Nienhuis, A. W. (1987). Human globin gene promoter sequences are sufficient for specific expression of a hybrid gene transfected into tissue culture cells. Mol. Cell. Biol. *7*, 398–402.

Smithies, O., Gregg, R. G., Boggs, S. S., Karalewski, M. A., and Kucherlapati, R. S. (1985). Insertion of DNA sequences into the human chromosomal β-globin locus by homologous recombination. Nature *317*, 230–234.

Southern, E. (1975). Detection of specific sequences among DNA fragments separated by gel electrophoresis. J. Mol. Biol. *98*, 503–517.

Taramelli, R., Kioussis, D., Vanin, E., Bartram, K., Groffen, J., Hurst, J., and Grosveld, F. (1986). γδβ-thalassaemias 1 and 2 are the result of a 100kb deletion in the human β-globin cluster. Nucl. Acids Res. *14*, 7017–7029.

Trudel, M., Magram, J., Bnjckner, L., and Costantini, F. (1987). Upstream ^Gγ-globin and downstream β-globin sequences required for stage-specific expression in transgenic mice. Mol. Cell. Biol. *7*, 4024–4029.

Tuan, D., Solomon, W., Qiliang, L. S., and Irving, M. L. (1985). The "β-like-globin" gene domain in human erythroid cells. Proc. Natl. Acad. Sci. USA *32*, 6384–6388.

van der Ploeg, L. H. T., Konings, A., Oort, M., Roos, D., Bernini, L., and Flavell, R. A. (1980). γβ-thalassaemia deletion of the γ- and δ genes influences δ-globin gene expression in man. Nature *283*, 637–642.

Wall, L., deBoer, E., and Grosveld, F. (1988). The human β-globin gene 3' enhancer contains multiple binding sites for an erythroid specific protein. Genes Dev. *2*, 1085–1100.

Wigler, M., Sweet, R., Sim, G. K., Wold, B., Pellicer, A., Lacy, E., Maniatis, T. Silverstein, S., and Axel, R. (1979). Transformation of mammalian cells with genes from procaryotes and eucaryotes. Cell *16*, 777–785.

Wright, S., deBoer, E., Grosveld, F. G., and Flavell, R. A. (1983). Regulated expression of the human β-qlobin gene family in murine erythroleukaemia cells. Nature *305*, 333–336.

Wrighton, N., and Grosveld, F. (1988). A novel *in vitro* transcription assay demonstrates the presence of globin-inducing trans-acting factors in uninduced Murine Erythroleukemia Cells. Mol. Cell. Biol. *8*, 130–137.

Purification and cloning of transcription factors

Experiments of the type described in Sections 6 and 7 have allowed the identification of a number of different DNA sequence elements that are involved in producing specific patterns of gene expression. Similarly, in many cases it has been possible to identify specific proteins that bind to these sequences, as described in the paper by Parker and Topol presented in Section 6.

In order to extend observations of this type and to characterize fully the DNA binding transcription factors involved, however, it is necessary to isolate cDNA clones encoding these transcription factors, so facilitating their detailed analysis. Although a number of methods for cloning the genes encoding transcription factors are now available (for a review, see Latchman, 1993), the earliest method for doing this involved the purification of the transcription factor protein. As with any other purified protein, a partial protein sequence could then be obtained and used to prepare specific oligonucleotide probes containing the DNA sequences capable of encoding small regions of the protein. These could then be used to screen cDNA libraries in order to isolate the appropriate cDNA clones (Figure 8.1).

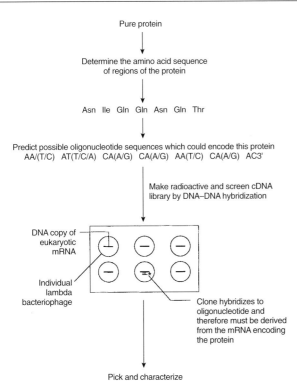

Figure 8.1 Isolation of cDNA clones by screening with short oligonucleotides predicted from the amino acid sequence of the purified protein. Since the multiple oligonucleotides contain all the possible sequences capable of encoding the amino acid sequence of the protein, one of these oligonucleotides will hybridize to a cDNA clone derived from the mRNA encoding the protein.

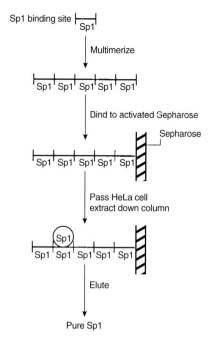

Sp1 binding site

Multimerize

Bind to activated Sepharose

Sepharose

Pass HeLa cell
extract down column

Elute

Pure Sp1

Figure 8.2 Purification of the Sp1 transcription factor. Sp1 can be purified on an affinity column in which multiple copies of the DNA binding site for Sp1 have been coupled to a Sepharose support.

Unfortunately, however, this method was extremely difficult to apply to transcription factors. Such factors are present in very small amounts within cells, necessitating the use of enormous amounts of material to obtain sufficient pure factor. More seriously, conventional chromatographic methods such as that employed by Parker and Topol to obtain preparations of the heat-shock factor (as described in Section 6) yield relatively impure preparations of transcription factor protein which are often only 1–2% pure. Although such preparations can be used to assay the activity of a transcription factor protein or its DNA binding ability, they are useless for obtaining partial protein sequences, since this requires highly purified protein.

This problem was solved by Kadonaga and Tjian in the paper presented here. In order to purify the transcription factor Sp1, these authors took advantage of its ability to bind specifically to the DNA sequence illustrated in Fig. 2(A) of the paper. They therefore prepared a DNA affinity column which contained multiple copies of this sequence coupled to an inert Sepharose resin. The material applied to this Sp1 affinity resin was a partially purified preparation, such as would be obtained by conventional chromatographic methods, which contained only 0.1% Sp1. When this material was passed down the column, the Sp1 bound specifically to its DNA recognition sequence, while the remaining protein passed directly through the column without binding (Figure 8.2). By eluting the bound Sp1 from the column, Kadonaga and Tjian were able to obtain a highly purified preparation of Sp1 which could be purified further by another passage through the column.

The highly purified Sp1 obtained in this way, unlike the material originally applied to the column, did not appear to be contaminated with any other proteins when subjected to gel electrophoresis, as illustrated in Fig. 3(B) of the paper. Similarly, this material retained its ability to bind to the specific Sp1 binding site, as assayed by the DNase I footprinting analysis illustrated in Fig. 3(A) of the paper, and was able to stimulate transcription *in vitro*, as illustrated in Fig. 3(C) of the paper.

A critical aspect of the success of Kadonaga and Tjian was the addition of a non-specific DNA competitor to the material applied to the Sp1 affinity column. This material serves to remove any DNA binding protein that can bind to any random sequence and would otherwise bind to the Sp1 binding sites on the column and contaminate the purified Sp1 preparation. As illustrated in Fig. 4 of the paper, the amount of non-specific competitor added is highly critical. If too little competitor is added, the other DNA binding proteins will bind to the Sp1 sites on the column and contaminate the Sp1 preparation. In contrast, however, if too much non-specific carrier is added, Sp1 will begin to bind to it, since it can bind to all DNA sequences with low affinity. The yield of Sp1 obtained will thus be reduced (Figure 8.3).

Hence by a combination of an Sp1 affinity column and the use of the correct amount of non-specific competitor, Kadonaga and Tjian were able to successfully purify Sp1, and this method has subsequently been employed by a variety of other groups interested in purifying specific transcription factors. Although such purified factors are of interest in themselves for assays of functional activity, the most important use of these factors is to generate partial protein sequences in order to obtain the corresponding cDNA clones. This was achieved by Kadonaga et al. in the second paper presented here. They used the method previously described in order to purify 300 μg of Sp1 to greater than 95% homogeneity. It is noteworthy that, as described in their paper, they had to begin with 800 g of cellular material in order to obtain an amount of Sp1 suitable for protein sequencing, given the low abundance of the protein in cellular material. As illustrated in Figure 1 of the paper, the amino acid sequences of short peptides within Sp1 were then obtained from this purified material. In turn the peptide sequences were used to predict corresponding oligonucleotides comprising all the possible DNA sequences that could encode these peptides, given the redundancy of the genetic code. These oligonucleotides were then used to screen a HeLa cell cDNA library and cDNA clones corresponding to Sp1 were isolated. DNA sequence analysis of these clones revealed that they did indeed contain DNA sequences capable of encoding the peptide sequences present in purified Sp1, confirming their identification as Sp1 cDNA clones.

Having isolated cDNA clones encoding Sp1, it was possible for Kadonaga et al. to perform a number of experiments aimed at characterizing this transcription factor and the gene which encodes it. Thus in the experiment illustrated in Figure 1(C) of their paper, they were able to use the cDNA clone in a Northern blot analysis to detect the Sp1 mRNA in HeLa cells. In addition, it was also possible to obtain the complete DNA sequence of the Sp1 cDNA clone and use it to predict the corresponding complete protein sequence of Sp1, as opposed to the very small amounts of partial protein sequence obtained by analysis of the purified protein itself. Analysis of the complete Sp1 sequence revealed the existence of three zinc-finger motifs, in each of which two cysteine and two histidine amino acid residues co-ordinate a molecule of zinc (Figure 8.4). As described in the paper by Miller et al. presented in Section 9, this motif was originally described within the RNA polymerase III transcription factor TFIIIA, which contains nine zinc fingers arranged in tandem. Although TFIIIA plays a critical role in the transcription of the 5 S rRNA genes, it is unusual among transcription factors in that it remains associated with the 5 S rRNA molecule after transcription. Thus immature *Xenopus* oocytes contain very large numbers of 5 S rRNA molecules, each with an associated TFIIIA molecule. This provides a very convenient source of large amounts of TFIIIA and resulted in it being one of the earliest transcription factors to be studied in detail.

The identification of zinc fingers within Sp1 thus indicated that such motifs are not unique to the relatively special case of TFIIIA, but are also found in a transcription factor that is expressed at a much lower level typical of most transcription factors. Moreover, the finding of such zinc fingers in Sp1 also indicated that they are not confined to transcription factors mediating transcription by RNA polymerase III, but could also be found in transcription factors involved in transcription by RNA polymerase II, which transcribes protein-coding genes (for a review, see Sentenac, 1985). Indeed, following their identification in Sp1, zinc fingers have been described in a number of different RNA polymerase II transcription factors (for reviews, see Evans and Hollenberg, 1988; Struhl, 1989).

Initial work on the TFIIIA molecule had indicated that these zinc-finger motifs are likely to be involved in mediating DNA binding by TFIIIA. In the paper presented here, Kadonaga et al. were able to show directly that this was also the case for Sp1. Thus, as illustrated in Figure 3 of the paper, they showed that DNA binding by Sp1 requires the presence of zinc ions, as would be expected if it is

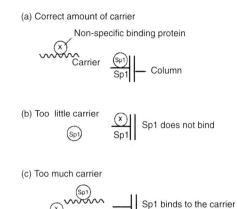

(a) Correct amount of carrier

(b) Too little carrier

(c) Too much carrier

Figure 8.3 Importance of the amount of non-specific carrier DNA in the purification of the Sp1 transcription factor.
(a) Purification of the Sp1 transcription factor in the presence of the correct amount of non-specific carrier DNA results in non-sequence-specific DNA binding proteins (X) binding to the carrier, allowing Sp1 to bind to the column and to be effectively purified. (b) If too little carrier is added, these non-specific binding proteins will bind to the Sp1 sites on the column, preventing the purification of Sp1. (c) If too much carrier is added, both the non-specific proteins and Sp1 will bind to the carrier, again preventing the binding of Sp1 to its sites on the column and hence preventing its purification.

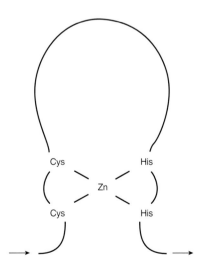

Figure 8.4 The zinc-finger motif, in which two cysteine and two histidine amino acid residues co-ordinate a molecule of zinc

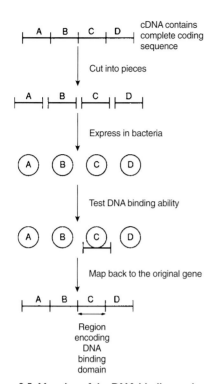

cDNA contains complete coding sequence

Cut into pieces

Express in bacteria

Test DNA binding ability

Map back to the original gene

Region encoding DNA binding domain

Figure 8.5 Mapping of the DNA binding region of a transcription factor by testing the ability of different regions to bind to DNA when expressed in bacteria. In this example, region C encodes a portion of the protein which is capable of binding to DNA, and this is therefore the DNA binding domain of the molecule.

dependent upon the zinc-containing finger motifs. Moreover, by expressing various regions of Sp1 in *Escherichia coli* they were able to show that the DNA binding ability of Sp1 did not require the entire protein molecule. Rather, as illustrated in Figure 5 of the paper, the C-terminal 168 amino acids of Sp1 were able to bind to DNA with the same sequence specificity as the intact molecule.

As these C terminal 168 amino acids contain the three zinc-finger motifs, this provides further evidence for the importance of these motifs in DNA binding. Moreover, the approach used by Kadonaga et al. to characterize the DNA binding region of Sp1 represents a model for mapping such regions within cloned transcription factors, even where no obvious DNA binding motifs are present. Thus different fragments of the cDNA clone can be expressed in bacteria to produce the corresponding regions of the protein. The ability of these different protein regions to bind to DNA can then be assessed (Figure 8.5). In this way the region of the protein that mediates DNA binding can be identified and its structure compared with that in previously described transcription factors, thereby indicating whether the DNA binding domain of the factor is of a novel type or corresponds to a previously characterized DNA binding domain.

The papers presented here from the Tjian group are therefore of major importance in the study of transcription factors. They provide a means by which a transcription factor present at typically low abundance in the cell can be obtained in a highly purified form, allowing protein sequencing and ultimately the isolation of cDNA clones derived from it. Such cDNA cloning in turn allows the analysis of features such as DNA binding, transcriptional activation, protein dimerization and regulation of factor activity, which are presented in subsequent sections.

Kadonaga & Tjian (1986) Proc. Natl. Acad. Sci. U.S.A. **83**, 5889–5893

Affinity purification of sequence-specific DNA binding proteins

(DNA-agarose affinity chromatography/competitor DNA/transcription factor Sp1/synthetic oligodeoxynucleotides)

JAMES T. KADONAGA AND ROBERT TJIAN

Department of Biochemistry, University of California, Berkeley, CA 94720

Communicated by Bruce M. Alberts, April 30, 1986

ABSTRACT We describe a method for affinity purification of sequence-specific DNA binding proteins that is fast and effective. Complementary chemically synthesized oligodeoxynucleotides that contain a recognition site for a sequence-specific DNA binding protein are annealed and ligated to give oligomers. This DNA is then covalently coupled to Sepharose CL-2B with cyanogen bromide to yield the affinity resin. A partially purified protein fraction is combined with competitor DNA and subsequently passed through the DNA-Sepharose resin. The desired sequence-specific DNA binding protein is purified because it preferentially binds to the recognition sites in the affinity resin rather than to the nonspecific competitor DNA in solution. For example, a protein fraction that is enriched for transcription factor Sp1 can be further purified 500- to 1000-fold by two sequential affinity chromatography steps to give Sp1 of an estimated 90% homogeneity with 30% yield. In addition, the use of tandem affinity columns containing different protein binding sites allows the simultaneous purification of multiple DNA binding proteins from the same extract. This method provides a means for the purification of rare sequence-specific DNA binding proteins, such as Sp1 and CAAT-binding transcription factor.

Many important cellular processes, such as transcription, replication, and recombination, involve the action of DNA binding proteins. For example, sequence-specific DNA binding proteins are directly involved in the regulation of mRNA transcription initiation in higher organisms (for reviews, see refs. 1–3). To study the biochemical properties of these transcription factors, it is necessary to purify the proteins to homogeneity. This would enable the factors to be characterized, facilitate the raising of antibodies, and ultimately provide a means for cloning the genes encoding these regulatory proteins. It has generally been very difficult, however, to obtain homogeneous preparations of these transcription factors because they typically constitute only 0.001% of the total cellular protein (M. R. Briggs, J.T.K., and R.T., unpublished data).

It has long been predicted that sequence-specific DNA binding proteins can be purified by chromatography through affinity resins that contain the proper DNA recognition sites attached to an immobile support (4–7). In the past, a number of DNA binding proteins, including various RNA and DNA polymerases, hormone receptors, and repressors, have been purified by nonspecific DNA-cellulose and DNA-agarose affinity chromatography (4, 5, 8). However, sequence-specific purification of DNA binding proteins has been performed only in a few cases (9, 10), and our attempts to purify eukaryotic promoter-specific transcription factors by the published methods have not been successful.

In the course of our studies on the regulation of mRNA synthesis by RNA polymerase II, we have found that conventional chromatography and HPLC of promoter-specific transcription factors resulted in preparations of only 1–2% purity. We were therefore prompted to develop an affinity chromatography method that could be successfully used for purification of low abundance sequence-specific DNA binding proteins, such as Sp1, CAAT-binding transcription factor (CTF; for reviews, see refs. 1 and 3), and activator protein 1 (AP-1; AP-1 is an RNA polymerase II transcription factor similar to Sp1 and CTF; W. Lee, P. Mitchell, and R.T., unpublished data). These proteins, which are typically derived from HeLa (human) cells, activate transcription by RNA polymerase II from a select group of promoters, such as the simian virus 40 (SV40) early and herpes simplex virus thymidine kinase promoters, that contain at least one properly positioned recognition site for Sp1, CTF, or AP-1. The complete purification and biochemical characterization of Sp1, CTF, and AP-1 will be described elsewhere (M. R. Briggs, J.T.K., and R.T., unpublished data; K. A. Jones, J.T.K., and R.T., unpublished data; W. Lee, P. Mitchell, and R.T., unpublished data). Here we report a simple and effective DNA affinity chromatography method that has allowed us to purify these transcription factors to homogeneity. This technique should be generally applicable for the purification of other sequence-specific DNA binding proteins.

MATERIALS AND METHODS

Materials. Sepharose CL-2B and T4 polynucleotide kinase were obtained from Pharmacia. T4 DNA ligase was from either New England Biolabs or Promega Biotec (Madison, WI). Econo-Columns were the products of Bio-Rad (no. 731-1550). Cyanogen bromide (CNBr; 97%) was purchased from Aldrich. Oligodeoxynucleotides were prepared with an Applied Biosystems 380A DNA synthesizer.

Preparation of DNA for Coupling to Sepharose. A scheme for the preparation of a sequence-specific DNA affinity resin is shown in Fig. 1. First, chemically synthesized complementary oligonucleotides, such as X and Y (see Fig. 2A), are annealed, 5′-phosphorylated, and ligated as follows. Gel-purified oligodeoxynucleotides (220 μg of each) are combined in 67 mM Tris·HCl buffer (pH 7.6) containing 13 mM $MgCl_2$, 6.7 mM dithiothreitol, 1.3 mM spermidine, and 1.3 mM EDTA in a total volume of 75 μl. This mixture is incubated at 88°C for 2 min, 65°C for 10 min, 37°C for 10 min, and room temperature for 5 min. ATP (20 mM) containing 5 μCi of [γ-^{32}P]ATP (pH 7; 15 μl; 1 Ci = 37 GBq) and T4 polynucleotide kinase (100 units; 10 μl) are then added to give a final volume of 100 μl, and the resulting solution is incubated at 37°C for 2 hr. This reaction is stopped by the addition of 5 M NH_4OAc, pH 5.5 (100 μl)/100 mM $MgCl_2$ (25 μl)/TE buffer (25 μl; TE is 10 mM Tris·HCl, pH 7.6/1 mM EDTA), and the mixture is heated at 65°C for 15 min to inactivate the kinase. The DNA is ethanol-precipitated; resuspended in TE buffer (200 μl), 3 M NaOAc (25 μl), and 100 mM $MgCl_2$ (25 μl);

Abbreviations: CTF, CAAT-binding transcription factor; AP-1, activator protein 1; NFI, nuclear factor I; SV40, simian virus 40.

reprecipitated with ethanol; washed with 70% ethanol; and dried *in vacuo*. The DNA is then dissolved in 88 mM Tris·HCl, pH 7.5/13.3 mM MgCl$_2$/20 mM dithiothreitol/1.3 mM spermidine (75 μl), and the ligation reaction is initiated by the addition of 20 μl of 20 mM ATP (pH 7) and 5 μl of T4 DNA ligase (10–30 Weiss units) to give a final volume of 100 μl. This mixture is incubated at 16°C for 4 hr (depending on the sequence and the length of the oligodeoxynucleotides, the optimal temperature for ligation may vary from 4°C to 16°C), and the DNA is then phenol-extracted, precipitated with ethanol, dried *in vacuo*, and dissolved in water (100 μl). (Note: do not dissolve the DNA in TE buffer—it will interfere with the coupling reaction.) Analysis of the resulting DNA by agarose gel electrophoresis typically shows oligomers of the basic oligodeoxynucleotide unit ranging from 3-mers to 75-mers (Fig. 2*B*).

Coupling of DNA to Sepharose. The DNA oligomers are covalently attached to Sepharose CL-2B by slight modification of the method of Arndt-Jovin *et al.* (8). Sepharose CL-2B (settled volume, 10 ml) is extensively washed with 250 ml of water, suspended in water to give a 20-ml slurry, and then equilibrated to 15°C in a water bath. CNBr (1.1 g; 10 mmol) is dissolved in *N,N*-dimethylformamide (2 ml) and added dropwise over 1 min to the Sepharose, which is mixed by magnetic stirring. Then, 5 M NaOH (1.8 ml; 9 mmol) is slowly added dropwise to the resin over 10 min. The pH of the reaction, which generates HBr as a by-product, should not exceed pH 10. The reaction is stopped by the addition of ice-cold water (100 ml) followed by gentle suction filtration of the resin on a coarse sintered-glass funnel. It is very important that the activated Sepharose is not suction-filtered into a dry cake. The CNBr-derivatized resin is then extensively washed with ice-cold water (300 ml) and 10 mM potassium phosphate (pH 8.0; 100 ml).

The activated Sepharose is immediately used for coupling to DNA as follows. The resin is transferred to a 15-ml polypropylene screw-cap tube, and 10 mM potassium phosphate (pH 8.0, 4 ml) is added to give a thick slurry. The ligated DNA (100 μl in water) is then added to this mixture, and the coupling reaction is carried out at room temperature for 16 hr on a rotary shaker. The resin is collected on a sintered-glass funnel, washed with water (200 ml) and 1 M ethanolamine·HCl (pH 8.0; 100 ml), and suspended in 1 M ethanolamine·HCl (pH 8) to give a final volume of 14 ml. This inactivation of unreacted CNBr-derivatized Sepharose by ethanolamine is carried out at room temperature for 4–6 hr on a rotary shaker. The resin is collected on a sintered-glass funnel and washed with 10 mM potassium phosphate (pH 8.0; 100 ml), 1 M potassium phosphate (pH 8.0, 100 ml), 1 M KCl (100 ml), water (100 ml), and 10 mM Tris·HCl (pH 7.6) containing 0.3 M NaCl, 1 mM EDTA, and 0.02% (wt/vol) NaN$_3$ (100 ml). The resin is stored at 4°C in 10 mM Tris·HCl (pH 7.6) containing 0.3 M NaCl, 1 mM EDTA, and 0.02% (wt/vol) NaN$_3$.

Because the DNA is labeled with ^{32}P, the efficiency of DNA attachment to the Sepharose can be crudely estimated by comparing the amount of radioactivity that is retained on the resin with the amount of radioactivity that remains in solution after the coupling reaction. The efficiency of DNA incorporation into the resin is usually 40–70%. Thus, the concentration of covalently bound DNA in the affinity resin is 20–30 μg of DNA per ml of resin.

DNA Affinity Chromatography. This procedure, which is described here for the purification of Sp1, should work, with only minor modification, for any high-affinity sequence-specific DNA binding protein. In one affinity chromatography step, the recovery of Sp1, as measured by a DNase I footprint assay, is typically 50–60%. All operations are performed at 4°C. Sp1 DNA affinity resin (1 ml) is equilibrated in a Bio-Rad Econo-Column with buffer Z containing

0.1 M KCl [20 ml; buffer Z is 25 mM Hepes (K$^+$), pH 7.8/12.5 mM MgCl$_2$/1 mM dithiothreitol/20% (vol/vol) glycerol/0.1% (vol/vol) Nonidet P-40]. A crude fraction of Sp1 (total protein, 5–10 mg, derived from 50 g of HeLa cells; $\approx$0.1% Sp1; 25 ml) in buffer Z containing 0.1 M KCl is combined with sonicated calf thymus DNA (440 μg; 200 μl of a 2.2 mg/ml solution in TE) and allowed to stand for 10 min. The protein-DNA mixture is passed through the affinity resin by gravity flow ($\approx$15 ml/hr), and the resin is washed four times with 2 ml of buffer Z containing 0.1 M KCl. The passage of buffer through the column is stopped, buffer Z containing 1.0 M KCl (1.2 ml) is added to the column, and the resin is thoroughly mixed with the buffer by using a narrow siliconized glass rod. The resin is allowed to stand for 10 min. and the protein is eluted to give the major Sp1 eluate ($\approx$1.2 ml). The column is then washed with buffer Z containing 0.5 M KCl (1.2 ml) to give the minor Sp1 eluate ($\approx$1.2 ml). Alternatively, Sp1 could be recovered by using a 0.1 M to 1.0 M KCl gradient, where the protein typically elutes from the resin from 0.4 M to 0.6 M KCl. Also, if further purification is desired, the Sp1 fractions can be diluted to 0.1 M KCl with buffer Z without KCl, mixed with competitor DNA, and reapplied to the affinity resin.

The affinity columns are regenerated by washing with 5 mM Tris·HCl, pH 7.6/2.5 M NaCl/0.5 mM EDTA (25 ml), and 10 mM Tris·HCl, pH 7.6/0.3 M NaCl/1 mM EDTA/0.02% (wt/vol) NaN$_3$ (10 ml). The columns are then stored at 4°C in 10 mM Tris·HCl, pH 7.6/0.3 M NaCl/1 mM EDTA/0.02% (wt/vol) NaN$_3$.

Other Methods. The partial purification of Sp1 has been described (11, 12). The complete purification of Sp1 to homogeneity as well as biochemical identification and characterization of the protein will be described elsewhere (M. R. Briggs, J.T.K., and R.T., unpublished data). Oligodeoxynucleotides were synthesized, separated from contaminants by electrophoresis on a denaturing 20% polyacrylamide gel, and then recovered by ethanol precipitation. The A_{260nm} and A_{280nm} of each oligodeoxynucleotide was measured, and the concentration of each sample was estimated by assuming that 1 A_{260nm} absorbance unit corresponds to 40 μg of DNA per ml. DNase I footprinting was carried out as described by Galas and Schmitz (13) and Dynan and Tjian (12). Reconstituted *in vivo* transcription reactions in the presence or absence of Sp1 were performed as described (11). DNA techniques were carried out as recommended by Maniatis *et al.* (14). Polyacrylamide gel electrophoresis of proteins in the presence of NaDodSO$_4$ was done according to the method of Laemmli (15).

RESULTS

Preparation of the DNA Affinity Resin. A scheme for the preparation of an affinity resin for the purification of sequence-specific DNA binding proteins is shown in Fig. 1. The DNA affinity resin was designed to have the following properties. First, to maximize the recovery of the desired protein and to minimize contamination by other proteins, the DNA that is attached to the resin consists only of tandem repeats of a strong binding site for the desired protein. We have found that it is best to survey different recognition sequences to determine a high-affinity binding site. Second, to ensure that the DNA is accessible to protein and stably bound to the resin, each DNA molecule is attached to an agarose bead by an average of one covalent bond. These aims were realized by the use of chemically synthesized oligodeoxynucleotides, which can be quickly and easily prepared in micromole quantities. Complementary oligodeoxynucleotides [such as X and Y (Fig. 2*A*), which, when annealed, possess complementary 5′-protruding ends] are annealed, 5′-phosphorylated, and ligated to give oligomers of the basic

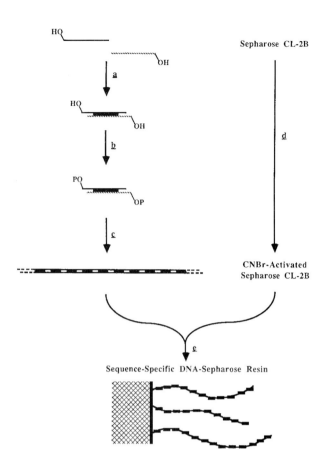

FIG. 1. Preparation of a sequence-specific DNA-Sepharose resin. Solid black rectangles represent the recognition site of a sequence-specific DNA binding protein. (*a*) Annealing of complementary oligodeoxynucleotides; (*b*) phosphorylation of 5′ protruding ends with T4 polynucleotide kinase and ATP; (*c*) polymerization of the complementary oligodeoxynucleotides with T4 DNA ligase and ATP; (*d*) activation of Sepharose CL-2B with CNBr; (*e*) coupling of the ligated DNA to the activated Sepharose to give the affinity resin. It is assumed that the DNA is covalently bound to the resin by the primary amino groups of the residues in the 5′ protruding ends; however, the exact position(s) where the DNA is attached to the resin is not known.

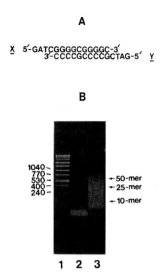

FIG 2. Polymerization of complementary 5′-phosphorylated oligodeoxynucleotides. (*A*) DNA sequence of complementary oligodeoxynucleotides, X and Y. When annealed, oligodeoxynucleotides X and Y create the high-affinity Sp1 binding site, 5′-GGGGCGGGGC-3′ (1, 3). (*B*) Analysis of the ligated DNA by 2% agarose gel electrophoresis. The DNA was visualized by ethidium bromide staining and ultraviolet light fluorescence. Lanes: 1, linear double-stranded DNA molecular size markers (sizes of selected fragments are given in base pairs); 2, oligodeoxynucleotides X and Y before 5′-phosphorylation and ligation (1 μg); 3, oligodeoxynucleotides X and Y after 5′-phosphorylation and ligation (2 μg). The estimated migration of multimers is indicated by arrows.

oligodeoxynucleotide unit that range from 3-mers to 75-mers (Fig. 2*B*). The ligated DNA is then covalently coupled to Sepharose CL-2B with CNBr. The efficiency of coupling of the DNA to the Sepharose is usually between 40% and 70%, and most of the DNA molecules are probably attached to the resin by a single covalent bond (8). We have successfully used oligodeoxynucleotides that range in length from 14 to 50 nucleotides. The concentration of DNA bound to the Sepharose is typically 20–30 μg/ml, which corresponds to a theoretical protein binding capacity of 2–3 nmol/ml if there is one recognition site per 15 base pairs. We have also observed that the resin is stable for at least 1 year and can be reused >30 times without any detectable loss of protein binding capacity.

DNA Affinity Chromatography. We have purified transcription factor Sp1 to an estimated 90% homogeneity by using any of three affinity resins that contain different Sp1 binding sites as well as different flanking oligodeoxynucleotide sequences. Similarly, we have purified CTF (16) and AP-1 to near homogeneity (K. A. Jones, J.T.K., and R.T., unpublished data; W. Lee, P. Mitchell, and R.T., unpublished data).

The results of a typical affinity chromatography experiment with Sp1 as a model protein are presented in Fig. 3. A partially purified preparation of Sp1 (total protein, 5–10 mg, derived from 50 g of HeLa cells; ≈0.1% purity; see *Materials*

and Methods) is combined with sonicated calf thymus competitor DNA, allowed to stand for 10 min, and then applied to an Sp1 affinity resin. The protein-DNA mixture is passed through the resin by gravity flow, the column is washed with buffer, and the Sp1 is eluted with 0.5 M KCl. The Sp1 DNA binding activity was monitored by DNase I footprinting of the 21-base-pair repeat sequences in SV40 (12). As displayed in Fig. 3*A*, >20% of the initial Sp1 activity is in the flowthrough fraction, and the recovery of Sp1 is 50–60%. Fig. 3*B* is a silver-stained NaDodSO₄/polyacrylamide gel that shows Sp1 before affinity purification and after one and two affinity chromatography steps, and Fig. 3*C* shows activation of transcription from the SV40 early promoter by purified Sp1 in a reconstituted *in vitro* transcription assay. Sp1 is purified 500- to 1000-fold with 30% yield to an estimated 90% homogeneity by two sequential affinity chromatography steps.

As shown in Fig. 4, the addition of competitor DNA to the crude protein fraction can be critical for successful affinity chromatography. In a typical experiment with a 1-ml affinity column that contains ≈20 μg of covalently bound DNA, 440 μg of sonicated calf thymus DNA, which is greater than a 20-fold excess over the synthetic DNA bound to the resin, is added to the crude protein fraction before chromatography. Thus, each passage through the affinity resin can give up to a 20-fold enrichment of Sp1 relative to other DNA binding proteins as well as separation from proteins that do not bind to DNA. In the absence of sufficient competitor DNA, nonspecific proteins will bind to the resin and subsequently contaminate the Sp1. On the other hand, a large excess of competitor DNA can result in a low recovery of Sp1, presumably due to weak binding of the protein to the competitor DNA (data not shown). Interestingly, the use of competitor DNA is important for affinity chromatography of Sp1 from crude extracts of ≈0.1% purity (Fig. 4*A*) but not for isolation of Sp1 from protein fractions of ≈1% purity (Fig. 4*B*). In the purification of a new protein, the composition and

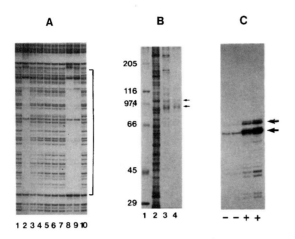

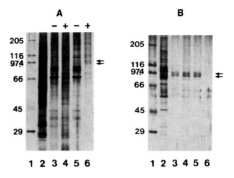

FIG. 3. Sequence-specific DNA affinity chromatography of transcription factor Sp1. (A) DNase I footprint assay of Sp1 binding to the SV40 21-base-pair repeat sequences. The degenerate 21-base-pair repeats in the SV40 genome contain six tandem Sp1 binding sites (12, 17, 18). Lanes 1 and 10, control DNase I digestion in the absence of protein. Lane 2, crude protein fraction before chromatography (25 μl of 6-ml fraction). Lane 3, affinity resin flowthrough (25 μl of 6-ml fraction). Lane 4, first column wash (25 μl of 2-ml fraction). Lane 5, second column wash (25 μl of 2-ml fraction). Lane 6, third column wash (25 μl of 2-ml fraction). Lane 7, fourth column wash (25 μl of 2-ml fraction). Lane 8, major Sp1 equate (10 μl of 1-ml fraction). Lane 9, minor Sp1 eluate (10 μl of 0.6-ml fraction). The region of DNA protected from DNase I digestion by Sp1 is indicated by a bracket. (B) NaDodSO$_4$/polyacrylamide gel electrophoresis of affinity purified Sp1. Protein samples were precipitated with trichloroacetic acid, subjected to electrophoresis on a NaDodSO$_4$/7% polyacrylamide gel, and stained with silver. Lanes: 1, protein molecular size markers; 2, crude Sp1 fraction (15 μg total protein); 3, Sp1 fraction after one affinity chromatography step (1.5 μg total protein); 4, Sp1 after two sequential affinity chromatography steps (0.3 μg total protein). The protein samples applied to lanes 3 and 4 contained roughly 10 times the Sp1 DNA binding activity as the protein sample applied to lane 2. The two polypeptides of 95 and 105 kDa, which are indicated by arrows, have been shown to be Sp1 (M. R. Briggs, J.T.K., and R.T., unpublished data). The size of each of the protein markers is also given in kDa. (C) Activation of in vitro transcription from the SV40 early promoter by affinity-purified Sp1. RNA synthesis was measured by a primer extension assay, and the arrows indicate the cDNA strands that derive from transcripts generated in vitro. The symbols + and − indicate the presence or absence of affinity-purified Sp1.

FIG. 4. Effects of competitor DNA and binding sites on the affinity purification of Sp1. Protein samples were precipitated with trichloroacetic acid, subjected to electrophoresis on a NaDodSO$_4$/8% polyacrylamide gel, and stained with silver. Sp1 is indicated by arrows, and the size of each of the protein markers is given in kDa. (A) Affinity chromatography of crude protein fraction containing 0.05–0.1% Sp1. Lanes: 1, protein molecular size markers; 2, crude protein fraction before chromatography; 3 and 4, Sp1 fractions after one affinity chromatography step; 5 and 6, Sp1 fractions after two affinity chromatography steps. The symbols + and − indicate the presence or absence of calf thymus competitor DNA in the protein fractions before chromatography. (B) Affinity chromatography of protein fraction containing ≈1% Sp1. Lanes: 1, protein molecular size markers; 2, protein fraction before chromatography; 3–5, eluate fractions from one passage through Sp1 affinity resins; 6, eluate fraction from one passage through an affinity resin that does not contain Sp1 binding sites. The relative amounts of calf thymus competitor DNA used in the affinity chromatography experiments shown in lanes 3, 4, 5, and 6 are 0, 0.2, 1, and 1, respectively. The two bands that migrate between the 45- and 66-kDa size markers derive from 2-mercaptoethanol in the electrophoresis sample buffer.

amount of competitor DNA must be carefully determined because the affinity of a test protein for different DNAs will vary. We have successfully used both sonicated calf thymus DNA and polydeoxyinosinic/polydeoxycytidylic acid as competitor DNAs.

In addition to variation of the competitor DNAs, the composition of the buffer is critical for successful DNA affinity chromatography. The most obvious constituents to vary are the concentration of mono- and divalent metal cations, such as sodium, potassium, and magnesium ions, as well as the buffer itself and the pH. If nucleases are present in the crude protein fraction, then divalent metal cations should be omitted and EDTA should be added to protect the resin. Also, specific ligands that affect the DNA binding properties of a protein could be used in some situations.

Finally, the affinity resins can assist in the identification of sequence-specific DNA binding proteins. For example, any protein that is enriched by chromatography through an affinity resin that contains Sp1 recognition sites is likely to be either Sp1 or a protein that is closely associated with Sp1. To eliminate the possibility that such proteins are not species that fortuitously bind either to DNA sequences other than the Sp1 binding sites or to the Sepharose CL-2B support, fractions containing Sp1 can be passed through DNA-

Sepharose resins that do or do not possess Sp1 recognition sequences. As shown in Fig. 4B, two polypeptides of 105 and 95 kDa, which have subsequently been shown to be Sp1 (M. R. Briggs, J.T.K., and R.T., unpublished data), are purified by affinity chromatography through a resin that contains Sp1 recognition sequences but are not enriched by passage through a similar resin that does not possess Sp1 binding sites.

DISCUSSION

In this paper, we describe a method for affinity purification of sequence-specific DNA binding proteins that is fast and effective. This technique provides a means for the purification of low abundance mammalian transcription factors, such as Sp1, CTF, and AP-1. In addition, multiple sequence-specific DNA binding factors can be simultaneously purified from a single protein fraction with the use of tandem affinity columns that contain the appropriate binding sites.

A general strategy for the purification of sequence-specific DNA binding proteins can now be outlined. (i) First, it is important that the protein is partially purified by conventional chromatography to remove nucleases and other contaminating activities that might adversely affect the affinity column. (ii) Next, the recognition sequence should be identified by various methods, such as DNase I footprinting (13), methidiumpropyl-EDTA·Fe(II) footprinting (19), and dimethyl sulfate methylation protection (20). (iii) If more than one binding site is known, the highest affinity site for the protein should be determined because low-affinity recognition sequences may not be useful for the purification of proteins (unpublished data). (iv) To assist in the identification of the desired protein, two or more DNA-Sepharose resins could be prepared that contain the highest affinity recognition site with different flanking oligodeoxynucleotide sequences in addition to a control resin that does not possess any recognition sites. If a protein is purified by passage through two different

Kadonaga & Tjian (1986) Proc. Natl. Acad. Sci. U.S.A. **83**, 5889–5893

affinity resins, then contamination by proteins that bind fortuitously to flanking oligodeoxynucleotide sequences is minimized. The control resin is used to identify proteins that bind nonspecifically to DNA-Sepharose. (*v*) Last, the conditions for successful affinity chromatography should be established by variation of such factors as competitor DNAs, pH, metal ion concentrations, ligands, recognition sequences, and the purity and quantity of the crude protein fraction.

Recently, affinity purification of nuclear factor I (NFI), a sequence-specific DNA binding protein from HeLa cells that stimulates adenovirus DNA replication, has been described (9). The NFI affinity resin consisted of plasmid DNA containing 88 copies of the NFI binding site adsorbed onto cellulose by the method of Alberts and Herrick (5). Highly purified NFI was obtained from a crude protein fraction by chromatography on a nonspecific *Escherichia coli* DNAcellulose column followed by two passages through the sequence-specific DNA-cellulose resin in the absence of any competitor DNA.

Sequence-specific DNA affinity chromatography has also been reported for SV40 large tumor antigen (T antigen), polyoma virus large T antigen, and *E. coli lac* repressor. SV40 large T antigen was purified in a similar manner as NFI by chromatography with plasmid DNA-cellulose (10), while a variation of the DNA-cellulose method was used to enrich for polyoma virus large T antigen (21). Last, the feasibility of sequence-specific affinity chromatography of E. *coli lac* repressor was investigated by comparison of the salt concentrations that were required to elute the protein from sequence-specific or nonspecific plasmid DNA resins (7).

Our previous attempts to purify Sp1 by sequence-specific DNA-cellulose chromatography with plasmid DNA containing multiple copies of a binding site failed to enrich for Sp1 relative to other DNA binding proteins (unpublished data). There are some significant differences between the previously described affinity chromatography procedures and the method we describe here that may account for the success of the synthetic DNA-Sepharose resins. First, the most important distinction is probably the addition of competitor DNA directly to the crude protein fraction before passage through the DNA-Sepharose resin. The presence of competitor DNA in solution minimizes retention of nonspecific DNA binding proteins on the affinity resin. Second, the use of synthetic oligodeoxynucleotides maximizes the specificity of the DNA-Sepharose resins. The oligodeoxynucleotides are not only more selective than plasmids containing multiple binding sites, but they can also be quickly and easily prepared. Finally, affinity chromatography has been previously performed with DNA-cellulose resins consisting of plasmid DNA adsorbed to cellulose, whereas, in this study, we have used DNA-Sepharose resins that were prepared by covalent coupling of synthetic DNA to an agarose support.

We are grateful to Kathy Jones, Mike Briggs, Bill Dynan, and Mark Biggin for their helpful suggestions. We also thank Bruce Malcolm for synthesizing the oligodeoxynucleotides that were used in this study. J.T.K. is a Fellow of the Miller Institute for Basic Research in Science. This work was funded by grants from the National Institutes of Health and by partial support from a National Institute for Environmental Health Sciences grant to R.T.

1. Dynan, W. S. & Tjian, R. (1985) *Nature (London)* **316**, 774–778.
2. Serfling, E., Jasin, M. & Schaffner, W. (1985) *Trends Genet.* **1**, 224–230.
3. Kadonaga, J. T., Jones, K. A. & Tjian, R. (1986) *Trends Biochem.* **11**, 20–23.
4. Alberts, B. M., Amodio, F. J., Jenkins, M., Gutmann, E. D. & Ferris, F. L. (1968) *Cold Spring Harbor Symp. Quant. Biol.* **33**, 289–305.
5. Alberts, B. & Herrick, G. (1971) *Methods Lnzymol.* **21**, 198–217.
6. Alberts, B. M. (1984) *Cold Spring Harbor Symp. Quant. Biol.* **48**, 1–12.
7. Herrick, G. (1980) *Nucleic Acids Res.* **8**, 3721–3728.
8. Arndt-Jovin, D. J., Jovin, T. M., Bahr, W., Frischauf, A.-M. & Marquardt, M. (1975) *Eur. J. Biochem.* **54**, 411–418.
9. Kosenfeld, P. J. & Kelly, T. J. (1986) *J. Biol. Chem.* **261**, 1398–1408.
10. Oren, M., Winocour, E. & Prives, C. (1980) *Proc. Natl. Acad. Sci. USA* **77**, 220–224.
11. Dynan, W. S. & Tjian, R. (1983) *Cell* **32**, 669–680.
12. Dynan, W. S. & Tjian, R. (1983) *Cell* **35**, 79–87.
13. Galas, D. & Schmitz, A. (1978) *Nucleic Acids Res.* **5**, 3157–3170.
14. Maniatis, T., Fritsch, E. F. & Sambrook, J. (1982) *Molecular Cloning: A Laboratory Manual* (Cold Spring Harbor Laboratory, Cold Spring Harbor, NY).
15. Laemmli, U. K. (1970) *Nature (London)* **227**, 680–685.
16. Jones, K. A., Yamamoto, K. R. & Tjian, R. (1985) *Cell* **42**, 559–572.
17. Gidoni, D., Dynan, W. S. & Tjian, R. (1984) *Nature (London)* **312**, 409–413.
18. Gidoni, D., Kadonaga, J. T., Barrera-Saldana, H., Takahashi, K., Chambon, P. & Tjian, R. (1985) *Science* **230**, 511–517.
19. Van Dyke, M. W. & Dervan, P. B. (1983) *Nucleic Acids Res.* **11**, 5555–5567.
20. Siebenlist, U. & Gilbert, W. (1980) *Proc. Natl. Acad. Sci. USA* **77**, 122–126.
21. Gaudray, P., Tyndall, C., Kamen, R. & Cuzin, F. (1981) *Nucleic Acids Res.* **9**, 5697–5710.

Kadonaga et al. (1987) Cell **51**, 1079–1090

Isolation of cDNA Encoding Transcription Factor Sp1 and Functional Analysis of the DNA Binding Domain

James T. Kadonaga,* Kristin R. Carner,* Frank R. Masiarzft,† and Robert Tjian *
* Howard Hughes Medical Institute
Department of Biochemistry
University of California
Berkeley, California 94720
†Chiron Corporation Research Laboratories
4560 Horton Street
Emeryville, California 94608

Summary

Transcription factor Sp1 is a protein present in mammalian cells that binds to GC box promoter elements and selectively activates mRNA synthesis from genes that contain functional recognition sites. We have isolated a cDNA that encodes the 696 C-terminal amino acid residues of human Sp1. By expression of truncated fragments of Sp1 in E. coli, we have localized the DNA binding activity to the C-terminal 168 amino acid residues. In this region, Sp1 has three contiguous Zn(II) finger motifs, which are believed to be metalloprotein structures that interact with DNA. We have found that purified Sp1 requires Zn(II) for sequence-specific binding to DNA. Thus, it is likely that Sp1 interacts with DNA by binding of the Zn(II) fingers. To facilitate the identification of mutant variants of Sp1 that are defective in DNA binding, we have also devised a bacterial colony assay for detection of Sp1 binding to DNA.

Introduction

Complex variations in gene expression are regulated, in part, at the level of transcription initiation, and one approach to the study of gene regulation has been to identify and characterize transcription factors that can distinguish between different genes (for reviews see Serfling et al., 1985; McKnight and Tjian, 1986; Maniatis et al., 1987). Many sequence-specific DNA binding proteins have been found to interact with promoter and enhancer elements, and several of these factors have also been shown to affect the level of RNA synthesis by in vitro transcription analysis. In addition, a variety of genes that appear to be involved in transcriptional regulation have been isolated, such as oncogenes, genes that encode steroid hormone receptors, and genes that govern developmental patterns of expression; but, in most cases, the proteins encoded by these genes have not yet been shown to modulate transcription in a biochemical assay.

Transcription factor Sp1 is a DNA binding protein that discriminates between different promoters and activates transcription of a subset of genes in animal cells (for reviews see Dynan and Tjian, 1985; Kadonaga et al., 1986; McKnight and Tjian, 1986). An important property of Sp1 is that the purified protein can activate RNA synthe-

sis in a reconstituted in vitro transcription reaction. Comparison of Sp1 activity in vitro and in vivo suggests that the mechanism by which the factor stimulates transcription in animal cells is similar to the means by which it functions in vitro (Jones et al., 1985; Lee et al., 1987a). Sp1 was originally identified as a protein from HeLa cells that binds to multiple GGGCGG sequences (GC boxes) in the 21 bp repeat elements of SV40 and activates in vitro transcription from the SV40 early promoter (Dynan and Tjian, 1983a, 1983b; Gidoni et al., 1984, 1985). Subsequently, a variety of cellular and viral promoters were shown to be activated by Sp1 in vitro. These studies revealed that Sp1-responsive promoters usually contain multiple GC box recognition sites, although a single binding site appears to be sufficient for a promoter to be stimulated by Sp1. Also, Sp1 recognition sequences are often found near binding sites for other transcription factors, such as CTF/NF-I (Jones et al., 1985) and AP-1 (Lee et al., 1987a), which suggests that these factors may act in conjunction with each other to modulate transcription. The Sp1 recognition sequence that is closest to the RNA initiation site is typically located 40 to 70 nucleotides upstream of the start site, and despite the asymmetry of Sp1 binding sites, they are functional in either orientation.

It is our hope that a detailed study of Sp1 will provide insight into the mechanisms of gene-specific transcriptional regulation in higher eukaryotes. Although the purification of Sp1 (Kadonaga and Tjian, 1986; Briggs et al., 1986) has facilitated studies of its transcriptional and DNA binding properties, many aspects of how Sp1 functions remain to be elucidated. We have not yet uncovered, for example, a unifying theme among Sp1-responsive genes. We would also like to characterize possible variations in the spatial and temporal expression of Sp1 and determine how the factor is regulated. Finally, we would like to determine the molecular interactions that allow Sp1 to participate in the initiation complex and stimulate transcription. To address these questions, which range from understanding the role of the factor in the cell to the biochemical mechanisms of Sp1 action, it is first necessary to isolate the Sp1 gene. In this paper, we describe cloning of a partial cDNA encoding Sp1 by determination of the amino acid sequence of Sp1 peptides and screening of a HeLa cDNA library with synthetic oligonucleotides that correspond to the coding sequence of Sp1. We report the structure of the 696 C-terminal amino acid residues of Sp1, as deduced from the cDNA sequence. Sp1 contains three contiguous Zn(II) finger structures, which are homologous to the repeated motifs that were first described in transcription factor IIIA (TFIIIA; Miller et al., 1985; Brown et al., 1985), and we show that the sequence-specific DNA binding activity of Sp1 requires Zn(II). By expression of truncated fragments of Sp1 in E. coli, we have localized the DNA binding region of the protein. In addition, we describe a means of detecting binding of Sp1 to DNA with a bacterial colony assay that will facilitate screening of Sp1 mutants that are defective in binding to DNA. These

A

Sp1 peptide #1:   NH₂-Val-Ser-Gly-Leu-Gln-Gly-Ser-⟦Gly⟧-Ala-Leu-Asn-Ile-Gln-Gln-Asn-Gln-Thr

Sp1 peptide #2:   NH₂-⟦Phe⟧-Asp-Glu-Leu-Gln-Arg

Sp1 peptide #3:   NH₂-Phe-Met-Arg

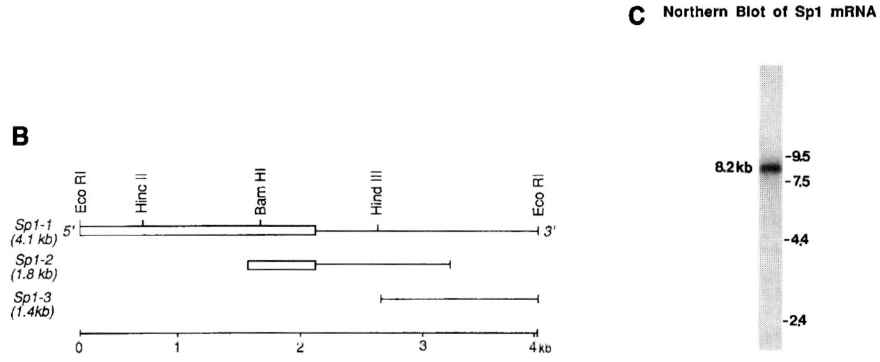

C **Northern Blot of Sp1 mRNA**

B

Figure 1. Isolation of cDNAs Encoding Transcription Factor Sp1

(A) Amino acid sequence of three tryptic peptides of Sp1. The boxes indicate residues that differ from the amino acid sequence deduced from the cDNA (see Figure 2). In these two instances, the correct amino acid residues were also detected, but in lower quantity than those shown in the figure.
(B) Restriction map of partial Sp1 cDNAs. The location of the three cDNA isolates relative to each other is shown. The open rectangles represent the coding sequence for Sp1. The DNA sequence of the 5′ end of the cDNA to the HindIII site is in GenBank (accession number J03133).
(C) Northern blot of Sp1 mRNA. The 2.7 kb EcoRI–HindIII fragment of Sp1-1 was randomly labeled with ³²p and used to probe 4 μg of poly(A)-enriched RNA from HeLa cells. The numbers indicate the size, in kb, of RNA molecular mass markers.

studies initiate a new phase of experiments that should improve our understanding of Sp1 function and provide information concerning the process of promoter-selective transcriptional activation.

Results

Isolation of a Partial Sp1 cDNA

To isolate the cDNA encoding transcription factor Sp1, we chose to determine a partial amino acid sequence of Sp1 and to screen a HeLa cDNA library with synthetic oligonucleotides that contain the coding sequence of the protein. Although the purified factor consists of two species of M_r 95 and 105 kd as determined by SDS–PAGE and silver staining (Kadonaga and Tjian, 1986; Briggs et al., 1986), these two forms appear to be variants of a single polypeptide because treatment of Sp1 with dilute ammonium hydroxide converts the M_r 95 and 105 kd species into a single form of M_r 105 kd (unpublished data). Hence, it seemed likely that a single gene encoded Sp1 and that any partial amino acid sequence obtained from purified Sp1 would be derived from a single polypeptide. Two attempts at N-terminal sequencing of Sp1 both failed, and consequently, we chose to generate and sequence Sp1 tryptic peptides. By using an abbreviated version of the procedure of Briggs et al. (1986), we purified 300μg of Sp1 to greater than 95% homogeneity from 800 g of HeLa cells. The cysteine residues were alkylated with 4-vinylpyridine under reducing conditions, and the protein was

digested with trypsin. The resulting peptides were purified by reverse phase HPLC and then subjected to sequence analysis to give the amino acid sequence of three Sp1 peptides, shown in Figure 1A.

A HeLa cDNA library in λgt10 was screened with synthetic oligonucleotides that contain the expected coding sequence of Sp1 peptide #1, and one positive clone, named Sp1-1, was isolated from 10⁶ plaques. The authenticity of the clone was verified by sequencing of the cDNA by the chain termination method of Sanger et al. (1977) after random shotgun cloning of DNA fragments into bacteriophage M13 (Bankier and Barrell, 1983). Sp1-1 is a 4.1 kb cDNA clone with a 2.1 kb open reading frame that contains the coding sequences for the three Sp1 peptides (see Figures 1 and 2). Each of the Sp1 peptides is also preceded by an arginine residue, a finding consistent with generation of the peptides by digestion with trypsin, which preferentially cleaves proteins at lysine and arginine residues. Two additional isolates of the Sp1 cDNA, named Sp1-2 and Sp1-3, were subsequently obtained by screening another 10⁶ plaques of the HeLa cDNA library. The location of the new clones relative to Sp1-1 was determined by DNA sequencing, and both Sp1-2 and Sp1-3 are contained within the larger Sp1-1 (Figure 1B). In addition, Northern blot analysis of poly(A)-enriched HeLa RNA was carried out and revealed that Sp1 mRNA appears to be a single species of approximately 8.2 kb (Figure 1C). Thus, the 4.1 kb Sp1-1 clone represents about one-half of the Sp1 mRNA.

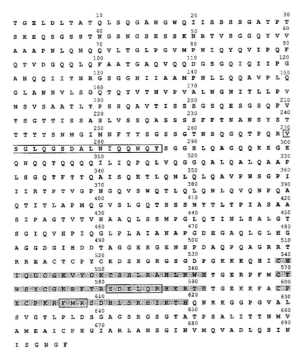

```
        10              20                  30
T G E L D L T A T Q L S Q G A N G W Q I I S S S S G A T P T
        40              50                  60
S K E Q S G S S T N G S N G S E S S K N R T V S G G Q Y V V
        70              80                  90
A A A P N L Q N Q Q V L T G L P G V M P N I Q Y Q V I P Q F
        100             110                 120
Q T V D G Q Q L Q F A A T G A Q V Q Q D G S G Q I Q I I P G
        130             140                 150
A N Q Q I I T N R G S G G N I I A A M P N L L Q Q A V P L Q
        160             170
G L A N N V L S G Q T Q Y V T N V P V A L N G N I T L L P V
        190             200                 210
N S V S A A T L T P S S Q A V T I S S S S G S Q E S G S Q P V
        220             230                 240
T S G T T I S S A S L V S S Q A S S S F F T N A N S Y S T
        250             260                 270
T T T T S N M G I M N F T T S G S S G T N S Q G Q T P Q R V
        280             290                 300
S G L Q G S D A L N I Q Q N Q T S G G S L Q A G Q Q K E G E
        310             320                 330
Q N Q Q T Q Q Q Q I L I Q P Q L V Q G G Q A L Q A L Q A A P
        340             350                 360
L S G Q T F T T Q A I S Q E T L Q N L Q L Q A V P N S G P I
        370             380                 390
I I R T P T V G P N G Q V S W Q T L Q L Q N L Q V Q N P Q A
        400             410                 420
Q T I T L A P M Q G V S L G Q T S S S N T T L T P I A S A A
        430             440                 450
S I P A G T V T V N A A Q L S S M P G L Q T I N L S A L G T
        460             470                 480
S G I Q V H P I Q G L P L A I A N A P G D H G A Q L G L H G
        490             500                 510
A G G D G I H D D T A G G E E G E N S P D A Q P Q A G R R T
        520             530                 540
R R E A C T C P Y C K D S E G R G S G D P G K K K Q H I C H
        550             560                 570
I Q G C G K V Y G K T S H L R A H L R W H T G E R P F M C T
        580             590                 600
W S Y C G K R F T R S D E L Q R H K R T H T G E K K F A C P
        610             620                 630
E C P K R F M R S D H L S K H I K T H Q N K K G G P G V A L
        640             650                 660
S V G T L P L D S G A G S E G S G T A T P S A L I T T N M V
        670             680                 690
A M E A I C P E G I A R L A N S G I N V M Q V A D L Q S I N

I S G N G F
```

Figure 2. Sequence of the 696 C-Terminal Amino Acid Residues of Sp1

The three tryptic peptides of Sp1 that were sequenced are enclosed in boxes. The shaded areas indicate the location of the three Zn(II) finger motifs in Sp1.

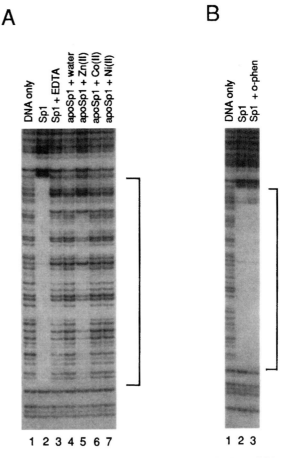

Figure 3. Sp1 Requires Zn(II) for Sequence-Specific Binding to DNA

(A) Inactivation of Sp1 by treatment with EDTA and recovery of DNA binding activity by addition of Zn(II). The footprinting reactions contained no Sp1 (lane 1) untreated Sp1 (lane 2), Sp1 + EDTA (lane 3) and apoSp1 incubated with water (as a negative control; lane 4) $ZnSO_4$ (lane 5), $CoSO_4$ (lane 6), or $NiSO_4$ (lane 7).

(B) Sp1 DNA binding activity is not affected by the presence of 1 mM 1,10-phenanthroline. The footprinting reactions contained no Sp1 (lane 1), untreated Sp1 (lane 2), and Sp1 + 1,10-phenanthroline (lane 3). Preparation of the protein fractions and conditions for DNAase I footprinting are described in Experimental Procedures.

Sequence of the C-Terminal 696 Amino Acid Residues of Sp1

The amino acid sequence of the C-terminal 696 amino acid residues of Sp1, as deduced from the cDNA sequence, is shown in Figure 2. The Sp1-1 cDNA probably does not contain the entire coding sequence for Sp1 because the calculated molecular mass of the 696 C-terminal amino acids encoded by the open reading frame (72 kd) is much less than the M_r of Sp1 from HeLa cells (95 and 105 kd) and the 2.1 kb open reading frame continues to the 5' end of the cDNA. Consequently, it is likely that cDNA clone Sp1-1 contains about 75% of the Sp1 coding sequence. At present, we are in the process of isolating additional Sp1 clones to obtain a full-length cDNA.

Sp1 Has Three Zn(II) "Fingers"

As a first step in the delineation of functional domains of Sp1, we focused upon identification and characterization of its sequence-specific DNA binding activity. By inspection of the amino acid sequence shown in Figure 2, we found that Sp1 has three Zn(II) "finger" motifs near its C-terminus (from Cys 539 to His 619). These structural motifs have the general form of Cys-$X_{2,4}$-Cys-X_3-Phe-X_5-Leu-X_2-His-X_3-His (Miller et al., 1985; Berg, 1986) and were first identified as nine tandemly repeated sequences in the RNA polymerase III transcription factor TFIIIA (Miller et al., 1985; Brown et al., 1985). The 2 Cys and 2 His residues are believed to co-ordinate a Zn(II) ion to give a metalloprotein structure that interacts with DNA. Except

for TFIIIA, however, proteins that contain finger motifs have not been demonstrated biochemically to require Zn(II) for sequence-specific binding to DNA.

The presence of three finger motifs in Sp1 prompted us to examine the dependence of Sp1 upon Zn(II) for sequence-specific binding to DNA (Figure 3). First, when a purified preparation of Sp1 is treated with 50 mM EDTA for 45 min at 4°C, the protein loses its ability to bind to the GC boxes in the SV40 21 bp repeat elements (Figure 3A, compare lanes 2 and 3). After dialysis and dilution of the mixture to a final EDTA concentration of 0.2 mM, various divalent metals were added to the apoprotein and recovery of specific DNA binding activity was assayed by DNAase I footprinting. Sp1 DNA binding activity is reconstituted, with approximately 10%–20% recovery, by the addition of Zn(II) (Figure 3A, lane 5), but not by the addition of metal-free water, Co(II), or Ni(II) (Figure 3A, lanes

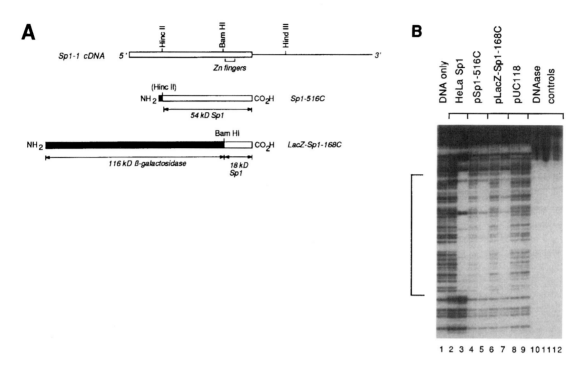

Figure 4. Expression of Sp1 in E. coli

(A) Structure of LacZ-Sp1 hybrid proteins synthesized in E. coli. Sp1-516C is encoded by pSp1-516C, and LacZ-Sp1-168C is encoded by pLacZ-Sp1-168C. The composition of the hybrid proteins and construction of the plasmids are described in the text.

(B) DNA binding activity in crude lysates of E. coli strain JK50 harboring Sp1-expressing plasmids. The footprinting reactions contained no Sp1 (lane 1), HeLa Sp1 (4 ng of protein, lane 2; 20 ng of protein, lane 3), lysate of JK50 (pSp1-516C) (10 μg of protein, lane 4; 50 μg of protein, lanes 5 and 10), lysate of JK50 (pLacZ-Sp1-168C,1 (10 μg of protein, lane 6; 50 μg of protein, lanes 7 and 11), and lysate of JK50 (pUC118) (10 μg of protein, lane 8; 50 μg of protein, lanes 9 and 12). The Sp1 footprint on the 21 bp repeat elements of SV40 is indicated by a bracket. To monitor the endogenous nucleases present in the extracts, control reactions (lanes 10 to 12) were carried out in which the DNA probe was incubated with the crude lysates without addition of DNAase I. The preparation of the E. coli extracts is described in Experimental Procedures. DNAase I footprinting of Sp1 was performed essentially as described by Dynan and Tjian (1983b).

4, 6, and 7). Thus, transcription factor Sp1 not only contains three finger motifs, but it also requires Zn(II) for sequence-specific binding to DNA. In contrast to TFIIIA, however, there is no detectable loss in Sp1 DNA binding activity after incubation of the protein with 1 mM 1,10-phenanthroline (Figure 3B), which is commonly used to remove Zn(II) from metalloproteins. Sp1 thus provides an example of a Zn(II) finger protein that is inert to inactivation by 1 mM 1,10-phenanthroline, and consequently, when testing Zn(II)-dependence of DNA binding proteins, it is necessary to examine a variety of chelating agents and conditions.

Expression of Sp1 in E. coli

To complement the findings that Sp1 has three finger motifs and requires Zn(II) for binding to DNA, we localized the DNA binding activity of Sp1 by expression of truncated fragments of the protein in E. coli, as outlined in Figure 4A. These experiments provide an independent means of mapping the region of Sp1 that is sufficient for DNA binding. We constructed two plasmids that express either 516 or 168 of the C-terminal amino acid residues of Sp1. The first construction, pSp1-516C, contains the 2112 bp HincII–HindIII cDNA fragment of Sp1-1 inserted into pUC118 digested with SmaI and HindIII. The hybrid protein encoded by this gene contains the first 10 amino acid residues of the *lacZ'* gene in pUC118 fused to the C-terminal 516 amino acid residues (54 kd) of Sp1. The other plasmid, pLacZ-Sp1-168C, contains the 1063 bp BamHI–HindIII fragment of Sp1-1 subcloned into pUR278 (Ruther and Muller-Hill, 1983) and encodes the entirety of β-galactosidase fused to the C-terminal 168 amino acid residues (18 kd) of Sp1. Both pSp1-516C and pLacZ-Sp1-168C contain the coding sequences for the three finger motifs, which comprise 71 amino acid residues near the C-terminus of the protein.

Crude lysates of E. coli strain JK50 containing pSp1-516C, pLacZ-Sp1-168C, or pUC118 (as a control) were tested for DNA binding activity by DNAase I footprinting with the SV40 21 bp repeat elements. The extracts prepared from JK50 (pSp1-516C) and JK50 (pLacZ-Sp1-168C), but not JK50 (pUC118), were found to protect the multiple GC box Sp1 binding sites in SV40 from digestion by DNAase I (Figure 4B). The Sp1 polypeptides encoded by pSp1-516C and pLacZ-Sp1-168C were purified to roughly 10% homogeneity (as determined by SDS–PAGE and silver-staining; unpublished data) by sequence-specific DNA affinity chromatography (Kadonaga and Tjian, 1986), and the binding of the protein fractions was tested on the SV40 21 bp repeats as well as the human

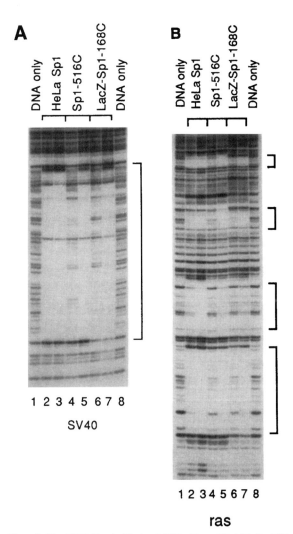

A

DNA only
HeLa Sp1
Sp1-516C
LacZ-Sp1-168C
DNA only

1 2 3 4 5 6 7 8

SV40

B

DNA only
HeLa Sp1
Sp1-516C
LacZ-Sp1-168C
DNA only

1 2 3 4 5 6 7 8

ras

Figure 5. The 168 C-Terminal Amino Acid Residues of Sp1 Bind to DNA with the Same Sequence Specificity as Full-Length HeLa Sp1

(A) Binding of affinity-purified Sp1-516C and LacZ-Sp1-168C to the SV40 21 bp repeat elements. The footprinting reactions contained no Sp1 (lanes 1 and 8), HeLa Sp1 (4 ng of protein, lane 2; 12 ng of protein, lane 3), Sp1-516C (40 ng of protein, lane 4; 120 ng of protein, lane 5), and LacZ-Sp1-168C (40 ng of protein, lane 6; 120 ng of protein, lane 7). The Sp1 footprint is indicated by a bracket. DNAase I footprinting of Sp1 on the SV40 21 bp repeat elements was carried out as described by Dynan and Tjian (1983b).

(B) Binding of affinity-purified Sp1-516C and LacZ-Sp1-168C to the human Harvey ras1 promoter. The footprinting reactions contained no Sp1 (lanes 1 and 8), HeLa Sp1 (4 ng of protein, lane 2; 12 ng of protein, lane 3), Sp1-516C (40 ng of protein, lane 4; 120 ng of protein, lane 5), and LacZ-Sp1-168C (40 ng of protein, lane 6; 120 ng of protein, lane 7). The Sp1 footprints are indicated by brackets. The binding of Sp1 to the Harvey ras1 promoter was described by Ishii et al. (1986).

Harvey ras1 promoter (Ishii et al., 1986). The polypeptides derived from pSp1-516C and pLacZ-Sp1-168C bind to both SV40 (Figure 5A) and the human Harvey ras1 promoter (Figure 5B) with the same sequence specificity as Sp1 from HeLa cells, although there are some differences in the patterns of Sp1-induced enhanced DNAase I cleavages. In addition, we have examined the DNA binding properties of a LacZ'-Spl hybrid protein that consists only

of the 11 N-terminal residues of the *lacZ'* gene of pUC118 fused to the C-terminal 168 residues of Sp1. When this truncated protein is expressed in E. coli, there is low, but detectable, levels of Sp1 DNA binding activity (approximately 10% of the activity obtained with pSp1-516C; unpublished data). It thus appears that the C-terminal 168 amino acid residues of Sp1, 71 of which constitute the three finger motifs, are sufficient for the sequence-specific binding of Sp1 to DNA. These experiments localize the region of Sp1 that is required for binding of the protein to DNA. In addition, they support the hypothesis that Zn(II) fingers are important for the binding of proteins to DNA and confirm the authenticity of the Sp1 cDNA clone.

Detection of Sequence-Specific Binding of Sp1 to DNA with a Bacterial Colony Assay

To analyze, in greater detail, the critical amino acid residues that are important for the binding of Sp1 to DNA, we constructed an E. coli strain in which the sequence-specific binding of Sp1 can be detected with a bacterial colony assay. This Sp1 test strain, JK901, should facilitate the screening and characterization of mutant variants of Sp1 that have altered DNA binding properties. The construction of JK901 is outlined in Figure 6A. This system for detection of sequence-specific binding of proteins to DNA in vivo in E. coli incorporates the method of Benson et al. (1986), which is a general approach to selection for DNA binding proteins in bacteria, and the cloning vectors of Simons et al. (1987), which enable insertion of desired DNA fragments into the E. coli chromosome in single copy. First, a plasmid was constructed in which a prokaryotic promoter that transcribes the *lacZYA* genes contains two tandem Sp1 recognition sequences at the transcriptional start site. The promoter was crossed into an appropriate λ phage in vivo, and then a Δ*lac* host strain was lysogenized with the recombinant phage. If Sp1 that is active for DNA binding is expressed in this lysogen, then the phenotype of the strain, which is normally Lac⁺, should become Lac⁻ due to repression of transcription of the *lacZYA* genes mediated by the binding of Sp1 to its "operator" sequence. Because the Sp1 recognition sequences are present in single copy on the E. coli chromosome as a λ lysogen, repression of *lacZYA* transcription requires only a few Sp1 binding sites per cell to be occupied. As shown in Figure 6B, the presence of Sp1 in JK901 transforms the phenotype of the strain from Lac⁺ to Lac⁻, as monitored with the indicator 5-bromo-4-chloro-3-indolyl β-D-galacto-pyranoside (X-Gal). Colonies of JK901 harboring pSp1-1, which is pUC118 that contains the Sp1-1 cDNA insert but does not express Sp1 protein, have a Lac⁺ phenotype (blue), whereas colonies of JK901 containing pSp1-516C, which expresses Sp1 that is active for binding to DNA, have a Lac⁻ phenotype (white). In contrast, strain JK31, which is identical to JK901 except that it does not contain Sp1 binding sites as an operator, is unaffected by the presence of Sp1 in the cell. Hence, in strain JK901, expression of Sp1 mutants that are defective in sequence-specific binding to DNA would yield a Lac⁺ phenotype, whereas expression of wild-type Sp1 would give a Lac⁻ phenotype.

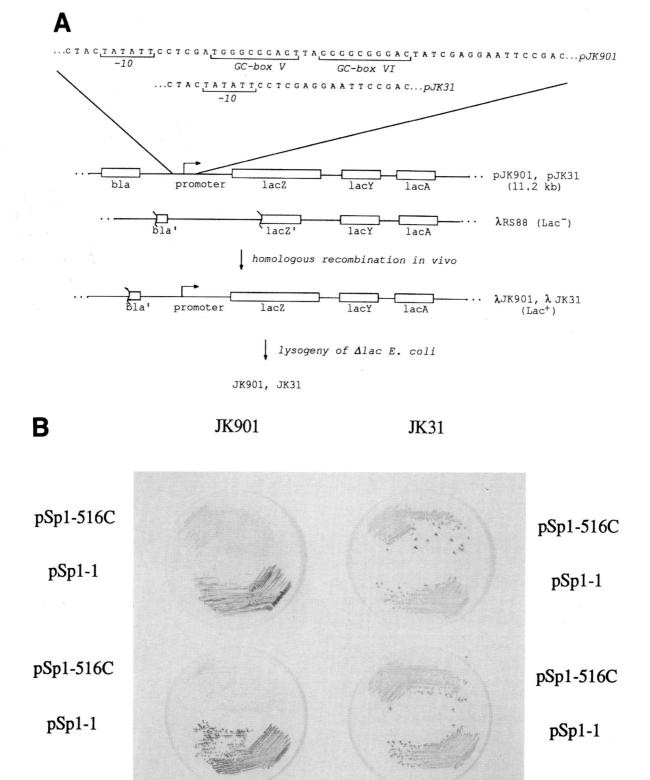

Figure 6. A Bacterial Colony Assay for Binding of Sp1 to DNA

(A) Construction of E. coli strains JK901 and JK31. The preparation of JK901 and JK31 is described in Experimental Procedures.

(B) The presence of Sp1 represses β-galactosidase synthesis in JK901, but not in JK31. Sp1 test strains containing either pSp1-516C, which expresses Sp1-516C protein, or pSp1-1, which contains the 4.1 kb cDNA insert from clone Sp1-1 and does not express Sp1 protein, were grown on nitrocellulose filters that were placed on LB plates containing 100 μg/ml ampicillin and 80 μl of a 25 mg/ml solution of X-Gal in dimethylformamide. JK901 contains Sp1 binding sites in the operator position, and the level of β-galactosidase activity in JK901 (pSp1-516C) is lower than that in JK901 (pSp1-1). JK31

Discussion

An important step at which gene expression can be regulated is the turning on and off of transcription initiation. Recent studies have identified protein factors that can discriminate between different promoters and activate RNA synthesis selectively in reconstituted transcription assays (for recent reviews see Dynan and Tjian, 1985; Serfling et al., 1985; McKnight and Tjian, 1986; Maniatis et al., 1987). Sp1 is an RNA polymerase II transcription factor from HeLa cells that activates RNA synthesis from promoters that contain properly positioned GC box Sp1 binding sites (Dynan and Tjian, 1985; Kadonaga et al., 1986). To study both the biochemical properties and the regulatory functions of Sp1, we had previously purified the protein to greater than 95% homogeneity and characterized its interaction with DNA (Kadonaga and Tjian, 1986; Briggs et al., 1986). In this work, we describe the isolation of cDNA clones encoding transcription factor Sp1, expression of Sp1 protein in E coli, and mapping of the region of Sp1 that is sufficient for sequence-specific binding to DNA.

The Sp1 cDNA was cloned by determining a partial amino acid sequence of protein purified from HeLa cells and then using synthetic oligonucleotides corresponding to the expected coding sequences of the peptides to screen a HeLa cDNA library. The largest clone, Sp1-1, is a 4.1 kb cDNA with a 2.1 kb open reading frame that contains the coding sequences for three Sp1 peptides. By Northern blot analysis, Sp1 mRNA appears to be a single species of approximately 8.2 kb, a length significantly larger than the 2.5 to 3 kb of message that is expected to encode Sp1, and the function of the estimated 5 kb of noncoding sequence in the Sp1 mRNA remains to be understood. Thus, at present, the largest cDNA represents roughly one-half of the Sp1 mRNA and contains an estimated 75% of the Sp1 coding sequence. In addition, Southern blot analysis of human genomic DNA indicates that Sp1 is encoded by a single gene (unpublished data). We have not yet, however, carried out low stringency Southern blots to examine whether there are other genes that are related to Sp1.

Structure of the C-Terminal 696 Amino Acid Residues of Sp1

The C-terminal 696 amino acids residues that are encoded by the 2.1 kb open reading frame in clone Sp1-1 have a calculated molecular mass of 72 kd. One of the most striking features of the protein is its unusually high glutamine content (83/696 = 11.9%). Sp1 also has relatively high levels of serine (11.2%), threonine (9.2%), and glycine (11.6%). The glutamine, serine, and threonine residues are unevenly distributed as the protein contains alternating stretches that are rich in either glutamine or serine/threonine (Figure 7A). It is possible that the serine/threonine-rich segments might be sites for phosphorylation, which would be a potential means for regulation of Sp1 activity, whereas it is difficult to imagine a function for

A

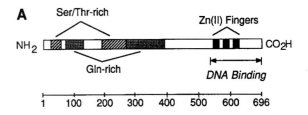

B

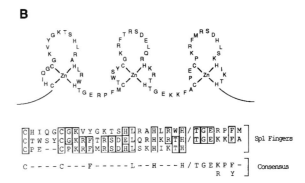

Figure 7. The 696 C-Terminal Amino Acid Residues of Transcription Factor Sp1

(A) Structure of the 696 C-terminal amino acid residues of Sp1. The serine/threonine-rich regions of the protein are indicated by striped boxes, and the glutamine-rich regions of Sp1 are depicted by shaded boxes. Residues 22 to 54 are 52% serine/threonine (17/33); residues 67 to 124 are 29% glutamine (17/58); residues 182 to 262 are 52% serine/threonine (42/81); and residues 263 to 391 are 29% glutamine (37/129). The three Zn(II) finger motifs are shown as solid boxes. The C-terminal 168 amino acid residues of the protein, which are sufficient for sequence-specific binding of Sp1 to DNA, are also indicated.
(B) Sequence of the three Zn(II) fingers in Sp1.

the glutamine-rich portions of the protein. Although Sp1 has a high glutamine content, its glutamine residues are not present as polyglutamine tracts, as it has been found, for example, in the rat glucocorticoid receptor (Miesfeld et al., 1986) and the protein encoded by the *Notch* gene in *Drosophila* (Wharton et al., 1985). Studies involving characterization of Sp1 mutants should reveal the importance of the serine/threonine-rich and glutamine-rich regions.

Other features of the C-terminal 696 residues of Sp1 are as follows. First, there are nine asparagine-X-serine/threonine motifs in the protein that are potential sites for N-linked glycosylation (Marshall, 1972). Second, because Sp1 is a large protein that functions in the nucleus, it is expected to have a nuclear location signal. Although there does not exist a consensus sequence for translocation of proteins to the nucleus, nuclear localization signals that have been characterized typically contain a cluster of 3 to 6 basic residues in a short peptide of roughly 4 to 9 amino acids (Dingwall and Laskey, 1986). Examination of the Sp1 sequence reveals only two small clusters of basic amino acid residues, Arg 508 to Arg 512 and Lys 533 to His 540, either of which might be components of a nuclear localiza-

---

is identical to JK901 except that it does not contain Sp1 binding sites in the operator position, and in contrast to JK901, the level of β-galactosidase activity in JK31 (pSp1-516C) is similar to that in JK31 (pSp1-1).

tion signal. Finally, the amino acid sequence of Sp1 was subjected to computer analysis with the Doolittle protein databank, and except for the three Zn(II) finger structures, which are discussed in the following section, there were no significant sequence homologies (R. Doolittle, personal communication).

Sp1 Binds to DNA with Zn(II) Fingers

Sp1 has three Zn(II) finger motifs, which are shown in Figure 7B. These putative DNA binding structures were initially identified as nine tandemly repeated sequences of approximately 30 amino acid residues in the RNA polymerase III transcription factor TFIIIA (Miller et al., 1985; Brown et al., 1985). TFIIIA has been analyzed by metal replacement studies (Hanas et al., 1983), atomic absorption spectroscopy (Hanas et al., 1983; Miller et al., 1985), and EXAFS (extended X-ray absorption fine structure) (Diakun et al., 1986); and based on these experiments, it has been postulated that the 2 Cys and 2 His residues in each finger bind a Zn(II) ion and that the resulting metalloprotein structures are involved in the sequence-specific binding of the factor to DNA. Similar motifs that are homologous to the fingers in TFIIIA have also been identified in other genes, including the Drosophila genes *Kruppel* (Rosenberg et al., 1986), *hunchback* (Tautz et al., 1987), *serendipity* (Vincent et al., 1985), and *terminus* (Baldarelli et al., 1988) and the yeast transcription factor ADR1 (Hartshorne et al., 1986). The fingers in these proteins have a consensus sequence of $Cys-X_{2,4}-Cys-X_3-Phe-X_5-Leu-X_2-His-X_3-His$, and thus, these motifs are referred to as Cys_2/His_2 fingers. With the exception of TFIIIA, however, finger-containing proteins have not yet been shown by direct biochemical analysis to depend on Zn(II) for binding to DNA. Here we have found that upon incubation of Sp1 with EDTA, DNA binding activity is lost. Then, after dialysis to remove the EDTA, binding activity is restored with Zn(II), but not with Co(II), Ni(II), or metal-free water (as a control). Hence, Sp1 appears to require Zn(II) for sequence-specific binding to DNA.

How do fingers interact with DNA? Klug and co-workers observed that TFIIIA, which has nine fingers, interacts with a 50 bp region in the 5S RNA gene (Miller et al., 1985), and they subsequently identified a $5^1/_2$ bp periodicity of guanine residues in the TFIIIA binding site (Rhodes and Klug, 1986). Based upon these findings, it was postulated that each finger motif binds to half a turn of DNA and that this $5^1/_2$ bp periodicity in the binding site correlates with the interaction of contiguous finger structures to DNA. Furthermore, Rhodes and Klug (1986) observed that Sp1 recognition sites, which are 10 bp in length, have a strong 5 bp sequence periodicity and suggested, on that basis, that Sp1 binds to DNA with fingers similar to those in TFIIIA. In this paper, we have shown that Sp1 has three finger motifs rather than the two finger structures that would be consistent with the Rhodes and Klug model. Assuming that all three fingers of Sp1 are essential for sequence-specific binding of the protein to DNA, Sp1 provides an example in which the number of finger motifs does not correlate with the number of $5^1/_2$ bp periodicities in the recognition site. Recent studies on the yeast tran-

scription factor ADR1, which has two Cys_2/His_2 fingers, also examine the binding of a finger-containing protein to DNA. In this work, Young and co-workers have found that ADR1 synthesized in E. coli binds to a 22 bp inverted repeat element (T. Young, personal communication); however, this binding site does not appear to possess, in half of the palindrome, a $5^1/_2$ bp sequence periodicity. Hence, although it appears that Zn(II) is an essential component in the binding of finger-containing proteins to DNA, the mode by which fingers interact with DNA will probably require direct structural analyses, such as X-ray crystallography and two-dimensional NMR studies.

A second type of finger motif, which we refer to as Cys_2/Cys_2 fingers, has been found in factors that either have been shown to be sequence-specific DNA binding proteins or are believed to be DNA binding proteins. These proteins include the glucocorticoid receptors from human (Hollenberg et al., 1985), rat (Miesfeld et al., 1986), and mouse (Danielsen et al., 1986), the estrogen receptors from human (Green et al., 1986; Greene et al., 1986) and chicken (Krust et al., 1986), the chicken progesterone receptor (Jeltsch et al., 1986; Conneely et al., 1986), the c-*erb*-A protein from human (Weinberger et al., 1986) and chicken (Sap et al., 1986), the chicken vitamin D receptor (McDonnell et al., 1987), the human mineralocorticoid receptor (Arriza et al., 1987), and the proteins encoded by the yeast genes *GAL4* (Laughon and Gesteland, 1984), *PPR1* (Kammerer et al., 1984), and *ARGRII* (Messenguy et al., 1986). The Cys_2/Cys_2 fingers have the form of $Cys-X_2-Cys-X_{13}-Cys-X_2-Cys$, and both the Cys_2/His_2 and the Cys_2/Cys_2 fingers are highly conserved from yeast to humans.

Further examination of the Sp1 fingers reveals that the second and third fingers (residues 569 to 591 and 599 to 619) are similar in the region between the 2 Cys and 2 His residues that delimit each finger. Seven out of 12 amino acid residues are identical, and it is possible that these two fingers interact with similar DNA sequences that each constitute a portion of an Sp1 binding site. In addition, there is a conserved T G E K/R P F/Y X sequence between the Sp1 finger motifs that is also present between contiguous fingers in TFIIIA, *Kruppel*, and ADR1. In contrast, the Cys_2/His_2 fingers in *hunchback* and *serendipity* are not separated by a similar sequence. The conservation of this 7 amino acid stretch suggests that it is important for the interaction of some finger-containing proteins with DNA but that it is also probably not essential for the DNA binding of all finger motifs.

Sp1 and TFIIIA are transcription factors that contain Cys_2/His_2 fingers. Given these examples, it seems likely that other promoter and enhancer binding proteins will have similar finger motifs. We have recently isolated the genes encoding transcription factors CTF/NF-I (C. Santoro et al., unpublished data; for CTF/NF-I references see Jones et al., 1985, 1987) and *jun*/AP-1 (Bohmann et al., unpublished data; for AP-1 references see Lee et al., 1987a, 1987b), but neither CTF/NF-I nor *jun*/AP-1 appears to have finger motifs. On the other hand, genes from Drosophila and mouse have been isolated on the basis of their homology to the Cys_2/His_2 finger motifs in *Kruppel* (Schuh et al., 1986; Chowdhury et al., 1987). These and other finger-containing genes may be transcription factors, but characterization of the proteins encoded by these genes will be necessary to demonstrate their function.

Localization of the DNA Binding Activity of Sp1

To localize the region of Sp1 that is sufficient for DNA binding activity, we have expressed fragments of Sp1 in E. coli and examined the binding of the resulting proteins to DNA. We have found that a LacZ-Sp1 hybrid protein that contains only the 168 C-terminal amino acid residues of Sp1 (LacZ-Sp1-168C) binds to DNA with the same sequence specificity as Spa purified from HeLa cells. Hence, these experiments localize the region of Spa that is sufficient for sequence-specific binding to DNA to the C-terminal 168 residues of the protein, which contain the 71 amino acid residues of the three Zn(II) finger motifs (Figure 7A). Preliminary in vitro transcription experiments with Sp1 synthesized in E. coli indicate that the LacZ-Sp1-168C protein does not stimulate RNA synthesis from Sp1-responsive promoters (A. Courey, J. T. K. and R. T., unpublished data). This finding, if confirmed, would indicate that binding of Sp1 to DNA is not sufficient for activation of transcription. This result is also consistent with the previous observation that a 40 kd proteolytic fragment of Sp1 isolated from human placenta binds to DNA but does not activate transcription (A. Axelrod, J. T. K., and R. T., unpublished data; Briggs et al., 1986). Sp1-mediated transcriptional stimulation is currently under investigation by assaying the factor both in vivo and in vitro (A. Courey, J. T. K., and R. T., unpublished data).

A Bacterial Colony Assay for Sp1 Binding to DNA

Finally, we have constructed a test strain, JK901, in which the sequence-specific DNA binding of Sp1 synthesized in E. coli can be detected with a bacterial colony assay. This strain will be used to screen a large number of mutants and to characterize variants of Sp1 that are defective in binding to DNA. JK901 contains a single copy of a modified *lacZ* gene in the E. coRi chromosome that has two Sp1 recognition sequences at the transcriptional start site, and thus, expression of β-galactosidase in this strain can be repressed by the presence of Sp1. Consequently, JK901 normally has a Lac+ phenotype that becomes Lac- upon expression of Sp1. In the presence of the β-galactosidase indicator X-Gal, JK901 colonies containing a plasmid that expresses Sp1 are white, whereas JK901 colonies harboring a plasmid that contains the Sp1 cDNA but does not express the protein are blue. Although this system of in vivo transcriptional repression works well for Sp1, similar studies carried out with SV40 large T antigen did not detect T antigen-mediated inhibition of *lacZ* expression (P. Kaufman, J. T. K., and R. T., unpublished data). Furthermore, β-galactosidase expression in strain JK605, which is identical to JK901 except that the asymmetric Sp1 binding sites are in the opposite orientation, is not inhibited in the presence of Sp1 (unpublished data). The latter result might be due to preferential binding of Sp1 to the G-rich strand of its recognition site (Gidoni et al., 1984). However, it appears to be difficult to predict whether or not this ap-proach for detecting DNA binding in vivo will be successful for other proteins, and as a consequence, we believe that this method may prove to be useful for characterization of cloned genes, but not as a general method for screening libraries with the aim of isolating genes that encode sequence-specific DNA binding proteins.

In this work, we have isolated a cDNA encoding Sp1 and characterized different aspects of the binding of the protein to DNA. The Sp1 clone can now be used to perform new experiments that will lead to a better understanding of the biological role of Sp1 in the cell as well as the molecular mechanisms by which the factor activates RNA polymerase II transcription. These studies will include systematic mutagenesis of the gene to map functional regions of the protein, high level expression of Sp1 in both E. coli and eukaryotic cells, raising of antibodies against Sp1, and examination of the spatial and temporal regulation of Sp1 expression. Sp1 is a promoter-selective RNA polymerase II transcription factor that appears to be important for gene regulation in mammalian cells, and thus, we hope that a detailed study of Sp1 will reveal a general theme for gene-specific modulation of transcription in higher eukaryotes.

Experimental Procedures

Isolation of cDNA Encoding Sp1

Sp1 (300 μ ≈ 3 nmol) was purified from HeLa cells (800 g) to obtain protein for amino acid sequence analysis. In each preparation, nuclear extract was prepared from 60 g of HeLa cells and subjected to Sepha-cryl S-300 gel filtration chromatography, as described by Briggs et al. (1986). The fractions that contained Sp1 DNA binding activity were pooled and subsequently applied to sequence-specific DNA affinity columns for Sp1 (four 1 ml columns) according to the procedure of Kadonaga and Tjian (1986). Three consecutive passes through the affinity resins gave Sp1 (20–25 μg) of greater than 95% purity with an overall yield of about 15%. The protein was treated with 4-vinylpyridine under reducing conditions to alkylate the cysteine residues (Enfield et al., 1980) and then digested exhaustively with trypsin (2% [w/w]; TPCK-treated; Sigma) in the presence of 2 M urea. This mixture was applied to a C-18 reverse phase column (Vydac), and the peptides were eluted with a 1-propanol, trifluoroacetic acid gradient and monitored by fluorescence after derivatization of one-tenth of the sample with fluorescamine (Stein and Moschera, 1981). The purified Sp1 peptides (ranging from 250 pmol to 1 nmol of each) were applied to an Applied Biosystems 470A protein sequencer, and amino acids were identified with an Applied Biosystems 120A PTH-amino acid analyzer to give the sequences shown in Figure 1A. A HeLa cell cDNA library in λgt10 (courtesy of Dr. C. Hauser) was then screened with the oligonucleotides 5'-GGCCTICAGGGCTCIGGCGCCCTIAACATCCAGCAGAAC-CAGACC-3' (Sp1 probe #1; I represents inosine) and 5'-AA(T/C)AT (T/C/A)CA(A/G)CA(A/G)AA(T/C)CA(A/G)AC-3' (Sp1 probe #2; degenerate positions are indicated within parentheses), which are based on the coding sequence of Sp1 peptide #1 (Figure 1A). Sp1 probe #1 was designed and used according to the recommendations of Lathe (1985) and Martin et al. (1985), and the final conditions for washing filters hybridized with Sp1 probe #1 were 59°C in 2× SST. The degenerate Sp1 probe #2 was used as recommended by Wood et al. (1985), and the final wash conditions were 57°C in 3 M tetramethylammonium chloride. The cDNA insert in isolate Sp1-1 was sequenced by the chain termination method of Sanger et al . (1977) after shotgun subcloning of the DNA into a bacteriophage M13 vector, as described by Bankier and Barrell (1983). However, because of nonrandom subcloning of Sp1-1 DNA fragments into M13, there is a region of approximately 500 bp between the HindIII site and the 3′ end of the cDNA that has not yet been sequenced.

Metal Replacement Studies

All equipment (pipet tips, plastic tubes, stir bars, dialysis tubing, beakers, etc.; glassware was not used) was soaked in 10 mM EDTA for 24 hr and then rinsed with metal-free water, which was prepared by treatment of glass-distilled water With Chelex 100 resin (Bio-Rad). Puratronic grade metal sulfate salts (Johnson Matthey Chemicals) were used in the metal reconstitution experiments. Buffer Z/0.6 M KCl is 25 mM HEPES (K+), pH 75, containing 0.6 M KCl, 12.5 mM MgCl₂, 20% (v/v) glycerol, 0.1% (v/v) NP40, and 1 mM DTT. Buffer zᵉ is 25 mM HEPES (K+), pH 7.5, containing 100 mM KCl, 20% (v/v) glycerol, and 0.01°/0 (v/v) NP40. All operations were carried out at 4°C.

Sp1 (240 µl; approximately 20 µg/ml protein of 90% purity in buffer Z/0.6 M KCl) was combined with 250 mM EDTA, pH 7, to a final EDTA concentration of 50 mM. This mixture was incubated for 45 min to give the "Sp1 + EDTA" fraction and then dialyzed twice (1 hr each) against 500 ml of buffer zᵉ containing 0.4 mM EDTA. The resulting protein (20 µl) was combined with solutions of metal sulfate salts (ZnSO₄,CoSO₄,NiSO₄, or water only as a control) in water to a final volume of 40 µl and concentrations of 0.5 mM metal sulfate salts and 0.2 mM EDTA. This mixture was incubated for 30 min to allow reconstitution of Sp1. These fractions were named "apoSp1 + M(II)," where M(II) is Zn(II), Co(II), Ni(II), or metal-free water. In the experiments involving treatment of Sp1 with 1,10-phenanthroline, Sp1 (48 µl; approximately 20 µg/ml protein of 90% purity in buffer Z/0.6 M KCl) was combined with 25 mM 1,10-phenanthroline in buffer Zᵉ (2 µl) to a final concentration of 1 mM 1,10-phenanthroline. This mixture was incubated for 30 min and named "Sp1 + o-phen." DNAase I footprinting of Sp1 on the SV40 21 bp repeat elements was carried out essentially as described by Dynan and Tjian (1983b), except that buffer Zᵉ was used instead of TM buffer Each footprint reaction contained about 0.03 pmol of ³²P-labeled DNA probe and approximately 15 ng of Sp1.

Expression of Sp1 in E. coli

The plasmids pSp1-516C and pLac.7-Sp1-168C were constructed by standard recombinant DNA techniques, such as those described by Maniatis et al. (1982). E. coli strain JK50 (hfl/A::Tn5 ΔlacX74 ara str ⌐m +/F' lacIᑫLₐZ::Tn10 [transposition deficient] proA proB) was used as the host for expression of Sp1. Crude cell lysates of JK50 containing pSp1-516C, pLacZ-Sp1-168C, or pUC118 were prepared by modification of the method of Desplan et al. (1985). Except when stated otherwise, all operations were performed at 4°C.

JK50 (500 ml in LB containing 100 µg/ml ampicillin; inoculated with 10 ml of an overnight culture) was grown at 37°C to A₅₅₀ₙₘ ≈ 0.1, and a solution of 100 mM isopropyl β-D-thiogalactopyranoside was added to 1 mM final concentration. The cells were grown at 37°C for approximately 2 hr until A₅₅₀ₙₘ ≈ 0.4. The cells were pelleted at 6000 × g for 10 min. washed in 20 ml of LB, repelleted at 6000 × g for 10 min, and suspended in buffer A (1.3 ml; buffer A is 40 mM Tris–HCl buffer, pH 7.7, containing 25% [w/v] sucrose, 0.2 mM EDTA, 1 mM DTT, 1 mM PMSF, and 1 mM sodium metabisulfite). Buffer A containing 1 mg/ml lysozyme (0.5 ml) was added, and the mixture was incubated for 1 hr. Then a two-thirds volume of 10 M urea was added to give a final concentration of 4 M urea. The mixture was incubated for 1 hr and subsequently centrifuged at 63,000 × g for 1 hr. The supernatant was dialyzed against buffer B containing 1 M urea (200 ml; buffer B is 20 mM Tris–HCl, pH 7.7, containing 50 mM KCl, 10 mM MgCl₂, 1 mM EDTA, 10 µM ZnSO₄, 20% [v/v] glycerol, 1 mM DTT, 0.2 mM PMSF, and 1 mM sodium metabisulfite) for 90 min. The mixture was then dialyzed against buffer B twice (500 ml each) for 2 hr and overnight, and the protein was stored at −70°C. This procedure typically gives 3 ml of a slightly cloudy mixture that contains 5 to 10 mg/ml protein, as determined by the method of Bradford (1976) by using bovine gamma globulin as the standard. These extracts can be used in DNAase I footprinting experiments (5 µl of extract will completely protect 0.03 pmol of probe DNA), or alternatively, the mixture can be directly applied to sequence-specific DNA affinity resins (Kadonaga and Tjian, 1986) to purify the E. coli-synthesized Sp1.

Construction of E. colt Strains JK901 and JK31

Recombinant DNA techniques were carried out essentially as described by Maniatis et al. (1982); growth and manipulation of E. coli and bacteriophage λ were performed as recommended by Simons et al. (1987) and Miller (1972). The plasmid pPY140 was constructed by Benson et al. (1986), and λ RS88 and pRS415 are described by Simons et al. (1987). The complementary oligonucleotides 5′-TCGATGGGC-GGAGTTAGGGGCGGGACTA-3′ and 5′-TCGATAGTCCCGCCCCTA-ACTCCGCCCA-3′, which represent GC boxes V and VI in the SV40 21 bp repeat elements (Gidoni et al., 1985), were annealed and inserted into the unique XhoI site in pPY140. The 371 bp EcoRI fragment that encompasses the inserted sequences was subcloned into pRS415 to give pJK901. In addition, the 343 bp fragment that encompasses the XhoI site in pPY140 was subcloned into pRS415 to give pJK31. Both pJK901 and pJK31 are shown in Figure 6A. The appropriate fragments of pJK901 and pJK31 were crossed into λ RS88 by genetic recombination in vivo to give λ JK901 and λ JK31, which were lysogenized into E. coli strain NK5442 (Δlac) to yield JK901 and JK31. The levels of β-galactosidase in the Sp1 test strains were monitored with the indicator 5-bromo-4-chloro-3-indolyl β-D-galactopyranoside (X-Gal).

Acknowledgments

We thank Al Courey, Craig Hauser, Mike Briggs, Joe Ladika, Amy Axelrod, and Jim Merryweather for their contributions to this work; Phil Youderian and Bob Simons for their generous gifts of phage and bacteria; and Russ Doolittle, Ted Young, Judy Lengyel, and Claude Desplan for communication of unpublished data. We are indebted to Mark Biggin, Pam Mitchell, Kathy Jones, and other members of the Tjian lab for their invaluable suggestions in the course of this work. We also thank Karen Ronan for assistance in the preparation of the manuscript. J. T. K. is a Lucille P. Markey Scholar in Biomedical Science. This work was supported in pan by a research grant from the National Cancer Institute to R. T.

The costs of publication of this article were defrayed in part by the payment of page charges. This article must therefore be hereby marked "*advertisement*" in accordance with 18 U.S.C. Section 1734 solely to indicate this fact.

Received October 16, 1987.

References

Arriza, J. L., Weinberger, C., Cerelli, G., Glaser, T. M., Handelin, B. L., Housman, D. E., and Evans, R. M. (1987). Cloning of human mineralocorticoid receptor complementary DNA: structural and functional kinship with the glucocorticoid receptor. Science *237*, 268–275.

Baldarelli, R. M., Mahoney, P. A., Salas, F,. Gustavson, E., Boyer, P. D., Chang, M.-F, Roark, M., and Lengyel, J. A. (1988). Transcripts of the *Drosophila* blastoderm-specific locus, *terminus*, are concentrated posteriorly and encode a potential DNA-binding finger. Dev. Biol., in press.

Bankier, A. T. and Barrels B. G. (1983). Shotgun DNA sequencing. In Techniques in the Life Sciences, Techniques in Nucleic Acid Biochemistry, B508. Vol. B5, R. A. Flavell, ed. (Ireland: Elsevier Scientific Publishers), pp. 1–34.

Benson, N., Sugiono, P., Bass, S., Mendelman, L. V., and Youderian, P. (1986). General selection for specific DNA-binding activities. Genetics *114*, 1–14.

Berg, J. (1986). Potential metal-binding domains in nucleic acid binding proteins. Science *232*, 485–487

Bradford, M. M. (1976). A rapid and sensitive method for the quantitation of microgram quantities of protein utilizing the principle of protein-dye binding. Anal. Biochem. *72*, 248–254.

Briggs, M. R., Kadonaga, J. T., Bell, S. R. and Tjian, R. (1986). Purification and biochemical characterization of the promoter-specific transcription factor Sp1. Science *234*, 47–52.

Brown, R. S., Sander, C., and Argos, P. (1985). The primary structure of transcription factor TFIIIA has 12 consecutive repeats. FEBS Lett. *186*, 271–274.

Chowdhury, K., Deutsch, U., and Gruss, P. (1987). A multigene family encoding several "finger" structures is present and differentially active in mammalian genomes. Cell *48*, 771–778.

Conneely, O. M., Sullivan, W. P., Toft, D. O., Birnbaumer, M., Cook, R. G., Maxwell, B. L., Zarucki-Schulz, T. Greene, G. L., Schrader, W. T., and O'Malley, B. W. (1986). Molecular cloning of the chicken progesterone receptor. Science *233*, 767–770.

Danielsen, M., Northrop, J. P., and Ringold, G. M. (1986). The mouse glucocorticoid receptor: mapping of functional domains by cloning, sequencing and expression of wild-type and mutant receptor proteins. EMBO J. *5*, 2513–2522.

Desplan, C., Theis, J., and O'Farrell, P. H. (1985). The *Drosophila* developmental gene, *engrailed*, encodes a sequence-specific DNA binding activity. Nature *318*, 630–635.

Diakun, G. P., Fairall, L., and Klug, A. (1986). EXAFS study of the zinc-binding sites in the protein transcription factor IIIA. Nature *324*, 698–699.

Dingwall, C., and Laskey, R. A. (1986). Protein import into the cell nucleus. Ann. Rev. Cell Biol. *2*, 367–390.

Dynan, W. S., and Tjian, R. (1983a). Isolation of transcription factors that discriminate between different promoters recognized by RNA polymerase II. Cell *32*, 669–680.

Dynan, W. S., and Tjian, R. (1983b). The promoter-specific transcription factor Sp1 binds to upstream sequences in the SV40 early promoter Cell *35*, 79–87.

Dynan, W. S., and Tjian, R. (t985). Control of eukaryotic messenger RNA synthesis by sequence-specific DNA-binding proteins. Nature *316*, 774–778.

Enfield, D. L., Ericsson, L. H., Fujikawa, K., Walsh, K. A., Neurath, H., and Titani, K. (1980). Amino acid sequence of the light chain of bovine factor X$_1$ (Stuart factor). Biochemistry *19*, 659–667.

Gidoni, D., Dynan, W. S., and Tjian, R. (1984). Multiple specific contacts between a mammalian transcription factor and its cognate promoters. Nature *312*, 409–413.

Gidoni, D., Kadonaga, J. T., Barrera-Saldana, H., Takahashi, K., Chambon, P. and Tjian, R. (1985). Bidirectional SV40 transcription mediated by tandem Sp1 binding interactions. Science *230*, 511–517.

Green, S., Walter, P., Kumar, V., Krust, A., Bornert, J.-M., Argos, P., and Chambon, P. (1986). Human oestrogen receptor cDNA: sequence, expression and homology to v-*erb*A. Nature *320*, 134–139.

Greene, G. L., Gilna, P., Waterfield, M., Baker, A., Hort, Y., and Shine, J. (1986). Sequence and expression of human estrogen receptor complementary DNA. Science *231*, 1150–1154.

Hanas, J. S., Hazuda, D. J., Bogenhagen, D. F., Wu, F. Y.-H., and Wu, C.-W. (1983). *Xenopus* transcription factor A requires zinc for binding to the 5S RNA gene. J. Biol. Chem. *258*, 14120–14125.

Hartshorne, T. A., Blumberg, H., and Young, E. T. (1986). Sequence homology of the yeast regulatory protein ADR1 with *Xenopus* transcription factor TFIIIA. Nature *320*, 283–287.

Hollenberg, S. M., Weinberger, C., Ong, E. S., Cerelli, G., Oro, A., Lebo, R., Thompson, E. B., Rosenfeld, M. G., and Evans, R. M. (1985). Primary structure and expression of a functional human glucocorticoid receptor cDNA. Nature *318*, 635–641.

Ishii, S., Kadonaga, J. T., Tjian, R., Brady, J. N., Merlino, G. T., and Pastan, I. (1986). Binding of the Sp1 transcription factor by the human Harvey *ras1* proto-oncogene promoter. Science *232*, 1410–1413.

Jeltsch, J. M., Krozowski, Z., Quirin-Stricker, C., Gronemeyer, H., Simpson, R. J., Garnier, J. M., Krust, A., Jacob, F., and Chambon, P. (1986). Cloning of the chicken progesterone receptor. Proc. Natl. Acad. Sci. USA *83*, 5424–5428.

Jones, K. A., Yamamoto, K. R., and Tjian, R. (1985). Two distinct transcription factors bind to the HSV thymidine kinase promoter in vitro. Cell *42*, 559–572.

Jones, K. A., Kadonaga, J. T., Rosenfeld, P. J., Kelly, T J., and Tjian, R. (1987). A cellular DNA-binding protein that activates eukaryotic transcription and DNA replication. Cell *48*, 79–89.

Kadonaga, J. T., and Tjian, R. (1986). Affinity purification of sequence-specific DNA binding proteins. Proc. Natl. Acad. Sci. USA *83*, 5889–5893.

Kadonaga, J. T., Jones, K. A., and Tjian, R. (1986). Promoter-specific activation of RNA polymerase II transcription by Sp1. Trends Biochem. *11*, 20–23.

Kammerer, B., Guyonvarch, A., and Hubert, J. C. (1984). Yeast regulatory gene *PPR1*: I . Nucleotide sequence, restriction map and codon usage. J. Mol. Biol. *180*, 239–250.

Krust, A., Green, S., Argos, P., Kumar, V., Walter, P., Bornert, J.-M., and Chambon, P. (1986). The chicken oestrogen receptor sequence: homology with v-*erb-A* and the human oestrogen and glucocorticoid receptors. EMBO J. *5*, 891–897.

Lathe, R. (1985). Synthetic oligonucleotide probes deduced from amino acid sequence data: theoretical and practical considerations. J. Mol. Biol. 183, 1–12.

Laughon, A., and Gesteland, R. R (1984). Primary structure of the *Saccharomyces cerevisiae GAL4* gene. Mol. Cell. Biol. *4*, 260–267.

Lee, W., Haslinger, A., Karin, M., and Tjian, R. (1987a). Activation of transcription by two factors that bind promoter and enhancer sequences of the human metallothionein gene and SV40. Nature *325*, 368–372.

Lee, W., Mitchell, P. and Tjian, R. (1987b). Purified transcription factor AP-1 interacts with TPA-inducible enhancer elements. Cell *49*, 741–752.

Maniatis, T., Fritsch, E. F., and Sambrook, J. (1982). Molecular Cloning: A Laboratory Manual (Cold Spring Harbor, New York: Cold Spring Harbor Laboratory).

Maniatis, T., Goodbourn, S., and Fischer, J. A. (1987). Regulation of inducible and tissue-specific gene expression. Science *236*, 1237–1245.

Marshall, R. D. (1972). Glycoproteins. Ann. Rev. Biochem. 41, 673-702.

Martin, F. H., Castro, M. M., Aboul-ela, F., and Tinoco, I., Jr. (1985). Base pairing involving deoxyinosine: implications for probe design. Nucl. Acids Res. *13*, 8927–8938.

McDonnell, D. P., Mangelsdorf, D. J., Pike, J. W., Haussler, M. R., and O'Malley, B. W. (1987). Molecular cloning of complementary DNA encoding the avian receptor for vitamin D. Science *235*, 1214–1217.

McKnight, S., and Tjian, R. (1986). Transcriptional selectivity of viral genes in mammalian cells. Cell *46*, 795–805.

Messenguy, F., Dubois, E., and Descamps, F. (1986). Nucleotide sequence of the *ARGRII* regulatory gene and amino acid sequence homologies between *ARGRII, PPR1* and *GAL4* regulatory proteins. Eur. J. Biochem. *157*, 77–81.

Miesfeld, R., Rusconi, S., Godowski, P. J., Maler, B. A., Okret, S., Wikström, A.-C., Gustafsson, J.-Å., and Yamamoto, K. R. (1986). Genetic complementation of a glucocorticoid receptor deficiency by expression of cloned receptor cDNA. Cell *46*, 389–399.

Miller, J. H. (1972). Experiments in Molecular Genetics (Cold Spring Harbor, New York: Cold Spring Harbor Laboratory).

Miller, J., McLachlan, A. D., and Klug, A. (1985). Repetitive zinc-binding domains in the protein transcription factor IIIA from *Xenopus* oocytes. EMBO J. *4*, 1609–1614.

Rhodes, D., and Klug, A. (1986). An underlying repeat in some transcriptional control sequences corresponding to half a double helical turn of DNA. Cell *46*, 123–132.

Rosenberg, U. B., Schroder, C., Preiss, A., Kienlin, A., Cote, S., Riede, I., and Jackle, H. (1986). Structural homology of the product of the *Drosophila Kruppel* gene with *Xenopus* transcription factor IIIA. Nature *319*, 336–339.

Ruther, U., and Muller-Hill, B. (1983). Easy identification of cDNA clones. EMBO J. *2*, 1791–1794.

Sanger, F., Nicklen, S., and Coulson, A. R. (1977). DNA sequencing with chain-terminating inhibitors. Proc. Natl. Acad. Sci. USA *74*, 5463–5467.

Sap, J., Munoz, A., Damm, K., Goldberg, Y., Ghysdael, J., Leutz, A., Beug, H., and Vennstrom, B. (1986). The c-*erb-A* protein is a high-affinity receptor for thyroid hormone. Nature *324*, 635–640.

Schuh, R., Aicher, W., Gaul, U., Coté, S., Preiss, A., Maier, D., Seifert, E., Nauber, U., Schröder, C., Kemler, R., and Jäckle, H. (1986). A conserved family of nuclear proteins containing structural elements of the finger protein encoded by *Krüppel*, a Drosophila segmentation gene. Cell *47*, 1025–1032.

Serfling, E., Jasin, M., and Schaffner, W. (1985). Enhancers and eukaryotic gene transcription. Trends Genet. *1*, 224–230.

Simons, R. W., Houman, F., and Kleckner, N. (1987). Improved single and multicopy *lac*-based cloning vectors for protein and operon fusions. Gene *53*, 85–96.

Stein, S., and Moschera, J. (1981). High-performance liquid chromatography and picomole-level detection of peptides and proteins. Meth. Enzymol. *79*, 7–16.

Tautz, D., Lehmann, R., Schnurch, H., Schuh, R., Seifert, E., Kienlin, A., Jones, K., and Jackle, H. (1987). Finger protein of novel structure encoded by *hunchback*, a second member of the gap class of Drosophila segmentation genes. Nature *327*, 383–389.

Vincent, A., Colot, H. V, and Rosbash, M. (1985). Sequence and structure of the *serendipity* locus of *Drosophila melanogaster:* a densely transcribed region including a blastoderm-specific gene. J. Mol. Biol. *186*, 149–166.

Weinberger, C., Thompson, C. C., Ong, E. S., Lebo, R., Gruol, D. J., and Evans, R. M. (1986). The c-*erb-A* gene encodes a thyroid hormone receptor. Nature. *324*, 641–646.

Wharton, K. A., Yedvobnick, B., Finnerty, V. G., and Artavanis-Tsakonas, S. (1985). *opa*: a novel fannily of transcribed repeats shared by the *Notch* locus and other developmentally regulated loci in D. melanogaster. Cell *40*, 55–62.

Wood, W. I., Gitschier, J., Lasky, L. A., and Lawn, R. M. (1985). Base composition-independent hybridization in tetramethylammonium chloride: a method for oligonucleotide screening of highly complex gene libraries. Proc. Natl. Acad. Sci. USA *82*, 1585–1588.

Note Added in Proof

The article by Bohmann et al. (cited as unpublished data) is currently in press (Science, 1987).

DNA binding domains of transcription factors

9

As described in previous sections, it has been possible to identify different DNA sequences that are required for specific patterns of gene expression (Sections 6 and 7), while specific transcription factors binding to these sequences have been identified and purified (Sections 6 and 8). Such characterization and purification of transcription factors that bind to specific DNA sequences raises the question of the mechanism by which these factors bind to DNA.

One of the earliest identifications of such a DNA binding domain was achieved by Miller et al. in the paper presented here. They took advantage of the very high abundance of the TFIIIA transcription factor in *Xenopus* oocytes. Thus, as described in Section 8, TFIIIA plays a critical role in the transcription of the 5 S rRNA genes. Most importantly, it remains associated with the 5 S rRNA molecule after transcription, and immature *Xenopus* oocytes contain very large numbers of TFIIIA molecules, each of which is associated with 5 S rRNA to form a 7 S ribonucleoprotein particle. Miller et al. took advantage of this and purified the 7 S particle under conditions where it remained intact. As described in the paper, this led to the most important finding that each 7 S particle contained between seven and eleven molecules of zinc per particle. The presence of zinc was necessary for the integrity of the particle, which dissociated when exposed to treatments that removed the zinc.

Further analysis of the 7 S particle, illustrated in Fig. 1 of the paper, involved its digestion with various proteases. This study revealed that the particle was progressively digested into a series of periodically spaced intermediates which were then themselves digested. This suggested that the TFIIIA protein was composed of a series of similar, small, protein domains of size approx. 3 kDa. This immediately suggested that the presence of multiple zinc ions in the 7 S particle could be due to each of these domains binding a single molecule of zinc.

This possibility was confirmed when Miller et al. examined the amino acid sequence of TFIIIA, which had been deduced from a cDNA clone (Ginsberg et al., 1984). This revealed that TFIIIA was composed of nine repeated units, each of approx. 30 amino acid residues. As illustrated in Fig. 4 of the paper, each of these repeated units contained, at the same positions, two cysteine amino acids and two histidine amino acids. These cysteine and histidine amino acids would be capable of co-ordinating an atom of zinc. Thus the presence of nine Cys_2-His_2 repeated units, each capable of binding a zinc atom, accounted for the original estimate that the TFIIIA molecule within the 7 S ribonucleoprotein particle contained between seven and eleven zinc atoms.

Most importantly, as illustrated in Fig. 4 of the paper by Miller et al., it was possible to produce a structure for the repeated domain in which a loop of 12 amino acids, containing conserved leucine and phenylalanine residues as well as several basic residues that could potentially interact with the acidic DNA, projects from the surface of protein. This loop is anchored at its base by the conserved cysteine and histidine residues which co-ordinate an

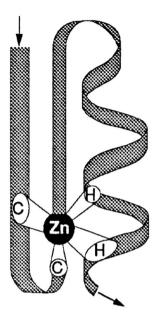

Figure 9.1 Alternative structure for the zinc finger, which is supported by structural analysis of this motif. Compare with the original model illustrated in Figure 8.4, and also in Fig. 4 of the paper by Miller et al.

atom of zinc. This structure was therefore referred to as a zinc finger (for reviews, see Klug and Rhodes, 1987; Rhodes and Klug, 1993; Klug and Schwabe, 1995). The proposed interaction of zinc with the conserved cysteine and histidine residues in this structure was subsequently confirmed by X-ray absorption spectroscopy of the purified TFIIIA protein (Diakun et al., 1986).

Indeed, the idea originally proposed by Miller et al. that the co-ordination of zinc via cysteine and histidine serves as a scaffold for the intervening region which makes direct contact with the DNA is now generally accepted. However, it is unlikely that the intervening amino acids form a simple loop structure as proposed in the original model. An alternative view of the structure of the zinc finger has been proposed based on the structures of other metallo-proteins (Figure 9.1; see Berg, 1988), and this structure is supported by NMR spectroscopy of the zinc fingers in the yeast SWI5 transcription factor (Neuhaus et al., 1990). In this structure the finger region forms a motif consisting of two anti-parallel β-sheets with an adjacent α-helix packed against one face of the β-sheet. Upon contact with the DNA, the α-helix lies in the major groove of the DNA and makes sequence-specific contacts with the DNA bases, while the β-sheets lie further away from the helical axis of the DNA and contact the DNA backbone.

Although the detailed structure of the zinc finger originally proposed by Miller et al. has been modified in the light of subsequent studies, the importance of this paper is that it identified for the first time the existence of the zinc-finger motif. This motif represents one of the earliest DNA binding domains to be specifically identified in a eukaryotic transcription factor and, as discussed in Section 8, it is present in many RNA polymerase II transcription factors, including Sp1 (for reviews, see Evans and Hollenberg, 1988; Struhl, 1989). Indeed, the link between the presence of this motif and the ability to bind to DNA and regulate gene expression is now so strong that it has been used as a probe to isolate genes encoding new regulatory proteins. For example, the zinc finger derived from the *Drosphila* kruppel transcription factor has been used to isolate the gene encoding Xfin, a protein containing 37 zinc fingers which is expressed in the early *Xenopus* embryo (Ruiz-I-Altaba et al., 1987).

The initial identification of the zinc finger was thus based on studies of purified proteins and on inspection of the amino acid sequence predicted from cDNA clones. Subsequently, as discussed above, this structure was refined on the basis of spectrographic analysis of purified proteins. This approach was also followed for a number of other DNA binding domains with characteristics different from those of the zinc finger. This occurred, for example, in the case of a number of *Drosophila* genes whose mutation results in altered development of the fly and which are referred to as homoeotic genes (for reviews, see Scott and Carroll, 1987; Ingham, 1988; see also Section 13). When the genes encoding these proteins were cloned, it was found that they each contained a short related DNA sequence of about 180 bp, capable of encoding 60 amino acids (Figure 9.2), which was flanked on either side by dramatically different sequences. This common element was named the homoeobox or homoeodomain (for reviews, see Gehring, 1987; Scott et al., 1989; Kornberg, 1993), and this region was shown to constitute the DNA binding domain of the protein, as described in the paper by Desplan et al. discussed in Section 13.

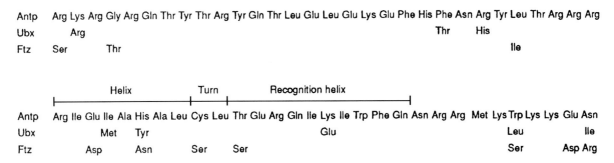

Antp	Arg	Lys	Arg	Gly	Arg	Gln	Thr	Tyr	Thr	Arg	Tyr	Gln	Thr	Leu	Glu	Leu	Glu	Lys	Glu	Phe	His	Phe	Asn	Arg	Tyr	Leu	Thr	Arg	Arg	Arg
Ubx		Arg																			Thr		His							
Ftz	Ser		Thr																				Ile							

	Helix			Turn		Recognition helix		

Antp	Arg	Ile	Glu	Ile	Ala	His	Ala	Leu	Cys	Leu	Thr	Glu	Arg	Gln	Ile	Lys	Ile	Trp	Phe	Gln	Asn	Arg	Arg	Met	Lys	Trp	Lys	Lys	Glu	Asn
Ubx			Met		Tyr										Glu										Leu				Ile	
Ftz			Asp		Asn		Ser		Ser																Ser			Asp	Arg	

Figure 9.2 Comparison of several *Drosophila* homoeodomains showing the conserved helical motifs. Differences between the sequences are indicated; a blank denotes identity in the sequence.

Analysis of the sequence of the homoeodomain revealed that it contained a so-called helix–turn–helix motif, which is highly conserved between the different homoeodomains (Figure 9.3; for a review, see Affolter et al., 1990). The existence of such a helix–turn–helix motif was confirmed by NMR spectroscopy of crystals prepared from the purified homoeodomain of the *Drosophila* factor Antennapedia (Quian et al., 1989). This study revealed that the Antennapedia homoeodomain does indeed contain a helix–turn–helix motif which is, however, preceded by another helical region. The helix–turn–helix motif thus constitutes helices II and III of the Antennapedia homoeodomain.

The existence of this helix–turn–helix motif in the crystal structure of the eukaryotic homoeodomain proteins was of particular interest, because such a helix–turn–helix structure had also been shown to act as a DNA binding motif in several prokaryotic regulatory proteins derived from bacteriophages, such as the λ cro protein and the phage 434 repressor. These prokaryotic proteins had been crystallized together with their corresponding DNA target site. X-ray crystallographic studies of such DNA–protein crystals had shown that the helix–turn–helix motif does indeed contact DNA, as predicted. One of the two helices lies across the major groove of the DNA, while the other lies partly within the major groove, where it can make sequence-specific contacts with the DNA bases. Thus it is this second helix (known as the recognition helix) that controls the sequence-specific DNA binding of these proteins (Figure 9.3).

Given the existence of this helix–turn–helix motif in the eukaryotic homoeodomain proteins and its DNA binding role in the prokaryotic proteins, it was important to determine whether this motif fulfilled a similar role in the eukaryotic proteins. This could only be achieved by extending previous studies involving the crystallization of the purified homoeodomain to allow the production of co-crystals of the homoeodomain bound to its DNA recognition site. This was achieved in the case of the homoeodomain of the engrailed protein in the paper by Kissinger et al. presented here. These authors obtained co-crystals of a 61-amino-acid peptide containing the engrailed homoeodomain bound to its 21 bp DNA recognition site. As indicated in Figure 3 of the paper, they found that helices II and III did indeed constitute a helix–turn–helix motif, with helix III fitting directly into the major groove of the DNA. Hence, as in the prokaryotic proteins, this helix does indeed constitute a recognition helix which can make sequence-specific contacts with the DNA.

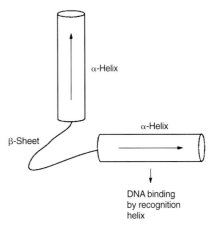

Figure 9.3 The helix–turn–helix motif

Interestingly, however, the crystal structure of the engrailed protein bound to DNA revealed certain differences in the alignment of the recognition helix within the major groove compared with that observed with the prokaryotic proteins. In the prokaryotic proteins, residues near the N-terminal end of the recognition helix are critical for sequence-specific contacts with the DNA. In contrast, as illustrated in Figure 7 of the paper by Kissinger et al., the critical base-specific contacts are made by amino acids nearer the C-terminus of the recognition helix within the eukaryotic proteins.

As well as being of importance in identifying differences between the prokaryotic and eukaryotic proteins, the studies of Kissinger et al. are valuable in that they provide a framework for understanding the effects of amino acid mutations within the homoeodomain. Thus alterations in the amino acids at the N-terminal end of the recognition helix had previously been shown to have no effect on the DNA binding specificity (Treisman et al., 1990). This was explained by the crystal structure which showed that, unlike the situation in the prokaryotic proteins, these residues are not in close contact with the DNA. In contrast, however, replacement of the lysine residue found at position 9 in the recognition helix in the bicoid homoeodomain with the glutamine residue found in the Antennapeadia protein resulted in a complete change in DNA binding specificity. Once again this is explained by the crystal structure of the molecule, which shows that the side chain of the glutamine residue present at position 9 in Antennapedia projects directly into the major groove, thereby allowing it to contribute specifically to sequence-specific binding to the bases of the DNA.

The paper by Kissinger et al. is thus of importance, since it represents one of the earliest examples in which the putative structure of a DNA binding domain (inferred from its protein sequence and crystallization of the purified protein) was confirmed by co-crystallization of the purified protein with its DNA recognition site. As indicated above, such an analysis can reveal differences between apparently similar stuctures, such as those found in the prokaryotic and eukaryotic helix–turn–helix proteins, as well as providing an important framework for understanding the effects of specific mutations on DNA binding. This type of study thus represents the ultimate aim which needs to be achieved following the initial identification of a DNA binding domain by the methods described in Section 8.

Repetitive zinc-binding domains in the protein transcription factor IIIA from *Xenopus* oocytes

J. Miller, A.D. McLachlan and A. Klug

MRC Laboratory of Molecular Biology, Hills Road, Cambridge CB2 2QH, UK

Communicated by A. Klug

The 7S particle of *Xenopus laevis* oocytes contains 5S RNA and a 40-K protein which is required for 5S RNA transcription *in vitro*. Proteolytic digestion of the protein in the particle yields periodic intermediates spaced at 3-K intervals and a limit digest containing 3-K fragments. The native particle is shown to contain 7–11 zinc atoms. These data suggest that the protein contains repetitive zinc-binding domains. Analysis of the amino acid sequence reveals nine tandem similar units, each consisting of approximately 30 residues and containing two invariant pairs of cysteines and histidines, the most common ligands for zinc. The linear arrangement of these repeated, independently folding domains, each centred on a zinc ion, comprises the major part of the protein. Such a structure explains how this small protein can bind to the long internal control region of the 5S RNA gene, and stay bound during the passage of an RNA polymerase molecule.
Key words: Xenopus laevis/transcription factor IIIA/zinc-binding domains/7S particle

Introduction

The 5S RNA genes of *Xenopus laevis*, which are transcribed by RNA polymerase III, have been the subject of intensive study in the last decade by the research groups of D.D. Brown and R.G. Roeder. The two types, oocyte and somatic, provide a system for studying the differential regulation of gene expression as well as transcription mechanisms (Brown, 1984). In the course of these studies it has been discovered that the correct initiation of transcription requires the binding of a 40-K protein factor, variously called factor A or transcription factor IIIA (TFIIIA), which has been purified from oocyte extracts (Engelke *et al.*, 1980). By deletion mapping it was found that this factor interacts with a region ~50 nucleotides long within the gene, called the internal control region (Bogenhagen *et al.*, 1980). This initiation complex is stabilised by the sequential binding of two further protein factors, called B and C (Segall *et al.*, 1980; Bieker *et al.*, 1985).

Immature oocytes store 20 000 5S RNA molecules in the form of 7S ribonucleoprotein particles (Picard and Wegnez, 1979), each containing a single protein which has been shown to be identical with transcription factor IIIA (Pelham and Brown, 1980; Honda and Roeder, 1980). TFIIIA therefore binds both 5S RNA and its cognate DNA and it was therefore suggested that it may mediate autoregulation of 5S gene transcription. Whether this autoregulation occurs *in vivo* or not, the dual interaction provides an interesting structural problem which can be approached because of the presence of large quantities of the protein TFIIIA in *Xenopus* oocytes. In this paper we report some results of our preliminary studies on TFIIIA which reveal a remarkable repeating structure within the protein.

Results

Zinc content of 7S particle

Because published methods for 7S purification involve ion exchange and high ionic strength buffers, which in our hands dissociates at least 30% of particle samples, we developed a new method for particle purification which causes no detectable dissociation. We found, in agreement with other workers (Denis and le Maire, 1983), that the 7S complex dissociated at salt concentrations >0.2 M, and that the isolated protein precipitated readily. We also observed that gel filtration of the complex in the presence of 0.1 mM dithiothreitol (DTT) invariably resulted in separate elution of protein and 5S RNA. This dissociation was also seen when the particle incubated with 0.1 mM DTT was electrophoresed on agarose gels, suggesting that the 20 cysteine groups per molecule (Picard and Wegnez, 1979) were somehow involved in particle stability. However, when we found that 25 mM $NaBH_4$ did not disrupt the complex, we suspected, by analogy with the results of Lewis and Laemmli (1982) on metaphase chromosomes, that metal binding might be involved. When the particle was incubated with 1,10-phenanthroline, DTT, EDTA or a number of other chelating agents and run on agarose gels, dissociation was observed which could only be prevented by prior addition of Zn^{2+}, and not by Mg^{2+}, Ca^{2+}, Mn^{2+}, Ni^{2+}, Fe^{2+}, Ca^{2+} or Co^{2+}. Cu^{2+} and Cd^{2+} induced dissociation by themselves, apparently by displacing bound Zn^{2+}.

Analysis of a 30–50% pure preparation of particle by atomic absorption spectroscopy revealed insignificant concentrations of Cd, Cu, Ni, Co or Fe, but a significant concentration of Zn, at a ratio of at least 5 mol Zn per mol. particle.

While these experiments were in progress, Hanas *et al.* (1983) reported the presence of Zn in the 7S RNP particle, and the requirement for Zn for the association of TFIIIA with the internal control region of the 5S gene. They found a Zn/particle ratio of about two in the presence of 5 mM EDTA and three in its absence, although the latter fell to two in samples purified by gel filtration.

We have now repeated the analysis with pure and undissociated particle preparations. To ensure that no contamination of the preparation could occur, all glass and plastic ware was washed in several changes of 10 mM EDTA for a time several times longer than the duration of the entire preparative procedure. The buffer was concentrated and submitted with particle sample for atomic absorption spectrophotometry. 7S particle at 65 μM contained Zn at 460 μM, giving a ratio of 7.0 ± 0.5 mol Zn per mol particle. The original buffer used was at most 60 nM in Zn, and particle was exposed to at most 1.5 litres of buffer during the purification. Since the yield of particle was 2 ml solution at 65 μM, containing 920 nmol of Zn, at most 90/920 = 10% could have been adsorbed at any time following homogenization of frog ovaries.

The buffers used by Hanas *et al.* (1983) contained 0.5 mM or 1 mM DTT, which has a large binding constant for Zn of ~10^{10} (Cornell and Corviro, 1972), and so their value for the Zn content may be an underestimate. Our buffer contained

20 mM MES, a weakly chelating buffer, which nevertheless, because of the large volumes used in the gel filtration and dialysis steps in the preparative procedure, might have reduced the Zn content of the particles, suggesting that the value we have obtained of seven Zn/particle may still be an underestimate. An experiment in which the particle was prepared in the presence of 10 μM Zn and then separated from unbound Zn on a gel filtration column gave a value of 11–12 mol Zn/mol particle. The sequence analysis described below suggests that there may be at least 9 mol Zn/mol particle.

The 7S particle is not unique in its requirement for Zn. The 42S particle of *X. laevis*, which Denis and le Maire (1983) have shown to contain, among other components, a 5S RNA and a 5S RNA-binding protein distinct from TFIIIA, also requires Zn for stability by agarose gel assay, has a very large molar Zn content, and contains no significant amounts of other metals (Miller, unpublished results).

Small domains in TFIIIA protein

Smith *et al.* (1984) have shown that, on treatment with proteolytic enzymes, the 40-K intact protein gives rise to a 30-K breakdown product, which is then converted to a 20-K product. These proteolytic fragments remain bound to the 5S RNA but Smith *et al.* (1984) have purified them and studied their interactions with the 5S gene by DNase I footprinting. From these experiments, Smith *et al.* (1984) concluded that TFIIIA consists of three structural domains which bind to different parts of the internal control region of the 5S gene: a 20-K protein which binds to the 3′ end of the coding strand, an adjacent 10-K domain which extends the binding to the 5′ end of the control region, and a third domain of 10-K which does not bind directly to the DNA but enables the intact protein to enhance transcription, presumably through interaction with RNA polymerase.

We have also carried out proteolysis of the 7S particle using trypsin, chymotrypsin, elastase, and also papain (D. Rhodes, personal communication) to determine whether smaller fragments of the protein might bind the 5S RNA and the 5S RNA gene. Our measurements of the tryptic breakdown products suggest that Smith's 40-K, 30-K and 20-K may be closer to 39-K, 33-K and 23-K (data not shown).

We have also observed that the 23-K fragment may be reduced to a 17-K fragment which remains bound to the 5S RNA. Further, the 17-K fragment, after prolonged proteolysis, can be reduced to a limit digest consisting of a mixture of 6-K, 4-K and 3-K fragments (Figure 1), which ultimately themselves disappear. Chymotrypsin produces higher multiples of these fragments as well. Doublets are seen at ~11 and 9.5-K, 7.5 and 6-K and 4-K and 3-K, a spacing between them of 3–3.5-K (Figure 2c and d). Early time points in tryptic and chymotryptic digestion reveal metastable bands between the production of the major fragments described above (data not shown), which are also approximately periodically spaced.

The finding of periodic intermediates in the course of proteolytic digestion, and the persistence of small fragments even after prolonged digestion, suggests a periodic arrangement of small, compact protein domains of size ~ 3-K. If such repetitive domains existed they might account for the large number of cysteine and histidine residues and the multiple Zn binding.

We therefore investigated the newly published amino acid sequence (Ginsberg *et al.*, 1984) to see if any structural periodicity was manifested in the sequence.

Analysis of sequence of TFIIIA

Sequence repeats. The amino acid sequence of TFIIIA, which

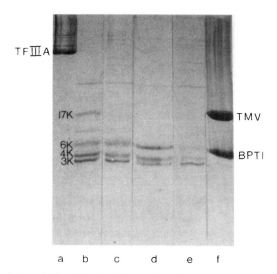

Fig. 1. Trypsin digestions. Particle samples (0.2 mg/ml) in 20 mM MES buffer, pH 6.0, 50 mM KCl were dialyzed against 50 mM Tris-Cl, pH 8.1, 50 mM KCl. 20 μl aliquots were digested with trypsin (20 μg/ml) for varying times and stopped with 2 mM benzamidine. Times were (**a**) 0; (**b**) 17 h; (**c**) 24 h; (**d**) 39 h; (**e**) 64 h. Electrophoresis was as described in Materials and methods.

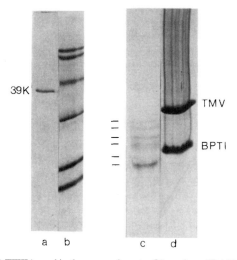

Fig. 2. (**a**) TFIIIA used in these experiments; (**b**) markers: 12.4 K, 17.2 K, 25.7 K, 45 K, 66.3 K, 78 K; (**c**) 17 h chymotrypsin (20 μg/ml) digest of sample prepared as described in legend to Figure 1. Doublet bands are marked (see Results). Acrylamide concentrations for (**a**) and (**b**) were 15%.

has been deduced from a cDNA clone (Ginsberg *et al.*, 1984), contains an unusually large number of Cys and His residues. At first sight these residues appeared to us to form roughly periodic groupings. We therefore made a systematic search for repeats in both the amino acid sequence and the cDNA, using the diagonal comparison matrix method and the damped Needleman and Wunsch method (see Materials and methods).

The protein comparison matrices (with short window lengths of 11, 22 and 30 residues) showed an exceptionally strong and regular pattern of 30-residue repeats in the sequence, with four repetitions evident in the first half of the molecule (residues 13–156) and two more clear repeats in the second half (residues 223–276). Further analysis of the protein by the damped Needleman and Wunsch method showed that the repeats were even more extensive, being partly obscured by a few gaps in the middle of the sequence. The final best alignment (Figure 3) shows that

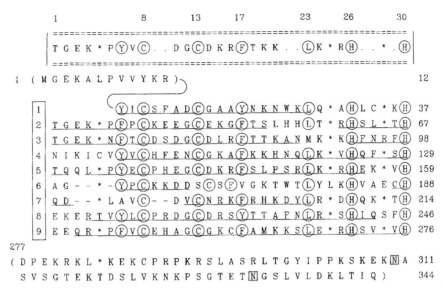

Fig. 3. Amino acid sequence of transcription factor IIIA from *X. laevis* oocytes, aligned to show the repeated units. The sequence is in one-letter code (Dayhoff, 1978). The molecule contains: an amino end region (residues 1–12); a lysine-rich zone (277–309) near the carboxyl end; a short tail region which may bind carbohydrate at Asn 310 and Asn 333, which are indicated by squares. The repeat units are numbered 1–9 on the left side of the diagram. The boxed-in consensus sequence at the top shows the characteristic features of a typical repeat unit, numbered as for a length of 30 residues. The end-point of each unit has been chosen arbitrarily after His-30. The best-conserved residues are ringed. Cys-8, Cys-13, His-26 and His-30 are believed to bind to a Zn²⁺ ion. Tyr-6, Phe-17 and Leu-23 are hydrophobic residues which may form an inner core for the proposed multiple-domain structure. Asterisks (*) mark positions where an insertion sometimes occurs in the normal pattern, and dots (.) mark variable positions in the sequence. In the main body of the repeats a dash (-) indicates an alignment gap. The underlined regions are those which show clear evidence of a relationship with at least one other unit (see Table I). Note that units 6 and 7 have diverged considerably from the usual pattern, and that residues 277-282 may form a fragmentary extension of repeat unit 9.

Table I. Overall similarities between the nine repeat units

Residues	Unit	Similarity values								
		5	9	2	8	3	4	1	7	6
130–159	5	*	23	20	15	18	14	6	8	2
247–276	9	23	*	23	16	18	12	10	1	0
38–67	2	20	23	*	11	17	16	6	1	8
215–246	8	15	16	11	*	17	13	11	5	5
68–98	3	18	18	17	17	*	6	6	6	2
99–129	4	14	12	16	13	6	*	16	7	0
13–37	1	6	10	6	11	6	16	*	0	0
189–214	7	8	1	1	5	6	7	0	*	1
160–188	6	2	0	8	5	2	0	0	1	*
Column totals		106	103	102	93	90	84	55	28	18

The value given to each pair is the number of overlapping 11-residue windows, in the correctly aligned comparison of the two units, for which the score exceeds the 0.1 % matching probability level. Since each unit is ~30 residues long, the maximum possible value is 30. The column totals give a measure of how close each unit is to the global consensus of them all. The underlinings below the sequence in Figure 1 show the parts of each unit that lie in the centre of a window, where the comparison score, with at least one other unit, reaches the 0.1 % level.

residues 13–276 form a continuous run of a repeated motif, of nine similar units, each of ~30 residues. Since the sequence repeats itself cyclically, the position of the boundary of the motif is uncertain, but the choice of an end-point shortly after His-30 of the consensus allows the largest number of complete units to be assigned. With this choice, most of the insertions and deletions occur near the ends of the units rather than in the central loop (consensus position numbers 14–25).

Our results suggest strongly that residues 13–276 of TFIIIA have evolved from a primitive ancestral unit of 30 amino acids by gene duplication or gene conversion (Hood *et al.*, 1975). It is not yet possible to make a proper analysis of the evolutionary relationships, because the end points of the units are not known; if the TFIIIA gene possesses intervening sequences these could be revealing. Table I gives a rough estimate of the degree of relatedness of each pair of units, derived from the numbers of high-scoring windows in the comparison matrix. The totals of the columns in the table give a measure of how closely each unit resembles the global consensus of the nine repeats.

It can be seen that units 5, 9 and 2, in that order, are clearly the most typical members of the family, and are very like one another. Common features include Pro-5, Gly-12, Lys-15 and Arg-25 (here we refer to the numbering of the consensus sequence in Figure 3). A second group comprises units 8, 3 and 4. These contain some single-residue insertions at positions 4A or 28A. Units 8 and 3 form a recognisable pair as shown by the identities Asp-11, Gly-12, Asp-15, Thr-18, Thr-19, Asn-22 and Phe-29. Lastly, units 1, 7 and 6 have diverged considerably from the normal pattern: in particular, units 6 and 7 are shortened at their amino ends with irregular (Cys-8)-(Cys-13) loops. We note that units 1 and 4 have Phe-10, Gly-14, Ala-16 and Lys-19 in common, while units 1 and 6 both have Trp-21.

Thus although all nine units belong to the same family they have diverged in detail and may have taken on specific individual DNA-binding functions as the 5S gene control system evolved. It would also not be surprising if the same 30-residue units were later found to occur in varying numbers in other related gene control proteins.

The evolutionary advantages of a repeating design are probably much the same as those in many other linear proteins, such as ovomucoid inhibitor (Laskowski *et al.*, 1980). Probably a single functional unit that binds to a half-turn of DNA was once evolved, and then became used in much more subtle and specialised ways when a large number of similar units were joined in series.

Characteristic structural features of the small domains

The well-marked repeat pattern in TFIIIA suggests strongly that each of the sequence repeats corresponds to a series of small structural elements, or domains, of ~30 residues, arranged linearly. In large globular proteins with repeated sequences the units most often form compact dimeric or tetrameric pseudo-symmetrical structures (Rossman and Argos, 1981; McLachlan, 1980). But there are also well-known examples of proteins with compact independently active structural units strung together in line, which can be separated by enzyme cleavage: the 'kringles' in plasminogen (Sottrup-Jensen *et al*, 1978) and the Kazal-type protease domains in the ovoinhibitor of Japanese quail (Laskow-ski *et al.*, 1980; Bolognesi *et al.*, 1982). In these molecules the structural units have lengths of ~80 and 63 residues. The TFIIIA unit is exceptional because of its small size and unique arrangement of Cys . . . Cys . . . His . . . His residues.

There is strong evidence, both that the repeat unit is a self-sufficient folded domain, and that it is stabilised by zinc ions. The first conclusion is suggested by the appearance of multiple small fragments on proteolysis, with quantized mol. wts. which are often multiples of 3 kd (see above). The second is supported by evidence that one TFIIIA molecule binds many zinc ions (7 – 11, see above) and is inactive in the absence of the metal (Hanas *et al.*, 1983). We therefore believe that most, if not all, of the nine units bind zinc.

Zinc is normally tetrahedrally coordinated in inorganic and metallo-organic compounds. The amino acids Cys and His are its most common ligands in enzymes (see Fersht, 1977), such as carbonic anhydrase, liver alcohol dehydrogenase, carboxy-peptidase A, Cu-Zn superoxide dismutase and thermolysin. Therefore we may picture each small domain as a compact unit formed round a central zinc ion to confer stability. Each zinc would be coordinated tetrahedrally to the two invariant pairs of Cys and His residues in each unit. The ends of the domain will then be pulled together round the zinc. In Figure 4 we have drawn one possible arrangement of the zinc ligands which fits this type of scheme. It is important to remember that the Cys . . . Cys and His . . . His loops probably cross over at an angle to form a tetrahedral box, and that the zinc may have more than four ligands. For example, a Cys side chain might be shared between zincs in two adjacent units.

Figure 5 shows how each domain contains three invariant hydrophobic groups (normally Tyr-6, Phe-17 and Leu-23) and how the acidic side chains are nearly all in or near the (Cys-8) . . . (Cys-13) loop.

The extended protein loop between residues 14 and 25 might form a DNA-binding region. The three-dimensional structures of gene repressor proteins (Ohlendorf and Matthews, 1983) show that the side chains of Lys, His, Asn, Gln and Thr often interact with the phosphate backbone of DNA, while Arg can form hydrogen bonds to the base pairs. Amino acids of this type are concentrated in the region 14–25 of TFIIIA (Figure 4).

A structural parallel for our proposed 30-residue metal-centred domain might be the 40-residue calcium-binding unit of the cal-modulin family (Moews and Kretsinger, 1975) or the 26-residue iron-sulphur half-domain of bacterial ferredoxin (Adman *et al.*, 1973). However, neither of these structures is an independent folding unit, but forms part of a linked pad of fragments. The metallothionein family of proteins (Boulanger *et al.*, 1983), which contain repeated Cys residues, are also not a good analogy for TFIIIA, but a rather peculiar group. The reason is that they bind metal ions in clusters, bridged by Cys residues, so that the ions are not separately packaged in independent protein cages, but

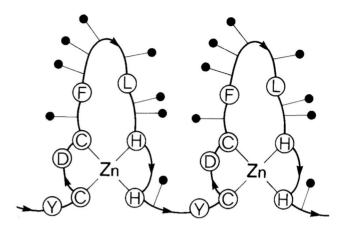

Fig. 4. Folding scheme for a linear arrangement of repeated domains, each centred on a tetrahedral arrangement of zinc ligands. Ringed residues are the conserved amino acids which include the Cys and His zinc ligands, the negatively charged Asp-11, and the three hydrophobic groups that may form a structural core. Black circles mark the most probable DNA-binding side chains (see Figure 5). In the scheme drawn here the metal ion draws the ends of each unit together, leaving the central residues 14–25 to form a potential DNA-binding loop or 'finger'. An alternative but much less likely position for the zinc is between the His residues of one unit and the Cys residues of an adjacent one.

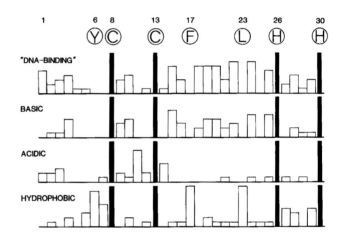

Fig. 5. Histograms for the average distribution of amino acids along the length of each 30-residue repeat unit. The height of each bar, in the range 0 – 9, gives the number of times that class of amino acid occurs at each position in the nine units. Positions are numbered as in Figure 3, with Cys-8, Cys-13, His-26 and His-30 treated as special positions (zinc ligands). Amino acids have been classed as follows. (a) DNA-binding: Lys, Arg, His, Asn, Gln, Thr (Ohlendorf and Matthews, 1983). (b) Basic: Arg, Lys, His. (c) Acidic: Asp, Glu. (d) Large hydrophobic: Leu, Ile, Val, Phe, Tyr, Trp. The strongest potential DNA-binding sites, with five or more DNA-binding amino acids in the nine units, are also marked in Figure 4.

instead built into a 'metal-sulphide' pseudocrystal bordered by the polypeptide chain.

TFIIIA thus appears to have an exceptionally compact molecular architecture for a self-sufficient structure; however, it may be that strong interactions between adjacent units are important for maintaining the structure. We have examined the two-helix DNA-binding motif of the Cro and Lac repressors, but it does not seem possible to accommodate the TFIIIA consensus sequence into this model: the loop 14 – 25 may instead fold into a long twisted ribbon of β-sheet which wraps in some way round the zinc-binding pocket, using the invariant hydrophobic groups as nuclei. The characteristic Cys-Cys-His-His consensus motif

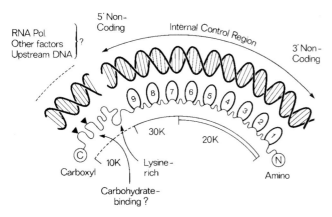

Fig. 6. An interpretation of the structural features of the protein TFIIIA and its interactions with DNA. The DNA is drawn curved, as if resting on a long beaded surface of the protein. The internal control region of the 5S RNA gene (bases 40–100) is drawn as six turns of DNA, with the 5′ end of the non-coding strand at the left. A separate piece of upstream DNA may be required to promote transcription. The protein sequence runs from right to left: the evidence for this orientation is given in the text. The amino end is followed by nine repeat units (residues 1–276) in contact with the control region. These units together form the 30-K proteolytic fragment of Smith *et al.* (1984). The first six repeat units (residues 13–188) probably constitute the 20-K proteolytic fragment: irregularities in the repeat pattern of units 6 and 7 may correspond to a susceptible cleavage region between the two fragments. The repeats are thought to be extended DNA-binding 'fingers' linked by flexible joints, each having a zinc ion centre. Residues 277–344 include a lysine-rich region, followed by two potential carbohydrate-binding sequences near the carboxyl end (marked with filled triangles), and together these parts from a separate domain, the 10-K proteolytic fragment. The relationship between the grooves of the DNA and the positions of the protein units is unknown and the drawing must not be taken literally, but each unit binds to about one half-turn of DNA. The tip of a unit could fit into a groove.

found in TFIIIA appears to be unique to this protein. We have searched for it without success in a large number of zinc enzymes, including those mentioned above, as well as in the metallothioneins. There is, therefore, no evidence yet that this motif, or indeed any other repetitive pattern, is a typical feature of zinc-binding proteins as a class.

Relation of large domains to the protein sequence. The nine small domains containing repeated pairs of cysteines and histidines comprise residues 1–276 or 1–280 (depending where the boundary is taken). The remaining 70 residues at the carboxyl end have no homology with the repeating units (Figure 3). The sequence suggests that this terminal region might be composed of two parts. The first half of ~30 residues is very rich in lysines and arginines, resembling in this respect a histone, although no significant sequence homology emerges. The second half lacks this enrichment and, as has been pointed out to us by Dr H-C. Thøgersen, contains two potential sites for carbohydrate addition at asparagines 310 and 333. [The characteristic sequence is N X S (Marshall, 1972).]

The marked difference in character of the last 70 residues suggest that this region might be identified with the 10-K 'domain' revealed at one end of the protein by the proteolysis studies of Smith *et al.* (1984). This 10-K domain is required for efficient RNA transcription, but, unlike the remaining 30-K fragment (and its smaller 20-K subfragment), not for binding to the internal control region of the 5S RNA gene. The amino and carboxyl orientation of the protein relative to the gene would then be as shown in Figure 6.

This orientation is consistent with the cyanogen bromide cleavage map of the protein shown by Smith *et al.* (1984, Figure

2). As pointed out by them, the small fragment CB2 stains quite differently with silver from the two other fragments. We therefore identify the fragments CB3, CB1 and CB2 with the residues 1–90, 91–266 and 267–344, respectively, since the latter is of different character from the first two (particularly in its high content of lysine). The lower mobility of the shorter peptide CB3 (78 residues) relative to CB3 (90 residues) could be accounted for if some of its residues were modified, as suggested above. The question of modifications is being investigated. The same amino-carboxyl orientation (Figure 6) has been found by the Carnegie group (D.D. Brown, personal communication).

Discussion

The experimental and theoretical studies described above have led to a picture in which the transcription factor IIIA consists mostly of a linear arrangement of nine repeated domains, each centred on a tetrahedral arrangement of zinc ligands. The repeats are thought to be extended DNA-binding fingers, linked by (flexible?) joints. This model is consistent with the highly asymmetric shape of the molecule indicated by its physico-chemical properties (Bieker and Roeder, 1984). A structure of this kind would explain how a relatively small protein of 40 K can bind to a long stretch of double-helical DNA: if each domain interacted with about half a DNA period, then, allowing for end effects, the nine domains could cover ~50–60 nucleotides, the length of the internal control region (Sakonju *et al.*, 1980). The DNA or RNA binding strength of each domain could be modulated, and specific recognition of short nucleic acid stretches established, by variations in the sequence at the finger tips.

One advantage of a many-fingered design for the transcription factor is that it binds to an internal control region of the gene in a system where a stable transcriptional complex is formed (Bogenhagen *et al.*, 1982; Gottesfeld and Bloomer, 1982; Lassar *et al.*, 1983). This complex can sustain many rounds of transcription during which the factor presumably remains bound to the gene. As the polymerase passes through the gene, the many-fingered protein could release those fingers bound ahead of the processing polymerase, but stay bound by its remaining fingers, whether to the intact DNA double helix or to the non-coding strand.

We have already mentioned above that the transcription factor appears to be a highly evolved version of a small molecule of the size of one of the contemporary 3-K domains stabilised by a metal ion. The primitive molecule could have simply assisted an early form of transcription. Evolution to the elaborate transcription apparatus found today could have taken place by gene duplication, with different repeats taking up extra functions. It is noteworthy, as remarked by Hanas *et al.* (1983) that RNA polymerase m contains zinc, and it could be that the initial activity of primitive TFIIIA promoted transcription in the absence of a polymerase molecule, which presumably only evolved later for greater efficiency.

Ginsberg *et al.* (1984) have noted a homology between a region of 5S DNA (or RNA) and the coding sequence of the TFIIIA gene, and have suggested that TFIIIA could interact with its own gene or the derived RNA to autoregulate expression at the transcriptional or translational levels. The evolutionary origin of this contemporary property may be as follows. RNA is widely believed to have preceded DNA, so that a small primitive RNA could have coded for a small protein (the precursor of the 3-K repeating unit) which bound back to the RNA and so stimulated its own production.

It may be that the deletions in units 6 and 7 of the contemporary protein (Figure 3) are necessary to enhance its flexibility and enable it to accommodate to the secondary or tertiary structure of the 5S RNA, as well as to the DNA.

The 10-K domain of Smith *et al.* (1984) does not bind to the internal control region of the gene and we have identified it with the 70 residues at the C-terminal end of the protein. Smith *et al.* (1984) have suggested that this 'transcription' domain may interact with other factors required to form a competent transcription complex or directly with RNA polymerase III. To this we would only add that, in view of its high lysine and arginine content, it might also be involved in binding to upstream elements of the DNA outside the coding region of the gene.

Materials and methods

7S particle purification

7S RNP particles were prepared from the ovaries of immature *X. laevis* (South African Snake Farm) by a method to be described elsewhere (Miller, in preparation). The method does not subject the particle to strong chelators, such as DTT, or to salt concentrations above 0.07 M. Proteins is at least 95% pure by Coomassie or silver stain, and >90% by amino acid analysis (data not shown). The particle is no more than 5% dissociated as measured by agarose gel electrophoresis (see below) and usually contains no detectable free 5S RNA.

For quantitation of metal content, glassware, tubing and any item to which buffer was exposed was treated for at least 48 h in 30% nitric acid, washed extensively with glass-distilled water and then treated for at least 72 h in 10 mM Na_2 EDTA (pH 4), the optimum pH for Zn^{2+} complexation by EDTA (West, 1969). Any items not resistant to nitric acid, such as column matrices or dialysis tubing, were rinsed in several changes of 10 mM Na_2 EDTA (pH 4) for at least 72 h. All materials were washed with glass-distilled water before use. Buffers were passed through a Chelex 100 ion-exchange column (BioRad) directly before use to remove divalent cations. Protein concentrations was measured by amino acid analysis.

Particle dissociation assay

7S particle preparations, usually at concentrations of 100 μg/ml, were loaded with an equal volume of 50% glycerol, 50 mM Tris-Cl, pH 7.5, 0.1 % xylene cyanol onto 0.7% agarose gels in 50 mM Tris-Cl, 90 mM boric acid pH 8.5 and run for ~1 h at 8 V/cm. Gels were stained with 1 mg/ml ethidium bromide and visualised under u.v. Experiments have confirmed that the fraction of RNA in particle preparations co-migrating with free 5S RNA correlates qualitatively with the fraction of 5S RNA eluting in a separate peak from 7S particle in gel filtration (Millerr, unpublished results), suggesting that it may be taken as a measure of free and bound 5S RNA in solution.

Polyacrylamide gel electrophoresis

Electrophoresis was as described by Laemmli (1970) with the following modifications. Stacking gel was 3% acrylamide, 0.15% bis-acrylamide, 0.125 M Tris-PO_4, pH 6.8, 0.1% SDS, 0.01% TEMED, 0.1% ammonium persulphate. Running gel was 22.5% acrylamide, 0.73% bis, 0.75 M Tris-Cl pH 8.8, 0.1% SDS, 0.01% TEMED, 0.1 ammonium persulphate. Samples were diluted with an equal volume of 100 mM NaPO₄ pH 6.8, 2% SDS, 100 M DTT and 50% glycerol and boiled for 3 min directly before loading. Approximate mol.wts were found from a logarithmic plot of mobility against the mol. wts. of two markers, tobacco mosaic virus protein (TMV) and bovine pancreatic trypsin inhibitor (BPTI). Gels were stained with 0.5% PAGE blue 83 (BDH Biochemicals).

Proteolysis

Proteolysis was always conducted at particle concentrations of 200–500 mg/ml in 50 mM Tris-Cl pH 8.1, 50 mM KCl, 0.5 mM $MgCl_2$, with 20 μg/ml trypsin, room temperature. Reactions were stopped either with 2 mM benzamidine, or by boiling for 2 min in loading buffer.

Sequence repeats – methods of analysis

We used two well-established methods. In the first, the diagonal comparison matrix (McLachlan, 1971, 1983), the sequence is divided into all its possible overlapping segments, or windows, of a given fixed length, without insertions or deletions, and every pair of segments is compared independently. The score for comparing each pair of amino acids in the window is derived from Dayhoff's mutation likelihood tables (Dayhoff , 1978) and is assessed against the exact calculated probability distribution of the window scores for random sequences with the same average composition as the whole protein (McLachlan, 1983). The cDNA sequences were scored by counting identical bases. In the second method, the damped Needleman-Wunsch alignment with gaps (Boswell and McLachlan, 1984; Needleman and Wunsch, 1970), sequences are aligned locally with penalties for insertions and deletions, but the scores are given weights which die away exponentially

with distance from the centre of each window. This newer method is more suitable for dealing with gaps, but is less susceptible to statistical analysis.

In the comparison matrices the highest observed score with a window of 30 corresponded to a double matching probability for the two peptide segments of only 0.53×10^{-10} (McLachlan, 1971) and was highly significant. We also calculated the highest expected score for the comparison of two random 344-residue sequences, i.e., the score which would be achieved on average just once with our protein. This score was exceeded not once, but 359 times in the natural sequence and showed that there are may significant repetitions.

In the damped Needleman-Wunsch method we used an effective range of 30 residues with rather low gap penalties: 2.0 to start each gap and 2.0 for each extension by one residue.

Acknowledgements

We thank Dr G.L. Everett (Johnson-Matthey Ltd., Royson) for determinations of the content of zinc and other metals, and Professor N. Hales and Dr G. Maguire (Department of Clinical Biochemistry, Cambridge University) for zinc determinations. We are grateful to Drs D.D. Brown and H. Pelham for advice, and our colleagues Drs D. Rhodes and H-C. Thørgersen for their help.

References

Adman, E.T., Sieker, L.C. and Jensen, L.H. (1973) *J. Biol. Chem.* **248**, 3987-3996.
Bieker, J.J., Martin, P.L. and Roeder, R.G. (1985) *Cell*, **40**, 119-127.
Bieker, J.J. and Roeder, R.G. (1984) *J. Biol. Chem.* **259**, 6158-6164
Bogenhagen, D.F., Sakonju, S. and Brown D.D. (1980) *Cell*, **19**, 27-35
Bogenhagen, D.F., Wormington, W.M. and Brown, D.D. (1982) *Cell*, **28**, 413-421.
Bolognesi, M., Gatti, G., Menegatti, E., Guarneri, M., Marquart, M., Papamokos, E. and Huber, R. (1982) *J. Mol. Biol.*, **162**, 839-868.
Boswell, D.R. and McLachlan, A.D. (1984) *Nucleic Acids Res.* **12**, 457-464.
Boulanger, Y., Goodman, C.M., Forte, C.P., Fesik, S.W. and Armitage, J.M. (1983) *Proc. Natl. Acad. Sci. U.S.A.* **80**, 1501-1505.
Brown, D.D. (1984) *Cell*, **37**, 359-365.
Cornell, N.W. and Crivaro, K.E. (1972) *Anal. Biochem.*, **47**, 203-208.
Dayhoff, M.O. (1978) *Atlas of Protein Sequence and Structure, Vol. 5, suppl. 3*, published by National Biomedical Research Foundation, Washington D.C.
Denis, H. and le Maire, M. (1983) in Roodyn, D.B. (ed), *Subcellular Biochemistry*, Vol. **9**, Plenum Publishing Co., pp. 263-297.
Engelke, D.R., Ng, S.Y., Shastry, B.S. and Roeder, R.G. (1984) *Cell*, **19**, 717-728.
Fersht, A. (1977) *Enzyme Structure and Mechanism*, published by W.H. Freeman & Co., San Francisco.
Ginsberg, A.M., King, B.O. and Roeder, R.G. (1984) *Cell*, **39**, 479-489.
Gottesfeld, J. and Bloomer, L.S. (1982) *Cell*, **28**, 781-791.
Hanas, J.S., Hazuda, D.J., Bogenhagen, D.F., Wu, F.Y.H. and Wu. C.W. (1983) *J. Biol. Chem.*, **258**, 14 120-14 125.
Honda, B. M. and Roeder, R.G. (1980) *Cell*, **22**, 119-126
Hood, L.M., Campbell, J.H. and Elgin, S.C.R. (1975) *Annu. Rev. Genet.*, **9**, 305-334.
Laemmli, U.K. (1970) *Nature*, **227**, 680-685.
Laskowski, M., Kato, I., Kohr, W.J., March, C.J. and Bodard, W.C. (1980) in Peeters, H. (ed)., *Protides of Biological Fluids*, Vol. **28**, Pergamon Press, Oxford, pp. 123-128.
Lassar, A.B., Martin, P.L. and Roeder, R.G. (1983) *Science*, **222**, 740-748.
Lewis, C.D. and Laemmli, U.K. (1982) *Cell*, **29**, 171-181.
McLachlan, A.D. (1971) *J. Mol. Biol.*, **61**, 409-421.
McLachlan, A.D. (1980) in Jaenicke, R. (ed.), *Protein Folding*, Elsevier/North Holland, Amsterdam, pp. 79-96.
McLachlan, A.D. (1983) *J. Mol. Biol.*, **169**, 15-30.
Marshall, R. D. (1972) *Annu. Rev. Biochem.* **41**, 673-702.
Moews, P.C. and Kretsinger, R.H. (1975) *J. Mol. Biol.*, **91**, 201-225.
Needleman, S.B. and Wunsch, C.D. (1970) *J. Mol. Biol.*, **48**, 443-453.
Ohlendorf, D.H. and Matthews, B.W. (1983) *Annu. Rev. Biophys. Bioeng.*, **12**, 259-284.
Pelham, H.R.B. and Brown, D.D. (1980) *Proc. Natl. Acad. Sci. U.S.A.* **77**, 4170-4174.
Picard, B. and Wegnez, M. (1979) *Proc. Natl. Acad. Sci. U.S.A.* **76**, 241-245.
Rossmann, M.G. and Argos, P.W. (1981) *Annu. Rev. Biochem.*, **50**, 497-532.
Sakonju, S., Bogenhagen, D.F. and Brown, D.D. (1980) *Cell*, **19**, 13-25.
Segall, J., Matsui, T. and Roeder, R.G. (1980) *J. Biol. Chem.*, **255**, 11986-11991.
Smith, D.R., Jackson, T.J. and Brown, D.D. (1984) *Cell*, **37**, 645-652.
Sottrup-Jensen, L., Claeys, H., Zajdel, M., Petersen, T.E. and Magnusson, S. (1978) *Prog. Chem. Fibrinolysis Thrombolysis*, **3**, 191-209.
West, T.S. (1969) *Complexometry with EDTA and Related Reagents*, published by BDH Chemicals Ltd., Poole, UK.

Received on 4 April 1985

Kissinger et al. (1990) Cell **63**, 579–590

Crystal Structure of an engrailed Homeodomain–DNA Complex at 2.8 Å Resolution: A Framework for Understanding Homeodomain–DNA Interactions

Charles R. Kissinger,* Beishan Liu,*
Enrique Martin-Blanco,[†] Thomas B. Kornberg,[†]
and Carl O. Pabo*
* Department of Molecular Biology and Genetics
and Howard Hughes Medical Institute
Johns Hopkins University School of Medicine
Baltimore, Maryland 21205
[†] Department of Biochemistry and Biophysics
University of California
San Francisco, California 94143

Summary

The crystal structure of a complex containing the engrailed homeodomain and a duplex DNA site has been determined at 2.8 Å resolution and refined to a crystallographic R factor of 24.4%. In this complex, two separate regions of the 61 amino acid polypeptide contact a TAAT subsite. An N-terminal arm fits into the minor groove, and the side chains of Arg-3 and Arg-5 make contacts near the 5′ end of this "core consensus" binding site. An α helix fits into the major groove, and the side chains of Ile-47 and Asn-51 contact base pairs near the 3′ end of the TAAT site. This "recognition helix" is part of a structurally conserved helix-turn-helix unit, but these helices are longer than the corresponding helices in the λ repressor, and the relationship between the helix-turn-helix unit and the DNA is significantly different.

Introduction

The homeodomain is a DNA binding motif that plays a central role in eukaryotic gene regulation (Scott et al., 1989). It was first discovered in a set of Drosophila proteins that regulate development, but it is now clear that the homeodomain occurs in a large family of proteins that regulate transcription in many higher organisms. Comparison of homeodomains from different genes and different organisms shows that their amino acid sequences are highly conserved (Scott et al., 1989). Although careful sequence comparisons allow the homeodomains to be grouped into subfamilies, it seems likely that all the homeodomains will have similar three-dimensional structures and use generally similar modes of DNA recognition. The structure of the Antennapedia homeodomain, a prototypical member of one of the largest sequence subfamilies, was recently determined by two-dimensional nuclear magnetic resonance (Qian et al., 1989). As anticipated from sequence comparisons (Laughon and Scott, 1984; Shepard et al., 1984), this homeodomain contains a helix-turn-helix (HTH) motif very similar to the HTH motif present in a number of prokaryotic repressors (Pabo and Sauer, 1984). Although high resolution crystal structures have been reported for several repressor–operator complexes (Otwinowski et al., 1988; Jordan and Pabo, 1988;

Aggarwal et al., 1988), there have not been any structural data available about homeodomain–DNA complexes. Intensive genetic and biochemical studies have elucidated some features of the protein–DNA interactions, but many puzzling questions remain about the specificity of homeodomain–DNA interactions, about the role of conserved residues in complex formation, and about the overall contribution of the homeodomain to site-specific binding and gene regulation.

We recently crystallized a 61 amino acid peptide that contains the homeodomain from the engrailed protein of Drosophila (a prototypic member of another major homeodomain subfamily) and also grew cocrystals with a 21 bp duplex DNA site (Liu et al., 1990; Figure 1). Using a strategy developed when cocrystallizing the λ repressor–operator complex (Jordan et al., 1985), we had tested a series of different DNA fragments with the engrailed homeodomain. It is not yet known which binding site(s) is functional in vivo, but excellent crystals were obtained with the DNA fragment shown in Figure 1, and gel mobility shift experiments confirmed that the engrailed homeodomain binds tightly to this site ($K_D = 1–2 \times 10^{-9}$ M in a buffer containing 100 mM KCl and 25 mM HEPES at pH 7.6). In this paper, we report the structure of this homeodomain–DNA complex and discuss the implications for our understanding of protein–DNA recognition and gene regulation.

Results and Discussion

Overall Arrangement of the Homeodomain–DNA Complex

In the crystal, two copies of the homeodomain bind to the 21 bp duplex. One binds near the center of the DNA fragment, and the other binds near the end (Figure 2). Superimposing the refined structures revealed that the conformations of the two protein monomers and the contacts they make with the DNA are nearly identical.

The engrailed homeodomain contains three α helices and an extended N-terminal arm. The structure of this motif provides the basis for understanding homeodomain–DNA interactions (Figure 3). Helix 1 (residues 10–22) and helix 2 (residues 28–37) pack against each other in an antiparallel arrangement. In the complex, each of these helices spans the major groove and is roughly perpendicular to the local direction of the DNA backbone. However, both of these helices are too far from the DNA to make many contacts. Helix 3 (residues 42–58) is roughly perpendicular to the first two helices. The exposed, hydrophilic face of helix 3 fits directly into the major groove, and side chains on this helix make extensive contacts with the DNA. The hydrophobic face of helix 3 packs against helices 1 and 2 to form the interior of the protein. In the engrailed cocrystal structure, we see no evidence for the kink (Qian et al., 1989) that has caused Wüthrich and his colleagues to describe residues 53–59 of the isolated Antennapedia homeodomain as a distinct helical segment

A

	Met	Asp	Glu	Lys	Arg	Pro	Arg	Thr	Ala	Phe	Ser
			1				5				

Helix 1

Ser	Glu	Gln	Leu	Ala	Arg	Leu	Lys	Arg	Glu	Phe	Asn	Glu	Asn	Arg
10					15					20				

Helix 2

Tyr	Leu	Thr	Glu	Arg	Arg	Arg	Gln	Gln	Leu	Ser	Ser	Glu	Leu	Gly
25					30					35				

Helix 3

Leu	Asn	Glu	Ala	Gln	Ile	Lys	Ile	Trp	Phe	Gln	Asn	Lys	Arg	Ala
40					45					50				

Lys	Ile	Lys	Lys	Ser
55				

B

```
1  2  3  4  5  6  7  8  9 10 11 12 13 14 15 16 17 18 19 20 21
T  T  T  T  G  C  C  A  T  G  T  A  A  T  T  A  C  C  T  A  A
A  A  A  C  G  G  T  A  C  A  T  T  A  A  T  G  G  A  T  T  A
```

Figure 1. Sequences of engrailed Homeodomain and Binding Site Used for Cocrystallization

(A) Sequence of the engrailed homeodomain. The fragment used for cocrystallization includes 60 amino acids from the Drosophila engrailed protein (Kornberg, 1981; Poole et al., 1985), and the cloning procedure adds a methionine at the N-terminal end of this peptide. Boxes mark the three α helices observed in the cocrystal. The numbering scheme corresponds to that used by Wüthrich and his colleagues in describing the structure of the Antennapedia homeodomain (Qian et al., 1989).

(B) DNA sequence used for cocrystallization. In the crystal, the overhanging 5' ends pair with those of neighboring duplexes to form a pseudocontinuous double helix. As discussed in the text, two copies of the homeodomain bind to this site. One protein makes its primary contacts near the TAAT site that includes base pairs 11–14 (shown in boldface). The other homeodomain binds at the end of the DNA (where neighboring duplexes overlap in the crystal). This homeodomain makes critical contacts in a region that includes a TAAA site on the lower strand (i.e., base pairs 1–3 and 21, shown in outline).

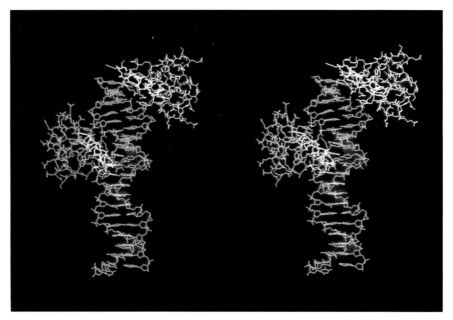

Figure 2. Two Homeodomains Bind to the 21 bp Duplex

One copy of the homeodomain (shown in orange) binds tightly to a site near the center of the DNA duplex. This is a higher affinity binding site, and the paper focuses on this complex. Another copy of the homeodomain (shown in purple) binds to a weaker site at the end of the duplex, and this homeodomain makes additional contacts with a contiguous duplex in the crystal. The overall arrangements of the two proteins with respect to the DNA are very similar, and nearly all of the protein–DNA contacts are identical. Helix 3 of each protein (highlighted in yellow) fits directly into the major groove of the double-helical DNA (shown in blue).

(helix IV). It is possible that this region changes conformation upon DNA binding, and crystallographic refinement of the isolated engraved protein is now in progress (N. Clarke, C. R. K., B. L., and C. O. P. unpublished data).

The first few residues of the homeodomain appear to be disordered in the crystal, but residues 3–9 form an extended N-terminal "arm" that fits into the minor groove and supplements the contacts made by helix 3. Helices 1 and 2 are connected by a relatively open loop (residues 23–27), and helices 2 and 3 are connected by a somewhat shorter "turn" (residues 38–41). The overall arrangement

of the homeodomain reveals a simple, functional design for DNA recognition.

There are no contacts between the two homeodomains seen in the crystal, and they appear to bind as independent monomers. Binding experiments using duplexes with the isolated subsites (see Experimental Procedures) indicate that the engrailed homeodomain binds to the central sequence with a K_D of $1–2 \times 10^{-9}$ M and binds to the terminal sequence with a K_D of 10^{-7} M. (Control experiments with a λ operator site show no binding; any interaction must be at least an order of magnitude weaker.) The

A

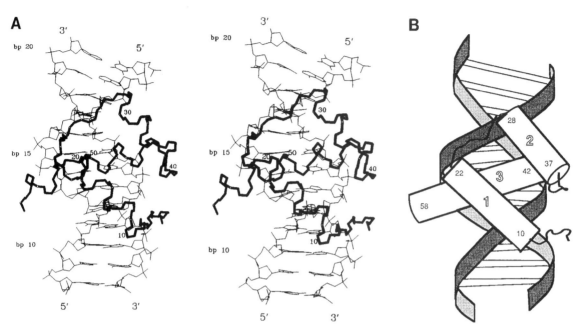

Figure 3. Overview of the Homeodomain–DNA Complex

(A) Stereo diagram showing how the α helices and the N-terminal arm are arranged in the homeodomain–DNA complex. To make it easier to see the overall relationship, this diagram shows only backbone atoms for the protein (N, C$_\alpha$, and C). Every tenth protein residue is numbered. The DNA shown here includes base pair 8 (bottom) through base pair 20 (top). Base pairs 10, 15, and 20 are labeled, along with the 3′ and 5′ termini of each DNA strand.

(B) Sketch summarizing the relationship of the α helices and the N-terminal arm with respect to the double-helical DNA. Cylinders are used to show the position of the α helices, ribbons are used to show the sugar-phosphate backbone of the DNA, and bars indicate base pairs.

two complexes seen in the crystal happen to be related by an approximate two-fold symmetry axis, but this does not appear to have any real significance for the engrailed–DNA interactions. Although the N-terminal arms have slightly different conformations, most of the contacts are the same in the two complexes. Our discussion will focus on the central site, since engrailed binds to this site with a higher affinity. Binding to the terminal sequence is inherently weaker, and interactions may also be perturbed because the protein is binding at the end of the duplex. (This homeodomain actually contacts two DNA duplexes, since the DNA fragments have overlapping 5′ ends and stack to form a pseudocontinuous helix in the crystal.)

Helix 3: Invariant Residues and Critical Contacts with the DNA

Helix 3 fits directly into the major groove and makes extensive contacts with the bases and with the sugar-phosphate backbone. It is interesting to note the critical roles played by Trp-48, Phe-49, Asn-51, and Arg-53. These residues occur in every one of the higher eukaryotic homeodomains compiled by Scott et al. (1989), and the crystal structure shows that these invariant residues occur in the section of helix 3 closest to the major groove (Figure 4). Trp-48 and Phe-49 form a key part of the hydrophobic core. They must play a major role in stabilizing the folded

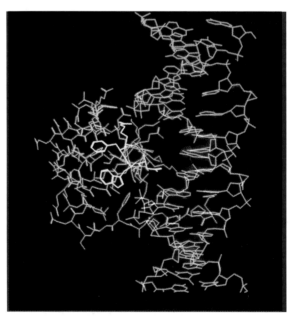

Figure 4. Photograph Highlighting the Position of Invariant Residues: Trp-48, Phe-49, Asn-51, and Arg-53

This view is looking down the axis of helix 3. The homeodomain is shown in orange with invariant residues highlighted in yellow. The DNA is shown in blue.

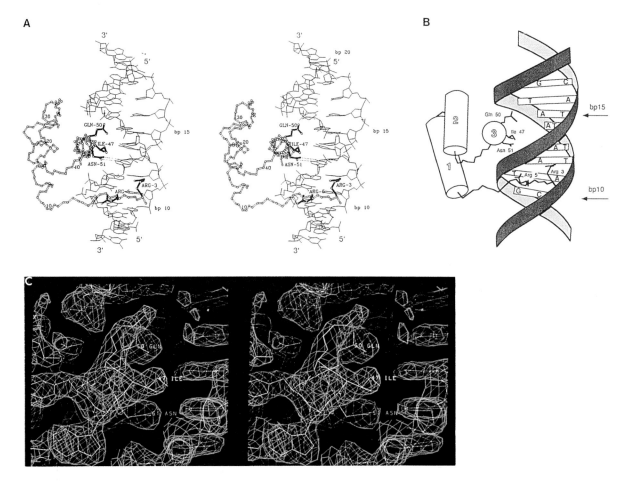

Figure 5. Key Contacts with the Base Pairs at the TAAT Subsite

(A) Stereo diagram showing a view along helix 3. Backbone atoms are shown for residues 3–59 of the homeodomain, and side chains are shown for residues that contact the base pairs: Arg-3, Arg-5, Ile-47, Gln-50, and Asn-51. The segment of DNA shown here includes base pair 8 (at the bottom) through base pair 20 (at the top) and thus includes the critical TAAT subsite. Base pairs 10, 15, and 20 are labeled, along with the 3' and 5' termini of each DNA strand. This diagram shows that the arm

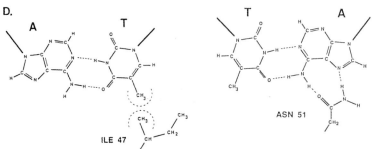

makes minor-groove contacts near the 5' end of the TAAT subsite and that helix 3 makes major-groove contacts near the 3' end of the TAAT subsite.
(B) Sketch summarizing the critical contacts.
(C) Stereo view of the calculated electron density from a $2F_o - F_c$ map in the vicinity of helix 3. The protein is shown in yellow, and the side chains of Ile-47, Gln-50, and Asn-51 are labeled. The DNA is shown in red and the electron density is shown in blue. The electron density is contoured at a level of one rms deviation above the average density
(D) Sketch showing the base contacts made by Ile-47 and Asn-51.

structure and in controlling how helix 1 packs against helix 3. (This will be critical for DNA recognition because it affects the spatial relationship between contacts made by the N-terminal arm and contacts made by helix 3.)

The invariant hydrophilic residues – Asn-51 and Arg-53 – make critical contacts with the DNA. Asn-51 makes a pair of hydrogen bonds with the adenine at base pair 13, donating a hydrogen bond to the N7 position and accepting a hydrogen bond from the N6 position (Figure 5). Arg-

53 hydrogen bonds with two phosphate groups on the other strand of the DNA (Figures 6 and 7).

Several of the neighboring residues in helix 3 make critical contacts with the DNA and are conserved within subsets of the homeodomain proteins. The side chains of Ile-47 and Gln-50 appear to be especially important for DNA recognition. Ile-47 provides a sequence-specific interaction by making hydrophobic contacts with the methyl group of the thymine at base pair 14. Valine, which occurs

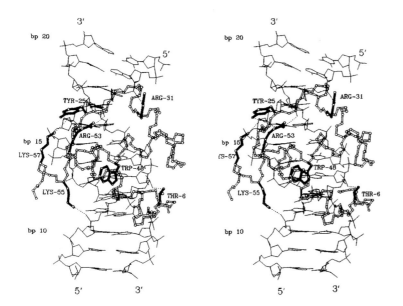

Figure 6. Contacts with the Phosphates

Stereo diagram showing side chains that contact the phosphodiester oxygens. The DNA segment is the same as that in Figures 3A and 5A. The view is roughly perpendicular to helix 3 and thus is similar to the view in Figure 3. Critical residues include Thr-6, Tyr-25, Arg-31, Trp-48, Arg-53, Lys-55, and Lys-57. These contacts also are summarized in Figure 7.

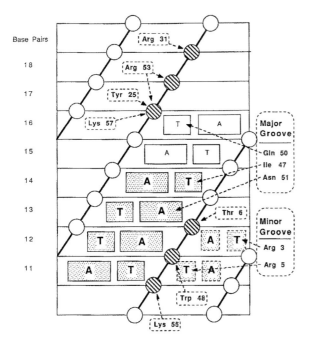

at this position in other homeodomains, would be able to

Figure 7 Sketch Summarizing All the Contacts Made by the Homeodomain

The DNA is represented as a cylindrical projection, and the shading emphasizes the TAAT subsite. Phosphates are represented with circles; hatched circles show phosphates that are contacted by the homeodomain.

make a similar hydrophobic contact. The side chain of Gln-50 projects into the major groove and clearly is in a position to make sequence-specific contacts. In the current structure, Gln-50 makes van der Waals contacts with the methyl group of the thymine at base pair 16. A modest rotation of the side chain would allow it to hydrogen bond with the O4 of the thymine at base pair 14, but the -NH$_2$ of the Gln side chain and the O4 of the thymine are about

4 Å apart in the current structure. It also appears that small changes in the DNA conformation could allow Gln-50 to hydrogen bond with the adenine at base pair 15. (The DNA geometry is discussed in a later section.) Residues on helix 3 also make an extensive set of contacts with the sugar-phosphate backbone (Figure 6); these contacts are discussed in a later section.

Contacts Made by the N-Terminal Arm

The crystal structure shows that the N-terminal arm binds to the minor groove of the DNA and reveals that conserved residues from the arm contact base pairs adjacent to the ones contacted by helix 3. The poor electron density for residues 1 and 2 of the engrailed homeodomain indicates that this region is disordered in the crystal. However, some backbone density can be seen for residue 2, and residues 3–5 form a well-defined region of extended chain that fits into the minor groove. The side chain of Arg-5, the most highly conserved residue in this portion of the homeodomain, reaches directly into the minor groove and hydrogen bonds with the O2 of the thymine at base pair 11. The electron density for Arg-3 is not nearly as well defined as the density for other side chains. However, it appears that the side chain of Arg-3 hydrogen bonds with the O2 of the thymine at base pair 12 and/or hydrogen bonds with the sugar oxygen from the adenosine at base pair 13.

Further studies are needed to understand what degree of sequence specificity is provided by these minor-groove contacts. Clearly these contacts explain a preference among homeodomains for AT-rich sites, since these have hydrogen bond acceptors at appropriate positions in the minor groove. Presumably a GC or a CG base pair would interfere with binding since the arginine side chain would have unfavorable steric or electrostatic interactions with the -NH$_2$ of the guanine. However, it is not clear whether these contacts could distinguish an AT from a TA base pair, since the N3 of adenine and the O2 of thymine occupy similar positions in the minor groove (Seeman et al., 1976).

Contacts with the DNA Backbone

The homeodomain makes an extensive set of contacts with the sugar-phosphate backbone, and we presume that these are critical for DNA binding. The full set of backbone contacts is shown in Figure 6 and summarized (along with the base contacts) in Figure 7. Most of these contacts are clustered in two regions flanking the sites where helix 3 makes contacts in the major groove. (In the sequence, each of these regions is offset in the 5′ direction along the DNA backbone; in the three-dimensional structure, these contacts surround the regions where helix 3 contacts specific base pairs.) The engrailed homeodomain makes one set of contacts with phosphates just "above" the region where helix 3 contacts bases in the major groove (Figures 6 and 7). At the edge of the site, Arg-31 contacts the phosphate that lies between the adenine at base pair 19 and the guanine at base pair 18. Proceeding in the 5′ to 3′ direction along this strand, we see that Arg-53 contacts the next phosphate. The phosphate after this appears especially critical and has contacts from Arg-53, Tyr-25, and Lys-57. (It is interesting that helix 1 does not make any contacts with the DNA and that Tyr-25 and Arg-31 provide the only DNA contacts from the loop or from helix 2.)

On the other strand, there are several contacts with phosphates just "below" the region where helix 3 fits into the major groove (Figures 6 and 7). Lys-55 contacts a phosphodiester oxygen from the thymidine at base pair 11. Proceeding in a 5′ to 3′ direction, we see that Trp-48 is very close to the next phosphate. The ring nitrogen is not in a favorable position for hydrogen bonding to the DNA, but the partial positive charge on the edge of the aromatic ring (Burley and Petsko, 1985) may provide a favorable electrostatic interaction with the phosphodiester oxygen. Thr-6 makes side chain and main chain hydrogen bonds to a phosphodiester oxygen of the adenosine at base pair 13.

DNA Conformation

The DNA duplex in the crystal is a relatively straight segment of B-DNA. The average helical twist of 34.2° (Table 1) corresponds to 10.53 bp per turn. This is very close to the average expected for B-DNA (Wang, 1979), and this suggests that protein binding and crystallization have not resulted in any large overall distortion of the DNA structure. As observed in single-crystal studies of B-DNA (Dickerson and Drew, 1981), the base pairs have significant propeller twist (average = 13.3°). The individual helical twists range from 28.1° to 44.7°, and the individual propeller twists range from 5.4° to 21.7° (Table 1). Base pairs at one end of the duplex (base pairs 13–14) have a significant tilt. The change in tilt occurs near base pairs 17 and 18, and the minor groove is unusually wide in this region.

Superimposing B-DNA on the complex confirms that there are no drastic distortions in the binding site, but it is clear that the major groove is several angstroms wider than normal in the region where helix 3 binds. Most of the changes seem to occur in the DNA strand that contacts the C-terminal portion of helix 3. The slightly lower helical twists between base pairs 14 and 15 and between base pairs 13 and 14 may contribute to the widening of the ma-

Table 1. Local Helical Parameters for the DNA Site				
		Twist (Degrees)		
Position	Base Pair	Helical	Propeller	Tilt (Degrees)
2	T·A		15.8	4.1
		35.3		
3	T·A		9.2	0.8
		32.1		
4	T·A		15.3	5.5
		33.1		
5	G·C		10.4	2.2
		37.4		
6	C·G		6.9	1.5
		27.9		
7	C·G		5.5	3.9
		44.7		
8	A·T		6.7	6.1
		29.5		
9	T·A		10.0	4.5
		40.6		
10	G·C		12.6	2.9
		28.1		
11	T·A		5.6	6.1
		39.8		
12	A·T		18.3	5.4
		35.1		
13	A·T		21.5	6.2
		30.4		
14	T·A		15.2	5.4
		32.2		
15	T·A		12.9	3.9
		34.5		
16	A·T		14.6	4.6
		31.0		
17	C·G		21.8	7.7
		40.6		
18	C·G		17.4	16.6
		31.9		
19	T·A		12.9	18.8
		33.6		
20	A·T		13.8	21.7
		32.6		
21	A·T		20.2	18.7

jor groove. The base pairs also have a significant tilt in this region, and this may affect the groove width.

TAAT Subsite Allows Alignment with Other Binding Sites

A deeper understanding of homeodomain–DNA interactions requires that we integrate the structural data with results from genetic and biochemical studies. To proceed with any detailed comparison, we need to align the binding site used in the crystal with binding sites used in other studies. Obtaining the correct alignment is complicated by the fact that individual homeodomain proteins can recognize a variety of different binding sites. However, the subsequence TAAT occurs in most homeodomain binding sites (Scott et al., 1989), and recent experiments with Ultrabithorax have emphasized the importance of a TAAT core (S. C. Ekker, K. E. Young, and R. A. Beachy, submitted). Aligning this subsite should provide the best prospect for correlating our structural data with results from other studies.

The most plausible alignment uses the TAAT subsite

that includes base pairs 11–14 on the upper strand. This is satisfying from a structural perspective, since the engrailed homeodomain contacts each of these base pairs. Another TAAT sequence occurs at base pairs 16–13 on the lower strand, but the engrailed homeodomain makes fewer contacts with this TAAT sequence and it seems less plausible that this could constitute the "core consensus" binding site. Other data confirm that the TAAT site at base pairs 11–14 is the appropriate one to use when aligning sequences:

First, this alignment is consistent with the model that S. D. Hanes and R. Brent (submitted) derived from an elegant genetic analysis of the homeodomain–DNA contacts. Their data allowed them to infer that the ninth residue of the recognition helix (i.e., Gln-50 of engrailed) makes critical contacts just to the 3′ side of the TAAT subsite. Our alignment is fully consistent with their model, since Gln-50 contacts base pair 16. Their data exclude the alternative assumption (that base pairs 16–13 provide the critical TAAT) since this would leave Gln-50 near the 5′ end of the TAAT subsite.

Second, ethylation interference experiments have been done with an Antennapedia binding site (Affolter et al., 1990) that contains a TAAT sequence. Our preferred alignment matches every phosphate contact seen in the engrailed cocrystal structure with a phosphate contact inferred from these ethylation interference experiments. (It seems safe to assume that the Antennapedia and engrailed homeodomains make generally similar contacts with the DNA backbones, and studies of the bacterial repressors have demonstrated that there is a very close correlation between the contacts seen in a cocrystal and contacts inferred from ethylation interference experiments [Johnson, 1980; Jordan and Pabo, 1988].)

Third, footprinting experiments using fragments of Oct-1 and complexes of Oct-1 with other proteins show that the Oct-1 homeodomain binds to a TAAT subsite and indicate that helix 2 is on the 3′ side of this TAAT site (T. M. Kristie and P. A. Sharp, submitted).

Comparison with Biochemical and Genetic Data about Homeodomain–DNA Contacts

The structure reported here is consistent with a vast body of data about homeodomain–DNA interactions. One obvious and satisfying aspect of the structure is the important role played by the most highly conserved residues of the homeodomain. Our structure shows that each of the invariant residues plays a critical role in folding and/or recognition. The structure also helps us understand why genetic experiments have pinpointed residue 50 as a critical residue for controlling specificity of the homeodomain–DNA interactions (Hanes and Brent, 1989; Treisman et al., 1989; S. D. Hanes and R. Brent, submitted). This side chain points directly into the major groove and clearly is in an excellent position to contribute to the specificity of binding and recognition. In the current complex, it forms a van der Waals contact with the methyl group of a thymine, but we were surprised that Gln-50 does not hydrogen bond to the adenine at base pair 15 or to some other position in the major groove. Studies are in progress to

see whether Gln-50 can hydrogen bond to a base under other circumstances. For example, it is possible that binding of the neighboring homeodomain distorts the DNA enough to prevent Gln-50 from making its preferred contact. The neighboring homeodomain contacts the phosphates between bases 16 and 19 on the upper strand. The high propeller twist at base pair 17, the high helical twist between base pairs 17 and 18, and the unusual tilt angles in this region (Table 1) all suggest that the DNA is somewhat distorted in this region. To address this issue, we have recently grown cocrystals in which the upper strand has been synthesized as two separate segments. After annealing, this leaves a "nick" between nucleotides 16 and 17 on the upper strand, and this may prevent the binding of one homeodomain from distorting the binding site of the other. In vitro selection schemes are also being used to find the optimal sequence for engrailed binding (E. M.-B. and T. K., unpublished data) and these experiments should help determine whether Gln-50 might prefer other bases at positions 15 and 16.

Residue 50 plays an important role in distinguishing one homeodomain binding site from another (Hanes and Brent, 1989; Treisman et al., 1989), but the structure makes it clear that Ile-47 and Asn-51 also play very important roles in recognition. The genetic experiments focused on differences in amino acid sequence that were responsible for differences in specificity. Asn-51 is invariant, and Ile or Val (which could make a similar sequence-specific contact) is almost always present at position 47. Contacts made by these residues will have a central role in recognizing the TAAT subsite and in distinguishing this from non-specific DNA.

The structure reported here readily explains the other highly conserved residues and regions of the homeodomain. A number of conserved residues — for example, Ile-16 and Phe-20 - help to form the hydrophobic interior in engrailed and Antennapedia (Qian et al., 1989). The crystal structure of the complex also reveals important contacts made by highly conserved residues (such as Arg-3 and Arg-5) near the N-terminal end of the homeodomain and by highly conserved residues near the C-terminal end of the homeodomain (Figures 5, 6, and 7).

The Homeodomain Uses the HTH Motif in a Novel Way

Sequence comparisons had suggested (Laughon and Scott, 1984; Shepard et al., 1984) and nuclear magnetic resonance studies of the Antennapedia homeodomain had confirmed (Qian et al., 1989) that the homeodomain contains an HTH motif that is structurally similar to that observed in the prokaryotic repressors (Pabo and Sauer, 1984). The engrailed structure also confirms that these HTH motifs are quite similar. The C_αs for residues 33–52 of λ repressor can be superimposed on the C_αs for residues 31–50 of the engrailed homeodomain with a root-mean-square (rms) distance of only 0.84 Å between corresponding atoms (Figure 8A). This rms distance is only slightly larger than the distances typically obtained when superimposing the HTH units of two prokaryotic repressors. Some of the aligned residues in the two proteins

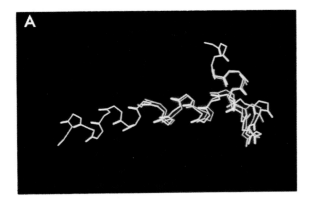

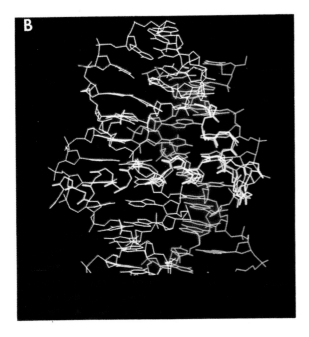

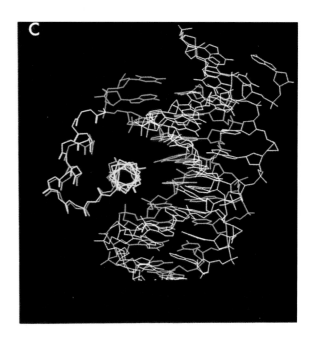

Figure 8. Comparison of the HTH Units in the λ Repressor–Operator Complex and the Homeodomain–DNA Complex

Helices 2 and 3 are shown for each protein (residues 33–52 for λ repressor; residues 28–58 for engrailed), but the complexes were aligned by superimposing corresponding C$_\alpha$s of the HTH units (residues 33–52 for λ repressor; residues 31–50 for engrailed).
(A) View of the HTH units that is roughly perpendicular to the recognition helixes. Residues from the λ repressor are shown in orange; residues from engrailed are shown in yellow.
(B) Complexes seen from the same perspective as (A). The λ DNA is shown in blue; the engrailed binding site is shown in purple.
(C) View of the complexes looking along the recognition helix.

(e.g., Ile-45 of engraved and Val-47 of λ repressor) play similar roles in stabilizing the HTH units, and these are precisely the positions that are most diagnostic in sequence searches for the HTH unit.

In spite of the similar backbone structures for the HTH units, the engrailed cocrystal structure reveals that the HTH units are used in significantly different ways in the homeodomain–DNA and repressor–operator complexes. Superimposing the two HTH units allows one to establish a common frame of reference for the two complexes, and it is clear that the DNA duplexes have very different positions in this reference frame (Figure 8B). If we take the arrangement seen in the λ repressor–operator complex as a starting point, it appears that the DNA in the homeodomain complex has been shifted toward the C-terminal end of the second helix in the HTH unit. These differences also are apparent if we compare the positions of residues that make critical contacts with the base pairs. Residues near the N-terminal end of the recognition helix make critical contacts in the λ repressor–operator complex. The critical residues in engrailed are near the center of an extended recognition helix.

A side view reveals other differences between the two arrangements (Figure 8C). In the λ repressor–operator complex, the first helix of the HTH unit (helix 2) fits partway into the major groove, and the N-terminal end of this helix contacts the sugar-phosphate backbone of the DNA. In contrast, the first helix of the homeodomain's HTH unit (helix 2) lies above the major groove. Relative to the framework provided by the superimposed HTH units, it appears that the DNA for the homeodomain has been rotated away from the N-terminal end of this helix. Although the position of helix 2 is rather different in the λ and in the engrailed complexes, it clearly plays related roles in the two complexes. Since it packs against helix 3, we expect that it will stabilize folding of the recognition helix. It also seems to serve as an "outrigger" that will prevent helix 3 from rocking in the major groove. (Moving in one direction would cause helix 2 to collide with the sugar-phosphate backbone; moving in the other direction would break the hydrogen bonds that helix 2 makes with the DNA backbone.) It is interesting that Arg-31, which provides a phosphate contact in the engraved complex, is aligned with Gln-33, which provides a corresponding contact in the λ complex.

Even though helix 2 of engrailed is longer and has a different orientation with respect to the DNA, a structurally analogous residue contacts the DNA backbone.

Implications for Understanding Protein–DNA Interactions

These differences in the arrangement of the HTH unit appear dramatic and certainly help us understand why the homeodomain proteins constitute a distinct subfamily of HTH proteins. However, the prokaryotic and eukaryotic complexes are similar in many fundamental ways. Although the homeodomain–DNA complex reveals a number of new and interesting features, it supports most of the fundamental ideas that have been developed from structural studies of the prokaryotic HTH proteins. In this section we try to place the homeodomain–DNA complex in a broader perspective. We discuss the role of α helices in recognition and discuss the respective roles of contacts with the base pairs and contacts with the sugar-phosphate backbone.

Fundamentally, the HTH unit of the prokaryotic repressors provides a way of positioning an α helix in the major groove, and side chains on this helix play a major role in site-specific binding (Jordan and Pabo, 1988; Aggarwal et al., 1988). (The Trp repressor–operator complex does not fit this paradigm [Otwinowski et al., 1988], but recent experiments suggest that the Trp repressor may have been crystallized with a nonspecific binding site [Staacke et al., 1990].) Despite significant differences in the position and orientation of the helices, it is clear that the recognition helix of the homeodomain has a fundamentally similar role. In fact, there was no reason to suppose that the position of the recognition helix needed to be conserved in all HTH proteins. There are no strict constraints on this relationship because there is no strict relationship between the periodicity of an α helix and the periodicity of B-DNA, and at an atomic level the shape and appearance of an α helix will be dominated by the amino acid sequence of the helix. Since the sequence will change from one protein to the next, the recognition helix will look significantly different and may pack against the DNA in a significantly different way.

Structural studies of the prokaryotic repressors also made it clear that a single α helix would only be able to contact a few base pairs. This constraint has a relatively simple geometric basis: since the major groove has the shape of a "helical saddle," a straight α-helical segment can only contact a few adjacent base pairs. Even though the recognition helix of the homeodomain is much longer, we see a similar constraint in this complex. A dramatic kink – sufficient to generate independent helical segments – would be needed to keep a longer helical region in contact with the major groove. Although the solution structure of the Antennapedia homeodomain had suggested that there was a small kink (of about 30°) after residue 52, the corresponding region of engrailed forms a continuous helix in the homeodomain–DNA complex. In any case, modeling experiments show that a small kink – like that reported for the Antennapedia homeodomain – would not allow significant additional contacts in the major groove.

Structural studies of prokaryotic DNA binding proteins indicate that there is no "code" for protein–DNA interactions (i.e., that there are no rigid rules determining which amino acids can contact which base pairs). The structure of the engrailed–DNA complex supports the notion that recognition depends on the detailed structure of the protein–DNA interface, but we note that the contacts between Asn-51 and adenine are similar to the Gln–adenine contacts seen in the λ and 434 complexes (Jordan and Pabo, 1988; Aggarwal et al., 1988). This type of contact – which had been proposed by Seeman et al. (1976) – may be an especially favorable hydrogen bonding arrangement and may play a significant role in site-specific recognition. Comparisons of the λ and 434 complexes (Pabo et al., 1990) had also emphasized that amino acids tend to make similar contacts when they occur at similar positions within the HTH unit, and it will be interesting to see whether similar relationships hold among the set of homeodomain proteins. It may be possible, within the context of a highly conserved structure, to predict the bases that will be preferred by particular amino acids at particular positions of the HTH unit.

The homeodomain makes an extensive set of contacts with the DNA backbone, and we presume that these play an important role in binding and recognition. As in the λ repressor–operator complex, these contacts occur just on the 5' sides of the region where the α helix contacts the base pairs, and in three dimensions these contacts are on the closest edges of the major groove. Although these backbone contacts could have some role in "indirect readout" of sequence information, it seems unlikely since this region is relatively uniform B-DNA. It seems simpler to imagine that their primary role is to provide a set of "fiducial marks" that help to align the homeodomain as it approaches the DNA. Such contacts would help to control the position and orientation of the recognition helix and thus would serve to enhance the specificity of complex formation.

Another interesting aspect of the homeodomain–DNA interactions involves the minor-groove contacts that are made by the N-terminal arm. Although Arg-43 of the 434 repressor fits partway into the minor groove (Aggarwal et al., 1988), the engrailed structure is the first complex to show how several side chains from an extended chain can make base-specific contacts in the minor groove. The engrailed contacts may be very similar to the contacts made by the Hin recombinase, which has been studied by incorporating EDTA–Fe at particular amino acid residues and then mapping DNA cleavage patterns (Sluka et al., 1990). Like engrailed, the N-terminal portion of this DNA binding domain contains an Arg-Pro-Arg sequence, and high resolution chemical footprinting experiments show that this N-terminal arm binds in the minor groove.

The λ repressor also has an extended N-terminal arm that contributes to site-specific binding, and some of the critical residues have identical positions when the two protein sequences are aligned. However, the repressor's arm binds in the major groove (Jordan and Pabo, 1988). Helix 1 has a very different orientation in the two proteins, and this seems to determine which groove will be closest to the N-terminal arm. Future studies will be needed to understand the precise role of contacts that engrailed makes in the minor groove, and it will be important to see how this N-terminal region of the homeodomain is constrained when it is present in the context of the intact protein. It also will be interesting to determine how often similar interactions occur in other proteins and whether regions of extended polypeptide chain might make similar contacts in the major groove of A-DNA (which is too narrow to accommodate an α helix).

General Principles of Homeodomain–DNA Interactions

The structure of the engrailed–DNA complex suggests some general principles about specificity in homeodomain–DNA interactions. It appears that the highly conserved residues on helix 3 and a few residues on the N-terminal arm form a sort of core recognition unit that is responsible for many of the contacts with the TAAT subsite. However, there will be a vast number of TAAT sequences in the genome, and these contacts, even when supplemented by sequence-specific contacts from residue 50, would presumably not be sufficient to provide for the differential regulation of gene expression. Specificity could be enhanced in many ways:

First, since the homeodomain occupies both the major and minor grooves, it may not be able to bind to DNA within nucleosomes. This could drastically simplify the problem of finding the appropriate binding sites.

Second, other mechanisms could be involved in particular cases. Co-operative binding of homeodomains to neighboring sites, binding in conjunction with other proteins, and dimerization of homeodomain proteins may increase the effective specificity of binding in particular systems.

Third, it also is interesting that both the N-terminus and C-terminus of the homeodomain are close to the DNA and that particular subsets of the homeodomain proteins often have a cluster of conserved residues immediately preceding or immediately following this standard 60 amino acid homeodomain. Some of these conserved clusters of residues may serve to modulate site-specific binding. The cocrystal structure makes it clear that these neighboring regions would be in an excellent position to directly contact neighboring bases on the DNA or to influence the structure or orientation of the N-terminal arm and the C-terminal helix. Finally, they also could provide "attachment" or "targeting" sites for other proteins that would bind to neighboring regions and modulate the specificity and/or affinity of DNA binding. At this stage, little information is available about the precise role of neighboring regions. However, the POU domain is a particularly striking example of a nearby conserved region that occurs in a number of homeodomain proteins (Herr et al., 1988), and recent experiments suggest that the POU domain recognizes a neighboring subsite on the DNA (Ingraham et al., 1990; Phillip Sharp, personal communication).

Although our model is somewhat speculative and schematic at this stage, it provides a clear connection between the structure of the complex and the known biological roles of the homeodomain proteins. Differential or modulated recognition may be particularly important for the subtle regulatory controls involved in differentiation and development. These structural data also provide a simple picture of how the family of homeodomain proteins may have diverged. Recognition could be based on a set of contacts with a core consensus sequence, and modulating the interactions would generate new specificities and therefore new regulatory activities.

These ideas provide a framework for thinking about family–subfamily relationships in proteins containing homeodomains. It clearly will be necessary to get structural information about other homeodomain–DNA complexes (particularly about the intact proteins) and to use the structural data to design more incisive experiments about the roles that particular residues or regions of the protein play in sequence-specific binding.

Conclusions

This study reveals the following basic structural features of homeodomain–DNA interactions:

The homeodomain makes contacts in both the major and minor grooves, and the critical contacts are centered on a conserved TAAT subsite that biochemical studies have highlighted as the most important part of the homeodomain binding site (S. C. Ekker, K. E. Young, and P. A. Beachy, submitted).

The HTH unit plays a significant role in recognition, but the helices in the homeodomain are longer than the corresponding helices of λ repressor and the orientation of the HTH unit with respect to the DNA is rather different. In the homeodomain, the first helix of the HTH unit lies entirely above the major groove. The second helix of the HTH unit (the recognition helix) fits directly into the major groove, but the critical contacts occur in a region that would correspond to the C-terminal end of the canonical (20 residue) HTH unit. However, the recognition helix is much longer in the homeodomain–DNA complex and residues that contact the bases are near the center of this extended helix.

Each of the invariant residues (defined by comparing sets of homeodomain sequences) plays a central role in folding and/or recognition. Trp-48 and Phe-49 form a central part of the hydrophobic core, and Trp-48 may have favorable electrostatic interactions with a phosphodiester oxygen. Asn-51 makes a pair of hydrogen bonds with an adenine at the third position of the TAAT subsite (TA<u>A</u>T). Arg-53 hydrogen bonds with a pair of phosphate groups on the DNA backbone.

Two other residues contact bases in the major groove. Ile-47 makes hydrophobic contacts with a thymine at the fourth position of the TAAT subsite (TAA<u>T</u>). Gln-50 projects directly into the major groove. In the current structure, it makes van der Waals contacts with a thymine methyl group, but small motions would allow Gln-50 to interact with several different positions near the 3′ side of the TAAT subsite (TAA<u>TNN</u>).

Residues near the N-terminal end of the homeodomain make minor-groove contacts near the 5′ end of the conserved TAAT subsite. Argus hydrogen bonds to the thy-

Table 2. Statistics for Data and Derivatives

Item	Native	IdU$_{16}$	IdUn$_1$	IdU$_{1+16}$
Resolution (Å)	2.8	2.7	3.8	2.8
Measured reflections	21,130	26,820	12,900	14,403
Unique reflections	9,072	10,052	3,849	8,969
R$_{sym}$[a]	4.26	4.34	3.34	3.23
Mean isomorphous difference [b]		0.21	0.11	0.20
Phasing power [c]		1.71	1.32	2.04
Cullis R factor [d]		0.60	0.65	0.56

Designations for the derivative data sets indicate the base pair(s) at which 5-iodouracil was substituted for thymine in the DNA used for crystallization.

[a] $\Sigma_h\Sigma_i|I_{h,i}-I_h|/\Sigma_h\Sigma_iI_{h,i}$, where I_h is the mean intensity of the i observations of reflection h.

[b] $\Sigma|F_{PH}-F_P|/\Sigma F_{PH}$, where F_{PH} and F_P are the derivative and native structure factor amplitudes, respectively.

[c] $[(F_{H(calc)}^2/I(F_{PH(obs)}-F_{PH(calc)})^2]^{1/2}$.

[d] $\Sigma||F_{der}\pm F_{nat}|-/F_{H(calc)}|/\Sigma|F_{der}-F_{nat}|$ for centric reflections, where $F_{H(calc)}$ is the calculated heavy atom structure factor.

mine at the first position (TAAT). Arg-3 appears to hydrogen bond to the thymine of the second base pair (TAAT), although electron density for this side chain is not as clear as for the other contacts.

There are extensive contacts with the sugar-phosphate backbone, and many of these contacts are made by residues in the N-terminal arm and by residues in helix 3.

Helix 1 does not make any direct contacts with the DNA. Only one residue from the following loop and only one residue from helix 2 actually contact the DNA. The primary role of these helices is to help stabilize the folded structure and to help fix the relative orientation of the N-terminal arm and helix 3.

The limited number of base contacts in the complex suggests that the isolated homeodomain can provide only a modest amount of sequence specificity. Cooperative binding, with other homeodomains and/or with other regulatory proteins, may serve to enhance specificity. However, since the N-terminal and C-terminal regions of the homeodomain are near the DNA, it also is possible that neighboring regions of the intact proteins could modulate the affinity or specificity and thus allow differential regulation of gene expression.

Experimental Procedures

The cocrystals were grown from a solution that contained equimolar amounts of the engrailed homeodomain and duplex DNA in a buffer containing 30 mM Bis-Tris–HCl. When the crystallization drops were set up, the pH was raised to 8.0–9.0 by the addition of ammonium hydroxide; crystals grew as the ammonium hydroxide diffused out and the pH returned to 6.7. Precession photographs revealed that the cocrystals form in space group C2 with α = 131.2 Å, b = 45.5 Å, c = 72.9 Å, and β = 119.0° (Liu et al., 1990). The crystals diffract to 2.5 Å in all directions, but the current data set is weak beyond 2.7 or 2.8 Å.

Native diffraction data were collected on a Xentronics area detector (Table 2), and isomorphous derivatives were obtained by preparing duplex DNA that had 5-iodouracil substituted for thymine at specific positions. The first derivative had iodouracil substituted at base pair 16 on the lower strand, and the second derivative had iodouracil substituted on the 5′ end of the upper strand. A third derivative had iodouracil at both positions. Difference Pattersons revealed that there was one DNA duplex in the asymmetric unit, and the doubly substituted derivative allowed us to determine the relative y coordinates for the iodine sites.

Heavy-atom parameters were first refined using the program REFINE from the CCP4 (1979) package (S.E.R.C. [U.K.] Collaborative

Computing Project No. 4, a Suite of Programs for Protein Crystallography, distributed from Daresbury Laboratory, Warrington, UK) and these data were used to phase an initial MIR map. This first map was relatively noisy, but the phases were improved by refining the heavy atom parameters against solvent-flattened phases (Rould et al., 1989). This is a cyclic process involving the following: B.-C. Wang's protocol for iterative phase improvement (Wang, 1985); refinement, using the program PHARE, of the heavy-atom parameters against the Wang phases; and calculation of new MIR phases. After several cycles, inspection of the Wang map revealed left-handed α helices, so we inverted the coordinates of the heavy atom sites and proceeded with several more cycles of flattening and refinement. Phases for the last MIR map had a mean figure of merit of 0.67 for data from 10–3.5 Å resolution. After a final round of solvent flattening, this map gave excellent density for the protein, and the map immediately revealed that there were two copies of the homeodomain bound to the 21 bp DNA fragment. We assumed that the structure of the engrailed homeodomain would be quite similar to that of the Antennapedia homeodomain, and we were able to rapidly fit the protein density by using a model extracted from published stereo photographs of the Antennapedia homeodomain (Qian et al., 1989; Rossmann and Argos, 1980).

The electron density for the DNA was less clearly defined, but it was readily fit by using a model of uniform B-DNA and matching the refined iodine positions with the methyl groups of the appropriate thymines. The program X-PLOR was used for refinement of this initial model (Brünger et al., 1987; Brünger, 1990). The rigid-body refinement option was used for an initial adjustment of the overall position and orientation of the DNA and two protein molecules. The first cycle of simulated annealing gave R = 0.30 for data from 10–3.2 Å, but extensive rebuilding of the DNA was required. Subsequent cycles of rebuilding and refinement gave the current model, which has R = 0.244 for data from 10–2.8 Å resolution using a single overall temperature factor of 17.2 Å^2 and without any water molecules included in the model. This model has good stereochemistry. The rms deviation for bond lengths is 0.019 Å, and the rms deviation for bond angles is 3.7°. This model fits the MIR map extremely well. It also has been confirmed with a procedure developed by Axel Brünger (personal communication) that involves systematically deleting short segments of the structure, using simulated annealing to minimize model bias from the phases, and examining $2F_o - F_c$ maps that span the deleted region.

Because structural analysis revealed that two homeodomains bind to the duplex used for cocrystallization, we synthesized two other duplexes so that we could estimate the intrinsic affinity of the engrailed homeodomain for each of these subsites. To eliminate binding at the end, we synthesized a DNA duplex with the sequence CCATGTAATTACCTGG (and its complement). Gel mobility shift experiments (in a buffer containing 100 mM KCl and 25 mM HEPES at pH 7.6) revealed that engrailed bound to this site with a K$_D$ of 1–2 × 10^{-9} M. To estimate the intrinsic affinity for the sequences that overlap the ends of the 21 bp duplex, we permuted the sequence of the original duplex and changed some of the flanking bases. We used a site with the sequence CCGCCTAATTTTGCCA (and its complement), and gel mobility shift experiments indicated that this site had a K_D of 1×10^{-7} M. (Base pairs 4–16 of this site correspond to base pairs 17–21 and 1–8 of the duplex used for crystallization.)

Acknowledgments

This project was supported by grants from the National Institutes of Health to C. O. P. (GM31471) and to T. B. K. (GM24635). We thank Anatoli Collector and Cynth ia Wendling for synthesizing the DNA, Upul Obeysekare and Neil Clarke for helping to build a model of the engrailed structure from stereo photographs, and Mark Rould and Tom Steitz for their help with refinement or the heavy-atom parameters against solvent-flattened phases. We thank Phil Beachy, Roger Brent, Axel Brünger, Neil Clarke, Phil Sharp, and Cynthia Wolberger for helpful discussions. We thank Roger Brent, Steven Hanes, and Phil Sharp for communicating unpublished data that was central for deriving the correct alignment of the TAAT subsites. We gratefully acknowledge the National Cancer Institute for use of the Cray during molecular dynamics refinement. B. L., C. R. K., and C. O. P. are in the Howard Hughes Medical Institute.

The costs of publication of this article were defrayed in part by the payment of page charges. This article must therefore be hereby marked "*advertisement*" in accordance with 18 USC Section 1734 solely to indicate this fact.

Received August 31, 1990; revised September 24, 1990.

References

Affolter, M., Percival-Smith, A., Muller, M., Leupin, W., and Gehring, W. J. (1990). DNA binding properties of the purified *Antennapedia* homeodomain. Proc. Natl. Acad. Sci. USA *87*, 4093–4097.

Aggarwal, A. K., Rodgers, D. W., Drottar, M., Ptashne, M., and Harrison, S. C. (1988). Recognition of a DNA operator by the repressor of phage 434: a view at high resolution. Science *242*, 899–907.

Brünger, A. T. (1990). X-PLOR v2.1 Manual (New Haven, Connecticut: Yale University Press).

Brünger, A. T., Kuriyan, J., and Karplus, M. (1987). Crystallographic R factor refinement by molecular dynamics. Science *235*, 458–460.

Burley, S. K., and Petsko, G. A. (1985). Aromatic–aromatic interactions: a mechanism of protein structure stabilization. Science *229*, 23–28.

Dickerson, R. E., and Drew, H. R. (1981). Structure of a B-DNA dodecamer. II. Influence of base sequence on helix structure. J. Mol. Biol. *149*, 761–786.

Hanes, S. D., and Brent, R. (1989). DNA specificity of the Bicoid activator protein is determined by homeodomain recognition helix residue 9. Cell *57*, 1275–1283.

Herr, W., Sturm, R. A., Clerc, R. G., Corcoran, L. M., Baltimore, D., Sharp, P. A., Ingraham, H. A., Rosenfeld, M. G., Finney, M., Ruvkun, G., and Horvitz, H. R. (1988). The POU domain: a large conserved region in the mammalian *pit*-1, *oct*-1, *oct*-2, and *Caenorhabditis elegans* *unc*-86 gene products. Genes Dev. *2*, 1513–1516.

Ingraham, H. A., Flynn, S. E., Voss, J. W., Albert, V. R., Kapiloff, M. S., Wilson, L., and Rosenfeld, M. G. (1990). The POU-specific domain of Pit-1 is essential for sequence-specific, high affinity DNA binding and DNA-dependent Pit-1–Pit-1 interactions. Cell *61*, 1021–1033.

Johnson, A. (1980). Mechanism of action of λ-cro protein. Ph.D. thesis, Harvard University, Cambridge, Massachusetts.

Jordan, S. R., and Pabo, C. O. (1988). Structure of the lambda complex at 2.5 Å resolution: details of the repressor–operator interactions. Science *242*, 893–899.

Jordan, S. R., Whitcombe, T. V., Berg, J. M., and Pabo, C. O. (1985). Systematic variation in DNA length yields highly ordered repressor–operator cocrystals. Science *230*, 1383–1385.

Kornberg, T (1981). *engrailed*: a gene controlling compartment and segment formation in *Drosophila*. Proc. Natl. Acad. Sci. USA *78*, 1095–1099.

Laughon, A., and Scott, M. P. (1984). Sequence of a *Drosophila* segmentation gene: protein structure homology with DNA-binding proteins. Nature *310*, 25–31.

Liu, B., Kissinger, C., Pabo, C. O., Martin-Blanco, E., and Kornberg, T. B. (1990). Crystallization and preliminary diffraction studies of the engrailed homeodomain and of an engrailed homeodomain–DNA complex. Biochem. Biophys. Res. Commun. *171*, 257–259.

Otwinowski, Z., Schevitz, R. W., Zhang, R.-G., Lawson, C. L., Joachimiak, A., Marmorstein, R. Q., Luisi, B. E, and Sigler, P. B. (1988). Crystal structure of *trp* repressor/operator complex at atomic resolution. Nature *335*, 321–329.

Pabo, C. O., and Sauer, R. T. (1984). Protein–DNA recognition. Annu. Rev. Biochem. *53*, 293–321.

Pabo, C. O., Aggarwal, A. K., Jordan, S. R., Beamer, L. J., Obeysekare, U. R., and Harrison, S. C. (1990). Conserved residues make similar contacts in two repressor–operator complexes. Science *247*, 1210–1213.

Poole, S. J., Kauvar, L. M., Drees, B., and Kornberg, T. (1985). The *engrailed* locus of Drosophila: structural analysis of an embryonic transcript. Cell *40*, 37–43.

Qian, Y. Q., Billeter M., Otting, G., Müller, M., Gehring, W. J., and Wüthrich, K. (1989). The structure of the *Antennapedia* homeodomain determined by NMR spectroscopy in solution: comparison with prokaryotic repressors. Cell *59*, 573–580.

Rossmann, M. G., and Argos, P. (1980). Three-dimensional coordinates from stereodiagrams of molecular structures. Acta Crystallogr. *B36*, 819–823.

Rould, M. A., Perona, J. J., Soll, D., and Steitz, T. A. (1989). Structure of *E. Coli* glutaminyl-tRNA synthetase complexed with tRNA^{Gln} and ATP at 2.8 Å resolution. Science *246*, 1135–1142.

Scott, M. P., Tamkun, J. W., and Hartzell, G. W. (1989). The structure and function of the homeodomain. Biochim. Biophys. Acta *989*, 25–48.

Seeman, N. C., Rosenberg, J. M., and Rich, A. (1976). Sequence-specific recognition of double helical nucleic acids by proteins. Proc. Natl. Acad. Sci. USA *73*, 804–808.

Shepard, J. C., McGinnis, W., Carrasco, A. E., De Robertis, E. M., and Gehring, W. J. (1984). Fly and frog homoeo domains show homologies with yeast mating type regulatory proteins. Nature *310*, 70–71.

Sluka, J. P., Horvath, S. J., Glasgow, A. C., Simon, M. I., and Dervan, R. B. (1990). Importance of minor-groove contacts for recognition of DNA by the binding domain of Hin recombinase. Biochemistry *29*, 6551–6561.

Staacke, D., Walter, B., Kisters-Woike, B., Wilcken-Bergmann, B., and Müller-Hill, B. (1990). How Trp repressor binds to its operator. EMBO J. *9*, 1963–1967.

Treisman, J., Gönczy, P., Vashishtha, M., Harris, E., and Desplan, C. (1989). A single amino acid can determine the DNA binding specificity of homeodomain proteins. Cell *59*, 553–562.

Wang, B.-C. (1985). Resolution of phase ambiguity in macromolecular crystallography. Meth. Enzymol. *115*, 90–112.

Wang, J. C. (1979). Helical repeat of DNA in solution. Proc. Natl. Acad. Sci. USA *76*, 200–203.

Availability of Coordinates

Coordinates are being deposited with the Brookhaven Data Bank. While these are being processed by Brookhaven and prepared for distribution, interested scientists may obtain a set of coordinates either by sending an appropriate BITNET address to us at PABO@JHUIGF or by sending a ½" tape with a mailing envelope and sufficient return postage.

Note Added in Proof

A preliminary NMR analysis of the Antp–DNA complex has been published: Otting, G., Qian, Y. Q., Billeter, M., Müller, M., Affolter, M., Gehring, W. J., and Wüthrich, K. (1990). Protein–DNA contacts in the structure of a homeodomain–DNA complex determined by nuclear magnetic resonance spectroscopy in solution. EMBO J. *9*, 3085–3092. Twelve intermolecular NOEs from six residues were used to dock the homeodomain against a model of B-DNA; each of these NOEs is consistent (using a cutoff of about 5 Å for the H–H distances) with the detailed model from our crystallographic study.

Dimerization domains in transcription factors

10

As discussed in Sections 8 and 9, the zinc-finger and homoeobox DNA binding domains were identified by comparison of the protein coding sequences of several different transcription factors. Thus these domains were identified as common features of several different transcription factors and then shown to mediate DNA binding. Subsequently structural predictions based on the amino acid sequence and direct analysis of the protein structure using crystallographic methods were used to determine the structure of these domains.

A similar approach was used in the paper by Landschulz et al. presented here. However, rather than identifying a DNA binding domain, these authors used this approach to identify the first example of a domain which mediates the formation of dimers between two transcription factor molecules.

C/EBP	L T S D N D R L R K R V E Q L S R E L D T L R G I F R Q L
Jun B	L E D K V K T L K A E N A G L S S A A G L L R E Q V A Q L
Jun	L E E K V K T L K A Q N S E L A S T A N M L R E Q V A Q L
GCN 4	L E D K V E E L L S K N Y H L E H E V A R L K K L V G E R
Fos	L Q A E T D Q L E D E K S A L Q T E I A N L L K E K E K L
Fra 1	L Q A E T D K L E D E K S G L Q R E I I E L Q K Q K E R L
c-Myc	V Q A E E Q K L I S E E D L L R K R R E Q L K H K L E Q L
n-Myc	L Q A E E H Q L L L E K E K L Q A R Q Q Q L L K K I E H A
l-Myc	L V G A E K K M A T E K R Q L R C R Q Q Q L Q K R I A Y L

Figure 10.1 Alignment of the leucine-rich region in several transcription factors. Note the conserved leucine residues (L; boxed) that occur every seven amino acids.

In their initial analysis, Landschulz et al. focused on the C-terminal region of the C/EBP transcription factor for which they had previously isolated cDNA clones (Landschulz et al., 1988). As illustrated in Fig. 1 of the paper presented here, they displayed the amino acids in this region of C/EBP on a helical wheel, which indicates the relative positions of different amino acids within an α-helical region of a protein. They made the observation, shown in Fig. 1 of the paper, that one side of the helix is predominantly composed of hydrophobic amino acids whereas the other side of the helix is predominantly composed of residues with charged side chains (basic and acidic residues). Most strikingly, there exists a periodic repetition of leucine residues, such that a leucine residue appears at every seventh position over a region of 35 amino acids, resulting in the leucines being present at the same position on each turn of the helical wheel.

Moreover, similar runs of leucine residues occurring at every seventh position can be identified in other transcription factors which do not otherwise show significant sequence identity to C/EBP (Figure 10.1). As in C/EBP, these leucine residues can be organized on a helical wheel so that they occur in an identical position on each turn of the wheel; this is illustrated in Fig. 2 of the paper by Landschulz et al.

Based on this unusual arrangement of leucine residues, Landschulz et al. made the proposal that such residues would serve as an

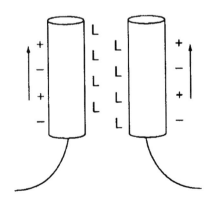

Figure 10.2 Model of the leucine zipper and its role in the dimerization of two molecules of a transcription factor. Note that the two leucine-rich regions associate in a parallel manner, with both helices orientated in the same direction and with charged amino acids on the outside of the dimer.

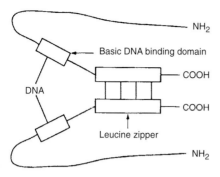

Figure 10.3 Model for the structure of the leucine zipper and the adjacent DNA binding domain following dimerization.

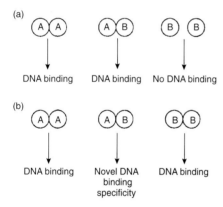

Figure 10.4 Regulation of transcription factor activity by dimerization. In (a), factor B cannot form a homodimer and can therefore only bind to DNA following heterodimerization with factor A, which can also bind to DNA as a homodimer. In (b), both factor A and factor B can form homodimers which bind to DNA. However, the heterodimer formed by A and B has a novel DNA binding specificity different from those of either factor alone.

interface for transcription factor dimerization. Thus the leucine-rich side of the helix would interface with a similar leucine-rich region in the partner transcription factor, with the highly charged regions remaining on the outside (Figure 10.2). The leucine-rich region would thus serve to mediate dimerization between two transcription factor molecules, and it was therefore named the 'leucine zipper' by Landschulz et al.

More recent studies of this leucine zipper motif (for reviews, see Hurst, 1994; Kerppola and Curran, 1995) have amply confirmed the original idea of Landschulz et al. that the leucine zipper constitutes a dimerization domain, and it thus represents the first example of such a domain to be identified. Although, as illustrated in Fig. 5(B) of their paper, Landschulz et al. were originally unclear as to whether the two leucine-rich helices interact in a parallel or anti-parallel manner, it is now clear that the arrangement shown in Figure 10.2 is the correct one, with the helices interacting in a parallel orientation. Such an interaction between the leucine zippers in two molecules of a transcription factor containing this motif provides the correct protein structure for DNA binding by the adjacent region of these proteins, which is rich in basic amino acids that can interact with the acidic DNA. The transcription factor dimer is rotationally symmetrical and contacts the DNA via the bifurcating basic regions (Figure 10.3; for further details, see Vinson et al., 1989; Glover and Harrison, 1995).

The importance of such transcription factor dimerization to allow DNA binding is seen in the case of the Fos and Jun transcription factors (for a review, see Lewin, 1991). Thus, while Jun can form a homodimer and therefore binds to DNA, the Fos protein cannot do so and does not bind to DNA when present in isolation. DNA binding by Fos requires the presence of the Jun molecule, so allowing a DNA binding Jun/Fos heterodimer to form.

Clearly, the requirement for some transcription factors to dimerize before they bind to DNA provides an additional level of gene regulation (for discussion, see Jones, 1990; Lamb and McKnight, 1991). In addition to homodimer formation, some factors may only bind to DNA as heterodimers with another factor, while some heterodimers may have novel DNA binding specificities or affinities distinct from those of either of the individual factors alone (Figure 10.4).

It is not suprising, therefore, that other motifs that can mediate transcription factor dimerization have also been described, such as the helix–loop–helix motif, which can also facilitate DNA binding by an adjacent basic DNA binding domain (for a review, see Littlewood and Evan, 1994). The leucine zipper originally described by Landschulz et al. was thus the first example of a class of motif which, rather than facilitating DNA binding directly, acts to facilitate the dimerization step which is required for subsequent binding, and thus plays a key role in the regulation of gene expression.

Landschulz (1988) Science **240**, 1759–1764

The Leucine Zipper: A Hypothetical Structure Common to a New Class of DNA Binding Proteins

WILLIAM H. LANDSCHULZ, PETER F. JOHNSON, STEVEN L. MCKNIGHT

A 30-amino-acid segment of C/EBP, a newly discovered enhancer binding protein, shares notable sequence similarity with a segment of the cellular Myc transforming protein. Display of these respective amino acid sequences on an idealized α helix revealed a periodic repetition of leucine residues at every seventh position over a distance covering eight helical turns. The periodic array of at least four leucines was also noted in the sequences of the Fos and Jun transforming proteins, as well as that of the yeast gene regulatory protein, GCN4. The polypeptide segments containing these periodic arrays of leucine residues are proposed to exist in an α-helical conformation, and the leucine side chains extending from one α helix interdigitate with those displayed from a similar α helix of a second polypeptide, facilitating dimerization. This hypothetical structure is referred to as the "leucine zipper," and it may represent a characteristic property of a new category of DNA binding proteins.

THE MOLECULAR BASIS OF SPECIFIC INTERACTION BETWEEN proteins and DNA has been the subject of intensive study. Proteins that regulate the various functions and metabolism of DNA can be fixed at an appropriate site of action by binding avidly to certain DNA sequences. Proteins that exhibit this property, termed sequence-specific DNA binding proteins, play integral roles in DNA replication, recombination, strand scission, and transcription. Moreover, additional roles for sequence-specific DNA binding proteins will probably be discovered in processes ranging from chromosome segregation during cytokinesis, to the ordered folding and unfolding of chromatin during successive transitions between interphase and mitosis.

The sequence-specific DNA binding proteins that have been studied most extensively are the gene activator and repressor proteins of bacteria. The active forms of these proteins are rotationally symmetric dimers, and their DNA binding sites are rotationally symmetric palindromes. X-ray crystallographic studies have revealed a structural motif common to the DNA binding domains of many bacterial repressors and activators (1). This DNA binding motif, termed "helix-turn-helix," is characterized by two successive α helices juxtaposed at approximately 90° by a turn of four amino acids (2). Dimerization of this class of bacterial gene regulatory

The authors are in the Department of Embryology, Carnegie Institution of Washington, 115 West University Parkway, Baltimore, MD 21210.

proteins arranges one α helix of each monomer with its analog so that they fit into successive major grooves of DNA.

A second DNA binding motif, common to many eukaryotic gene regulatory proteins, has recently been discovered. The founding member of this second family is a metalloprotein, termed TFIIIA, which acts as a positive regulator of 5S RNA gene expression (3). The amino acid sequence of TFIIIA has a repeating motif consisting of two closely spaced cysteines followed by two histidines (4). Biophysical studies indicate that a zinc ion is tetrahedrally coordinated by the cysteine-histidine motif, and that the coordination complex imparts both stability and DNA sequence-specificity to the TFIIIA polypeptide (5). The amino acid sequence diagnostic of this motif, termed the "zinc finger," is being observed in an ever increasing number of eukaryotic gene regulatory proteins (6, 7).

Two fundamental concepts have been established from studies on sequence-specific DNA binding proteins. The first, which satisfies long-standing predictions, holds that binding specificity results from direct atomic interaction between amino acid side chains and base pairs in minimally distorted B DNA. There seems, however, to be no general code that specifies the amino acid surface necessary to interact stably with a given DNA surface. In contrast, a second and less intuitively obvious concept has emerged. That is, a limited number of structural motifs "steer" the appropriate amino acid side chains of a protein into the grooves of double helical DNA where they can interact directly with base pairs. These structural motifs are composed primarily of amino acid residues that do not make direct contact with the DNA. Rather, they form three-dimensional "scaf-

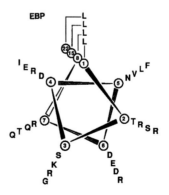

Fig. 1. Helical wheel analysis of a carboxyl-terminal portion of C/EBP. The amino acid sequence of a portion of the C/EBP polypeptide (11) is displayed end-to-end down the axis of a schematic α helix. The most amino-terminal residue included in the analysis is leucine residue 315 of C/EBP, and is placed at position number one of the idealized helix. The second position on the helical wheel, which is occupied by threonine 316, is located at an azimuth displaced by 100° from position number one. The helical wheel consists of seven "spokes", corresponding to the fit of seven amino acid residues into every two α-helical turns. A total of 28 amino acids of C/EBP are included in the analysis, with the most carboxyl-terminal residue being glutamine 342. The amino acid abbreviations are as follows: A, alanine; C, cysteine; D, aspartic acid; E, glutamic acid; F, phenylalanine; G, glycine; H, histidine; I, isoleucine; K, lysine; L, leucine; M, methionine; N, asparagine; P, proline; Q, glutamine; R, arginine; S, serine; T, threonine; V, valine; W, tryptophan; Y, tyrosine.

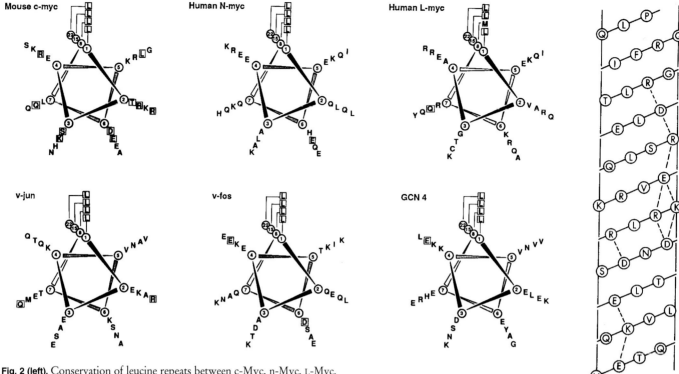

Fig. 2 (left). Conservation of leucine repeats between c-Myc, n-Myc, L-Myc, Jun, Fos, and GCN4. The amino acid sequence of a portion of each protein is displayed on a schematic α helix according to the conventions of Fig. 1. The region of c-Myc selected for display had been observed to share substantial sequence similarity with C/EBP (11), and starts at leucine residue 413 of the mouse c-Myc polypeptide (28). The regions of n-Myc and L-Myc selected for display correspond to the same carboxyl-terminal segment as was chosen for c-Myc. The analysis of n-Myc started at leucine residue 425, and that of L-Myc started at leucine residue 333 (29). A slight similarity in amino acid sequence has been observed among c-Myc, Jun, Fos, and the DNA binding domain of GCN4 (14). Helical wheel analyses were carried out on these segments starting at leucine residues 465 of Jun (30), 165 of Fos (31), and 253 of GCN4 (32). Amino acid residues that share identity with C/EBP (see Fig. 1) are boxed. **Fig. 3 (right).** Helical net analysis of a carboxyl-terminal region of C/EBP. The helical region containing the leucine repeats of C/EBP was flattened into two dimensions by splitting the helix lengthwise along the face opposite to the strip of leucines. The display starts in the lower left corner with valine residue number 306. The first helical turn is completed with glutamine residue 309, the second turn is initiated by glutamine residue 310, and so on until proline residue 344. The split along the helix bisects two residues, lysine 324 and glutamine 342. As such, each of these residues is displayed in duplicate, corresponding to the "last" and "first" residues of helical turns 5 and 6 and 10 and 11, respectively. Oppositely charged residues located in positions suitable for ion pairing ($i \pm 3$ or $i \pm 4$) are connected by a dashed line.

folds" that match the contour of DNA. These scaffolds dictate the appropriate positioning of the interacting protein surface, allowing atomic interaction between amino acid side chains and the base pairs that constitute a specific binding site on DNA. Indeed, it is the amino acid sequences of these scaffolds, rather than surface-contacting sequences, that exhibit salient protein-to-protein similarity (6, 8).

There are ample reasons to anticipate new and different structural motifs for DNA binding proteins. For example, crystallographic studies of the restriction endonuclease Eco RI show that its recognition specificity is not established by either of the aforementioned structural motifs (9). Moreover, the amino acid sequences of at least three newly identified sequence-specific DNA binding proteins have failed to show relatedness to either the "helix-turn-helix" motif or the "zinc finger" motif (10, 11). Finally, amino acid sequence analyses of nuclear oncogene products, which have been anticipated to modulate gene expression by binding DNA (12), have failed to show similarity to either of the established DNA binding motifs.

We now describe an amino acid sequence motif common to five DNA binding proteins: three nuclear transforming proteins, and two transcriptional regulatory proteins. The motif consists of a periodic repetition of leucine residues. We propose that the leucines extend from an unusually long α helix, and that the leucine side chains of one helix interdigitate with those of a matching helix from a second polypeptide to form a stable noncovalent linkage. Further-

more, we predict that the paired helices of this class of proteins play a fundamental role in arranging the contact surface for sequence-specific interaction with DNA.

A periodicity of leucines. Rat liver nuclei contain a heat-stable protein that is capable of binding in a sequence-specific manner to regulatory DNA sequences common to a number of different animal virus chromosomes (13). Two of the cis-regulatory DNA sequences to which this protein binds are the "CCAAT homology" common to many promoters of genes that encode messenger RNA, and the "enhancer core homology" common to many viral enhancers. Since it was not initially realized that the same protein accounted for both DNA binding activities, it was variously termed CBP (CCAAT binding protein) or EBP (enhancer binding protein). The physiological role of this protein is poorly understood; as such, we continue to designate the protein merely according to its binding specificity (C/EBP).

The gene that encodes C/EBP has been isolated as a recombinant clone, and the DNA binding domain of the protein has been localized to a 14-kD segment (11). The amino acid sequence of the C/EBP DNA binding domain contains an abundance of residues with charged side chains, especially lysines and arginines; however, no prolines occur within this region. Since proline residues are seldom found in α helices, we arranged the amino acid sequence of C/EBP on a schematic α helix. When analyzed in this way, a 28-amino-acid segment of the DNA binding domain exhibited notable amphipathy (Fig. 1). One side of the hypothetical helix was predominately composed of hydrophobic amino acids (particularly leucines), while the other was composed of residues with charged side chains (six basics and four acidics) and uncharged polar side chains (glutamines, threonines, and serines).

Most conspicuous was the periodic repetition of leucine residues. Leucines appear at every seventh position over a region of 35 amino acids within the DNA binding domain of C/EBP. This distribution is not simply a function of the abundance of this residue since the 28-amino-acid region contains only one other leucine. In a computer-assisted search for related protein sequences, we found substantial similarity between C/EBP and the product of the mouse c-myc proto-oncogene (11). The two proteins share 17 identities within a 30-amino-acid region localized near their respective carboxyl termini. The region of similarity between C/EBP and the transforming protein c-Myc coincides almost perfectly with the proposed α helix of C/EBP shown in Fig. 1. Having noted the unusual repetition of leucines in C/EBP, we arranged the c-Myc sequence on a schematic α helix and discovered the same motif (Fig. 2). The sequence of mouse c-Myc, starting 32 residues upstream from its carboxyl terminus, exhibits leucines at every seventh residue over eight hypothetical turns of α helix. The same periodic array of leucines occurs in human c-Myc, a related human protein termed n-Myc, and at three out of four positions in human L-Myc (Fig. 2). The single deviation from this heptad periodicity of leucines was observed in human L-Myc, and consisted of a replacement by methionine.

Amino acid sequence similarities have been noted between c-Myc and two other nuclear transforming proteins, Fos and Jun (14). Moreover, both structural and functional evidence has indicated that Jun is related to GCN4, a DNA binding protein from yeast that plays a direct role in regulating transcription (14, 15). When displayed on schematic α helices, we noticed that the related sequences of each of these proteins exhibit at least four periodic repeats of leucine residues (Fig. 2).

The invariant occurrence of at least four leucine repeats in five different proteins is not reflective of general sequence similarity. Although any two of the proteins share substantial similarity within this intriguing region (that is, C/EBP is quite similar to c-Myc, and Jun is even more similar to GCN4), no single amino acid, other than the four leucines, is conserved among all four proteins. Indeed, Jun and GCN4 share only one other identity with C/EBP within this 28-residue window, and Fos shares only two other identities (see Fig. 2).

Prediction of unusual helix stability. The leucine repeat common to the aforementioned proteins was observed as a consequence of projecting amino acid sequences on an idealized α helix. If these segments of protein actually exist in an α-helical conformation, the projected helices would be unusually long. The distance from the first leucine to the fourth, 22 amino acids, would require a minimum of six helical turns. C/EBP, Jun, and Fos actually contain five leucine repeats, which would span at least eight α-helical turns. Two of the primary forces that stabilize α helices are the amphipathic arrangement of hydrophobic amino acids and the occurrence of oppositely charged amino acid pairs configured in a manner that allows formation of a salt bridge (16–19). The disposition of hydrophobic residues on one side of an α helix can provide a contiguous array of stabilizing interactions with the globular fold of a protein. In more pronounced cases, hydrophobic interactions of this nature facilitate the coiled-coil intertwining of very long α helices found in keratins, lamins, and paramyosin (20).

The stabilizing influence of ion pairs, which depend on the appropriate juxtaposition of acidic and basic amino acids within an α helix (21), has been inferred from two lines of evidence. On the analytical side, computer searches of the Brookhaven Data Bank of solved protein structures have shown that α helices are relatively rich in ion pairs. In a computer study of all identified α-helical structures, Sundaralingam et al. (18) noticed a proportional increase in the frequency of ion pairs as the size of the helix lengthened. The class including the largest α helices, which consisted of at least six helical turns, contained an average of 0.4 ion pair per turn. In a separate study, Sundaralingam also examined the incidence of ion pairs in a category of α helices common to calmodulin and troponin C (22). These unusual α helices, which connect dual globular domains within each protein, are exceptionally long, stable, and solvent-exposed; moreover, they are extremely rich in intrahelical ion pairs, containing an average density of 0.7 ion pair per helical turn. On the experimental side, short peptides with systematically varied amino acid sequences have been synthesized and tested for their propensity to adopt solvent-stable α helical structure. Using such an approach, Marqusee and Baldwin (19) found that the appropriate juxtaposition of oppositely charged residues fostered helix stability.

As noted above, C/EBP exhibits an amphipathic array of hydrophobic residues in the area that shares sequence similarity with c-Myc (see Figs. 1 and 2). Moreover, this same region of the protein is unusually rich in oppositely charged residues that are juxtaposed in a manner suitable for ion pairing. An α-helical display of the region containing the five leucine repeats of C/EBP (Fig. 3) indicates that eight ion pairs occur within the eight helical turns that separate the first leucine from the fifth leucine. Two additional pairs occur in the helical turns that precede the leucine closest to the amino terminus.

The density of ion pairs observed in C/EBP, roughly one per helical turn and evenly spaced throughout the 30-amino-acid region of interest, is exceptionally high. Only helices known to be long, stable, and solvent-exposed have ion pair densities approaching one per turn. On this basis, we predict that a substantial portion of the DNA binding domain of C/EBP, which includes the leucine repeat motif, exists in a solvent-stable, α-helical conformation. Indeed, since the sequence-specific DNA binding activity of C/EBP is unusually heat-stable (13), we suggest that the DNA binding surface may occur within or adjacent to this helix.

Interhelical interdigitation of leucines: The zipper model. Why do each of the nuclear proteins that we have considered contain leucine residues at every seventh position over at least six helical turns? The most obvious response to this question is that the leucines establish amphipathy, which helps to stabilize a long α helix. The "helical wheel" plots of Figs. 1 and 2 show that C/EBP, GCN4, and Jun contain hydrophobic residues at three out of four positions on the "spoke" immediately adjacent to that containing the four leucines. This arrangement establishes a continuous "spine" of hydrophobicity over six to eight helical turns. The opposite face of each putative helix is rich in amino acids with either charged side chains or uncharged polar side chains. According to our conventions, "spokes" 3 and 6 are opposite from the "spoke" containing leucine residues. Out of the 40 residues that occupy "spokes" 3 and 6 in the five proteins (C/EBP, c-Myc, Fos, Jun, and GCN4), 21 bear charged side chains and 11 bear uncharged polar side chains.

Although amphipathy is a common feature of each of these putative α helices, the degree of surface hydrophobicity along the nonpolar side is not remarkably high. Two of the proteins have almost no hydrophobic amino acids at "spoke" positions adjacent to the leucines; Fos has only a single isoleucine to abet hydrophobicity, and c-Myc has only a single leucine (both are located at position 3 of "spoke" 5; see Fig. 2). In these cases, surface hydrophobicity is limited to a very thin "ridge" consisting of one leucine residue every other turn of the helix. Furthermore, if hydrophobicity were the only property necessary along the nonpolar side of these helices, it is perplexing that leucines would be used to the virtual exclusion of all other hydrophobic amino acids.

We have used a computer program to generate a molecular model of C/EBP in the region including its leucine repeats (23). The side chains of C/EBP were programmed onto the α carbon backbone of an idealized α helix and displayed in three dimensions. The graphics program highlights van der Waals radii in color. We chose to display the periodically repeating leucine residues of C/EBP in blue and all remaining amino acids in red. Figure 4 shows the putative C/EBP helix as viewed from three different perspectives around the long axis of the helix, each differing by 90°.

Three-dimensional projections of C/EBP show that the van der Waals radii of the repeating leucine side chains are insufficiently close to provide stabilizing intrahelical interactions. This observation reinforced the need to find a hydrophobic surface to match the thin "ridge" of hydrophobicity. Such a surface could exist within the globular fold of the same polypeptide. Alternatively, as in the case of coiled-coil α helices, the complementary surface might be presented by a separate polypeptide. Our attention became focused on the latter possibility for several reasons. (i) The leucine motif maintains strict adherence to a heptad repeat. In no case was the continuity of the repeat shifted by even a single amino acid. The heptad repeat is the quintessential feature shared by all proteins that adopt a coiled coil structure (20). (ii) The hydrophobic surface of the leucine repeat motif seemed unusually thin. If the helix were to exist within a globular fold, the properties of that fold would have to accommodate the unusual aspects of the helix (the narrowness of its hydrophobic "ridge," an abundance of charged amino acids, and an abundance of uncharged polar amino acids).

If we assume that the "ridge" of hydrophobicity of each helix requires a matching surface, and consider that this surface might be donated by a separate polypeptide, perhaps the best candidate is the matching surface donated by the same helix from a second monomer. That is, the hydrophobic surfaces of two leucine helices might interact to form a dimer. Although based on the coiled-coil paradigm, this hypothetical structure is distinctive in several ways. First, the interaction surface is relatively short. Keratins, lamins and other proteins that adopt a coiled-coil structure rely on dozens or even hundreds of heptad repeats, resulting in extensively interwoven helices (20). Moreover, the interhelical hydrophobic interactions of those proteins can be established by almost any hydrophobic amino acid. The motif that we propose covers only four or five heptad repeats, and depends almost exclusively on leucine side chains.

Why is leucine used to the virtual exclusion of all other hydrophobic residues? An examination of the structural properties of hydrophobic amino acids may be instructive in this regard.

The leucine side chain has two methyl groups extending from its single γ carbon, and no methyl groups appended to its β carbon. As such, its side chain is long, symmetrical, and bulky at the tip. We predict that these properties allow the leucine residues from one α helix to interdigitate with those of a second α helix, forming a molecular zipper between two polypeptides (Fig. 5A). It is important to note that an ideal α helix contains slightly more than 3.5 residues per turn. Therefore, if our hypothesis is correct, the paired helices must be distorted in order to maintain maximum packing of the leucine side chains.

Our model of ordered interdigitation may account for the exclusive use of leucine. Like the tooth of a zipper, the leucine side chain is relatively long and bulky at its tip. Isoleucine also contain an extended side chain that is appended with two methyl groups. However, one of the methyl groups is attached to the β carbon of the isoleucine side chain. We propose that the projection of a methyl group from a β carbon would interfere sterically with the sequential interdigitation of hydrophobic side chains. According to these interpretations, valine would constitute a particularly ill-suited amino acid. Its side chain is short (relative to leucine) and is appended with two methyl groups extending from the β carbon atom. Rather than acting to lock two helixes together, valine would block interdigitation. Given these considerations, methionine might be expected to constitute the most suitable alternative to leucine. Like leucine, its β carbon is free of attached methyl groups. Moreover, it contains both a methyl group and a bulky sulfur atom at the tip of its side chain. Out of 31 heptad repeats analyzed in our study, we observed 30 leucines and one methionine (see Fig. 2).

The model that we have presented makes the explicit prediction that the "leucine zipper" would represent the dimerization domain of this class of DNA binding proteins. Although this prediction has not been tested directly, it is known that GCN4 and C/EBP exist in solution as stable dimers (24). Moreover, the domains of both proteins critical for dimerization map to regions coincident with DNA binding, which includes the periodic leucine repeats. The possibility that a stable dimerization interface can be generated from an α helix of less than ten turns has been demonstrated in model systems with synthetic peptides (25).

If our "leucine zipper" model is correct, then the α helices must interlock in one of two orientations, parallel or anti-parallel relative to the amino-to-carboxyl dipole of each helix. Crystallographic studies have shown that helix packing can occur in either orientation (17). However, all precedents for coiled-coil interaction between two polypeptides adopt a parallel conformation (20). Despite this fact, we offer several reasons to anticipate that the "leucine zipper" motif exists in an antiparallel conformation. First, the intimate degree of side chain packing implicit to our model might only be compatible with antiparallel helices. Amino acid side chains are disposed at an angle pointing toward the amino terminus of the helix (Fig. 4); as such, an antiparallel arrangement might be better suited for side chain interlocking. Second, an antiparallel conformation would allow the dipole of one helix to attract, rather than repel, the dipole of the matching helix.

We anticipate that the issue of helical orientation will be central to the mode by which these proteins bind DNA. Either conformation, parallel or antiparallel, results in a rotationally symmetric molecule (Fig. 5B). In an antiparallel conformation, the axis of rotation would be perpendicular to the paired helices, whereas in a parallel conformation, the axis would be in line with the helices. The relevance of helix orientation is underscored by the fact that the "leucine zipper" alone is not sufficient to confer sequence-specific interaction with DNA. Amino acid sequence analysis of C/EBP has shown that the protein contains a high proportion of basic residues in a 30-amino-acid region immediately adjacent to its "leucine zipper" (11). This highly basic region of C/EBP exhibits substantial sequence similarity to a region of the Fos-transforming protein that is juxtaposed identically to its "leucine zipper" (11). Moreover, the C/EBP basic region must remain intact in order for C/EBP to bind DNA (26). We predict that the "leucine zipper" juxtaposes the basic regions of two polypeptides in a manner suitable for sequence-specific recognition of DNA (Fig. 6).

By comparing the amino acid sequences of several DNA binding proteins we have discovered a repeating motif of leucine residues. We propose that these leucines project from comparatively long, stable, solvent-exposed α helices, and that the leucine residues that

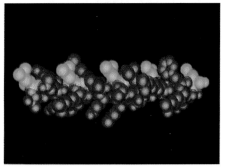 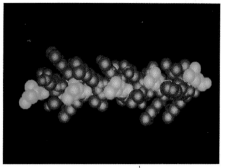 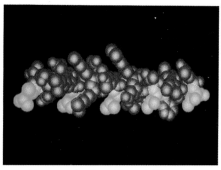

Fig. 4. Three-dimensional model of the helical leucine repeat region of C/EBP. The amino acid side chains of C/EBP between leucine 315 and lucine 343 were appended onto the α carbon backbone of an idealized α helix and displayed in three dimensions by means of a computer graphics program (*23*). Leucine residues that occur at every seventh residue are displayed in blue, all other residues are displayed in red. The amino-to-carboxyl dipole of the helix is arranged left to right. Each successive view of the helix is rotated by 90° around an axis parallel to the helix. Note that the distance between leucine residues exceeds the van der Waals radii of the side chain R groups and that side chains tilt toward the amino terminus of the helix.

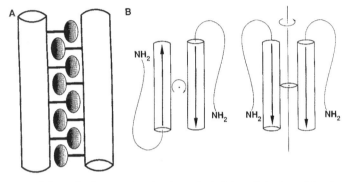

Fig. 5. Schematic diagram showing hypothetical interdigitation of leucine side chains between two α helices. (**A**) Two parallel tubes represent the approximate dimensions of the α carbon backbone of idealized α helices. Interdigitating protrusions symbolize leucine side chains. The sphere located at the tip of each residue represents the two methyl groups attached to the γ carbon atom of the leucine side chain. (**B**) C/EBP dimers disposed in either an antiparallel (left) or parallel array (right) with respect to the leucine repeat. Thick arrows within helical cylinders denote amino-to-carboxyl dipole. Thin arrows denote axis of rotational symmetry.

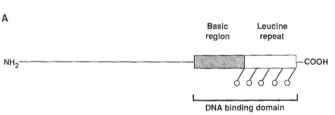

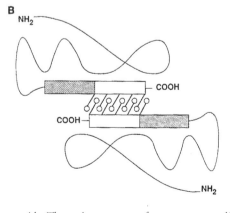

Fig. 6. Disposition of the leucine repeat and DNA binding domain within the intact C/EBP polypeptide. (**A**) Location of the leucine repeat within the intact C/EBP polypeptide. The region necessary for sequence-specific interaction with DNA extends beyond the leucine repeat toward the amino-terminus of the protein, and includes a 30-amino-acid region that is highly positively charged (stippled). (**B**) A hypothetical model of a C/EBP dimer established by the interdigitation of leucine repeat helices in an antiparallel conformation.

project from one α helix interdigitate with those of a second helix, causing the two molecules to dimerize. We believe that this motif, the "leucine zipper," represents a part of the scaffold that molds a protein to interact with its target site on DNA. Finally, we point out the possibility that the "leucine zipper" might allow dimerization of different polypeptides so long as each subunit contained the motif. Indeed, recent evidence has raised the possibility that the Fos and Jun transforming proteins, which have been shown to contain the leucine repeat motif, form a heterotypic complex (*27*). The notion of heterotypic interactions raises potentially important implications relating to the combinatorial action of gene regulatory proteins, and may facilitate a more direct attack on the function of nuclear oncogenes.

REFERENCES AND NOTES

1. W. F. Anderson, D. H. Ohlendorf, Y. Takeda, B. W. Mathews, *Nature* **290**, 754 (1981); D. B. McKay and T. A. Steit, *Nature ibid.*, p. 744; C. O. Pabo ard M. Lcwis, *Nature* **298**, 443 (1982); R. W. Schevitz, Z. Otwinowski, A. Joachimiak, C. L. Lawson, P. B. Sigler, *ibid.* **317**, 782 (1985).
2. T. A. Steitz, D. H. Ohlendorf, D. B. McKay, W. F. Anderson, B. W. Mathews, *Proc. Natl. Acad. Sci. U.S.A.* **79**, 3097 (1982).
3. D. R. Engelke, S.-Y. Ng, B. S. Shastry, R. G. Roeder, *Cell* **19**, 717 (1980); H. R. B. Pelham, and D. D. Brown, *Proc. Natl. Acad. Sci. U.S.A.* **77**, 4170 (1980); J. S. Hanas, D . J . Hazuda, D. F. Bogenhagen, F. Y.-H. Wu, C.-W. Wu, *J. Biol. Chem.* **258**, 14120 (1983).
4. R. S. Brown, C. Sander, P. Argos, *FEBS Lett.* **186**, 271 (1985); J. Miller, A. D. McLachlan, A. Klug, *EMBO J.* **4**, 1609 (1985).
5. G. P. Diakun, L. Fairall, A. Klug, *Nature* **18**, 698 (1986); A. D. Frankel J. M. Berg, C. O. Pabo, *Proc. Natl. Acad. Sci. U.S.A.* **84**, 4841 (1987).
6. J. M. Berg, *Science* **232**, 485 (1986).
7. T. A. Hartshorne, H. Blumberg, E. T. Young, *Nature* **320**, 283 (1986); U. B. Rosenberg *et al.*, *ibid.* **319**, 336 (1986); J. T. Kadonaga, K. R. Carner, F. R. Masiarz, R. Tjian, *Cell* **51**, 1079 (1987); D. C. Page *et al.*, *ibid.*, p. 109; R. M. Evans and S. M. Hollenberg, *ibid.* **52**, 1 (1988).
8. C. O. Pabo and R. T. Sauer, *Annu. Rev. Biochem.* **53**, 293 (1984).
9. J. A. McClarin *et al.*, *Science* **234**, 1526 (1986).
10. I. Hope and K. Struhl, *Cell* **46**, 885 (1986); J. L. Pinkham, J. T. Oleson, L. P. Guarente, *Mol. Cell. Biol.* **7**, 578 (1987).
11. W. H. Landschulz, P. F. Johnson, E. Adashi, B. J. Graves, S. L. McKnight, *Genes Dev.*, in press.
12. R. E. Kingston, A. S. Baldwin, Jr., P. A. Sharp, *Nature* **312**, 280 (1984); C. Setoyama, R. Frunzio, G. Liau, M. Mudryi, B. de Crombrugghe, *Proc. Natl. Acad. Sci. U.S.A.* **83**, 3213 (1986); R. Kaddurah-Daouk, J. M. Green, A. Baldwin, Jr., R. Kingston, *Genes Dev.* **1**, 347 (1987); R. J. Distel, H.-S. Ro, B. S. Rosen, D. L. Groves, B. M. Spiegelman, *Cell* **49**, 835 (1987); K. Loch, K. Anderson, R. Brent, *ibid.* **52**, 179 (1988).
13. B. J. Graves, P. F. Johnson, S. L. McKnight, *Cell* **44**, 565 (1986); P. F. Johnson, W. H. Landschulz, B. J. Graves, S. L. McKnight, *Genes Dev.* **1**, 133 (1987).
14. P. K. Vogt, T. J. Bos, R. F. Doolittle, *Proc. Natl. Acad. Sci. U.S.A.* **84**, 3316 (1987).
15. K. Struhl, *Cell* **50**, 841 (1987).
16. G. E. Schulz and R. H. Schirmer, *Principles of Protein Structure* (Springer-Verlag, New York, 1979).
17. C. Chothia, *Annu. Rev. Biochem.* **53**, 537 (1984).
18. N. Sundaralingam, Y. C. Sekharudu, N. Yathindra, V. Ravichandran, *Proteins* **2**, 64 (1987).

19. S. Marqusee and R. L. Baldwin, *Proc. Natl. Acad. Sci. U.S.A.* **84**, 8898 (1988).

20. F. H. C. Crick, *Nature* **170**, 882 (1952); D. A. D. Parry, W. G. Crewther, R. D. B. Fraser, T. P. MacRae, *J. Mol. Biol.* **113**, 449 (1977); F. D. McKeon, M. W. Kirschner, D. Caput, *Nature* **319**, 463 (1986); C. Cohen and K. C. Holmes, *J. Mol. Biol.* **6**, 423 (1963).

21. By ion pair we refer simply to the $i + 3$ and $i + 4$ juxtaposition of basic (lysine and arginine) and acidic (glutamate and aspartate) amino acids. Marqusee and Baldwin (*19*) have indicated that the $i + 4$ glutamate-lysine arrangement fosters helix stability better than the $i + 3$ arrangement, and that the stability imparted by pairs of oppositely charged residues is probably mediated by salt bridges.

22. M. Sundaralingam, W. Drendel, M. Greaser, *Proc. Natl. Acad. Sci. U.S.A.* **82**, 7944 (1985).

23. Computer graphic modeling was done with the Promodel software package (New England Biographics). We placed the side chains found in the leucine repeat region of C/EBP on the idealized α-helical backbone specified by the software. The helix mantains 3.6 residues per turn with uniformly constant phi and psi angles, helical pitch, and rise. Although the program uses the lowest average conformation for the side chains, it neither alters side chain conformation due to neighboring residues nor evaluates plausibility of a proposed structure; no steric hindrance for any residues was observed.

24. I. Hope and K. Struhl, *EMBO J.* **6**, 2781 (1987); W. H. Landschulz, P. B. Sigler, S. L. McKnight, unpublished data. The carboxyl-terminal 18 kD of C/EBP was expressed in bacteria and purified to homogeneity. A 1 μM solution of the protein was treated with 0.001 percent glutaraldehyde at room temperature for 1 to 3 hours, and the products were analyzed by SDS–polyacrylamide gel electrophoresis. Within 3 hours, more than 90 percent of the 18-kD species was cross-linked into a 35-kD species.

25. S. Y. M. Lau, A. K. Taneja, R. S. Hodges, *J. Biol. Chem.* **259**, 13253 (1984).

26. W. H. Landschulz and S. L. McKnight, unpublished data. A recombinant DNA clone encoding C/EBP was mutagenized in vitro and expressed in bacteria. Amino-carboxyl deletion mutants were capable of sequence-specific interaction with C/EBP binding sites on DNA so long as the leucine repeat and adjacent basic region remained intact.

27. B. R. Franza, Jr., F. J. Rauscher III, S. F. Josephs, T. Curran, *Science* **239**, 1150 (1988); F. J. Rausher III *et al., ibid.* **240**, 1010 (1988); R. Tjian and R. Turner, personal communication.

28. L. W. Stanton, P. D. Pahrlander, P. M. Tesser, K. B. Marcu, *Nature* **310**, 423 (1984).

29. F. Kaye *et al., Mol. Cell. Biol.* **8**, 186 (1988); R A. DePinho, D. S. Hatton, A. Tesfaye, G. D. Yancopoulos, F. W. Alt, *Genes Dev.* **1**, 311 (1987).

30. Y. Maki, T. J. Bos, C. Davis, M. Starbuck, P. K. Vogt, *Proc. Natl. Acad. Sci U.S.A.* **84**, 2848 (1987).

31. C. van Beveren, F. van Straaten, T. Curran, R. Mueller, I. M. Verma, *Cell* **32**, 1241 (1983)

32. A. G. Hinnebusch, *Proc. Natl. Acad. Sci. U.S.A.* **81**, 6442 (1984).

33. We thank R. Kretsinger and P. Sigler for invaluable input; H. Smith for assistance with computer graphics; K. LaMarco for comments on the manuscript; C. Jewel for technical assistance; and C. Pabo, R. Baldwin, K. Yamamoto, A. Klug, and J. Berg for critical comment. Supported by a Medical Scientist Scholarship of the Life and Health Insurance Medical Research Fund (W.H.L.), postdoctoral fellowship awards from the Damon Runyon/Walter Winchell Cancer Fund and the Carnegie Corporation (P.F.J.), a Faculty Research Award from the American Cancer Society (S.L.M.), and grant award (S.L.M.) from NIH.

22 March 1988; accepted 20 May 1988

Activation domains in transcription factors

Once a transcription factor has bound to DNA, it evidently needs to influence the rate of transcription of the gene to which it has bound. Initially, two possibilities were considered to explain how this might occur (Figure 11.1). In the first mechanism, the binding of the transcription factor to DNA would somehow alter the structure of the DNA so that it was more readily transcribed (Figure 11.1a). Alternatively, the activation of transcription could occur without a change in the structure of the DNA, but rather via the interaction of the factor with other proteins involved in transcription (Figure 11.1b).

These two possibilities were distinguished by the paper by Brent and Ptashne which is presented here. They fused the entire coding region of the yeast transcriptional activator GAL4 to the DNA binding region of a completely different factor, known as LexA, derived from bacteria. As indicated in Table 2 of the paper, the LexA–GAL4 fusion protein is capable of activating gene transcription, but only when the gene contains a binding site for the LexA protein (*lexA* op), and this effect occurs regardless of whether the LexA binding site is located at position −178 or −577 relative to the transcriptional start site. Moreover, as indicated in Table 5 of the paper, this effect can also be observed when the LexA binding site is located downstream of the normal transcriptional start site.

In all these cases (see, for example, Tables 2 and 5 in the paper by Brent and Ptashne), activation of transcription was not observed with the LexA protein in isolation, indicating that it was dependent on the GAL4 portion of the fusion protein. Hence, as illustrated schematically in Figure 6 of the paper, gene activation only occurs in the presence of the GAL4 region and not with LexA alone.

This indicates that GAL4 can activate transcription when delivered to the gene via the LexA DNA binding domain. Since in this situation the GAL4 protein does not bind directly to the DNA itself, this eliminates models of the type shown in Figure 11.1(a) in which binding of the transcription factor changes the DNA structure in such a manner as to allow increased transcription. Rather, it strongly indicates that activating transcription factors such as GAL4 act as illustrated in Figure 11.1(b) by interacting directly with other proteins involved in transcription so as to stimulate their activity.

Moreover, in an observation noted in the paper by Brent and Ptashne but not directly presented, it is indicated that the stimulation of transcription cannot be observed with a fusion protein containing only the first 74 amino acids of GAL4 and lacking the remaining portion of the protein. This indicates that the region of GAL4 that interacts with other proteins to stimulate transcription is likely to be located at the C-terminus of the molecule, whereas the N-terminus is involved in mediating DNA binding. The region of a transcription factor that allows it to activate transcription has been called the activation domain, and this generally constitutes a region of the molecule distinct from that which performs DNA binding (for reviews, see Latchman, 1995a,b). Indeed, a number of

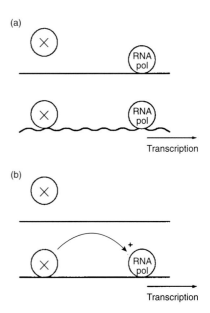

Figure 11.1 Possible mechanisms by which a transcription factor could influence transcription. In (a), the transcription factor (X) binds to DNA and alters its structure so that it can be more readily transcribed by RNA polymerase (RNA pol). In (b), the factor binds to DNA and interacts with the RNA polymerase by a protein–protein interaction to stimulate its activity and therefore increase transcription.

such activation domains have now been mapped within individual transcription factors by using the approach pioneered by Brent and Ptashne, in which different regions of a transcription factor are linked to the DNA binding domain of another factor and the ability of the fusion product to stimulate transcription is assessed (Figure 11.2).

The paper by Brent and Ptashne is thus of considerable importance, in that it both proved that transcription factors stimulate transcription by interacting with other proteins and also pioneered the method of identifying the activation domains that achieve this effect within the transcription factor molecule.

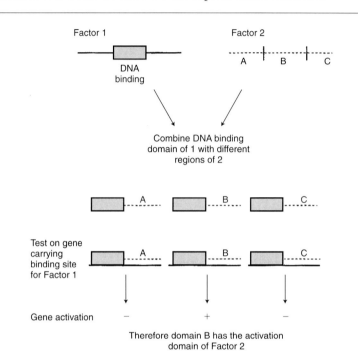

Figure 11.2 Identification of the activation domain of a transcription factor in a domain swapping experiment. The various regions of Factor 2 are combined with the DNA binding region of Factor 1, and the ability to activate transcription on a gene carrying the binding site for Factor 1 is assessed. Only region B can confer the ability to activate transcription following DNA binding, and this region therefore must contain the activation domain of Factor 2.

Brent & Ptashne (1985) Cell **43**, 729–736

A Eukaryotic Transcriptional Activator Bearing the DNA Specificity of a Prokaryotic Repressor

Roger Brent* and Mark Ptashne

Department of Biochemistry and Molecular Biology
Harvard University
7 Divinity Avenue
Cambridge, Massachusetts 02138

Summary

We describe a new protein that binds to DNA and activates gene transcription in yeast. This protein, LexA-GAL4, is a hybrid of LexA, an Escherichia coli repressor protein, and GAL4, a Saccharomyces cerevisiae transcriptional activator. The hybrid protein, synthesized in yeast, activates transcription of a gene if and only if a *lexA* operator is present near the transcription start site. Thus, the DNA binding function of GAL4 can be replaced with that of a prokaryotic repressor without loss of the transcriptional activation function. These results suggest that DNA-bound LexA-GAL4 and DNA-bound GAL4 activate transcription by contacting other proteins.

Introduction

In Saccharomyces cerevisiae, the protein GAL4 turns on transcription of the *GAL1* gene when bound to an upstream region called UAS_G. This region contains four 17 bp sites of related sequence, a near-consensus of which (the "17-mer") mediates GAL4 activity in vivo and binds GAL4 in vitro (Giniger et al., 1985; Keegan, personal communication). Both UAS_G and a single 17-mer function when placed at several positions within a region between 40 and 600 nucleotides from the *GAL1* transcription start site, or when placed upstream of a different gene, *CYC1* (Guarente et al., 1982b; West et al., 1984; Giniger et al., 1985). GAL4 is active in wild-type strains only when cells are grown on medium containing galactose, because, it is thought, growth on this medium leads to dissociation of GAL4 from an inhibitory protein, GAL80 (Oshima, 1982). GAL4 activity is reduced when cells are grown on medium that contains glucose and galactose (Oshima, 1982; Yocum et al., 1984), at least partly because GAL4 binds UAS_G inefficiently under these conditions (Giniger et al., 1985).

Upstream activation sites (UASs) have been found upstream of all RNA polymerase II-dependent yeast genes the regulatory regions of which have been carefully studied. For example, upstream of *CYC1* are two sites called UAS_{C1} and UAS_{C2}. Cellular gene products, probably encoded by the *HAP1*, *HAP2*, and *HAP3* genes (Guarente et al., 1984), presumably interact with these sites. If UAS_G is inserted upstream of *CYC1* in place of UAS_{C1} and UAS_{C2}, *CYC1* transcription becomes dependent on GAL4 and is regulated like *GAL1* transcription. Although the properties of UASs are similar in other respects to those of the enhancer sequences found in higher eukaryotes, UAS_G and the two UAS_Cs have been reported to be inactive when positioned downstream of the transcription start point of a gene (Struhl, 1984; Guarente and Hoar, 1984).

The current investigation was prompted by a consideration of two mechanisms by which GAL4 might turn on transcription. According to the first, GAL4 would bind to DNA in some way that would stabilize an unusual DNA structure (e.g., left-handed DNA), and the perturbed structure would then somehow be transmitted down the helix, where it would help proteins bind near the transcription start. According to the second idea, GAL4 would contact DNA without greatly perturbing the structure of the DNA around the binding site, and activation of transcription would occur when GAL4 touches other proteins. In Escherichia coli, lambda repressor acts as a positive regulator (of its own gene) by the second mechanism; repressor binds to a site adjacent to the RNA polymerase binding site and touches RNA polymerase. One line of evidence that led to this picture was the isolation of lambda repressor mutants called *pc* (for Positive Control) that bind DNA but fail to activate transcription (Guarente et al., 1982a). The amino acids changed in *pc* mutants are clustered in a region on the surface of the lambda repressor molecule (Hochschild et al., 1983) that is thought on the basis of other experiments to be that portion of the molecule that touches RNA polymerase.

Consideration of the lambda experiments led us to try to separate the ability of GAL4 to bind DNA from its ability to stimulate transcription. However, instead of seeking to preserve GAL4's DNA binding while eliminating its ability to activate transcription, we sought to confer a new DNA binding specificity on GAL4 while preserving its ability to stimulate transcription. To this end, we constructed a new protein called LexA-GAL4, the DNA binding specificity of which came from an E. coli repressor called LexA.

In E. coli, LexA represses many genes. Like the repressors of lambda-like phages, LexA probably binds as a dimer to its operators (R. Brent, Ph.D. thesis, Harvard University, Cambridge, Massachusetts, 1982). Moreover, the LexA monomer seems to have an overall organization similar to that of the phage repressors; an amino terminal domain that binds operator DNA and contains weak dimerization contacts, a carboxy-terminal domain that contains stronger dimer contacts, and a flexible hinge region that connects the two (Brent and Ptashne, 1981; R. Brent, Ph.D. thesis 1982; Little and Hill, 1985; Shnarr et al., 1985). The first 87 amino acids of LexA contain the information necessary for specific binding to the LexA operator (Brent, unpublished) and 16 amino acids of the putative hinge region (Little and Hill, 1985). If the cellular DNA is damaged, RecA protein and amino acids within the C-terminus of LexA catalyze cleavage of LexA within the hinge region.

* Present address: Department of Molecular Biology, Massachusetts General Hospital, 50 Blossom Street, Boston, Massachusetts 02114, and Department of Genetics, Harvard University Medical School, Boston, Massachusetts 02115.

1a.

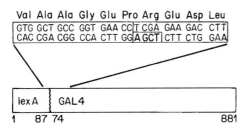

Val	Ala	Ala	Gly	Glu	Pro	Arg	Glu	Asp	Leu

GTG GCT GCC GGT GAA CC|T CGA| GAA GAC |CTT|
CAC CGA CGG CCA CTT GG|AGCT| CTT CTG |GAA|

lexA	GAL4	
1	87 74	881

b.

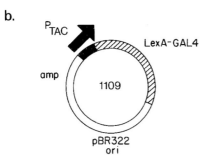

P_{TAC}

amp

1109

LexA-GAL4

pBR322
ori

c.

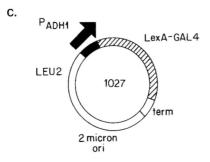

P_{ADH1}

LEU2

1027

LexA-GAL4

term

2 micron
ori

Figure 1. LexA-GAL4 and Plasmids That Direct Its Synthesis in Bacteria and in Yeast

(A) Sequence of LexA-GAL4 fusion junction. *lexA* coding sequence extending to the Xmn I site within the hinge region (vertical line), was abutted to a filled-in Xho I site in *GAL4*, as described in Experimental Procedures. Bases of the filled-in site are shown in boldface. This ligation recreated the Xho I site Amino acid sequence of the protein near the fusion junction is shown above the line. (B) Plasmid 1109, which directs synthesis of LexA-GAL4 in E coli. (C) Plasmid 1027, which directs synthesis of LexA-GAL4 in yeast.

Because it is unable to dimerize efficiently, the proteolytic amino-terminal fragment binds operator with greatly reduced efficiency, and transcription of LexA-repressed genes is induced (reviewed in Walker, 1984; and Little and Mount, 1982).

We have recently reported the synthesis of LexA in yeast. If a LexA operator is inserted upstream of *GAL1* between UAS_G and the transcription start, LexA represses *GAL1* transcription. We inferred from this that LexA enters the yeast nucleus and binds to the *lexA* operators upstream of *GAL1* (Brent and Ptashne, 1984). In this paper we describe the construction of a gene that encodes a hybrid protein, LexA-GAL4. When LexA-GAL4 is synthesized in E. coli, it binds to *lexA* operators. When LexA-GAL4 is synthesized in yeast, it activates transcription, if

Table 1. LexA-GAL4 Represses *lexA* Transcription

Plasmid	Regulatory Protein Produced	Units of β-galactosidase
pBR322	none	2000
pKB280	Lambda repressor	1800
PRB451	LexA	120
1109	LexA-GAL4	140

LexA-GAL4 represses transcription in E coli. The bacterial strain used, RB1003, lacks functional LexA, but contains a *lacZ* gene, the transcription of which is directed by a *lexA* promoter. It was transformed separately with the plasmids below, which encoded regulatory proteins under the control of the lac or tac promoters. In these experiments, IPTG was added to the culture medium to inactivate lac repressor and induce synthesis of the respective proteins. Cells were grown to a density of 1×10^6 cells/ml in LB medium that contained 60 μg/ml carbenicillin. At this point IPTG was added to the culture to a concentration of 5×10^{-4} M. Growth was resumed for 4 hr, and β-galactosidase was assayed, and numbers refer to units of β-galactosidase activity, computed as described in Miller, 1972, in different cultures.

and only if a *lexA* operator is present near the start point of transcription.

Results

The gene that encodes LexA-GAL4 was derived from two DNA fragments, one encoding the amino-terminal 87 residues of LexA, and the other encoding the carboxy-terminal 807 amino acids of GAL4. These fragments were ligated and were inserted into plasmids that directed synthesis of LexA-GAL4, in one case in bacteria under the control of the tac promoter, and in the other case in yeast under the control of the *ADH1* promoter (see Experimental Procedures and Figure 1).

LexA-GAL4 Recognizes *lexA* Operators in E. coli

Two experiments show that, in E. coli, LexA-GAL4 recognizes *lexA* operators. The first test exploits the fact that LexA represses transcription of its own gene (Brent and Ptashne, 1980). In the bacterial strain used in this experiment, the *lacZ* gene was fused to the *lexA* promoter, so that the amount of β-galactosidase in the strain was a measure of transcription from that promoter. The strain also carried a mutation that inactivated the chromosomal *lexA* gene. Table 1 shows that LexA-GAL4 repressed transcription from the LexA promoter by 16-fold. The second test is based on the fact that derepression of certain LexA-repressed genes is necessary for recovery from DNA damage. Cells that contain a mutant form of LexA, which recognizes operators but cannot be proteolysed, are especially sensitive to the lethal effect of ultraviolet irradiation (UV) (reviewed in Walker, 1984; and Little and Mount, 1982). Proteolysis of LexA in vivo is catalyzed by RecA, which apparently recognizes, in part, amino acids in the C terminus of LexA (Little, 1984). Since LexA-GAL4 lacks the C terminus of LexA, we expected that otherwise wild-type E. coli containing LexA-GAL4 would be UV-sensitive. Figure 2 shows that, as expected, cells containing LexA-GAL4 were profoundly UV-sensitive (Figure 2).

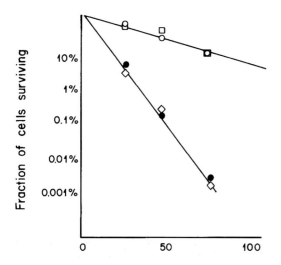

Figure 2. Effect of LexA-GAL4 Synthesis on Sensitivity of E. coli to Killing by UV Irradiation

The top line shows UV sensitivity of cells transformed with a plasmid that contains a promoter but does not direct the synthesis of a regulatory protein (open squares), or with a plasmid that directs the synthesis of lambda repressor (open circles). The bottom line shows the sensitivity of cells transformed with plasmids that direct the synthesis of LexA (closed circles) or of LexA-GAL4 (diamonds).

LexA-GAL4 Activates Transcription in Yeast when Bound to a *lexA* Operator

We transformed yeast with a plasmid that directs the synthesis of LexA-GAL4 (Figure 1) and separately transformed cells with a plasmid that directs synthesis of native LexA (Brent and Ptashne, 1984). In addition, we transformed these strains with plasmids that carried the constructs shown in Figure 3. As shown in the figure, these plasmids carry part of the *GAL1* or *CYC1* gene fused to *lacZ*. Upstream of the fusion gene the plasmids contained one of the following: UAS_G, the 17-mer, UAS_{C1}, and UAS_{C2}, a *lexA* operator, or none of these elements. To determine the amount of transcription of the *lacZ* fusion genes, we measured the amount of β-galactosidase activity in cultures of these doubly transformed cells.

LexA-GAL4 stimulated production of β-galactosidase directed by the *CYC1-lacZ* fusion gene if and only if the plasmid contained a *lexA* operator (Table 2). When the *lexA* operator was located 590 nucleotides upstream of the nearest *CYC1* transcription start site, β-galactosidase production was two-thirds that obtained when the *lexA* operator was positioned 178 nucleotides upstream. Compared with the amount obtained when cells were grown in medium that contained galactose as the only carbon source, the amount of β-galactosidase directed by LexA-GAL4 in cells grown in glucose medium was diminished by about a factor of three. Native LexA did not stimulate β-galactosidase production from these plasmids.

LexA-GAL4 also stimulated production of β-galactosidase from a plasmid that contained the *GAL1-lacZ* fusion gene if and only if the plasmid contained a *lexA* operator

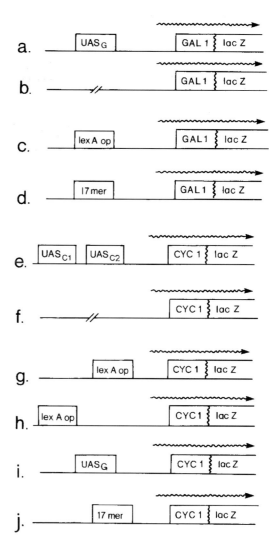

Figure 3. DNA Constructions Used in These Experiments

In each construction, the name of the plasmid, and the distance from the primary or most upstream transcription start site (for the *GAL1* and *CYC1* constructions, respectively) to the upstream element is indicated. The methods used to make these constructions are described in Experimental Procedures. (a) LR1Δ20B, 128 nucleotides; (b) LR1Δ1 lacks an upstream element; (c) 1145, 167 nucleotides; (d) pSV15, 128 nucleotides; (e) pLG669-Z, about 190 nucleotides; (f) pLG570Z lacks an upstream element; (g) 1155, 178 nucleotides; (h) 1057, 577 nucleotides; (i) pLGSD5, about 190 nucleotides; (j) pSV14, 178 nucleotides.

(Table 3). Table 3 also shows that, growth of cells in glucose medium decreased that stimulation by about half. Native LexA did not stimulate β-galactosidase production.

The experiments described in Table 2 and Table 3 were performed in a $GAL4^+$ host. We have performed similar experiments in two other strains, one of which carried a *gal4* point mutation, the other a *gal4* deletion. In these strains, LexA-GAL4 stimulated β-galactosidase production from the CYC1-lacZ and GAL1-lacZ plasmids if and only if they contained a *lexA* operator. In particular, LexA-GAL4 did not stimulate β-galactosidase production from plasmids that contained UAS_G but no *lexA* operator, nor from an integrated *Gal1-lacZ* fusion gene, nor did it complement the

Table 2. LexA-GAL4 Activates Transcription of a *CYC1-lacZ* Fusion Gene

Growth Medium	Upstream Element	Regulatory Protein	
		LexA	LexA-GAL4
Galactose	no UAS (f)	<1	<1
	lexA op at − 178 (g)	<1	590
	lexA op at − 577 (h)	<1	420
	UAS$_{C1}$ and UAS$_{C2}$ (e)	550	500
	UAS$_G$ (i)	950	950
	17-mer (j)	600	620
Glucose	no UAS (f)	<1	<1
	lexA op at − 178 (g)	<1	210
	lexA op at − 577 (h)	<1	140
	UAS$_{C1}$ and UAS$_{C2}$ (e)	180	160
	UAS$_G$ (i)	<1	<1
	17-mer (j)	<1	<1

The letter in parentheses in each case refers to the construction diagrammed in Figure 3 and described in Experimental Procedures. Numbers denote units of β-galactosidase activity measured in cultures of doubly transformed yeast as described in Experimental Procedures.

Table 3. LexA-GAL1 Activates Transcription of a *GAL-lacZ* Fusion Gene

Growth Medium	Upstream Sequence	Regulatory Protein	
		LexA	LexA-GAL4
Galactose	no UAS (b)	0	0
	lexA op at − 167 (c)	<1	520
	17-mer (d)	1050	1000
	UAS$_G$ (a)	1800	1800
Glucose	no UAS (a)	0	0
	lexA op at − 167 (c)	<1	190
	17-mer (c)	<1	<1
	UAS$_G$ (a)*	160	150

The letter in parentheses in each case refers to the construction diagrammed in Figure 3 and described in Experimental Procedures. Numbers denote units of β-galactosidase activity measured in cultures of doubly transformed cells.

* This particular plasmid, LR1Δ20B, directs a low level of GAL1 transcription and β-galactosidase production when cells containing it are grown on glucose medium (West et al., 1984).

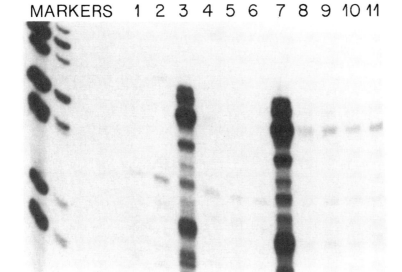

MARKERS 1 2 3 4 5 6 7 8 9 10 11

Figure 4. 5′ Ends of RNAs Made from *GAL1* Derivatives in the GAL4⁺ Strain Sc294

Cells were grown in glucose medium, RNA was mapped, and an autoradiogram was generated as described in Experimental Procedures. The lanes on the left contain size markers. In this experiment, Sc294 carried two plasmids, the relevant features of which are given below. The plasmid LR1Δ20B was used as a control because it directs the transcription of correctly 5′-ended GAL1 mRNA when cells harboring it are grown in glucose medium (West et al., 1984). Lane 1 contains the probe alone. Lane 2, RNA extracted from GAL4⁺ strain Sc294 transformed with LR1Δ1 and PRB500; LR1Δ1 lacks a UAS, and PRB500 directs the synthesis of LexA protein. Lane 3, LR1Δ20B, which contains UAS$_G$, and PRB500. Lanes 4 and 5, 1145, which contains a *lexA* operator but does not contain UAS$_G$, and pRB500. Lane 6, LR1Δ1 and 1027, which directs the synthesis of LexA-GAL4. Lane 7, LR1Δ20B, which contains UAS$_G$, and 1027. Lanes 8, 9, 10, and 11, 1145 and 1027.

inability of the *gal4⁻* strain to grow on galactose medium (not shown). As above, LexA-GAL4 directed synthesis of less β-galactosidase production from the *CYC1-lacZ* fusion gene when the *lexA* operator was located 590 nucleotides upstream of the nearest transcription start site than when it was 178 nucleotides upstream (not shown). In experiments using these strains, we in most cases estimated β-galactosidase levels from the color of colonies on indicator plates (see Experimental Procedures).

Figure 4 shows that, when LexA-GAL4 stimulated transcription of *GAL1* derivatives, the RNAs made had the same 5′ ends as the RNAs made from a plasmid that con-

tained wild-type UAS$_G$. LexA-GAL4 directed the synthesis of 5% as much *GAL1* transcript as was expected from the amount of β-galactosidase activity in these cultures. We do not yet know the cause of this apparent discrepancy between these two measures of the amount of transcription.

LexA-GAL4 May Interact with GAL80

Table 4 shows that, in a *GAL4⁺* strain, LexA-GAL4 induced synthesis of β-galactosidase from a *GAL1-lacZ* fusion gene that carried UAS$_G$ upstream but no *lexA* operator. In this experiment, cells were grown on glucose medium.

Table 4. LexA-GAL4 Induces *GAL1-lacZ* Expression

Plasmid	Regulatory Protein	Units of β-galactosidase
pAAH5	none	0
pRB500	LexA	0
C15	GAL4	80
1027	LexA-GAL4	75

In a *GAL4⁺* strain, synthesis of large amounts of LexA-GAL4 induces expression of a *GAL1-lacZ* fusion gene. 745::pRY171, which contains an integrated *GAL1-lacZ* fusion gene, was transformed with the plasmids below (described in Experimental Procedures). Cells were grown on glucose medium, and β-galactosidase levels were determined as described in Experimental Procedures.

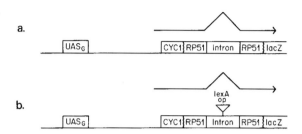

Figure 5. DNA Constructions Used in Downstream Activation Experiment

Constructions are described in more detail in Experimental Procedures. Arrow represents transcript when transcription is activated by UAS_G upstream. The distance from the Sal I site in the *RP51* intron to the start of the most upstream *CYC1* transcript is about 160 nucleotides. (a) the plasmid HZ18 (Teem and Rosbash, 1983). (b) the derivative 1146, which bears a *lexA* operator in the RP51 intron.

This aspect of the behavior of LexA-GAL4 is consistent with a model in which large amounts of the C-terminus of GAL4 titrate the negative regulator GAL80, so that the wild-type GAL4 present in the cell is free to activate transcription from UAS_G (see Discussion) (Laughon and Gesteland, 1982; Johnston and Hopper, 1982).

LexA-GAL4 Activates Transcription from a Downstream Site

To test whether LexA-GAL4 could activate transcription from a site downstream of the normal transcription start, we inserted a *lexA* operator into the intron of a spliced yeast gene, downstream of the normal transcription start site. We adopted an approach first used by Guarente and Hoar (1984), and inserted the *lexA* operator into the plasmid diagrammed in Figure 5 (Teem and Rosbash, 1983). This plasmid contains UAS_G upstream of the *CYC1* coding sequence, which is fused to a fragment containing a portion of *RP51*, a gene of which the transcript is spliced. The *RP51* fragment contains part of the first exon, the intron, and part of the second exon. The second exon of RP51 is fused to *lacZ*. Insertion of a *lexA* operator into the intron allowed us to test whether LexA-GAL4 activated transcription when bound downstream of the transcription start point, by measuring β-galactosidase produced by the plasmid.

Table 5 shows that LexA-GAL4 stimulated production of β-galactosidase if and only if the *RP51* intron contained a *lexA* operator. No β-galactosidase was produced when LexA was present instead of LexA-GAL4. This experiment was done in a *gal4⁻* strain to eliminate upstream activation by UAS_G. β-galactosidase activity in this experiment was about 4% of the level observed in a *GAL4⁺* strain when transcription was activated from UAS_G upstream (not shown).

Discussion

We have described a new protein LexA-GAL4, that activates transcription in yeast. Our most important conclusion is diagrammed in Figure 6. Although LexA-GAL4 and LexA both bind *lexA* operators in yeast (this paper, and Brent and Ptashne, 1984), LexA-GAL4 activates transcription, while LexA does not. LexA-GAL4 does not interact

with UAS_G. Activation of transcription by LexA-GAL4 is less effective when the *lexA* operator is far upstream of the transcription start point than when it is close. In the one case tested, the mRNAs for which synthesis is stimulated by LexA-GAL4 have the same 5′ ends as those generated by the wild-type promoter, but are present at an unexpectedly low level. LexA-GAL4 activates transcription, at further reduced efficiency, when its binding site lies downstream of the normal transcription start. In this case, we have not yet determined the location of the 5′ end of the RNA.

Since LexA-GAL4 stimulates transcription when bound to a *lexA* operator, but native LexA does not, our results argue against any model of GAL4 action that would posit that the sequence-specific contact GAL4 makes with UAS_G changes the structure of DNA and that this change is crucial to gene activation. If gene activation by DNA-bound GAL4 is not effected via a change in the structure of DNA, we infer that gene activation depends on the interaction of GAL4 with other cellular components, most likely proteins. This suggestion has received further support from the work of Keegan, Gill, and Ptashne (unpublished), which shows that a GAL4-β-galactosidase hybrid protein containing only the first 74 amino acids of GAL4 binds UAS_G but cannot stimulate transcription.

Indirect immunofluorescence using anti-lexA antibody shows that, LexA-GAL4, which lacks the portion of GAL4 thought to be necessary for its nuclear localization, is not concentrated to the nucleus, but is dispersed throughout the cell (P. Silver, R. Brent, and M. Ptashne, unpublished). We imagine that LexA-GAL4 (monomer weight 99,000 daltons) enters the nucleus by diffusion through the nuclear pores. This idea is not unprecedented, at least for a smaller protein; native LexA (monomer molecular weight 20,000 daltons) binds *lexA* operators in the nucleus even though it is not localized to the nucleus (Brent and Ptashne, 1984; Silver, Brent, and Ptashne, unpublished). LexA-GAL4 functions in cells that contain deletions of the GAL4 (this paper) and GAL80 genes (C. L. Anderson and Brent, unpublished), suggesting that its entry into the nucleus is not facilitated by an interaction with either of these gene products.

Table 5. LexA-GAL4 Stimulates Synthesis of β-galactosidase when sound to a Site Downstream of the Normal Transcription Start site

Downstream Element Found in First Plasmid	Regulatory Protein Produced by Second Plasmid	
	LexA	LexA-GAL4
No Operator	<1	<1
lexA Operator	<1	30

Constructions are shown in Figure 5 and are described in Experimental Procedures. β-galactosidase activity was measured in cultures of the gal4⁻ strain SHC22C grown in glucose medium, which were doubly transformed with the indicated plasmids. Numbers refer to units of β-galactosidase activity measured as described in Experimental Procedures.

GAL4 is thought to form oligomers (Oshima, 1983; Giniger et al., 1985). The fact that it recognizes a 17 bp sequence that has approximate 2-fold rotational symmetry is consistent with the idea that the DNA binding form of GAL4 is a dimer or tetramer (Giniger et al., 1985). The DNA binding form of LexA is also likely to be a dimer (R. Brent, Ph.D. thesis, 1982). We think it likely that the part of the LexA hinge region contained in LexA-GAL4 provides sufficient flexibility to allow the amino-terminal LexA moieties to assume a conformation identical with the one they have when they are part of a dimer of native LexA.

In contrast to GAL4 activity in a wild-type cell, the activity of LexA-GAL4 in our experiments did not depend on the presence of galactose in the medium. We can explain the galactose-independence of the activity of LexA-GAL4 by assuming that LexA-GAL4 retains the portion of GAL4 that interacts with GAL80, an assumption supported by the experiment shown in Table 4. Since, in our experiments, the synthesis of LexA-GAL4 was directed by the *ADH1* promoter, we suspect that GAL80 was titrated, and the excess, uncomplexed LexA-GAL4 was free to activate transcription. We interpret the relative insensitivity of LexA-GAL4 activity to the presence of glucose in the medium to mean that, under these conditions, LexA-GAL4 retains normal ability to bind DNA. We think, at least in the case of LexA-GAL4-dependent *GAL1* transcription, that the residual 2-fold glucose repression we observe arises from a different mechanism that depends for its action on a specific sequence upstream of *GAL1* that is present in our construction (West et al., 1984; M. Lamphier and M. Ptashne, unpublished).

Our experiments are thus consistent with a picture in which GAL4 is divided into distinct functional domains; an amino-terminal domain that directs nuclear localization (Silver et al., 1984) and binds UAS_G, and a C-terminal domain that stimulates transcription, interacts with GAL80 (Loughn and Gesteland, 1984), and directs oligomerization.

We have recently constructed a hybrid protein composed of LexA and a positive regulator of amino acid biosynthesis called GCN4 (Lucchini et al., 1984; Driscoll-Penn et al., 1983). LexA-GCN4 activates transcription from the *lexA* operator containing constructions used in

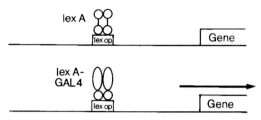

Figure 6 Activation of Transcription by LexA-GAL4

The first line shows that LexA protein is unable to activate transcription when bound to its operator upstream of a yeast gene. The second line shows that LexA-GAL4 activates transcripbon when bound to the same site.

this paper (Brent and C. L. Anderson, unpublished). This fact suggests that GCN4 activates transcription, if and only if it is bound to DNA. From this experiment, we cannot exclude the possibility that GCN4 normally interacts with some other protein that brings it to the DNA, and that fusion of LexA to GCN4 has circumvented this mechanism. However, GCN4 has recently been shown to interact with DNA directly (Hope and Struhl, 1985). Construction of analogous hybrid proteins may prove to be a useful tool for identifying and studying transcriptional activation function, in other eukaryotic regulatory proteins.

Experimental Procedures

Strains

DBY745 is α *leu2 ura3*. SHC22C, α *gal4 ura3 leu2*, was a gift of Susan Hanley. Sc294, α *(gal1-gal10) ura3 leu2*, was a gift of Jim Hopper. DBY745::pRY171 and SHC22C::pRY171 were made by cutting pRY171 (Yocum et al, 1984), which contains a *GAL1-lacZ* fusion gene, with Apa I in the *URA3* gene, transforming the strains with the linearized plasmid DNA, and selecting stable URA⁺ transformants. Bacterial strain JM101 (Messing et al., 1981) was the host for most plasmid constructions. XA90 (Amman et al., 1983), which contains the *lacI*^q1 mutation, and thus contains large amounts of lac repressor, was used as the host in the ultraviolet killing experiments. RB1003 F′ (*lacI*^q1 *lacZ*::Tn10)/*lexA*(def) Δ (*lac-pro*) *arg ile val thiA str*^r was constructed by standard bacterial techniques and lysogenized with RB230, an *imm*^21 phage that contains the *lexA* promoter fused to an operon that contains *trpA* and *lacZ*, similar to RB200 (Brent and Ptashne, 1980).

Plasmids

Plasmids were constructed by standard techniques (Maniatis et al., 1982).

lacZ Fusion Plasmids

All plasmids carry the *URA3*⁺ gene, a 2μ replicator, and relevant portions diagrammed in Figure 2. LR1Δ20B, which contains UAS_G and the *GAL1* gene fused to *lacZ*, and LR1Δ1, which does not contain UAS_G, have been described (West et al., 1984). 1155 was made from LR1Δ1 by inserting a single synthetic consensus *lexA* operator (Brent and Ptashne, 1984) into the Xho I site of the plasmid, 167 nucleotides from the primary transcription start site. pSV15 is a derivative of LR1Δ20B in which UAS_G has been deleted and replaced with a DNA fragment, 128 nucleotides upstream of the primary *GAL1* transcription start site, that contains the "17-mer" (the 17 bp near-consensus GAL4 site of action [Giniger et al., 1985]). pLG669-Z, which carries UAS_C1 and UAS_C2 upstream of a *CYC1-lacZ* fusion gene, and pLGSD5, which carries UAS_G upstream of the *CYC1-lacZ* gene, have been described (Guarente and Ptashne, 1981; Guarente et al., 1982b). pLG670Z, a gift of L. Guarente, is a derivative of pLG669-Z from which the Xho I–Xho I fragment that contains the upstream activation sites has been deleted. 1057 and 1155 were constructed from pLG670Z by inserting a single *lexA* operator at the Xho I site or the Sal I site, 178 and 577 nucleotides upstream of the most upstream *CYC1* transcription start site, respectively. pSV14 is a derivative of pLG670Z in which a fragment of DNA that contains the 17-mer has been inserted at the Xho I site 178 nucleotides upstream of the *CYC1-lacZ* fusion gene. HZ18, a gift of John Teem, was described by Teem and Rosbash (1983). It contains a Sal I site in the RP51 intron about 160 nucleotides downstream from the most upstream *CYC1* transcription start site. 1146 was constructed from HZ18 by inserting a *lexA* operator into the Sal I site.

LexA-GAL4 Hybrid Plasmids

Plasmid 1109, which directs the synthesis of LexA-GAL4 in E. coli, was constructed from three DNA fragments. In this construction, a Pst I–Xmn I piece was isolated from pRB451, a tac promoter derivative of pRB191 (Brent and Ptashne, 1981). This piece of DNA contained the tac promoter, a hybrid ribosome binding site, and *lexA* DNA from codon 1 to within codon 87. Plasmid C15, a gift of Liam Keegan, which contains the S. cerevisiae *GAL4* gene transcribed by its own promoter, was cut with Xho I. The Xho I ends were rendered flush by treatment with the Klenow fragment of DNA polymerase 1 and then were cut with Hind III to liberate a fragment of DNA that encoded the C-terminus of GAL4, extending from the filled-in Xho I site within codon 74 (Laughon and Gesteland, 1984) to about 250 nucleotides beyond the termination codon after amino acid 881. These two pieces were inserted into a Pst I–Hind III backbone fragment of pBR322 (Bolivar et al., 1977). Plasmid 1027, which directs the synthesis of LexA-GAL4 in S. cerevisiae, was constructed in two steps. In the first, we isolated a Hind III–Xmn I fragment from pRB500, a plasmid that directs the synthesis of *lexA* in yeast (Brent and Ptashne, 1984) . This fragment contained DNA from immediately upstream of the *lexA* coding sequence to within codon 87. We inserted this fragment, and the same *GAL4* fragment used in constructing 1109 above, into plasmid pi^4-8, a gift of R. R. Yocum, that confers resistance to tetracycline if Hind III-ended fragments are inserted into it. This ligation created plasmid 1002, which contained the *lexA-GAL4* hybrid gene flanked by Hind III sites. The Hind III-ended *lexA-GAL4* fragment was inserted into pAAH5, a plasmid made by Gustav Ammerer that contains the *ADH1* promoter, the *LEU2* selectable marker, and the 2 μ replicator, to create plasmid 1027. Like all plasmids from this construction that contained the gene encoding LexA-GAL4 in the correct orientation, purified 1027 transformed JM101 at 1% of the frequency expected from the amount of DNA used, and most transformed cells grew slowly. About 10% of the JM101 cells transformed to carbenicillin resistance from a preparation of 1027, grew quickly, and proved to contain plasmids that carried deletions of portions of the gene encoding LexA-GAL4 (not shown).

Cell Growth, Transformation, and Assay of β-galactosidase

Bacterial strains were grown in LB medium that contained, when appropriate, tetracycline at a concentration of 15 μg/ml, carbenecillin at 60 μg/ml, and isopropyl-β-D-thio-galactopyranoside (IPTG) at 5 × 10^{-4} M. Yeast were grown on YEPD medium, or, when they contained plasmids, on minimal medium containing either glucose 2% weight/volume ("glucose medium") or galactose at the same concentration ("galactose medium") but lacking either leucine or uracil or both (Sherman et al., 1983). Yeast were made competent by treatment with lithium acetate (Ito et al., 1983). For assay of LexA-GAL4 activity, cells were first transformed with 1027 or another *LEU2*+ plasmid. As determined by their ability to activate β-galactosidase production from a plasmid that contained a *lexA* operator near a *lacZ* fusion gene, about 90% of the 1027 transformants produced LexA-GAL4. The amount of β-galactosidase in liquid cultures of E. coli was measured according to Miller (1972) and determined for S. cerevisiae in liquid culture and from the degree of blue color on indicator plates containing 5-bromo-4-chloro-3-indolyl-β-D-galactopyranoside, as described (Brent and Ptashne, 1984; Yocum et al., 1984). In DBY745, results of liquid assays were quite consistent from day to day; in contrast, due to a tendency of cells in cultures of the *gal4*$^-$ strain SHC22C to form clumps on some days but not on others, liquid assays had to be repeated many times to generate reliable data. For all strains used, estimation of β-galactosidase activity from color on indicator plates was consistent from day to day.

Measurement of Sensitivity to Ultraviolet Radiation

In the experiment shown in Figure 2, E. coli XA90, which contains large amounts of *lac* repressor because of its *lacI*^{q1} mutation, was trans-

formed separately with four similar plasmids that contain a tac promoter as follows: ptac12, which does not direct the synthesis of a regulatory protein; pEA305, which directs the synthesis of bacteriophage lambda repressor (Amman et al., 1983); pRB451, which directs the synthesis of LexA; or 1109, which directs the synthesis of LexA-GAL4. Since the amount of wild-type LexA that can be disposed of by proteolysis after DNA damage is limited, cells that contain large amounts of wild-type LexA display a sensitivity to killing by ultraviolet radiation equivalent to that observed for cells that contain a mutant LexA that cannot be proteolysed (Brent and Ptashne, 1980; Little and Mount, 1982; Walker, 1984). Cells were grown at 37°C in LB medium that contained 60 μg/ml of carbenicillin to a density of 2 × 10^7 cells/ml. IPTG was added to these cultures to a concentration of 5 × 10^{-4} M to derepress the tac promoters, and growth was resumed for 90 min. Cells were then irradiated with ultraviolet light, and a series of 10-fold dilutions of the cultures was plated immediately on LB plates that contained carbenicillin at 60 μg/ml but lacked IPTG. The fraction of cells surviving was determined by counting colonies on plates that had been incubated in the dark at 37°C for 2 days.

RNA Mapping

A Bam HI–Xho I fragment from LR1Δ1, which contains DNA from the boundary between *GAL1* and *lacZ* and extends upstream to 167 nucleotides 5′ of the primary GAL1 transcription start, was inserted into the SP6 vector pSP64 (Melton et al., 1984) that had been cut with Sal I and Bam HI to create plasmid 1164. RNA made in vitro by SP6 polymerase from plasmid 1164 is complementary to RNA made in vivo from *GAL1*. Since any transcription from the chromosomal *GAL1* gene would produce message that could also hybridize with the SP6 probe, we used strain Sc294, which contains a deletion of the *GAL1* and *GAL10* genes, for these experiments. Cells contained the plasmids shown in the legend to Figure 4. Since *gal1*$^-$ strains cannot grow on galactose medium, we grew the cells on glucose medium and used cells transformed with the plasmid LR1Δ20B, which constitutively produces correctly 5′-ended RNA under these conditions (West et al., 1984), as a positive control. RNA was extracted from cells according to a procedure provided by K. Durban, very similar to that used by Carlson and Botstein (1982). Precise mapping of the RNAs made from the *GAL1* promoter was done after hybridization with probe and digestion with RNAases A and T1 according to Zinn et al. (1983). To generate size markers, plasmid pRB322 was digested with Hpa II (Sutcliffe, 1978), and the fragments were labeled with ^{32}P (Maniatis et al., 1982).

Acknowledgments

We are grateful to Susan Varnum and Catherine Anderson for technical assistance, to Kai Zinn for help with the RNA mapping, to the people cited in the text for gifts of plasmids, strains, and protocols, to Pam Silver, Liam Keegan, Grace Gill, and Marc Lamphier for permission to cite unpublished data, to many members of the Ptashne laboratory for useful discussions, and to Robin Wharton, Edward Giniger, Anne Ephrussi, Ann Hochschild, and Sharon Plon for comments on the manuscript. This work was supported by grants from the National Institutes of Health to M. P.

The costs of publication of this article were defrayed in part by the payment of page charges. This article must therefore be hereby marked "*advertisement*" in accordance with 18 U.S.C. Section 1734 solely to indicate this fact.

Received August 12,1985; revised October 16, 1985

References

Amman, E., Brosius, J., and Ptashne, M. (1983). Vectors bearing a hybrid *trp-lac* promoter useful for regulated expression of cloned genes in *Escherichia coli*. Gene *25*, 167–178.

Bolivar, F. Rodriguez, R. L., Greene, R. J., Betlach, M. C., Heynecker, H. L., and Boyer, H. W. (1977). Construction of useful cloning vectors. Gene *2*, 95–113.

Brent, R., and Ptashne, M. (1980). The *lexA* gene product represses its own promoter. Proc. Natl. Acad. Sci. USA *77*, 1932–1936.

Brent, R., and Ptashne, M. (1981). Mechanism of action of the *lexA* gene product. Proc. Natl. Acad. Sci. USA *78*, 4204–4208.

Brent, R., and Ptashne, M. (1984). A bacterial repressor protein or a yeast transcriptional terminator can block upstream activation of a yeast gene. Nature *312*, 612–615.

Carlson, M., and Botstein, D. (1982). Two differentially regulated mRNAs with different 5′ ends encode secreted and intracellular forms of yeast invertase. Cell *28*, 145–154.

Driscoll-Penn, M. D., Galgoli, B., and Greer, H. (1983). Identification of AAS genes and their regulatory role in general control of amino acid biosynthesis in yeast. Proc. Natl. Acad. Sci. USA *80*, 2704–2708.

Giniger, E., Varnum, S. M., and Ptashne, M. (1985). Specific DNA binding of GAL4, a positive regulatory protein of yeast. Cell *40*, 767–774.

Guarente, L., and Hoar, E. (1984). Upstream sites of the CYC1 gene of *Saccharomyces cerevisiae* are active when inverted but not when placed downstream of the "TATA box". Proc. Natl. Acad. Sci. USA *81*, 7860–7864.

Guarente, L., and Ptashne, M. (1981). Fusion of *Escherichia coli lacZ* to the cytochrome c gene of *Saccharomyces cerevisiae*. Proc. Natl. Acad. Sci. USA *78*, 2199–2203.

Guarente, L., Nye, J. S., Hochschild, A., and Ptashne, M. (1982a). A mutant lambda repressor with a specific defect in its positive control function. Proc. Natl. Acad. Sci. USA *79*, 2236–2239.

Guarente, L., Yocum, R. R., and Gifford, P. (1982b). A GAL10-CYC1 hybrid yeast promoter identifies the GAL4 regulatory region as an upstream site. Proc. Natl. Acad. Sci. USA, *79*, 7410–7414.

Guarente, L., Lalonde, B., Gifford, P. and Alani, E. (1984). Distinctly regulated tandem upstream activation sites mediate catabolite repression of the *CYC1* gene of S. cerevisiae. Cell *36*, 503–511.

Hochschild, A., Irwin, N., and Ptashne, M. (1983). Repressor structure and the mechanism of positive control. Cell *32*, 319–325.

Hope, I., and Struhl, K. (1985). GCN4 protein, synthesized in vitro, binds *HIS3* regulatory sequences: implications for general control of amino acid biosynthetic genes in yeast. Cell *43*, 177–188.

Ito, H., Fukuda, Y., Murata, K., and Kimura, A. (1983). Transformation of intact yeast cells treated with alkali cations. J. Bacteriol. *53*, 163–168.

Johnston, S., and Hopper, J. (1982). Isolation of the yeast regulatory gene GAL4 and analysis of its dosage effects on the galactose/melibiose region. Proc. Natl. Acad. Sci. USA *79*, 6971–6975.

Laughon, A., and Gesteland, R. (1982). Isolation and preliminary characterization of the GAL4 gene, a positive regulator of transcription in yeast. Proc. Natl. Acad. Sci. USA *79*, 6827–6831.

Laughon, A., and Gesteland, R. (1984). Primary structure of the *Saccharomyces cerevisiae* GAL4 gene. Mol. Cell. Biol *4*, 260–267.

Little, J. W. (1984). Autodigestion of *lexA* and lambda repressors. Proc. Natl. Acad. Sci. USA *81*, 1375–1379.

Little, J. W., and Hill, S. A. (1985). Deletions within a hinge region of a specific DNA binding protein. Proc. Natl. Acad. Sci. USA *82*, 2301–2305.

Little, J. W., and Mount, D. W. (1982). The SOS regulatory system of Escherichia coli. Cell *29*, 11–22.

Lucchini, G., Hinnebush, A. G., Chan, C., and Fink, J. R. (1984). Mol. Cell. Biol. *4*, 1326–1333.

Maniatis, T., Fritsch, E., and Sambrook, J. (1982). Molecular Cloning: A Laboratory Manual. (Cold Spring Harbor, New York: Cold Spring Harbor Laboratory).

Melton, D. A., Krieg, R A., Rebagliati, M. R., Maniatis, T., Zinn, K., and Green, M. R. (1984). Efficient *in vitro* synthesis of biologically active RNA and RNA hybridization probes from plasmids containing a bacteriophage SP6 promoter. Nucl. Acids Res. *12*, 7035–7056.

Messing, J., Crea, R., Seeburg, P. H. (1981). A system for shotgun DNA sequencing. Nucl. Acids Res. *9*, 309–318.

Miller, J. (1972). Experiments in Molecular Genetics. (Cold Spring Harbor, New York: Cold Spring Harbor Laboratory).

Oshima, Y. (1982). Regulatory circuits for gene expression: the metabolism of galactose and phosphate. In Molecular Biology of the Yeast *Saccharomyces cerevisiae*, J. N. Strathern, E. W. Jones, and J. R.

Broach, eds. (Cold Spring Harbor, New York: Cold Spring Harbor Laboratory), pp. 159–180.

Sherman, F., Fink, G. R., and Lawrence, C. W. (1983). Methods in Yeast Genetics. (Cold Spring Harbor, New York: Cold Spring Harbor Laboratory).

Shnarr, M., Pouyet, J., Granger-Schnarr, M., Daune, M. (1985). Large-scale purification, oligomerization, equilibria, and specific interaction of the LexA repressor of Escherichia coli. Biochemistry *24*, 2812–2818.

Silver, P. A., Keegan, L. P. and Ptashne, M. (1984). Amino terminus of the yeast *GAL4* gene product is sufficient for nuclear localization. Proc. Natl. Acad. Sci. USA *81*, 5951–5955.

Struhl, K. (1984). Genetic properties and chromatin structure of the yeast *gal* regulatory element: an enhancer-like sequence. Proc. Natl. Acad. Sci. USA *81*, 7865–7869.

Sutcliffe, J. G. (1978). Sequence of the plasmid pBR322. Cold Spring Harbor Symp. Quant. Biol. *43*, 77–90.

Teem, J. L., and Rosbash, M. (1983). Expression of a β-galactosidase gene containing the ribosomal protein 51 intron is sensitive to the rna2 mutalion of yeast. Proc. Natl. Acad. Sci. USA *80*, 4403–4407.

Walker, G. (1984). Mutagenesis and inducible responses to deoxyribonucleic acid damage in *Escherichia coli*. Microbiol. Rev. *48*, 60–93.

West, R. W., Jr., Yocum, R. R., and Ptashne, M. (1984). *Saccharomyces cerevisiae* GAL1-GAL10 divergent promoter region: location and function of the upstream activator sequence UAS$_G$. Mol. Cell. Biol. *4*, 2467–2478.

Yocum, R. R., Hanley, S., West, R., Jr., and Ptashne, M. (1984). Use of *lacZ* fusions to delimit regulatory elements of the inducible divergent GAL1-Gal10 promoter in *Saccharomyces cerevisiae*. Mol. Cell. Biol. *4*, 1985–1998.

Zinn, K., DiMaio, D., and Maniatis, T. (1983). Identification of two distinct regulatory regions adjacent to the human β-interferon gene. Cell *34*, 865–879.

Regulation of transcription factor activity

12

The experiments described in previous sections have identified regions of specific transcription factors that allow them to dimerize with one another (Section 10), bind to DNA (Section 9) and activate transcription (Section 11). Similarly, I have discussed how the interaction of specific transcription factors with particular DNA sequences allows such sequences to render a gene active in a particular cell type or inducible following exposure to a specific stimulus (Sections 6 and 7).

Obviously, this mechanism for producing tissue-specific or inducible gene transcription requires that the activity of the transcription factor be regulated in some way, so that its binding to the particular DNA sequence and/or the activation of transcription only occurs in the appropriate cell type or following exposure to the appropriate stimulus. Two basic mechanisms by which this can be achieved have been described (Figure 12.1). Thus it is possible to control the synthesis of the transcription factor so that it is made only when appropriate (Figure 12.1a). In this situation, the factor is only synthesized in the appropriate cell type or following exposure to the appropriate stimulus, and can therefore bind to its specific DNA recognition site and activate transcription only in these situations.

An example of this type of regulation is provided by the MyoD transcription factor, which plays a critical role in regulating gene expression in skeletal muscle cells. Thus, as described in the paper by Davis et al. presented in Section 13, this transcription factor and the mRNA encoding it are only synthesized in skeletal muscle cells and not in other cell types. Indeed, the artificial expression of MyoD in undifferentiated fibroblast cells is sufficient to convert them into muscle cells. Hence, in this case, the ability of MyoD to activate transcription only in the appropriate cell type is controlled by ensuring that the factor is synthesized only in that cell type.

It is also possible, however, for a transcription factor to be synthesized in all cell types but in a form that is incapable of activating gene expression (Figure 12.1b). This could occur, for example, because the inactive factor cannot bind to DNA or, alternatively, because it cannot stimulate transcription following DNA binding. In response to an appropriate stimulus or in a particular cell type, the inactive factor becomes converted into an active form, allowing it to mediate transcriptional activation of specific genes in response to a particular stimulus or in a particular cell type. The papers presented in this section deal with some of the mechanisms by which pre-existing inactive transcription factors can become activated.

One of the best examples of this is provided by the CREB factor described in the paper by Gonzalez and Montminy presented here. The CREB factor mediates the transcriptional activation of specific genes in response to cyclic AMP (cAMP), and achieves this effect by binding to a DNA sequence known as the cAMP response element (CRE) which is present in cAMP-inducible genes (Montminy et al., 1986). CREB is present in all cell types, but only activates transcription via the CRE following exposure of the cell

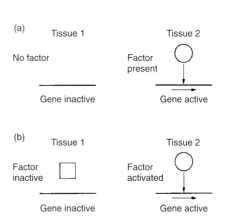

Figure 12.1 Gene activation can be mediated by the synthesis of a specific factor only in a particular tissue (a) or by its activation only in a specific tissue (b).

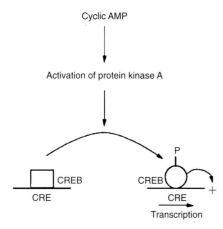

Figure 12.2. cAMP activates protein kinase A, which phosphorylates the CREB transcription factor, increasing its ability to stimulate transcription. This therefore results in the activation of cAMP-inducible genes that contain the CRE to which CREB binds.

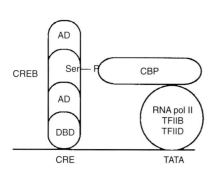

Figure 12.3 Mechanism by which phosphorylation of the CREB factor enhances its ability to stimulate transcription.
Phosphorylation of the serine (ser) residue located between the activation domains (AD) of CREB results in enhanced binding of the CBP factor. This factor also interacts with the basal transcriptional complex of RNA polymerase (RNA pol) and associated factors, such as TFIIB and TFIID, bound to the TATA box and stimulates the activity of the complex. Hence phosphorylation of CREB results in enhanced recruitment of CBP, leading to increased transcription. DBD indicates the DNA binding domain of CREB bound to the CRE DNA binding site.

to cAMP. The experiments in the paper by Gonzalez and Montminy elucidated the mechanism by which CREB becomes activated following exposure to cAMP. As illustrated in Figure 1 of the paper, they showed that, following exposure of PC12 cells to cAMP, the CREB protein becomes phosphorylated on a specific serine residue present at position 133. In addition, they showed that this effect did not occur in A126-1B2 cells, which lack the enzyme protein kinase A that is stimulated by cAMP. This therefore establishes a pathway by which cAMP stimulates the activity of protein kinase A, which in turn phosphorylates CREB on serine-133 (Figure 12.2).

The importance of this phosphorylation for CREB activation was demonstrated by Gonzalez and Montminy in two ways. Thus, as illustrated in Figure 2 of the paper, they investigated the reasons why the F9 teratocarcinoma cell line was not responsive to cAMP treatment. These cells had previously been shown to have only low levels of CREB and of protein kinase A. Gonzalez and Montminy showed that the known unresponsiveness of these cells to cAMP could only be overcome by overexpressing both CREB and the catalytic subunit of protein kinase A, and not by overexpressing one of these factors without the other. This is in accordance with a model in which the ability of CREB to stimulate transcription in response to cAMP is itself dependent upon CREB being phosphorylated by protein kinase A.

Moreover, in the experiment presented in Figure 4 of the paper, Gonzalez and Montminy investigated the effect of changing the critical serine residue within CREB to other amino acids that could not be phosphorylated (see Figure 3 of the paper for details of these amino acids). They showed that these mutant forms of CREB that could not be phosphorylated were incapable of activating transcription in response to cAMP. This experiment thus directly indicates that the phosphorylation of CREB on serine-133 is essential for its ability to activate transcription.

Thus the paper by Gonzalez and Montminy represents one of the earliest examples elucidating the mechanism by which a transcription factor becomes activated in response to a particular stimulus. The mechanism by which phosphorylation of CREB on serine-133 allows it to stimulate transcription has now been analysed in great detail (for reviews, see Lalli and Sassone-Corsi, 1994; Sassone-Corsi, 1994). As indicated in Figure 12.2, the CREB factor is bound to the CRE even before cAMP treatment. Exposure to cAMP and phosphorylation on serine-133 does not affect the DNA binding ability of CREB. Rather, the serine at position 133 is located between the two activation domains of the CREB factor, and its phosphorylation enhances the ability of these activation domains to stimulate transcription following DNA binding (Figure 12.2).

Most interestingly, it has been shown that the phosphorylation of serine-133 allows CREB to interact with another protein, CBP, which does not bind to non-phosphorylated CREB (Figure 12.3) (Chrivia et al., 1993). It appears that the CBP protein, as well as interacting with CREB, also interacts with the basal transcriptional complex of RNA polymerase and associated factors stimulating its activity (Figure 12.3). Hence CBP acts as a bridge between CREB and the basal transcriptional complex, and its binding to CREB is necessary for CREB to stimulate transcription (Kwok et al., 1994). The important role of CBP and its binding only to the phosphorylated form of CREB thus explains why phosphorylation of CREB is necessary for transcriptional activation.

In this example, therefore, the activation of a transcription factor is controlled directly by its phosphorylation, which in turn affects its ability to associate with another protein that is essential for its ability to stimulate transcription. The second paper presented in this section, by Baeuerle and Baltimore, also involves the regulation of transcription factor activity by phosphorylation and by protein–protein association. In this case, however, the protein–protein association produces an inhibition of transcription factor activity rather than a stimulation, while the phosphorylation functions to dissociate the protein–protein complex rather than to promote its formation.

Baeuerle and Baltimore studied nuclear factor κB (NF-κB). This factor is active in mature B-lymphocytes, where it stimulates immunoglobulin κ light chain gene expression, and is also active in activated T-lymphocytes, where it stimulates other genes such as that encoding the interleukin-2α receptor (for reviews, see Blank et al., 1992; Grimm and Baeuerle, 1993). In a previous paper (Baeuerle and Baltimore, 1988) it had been shown that active NF-κB capable of binding to DNA was, as expected, detectable in the nuclei of mature B-lymphocytes and activated T-cells. In contrast, however, no such active NF-κB was detectable in the nuclei of other cell types, including immature B-lymphocytes and resting T-lymphocytes. However, treatment of the cytoplasm of other cell types with the detergent sodium deoxycholate, which dissociates protein–protein complexes, resulted in the detection of NF-κB within the cytoplasm (Figure 12.4). This immediately suggested that NF-κB was present in an inactive form in all cell types but was complexed with an inhibitory protein that prevented it from

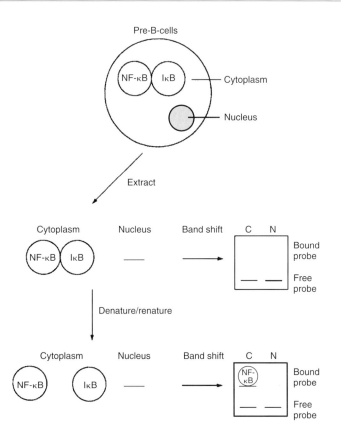

Figure 12.4 Detection of active NF-κB in the cytoplasm of pre-B-cells when the cytoplasmic extract is first denatured and renatured, resulting in the release of NF-κB from its inactive complex with IκB.

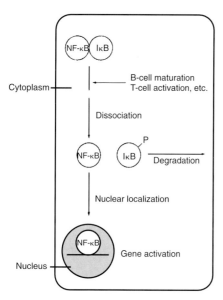

Figure 12.5 Activation of NF-κB by dissociation of the inhibitory protein IκB.
Phosphorylation (P) promotes the dissociation of IκB from NF-κB and its subsequent degradation. Removal of NF-κB allows IκB to move to the nucleus and switch on gene expression.

entering the nucleus and binding to DNA (Figure 12.5). The experiments described in the paper by Baeuerle and Baltimore presented here provided the evidence that directly demonstrated the existence of such an inhibitor, which they named IκB.

To identify this factor, Baeuerle and Baltimore utilized immature B-cells which they had previously shown did not contain active NF-κB in the nucleus, although NF-κB activity could be demonstrated in these cells following deoxycholate treatment of their cytoplasm. Having released such latent NF-κB activity, Baeuerle and Baltimore then purified from the cytoplasm another factor which, when added back to this NF-κB, could inhibit its DNA binding activity. The purification of this inhibitory factor is illustrated in Fig. 1 of the paper, while Fig. 2(A) of the paper demonstrates that the factor is approx. 70 kDa in size. Subsequent experiments shown in Fig. 2(B) of the paper established that the inhibitory factor is a protein, since it can be digested with the protease trypsin.

Baeuerle and Baltimore also demonstrated directly that addition of this IκB protein to active NF-κB was able to inhibit the DNA binding activity of the latter, as illustrated in Fig. 3 of the paper, and that this effect was specific for NF-κB, since IκB had no effect on the DNA binding ability of other transcription factors.

These findings thus lead to a model whereby, in most cell types, the NF-κB protein is retained in the cytoplasm by a protein–protein interaction with IκB, whereas in mature B-lymphocytes and activated T-lymphocytes NF-κB is released from IκB and therefore moves to the nucleus, where it can activate gene expression (Figure 12.5). Moreover, as indicated in Fig. 4(B) of the Baeuerle and Baltimore paper, active NF-κB purified from the nuclei of these cells can be inhibited by the addition of IκB purified from other cell types. These results indicate that, during the maturation of B-lymphocytes or the activation of T-lymphocytes, IκB disappears, thereby releasing NF-κB (for reviews, see Beg and Baldwin, 1993; Thanos and Maniatis, 1995). It is now clear that this effect is dependent upon the phosphorylation of IκB, which results in its dissociation from NF-κB and its rapid degradation (Brown et al., 1995) (Figure 12.5).

Taken together, therefore, the CREB and NF-κB examples described in this section illustrate the manner in which both phosphorylation events and the modulation of protein–protein interactions can control the activity of transcription factors, allowing them to be activated in response to a specific signal. In addition, they indicate how phosphorylation events can either promote or inhibit the interactions between two different proteins, as well as illustrating that such protein–protein interactions can both stimulate and inhibit the activity of an individual transcription factor. Such regulatory mechanisms have now been described for a wide variety of transcription factors (for reviews, see Latchman, 1995a,b) and are central to the ability of specific transcription factors to stimulate transcription only in a specific cell type or following exposure to a particular stimulus. The papers presented in this section are thus of critical importance in providing two of the earliest examples analysing the mechanisms of transcription factor activation which are essential for the appropriate regulation of gene expression in different cell types or in response to specific stimuli.

Cyclic AMP Stimulates Somatostatin Gene Transcription by Phosphorylation of CREB at Serine 133

Gustavo A. Gonzalez and Marc R. Montminy

The Clayton Foundation Laboratories for
   Peptide Biology
The Salk Institute
10010 North Torrey Pines Road
La Jolla, California 92037

Summary

In this paper, we demonstrate that phosphorylation of CREB at Ser-133 is induced 6-fold in vivo, following treatment of PC12 cells with forskolin. By contrast, no such induction was observed in the kinase A–deficient PC12 line A126-1B2 (A126). Using F9 teratocarcinoma cells, which are unresponsive to cAMP, we initiated a series of transient expression experiments to establish a causal link between phosphorylation of CREB and *trans*-activation of cAMP-responsive genes. Inactivating the kinase A phosphorylation site by in vitro mutagenesis of the cloned CREB cDNA at Ser-133 completely abolished CREB transcriptional activity. As CREB mutants containing acidic residues in place of the Ser-133 phosphoacceptor were also transcriptionally inactive, these results suggest that phosphorylation of CREB may stimulate transcription by a mechanism other than by simply providing negative charge.

Introduction

Cyclic AMP (cAMP) mediates the hormonal stimulation of a variety of eukaryotic genes through a conserved cAMP response element (CRE) (Montminy et al., 1986; Comb et al., 1986). Transcriptional induction by cAMP is rapid, peaking at 30 min and declining gradually over 24 hr (Sasaki et al., 1984; Lewis et al., 1987). This burst in transcription is resistant to inhibitors of protein synthesis, suggesting that cAMP may stimulate gene expression by inducing the covalent modification rather than de novo synthesis of specific nuclear factors. Since all of the known cellular effects of cAMP occur via the catalytic subunit (C-subunit) of cAMP-dependent protein kinase (kinase A), it appears likely that this enzyme mediates the phosphorylation of factors that are critical for the transcriptional response. Kinase A–deficient cell lines, for example, are unable to stimulate somatostatin gene transcription in response to forskolin (Montminy et al., 1986). Furthermore, microinjection of the C-subunit into cells can directly activate CRE-dependent transcription without simultaneous addition of cAMP (Riabowol et al., 1988).

We have previously characterized and isolated cDNAs for a 43 kd nuclear CRE binding protein, CREB, which binds to and stimulates transcription of the somatostatin gene in vitro (Montminy and Bilezikjian, 1987; Yamamoto et al., 1988; Gonzalez et al., 1989). Sequence analysis of the CREB cDNA predicts a single kinase A consensus

recognition site (R-R-P-S) from amino acids 130–133. Indeed, CREB is efficiently phosphorylated at Ser-133 by the C-subunit in vitro, suggesting that this site may be critical for cAMP responsiveness (Yamamoto et al., 1988). In this report we show that CREB is phosphorylated in vivo at Ser-133 in response to forskolin and that phosphorylation of Ser-133 is critical to the activation of gene transcription by cAMP.

Results and Discussion

Forskolin stimulates transient expression of the somatostatin–CAT fusion gene $\Delta(-71)$ 10- to 20-fold in PC12 cells, but has no effect in kinase A–deficient A126-1B2 cells (Montminy et al., 1986). To test the hypothesis that cAMP enhances transcription of the somatostatin gene by inducing the phosphorylation of CREB through kinase A, we labeled PC12 and A126 cells with inorganic ^{32}P and then treated them with forskolin or ethanol vehicle for 20 min. Cells were harvested in standard SDS lysis buffer and immunoprecipitated with W39 CREB antiserum (Gonzalez et al., 1989). Following SDS–PAGE, the ^{32}P-labeled 43 kd bands corresponding to CREB were eluted and digested with trypsin. Two-dimensional phosphopeptide maps of control and forskolin-treated samples showed a single ^{32}P-labeled spot (Figure 1A) migrating at the same relative position as CREB tryptic peptide W51 (Gonzalez et al.,1989). Using gas phase microsequencing, we had previously determined that the W51 peptide contains the kinase A phosphorylation motif (R-R-P-S). Forskolin treatment caused a 6-fold increase in the phosphorylation of this peptide in PC12 cells, but had no effect in A126-1B2 cells. The increased phosphorylation in PC12 cells does not appear to arise from de novo synthesis of CREB, as incorporation of [^{35}S]methionine into the protein was unaffected by forskolin treatment from 30 min (data not shown) up to 4 hr (Figure 1B).

To determine whether CREB is both necessary and sufficient to stimulate transcription of the somatostatin gene in response to cAMP, we initiated a series of transient expression experiments. Because PC12 cells have high levels of endogenous CREB activity (Montminy and Bilezikjian, 1987; Yamamoto et al., 1988), we did not consider these cells suitable for mutagenesis studies. F9 teratocarcinoma cells, in contrast, are unresponsive to cAMP, becoming inducible only after differentiation with retinoic acid (Rickles et al., 1989). Kinase A (Plet et al., 1982) and CREB levels (M. M., unpublished data) are reduced 3- to 5-fold in undifferentiated versus differentiated F9 cells, suggesting that a deficiency in both components may account for the lack of cAMP responsiveness. Consistently, $\Delta(-71)$ CAT activity was not stimulated by forskolin after transfection of F9 cells with this reporter plasmid (data not shown). When cotransfected with a Rous sarcoma virus (RSV) plasmid expressing CREB or a metallothionein vector expressing the C-subunit of kinase A (Mellon et al., 1989), $\Delta(-71)$ activity was only modestly induced by each

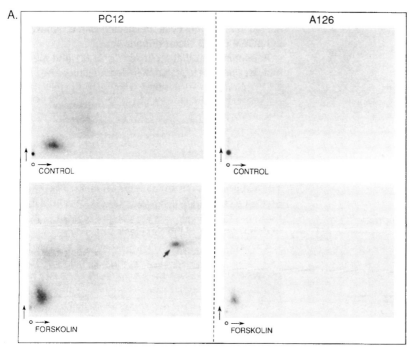

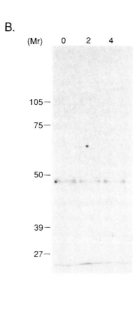

Figure 1. Phosphorylation and Synthesis of CREB in PC12 Cells

(A) Phosphorylation of CREB in response to forskolin treatment of PC12 and A126 cells. Two-dimensional phosphotryptic maps of ^{32}P-labeled CREB protein after immunoprecipitation wiht W39 CREB antiserum. Arrow points to the ^{32}P-labeled spot, which migrates at the same position as CREB tryptic peptide W51.

(B) Biosynthesis of CREB in PC12 cells in response to forskolin after 0.2, and 4 hr of forskolin treatment. Arrow points to ^{35}S-labeled CREB. Mr = relative molecular weight (in thousands). Cells were treated with forskolin (10 mM) for times indicated and then labeled with [^{35}S]methionine (0.1 mCi/ml) for 15 min. Cell extracts were immunoprecipitated with affinity-purified W39 CREB antiserum. Samples were then resolved by SDS–PAGE.

(Figure 2). Cotransfection of both the C-subunit and CREB genes, however, caused a dramatic 200-fold induction in Δ(−71) CAT activity, suggesting that both CREB and C-subunit activities are critical to the induction of gene transcription. No such induction was observed when CREB was cotransfected with an inactive mutant C-subunit plasmid pCaK72M (gift of M. Uhler) encoding a lysine to methionine substitution at position 72 (data not shown). Furthermore, C-subunit and CREB activities appear to be specific for the CRE sequence, as neither CREB nor the C-subunit, alone or in combination, could stimulate CAT expression when coexpressed with an RSV-CAT reporter gene (Figure 2).

To establish a causal link between C-subunit expression and CREB *trans*-activation, we prepared mutant forms of CREB that were defective in the kinase A phosphorylation motif by in vitro mutagenesis of the cloned CREB cDNA (Figure 3a). After insertion into RSV vectors, each mutant was evaluated for *trans*-activating potential in F9 cells by cotransfection with the C-subunit expression plasmid.

Mutant M1 contains a conservative serine to alanine substitution at position 133, which destroys the kinase A phosphorylation site while maintaining charge balance (Figure 3a). Combined transcription–translation of the M1 cDNA in vitro showed an immunoreactive protein that migrated identically to the wild-type CREB protein on SDS–PAGE (Figure 3b). When tested in F9 cells, however,

M1 was completely unable to activate transcription of the somatostatin gene, suggesting that Ser-133 was indeed critical to the *trans*-activation of CREB by the C-subunit (Figure 4a).

To test whether phosphorylation of Ser-133 activates transcription simply by providing negative charge at this residue, we constructed a mutant, M2, containing a Ser-133 to Asp-133 substitution (Figure 3a). As with M1, the M2 cDNA encoded a 43 kd immunoreactive CREB product (Figure 3b). When coexpressed with the C-subunit expression plasmid in F9 cells, however, the M2 protein was ineffective in stimulating Δ(−71) CAT activity (Figure 4a). A CREB mutant containing a Ser-133 to Glu-133 substitution was similarly inactive (G. G., unpublished data). These results suggest that phosphorylation of CREB at Ser-133 may activate transcription by a mechanism other than by providing negative charge.

In the event that the M1 and M2 mutants were inactive owing to reduced expression from their RSV promoters, we performed RNAase protection assays on total RNA from F9 cells transfected with wild-type or mutant (M1 and M2) RSV-CREB plasmids. Using an 800 base CREB antisense RNA probe, we observed a 745 nucleotide RNAase-protected fragment on denaturing polyacrylamide gels, the reduced size of which was due to RNAase digestion of plasmid polylinker sequences. The 745 base fragment was present at comparable levels in samples

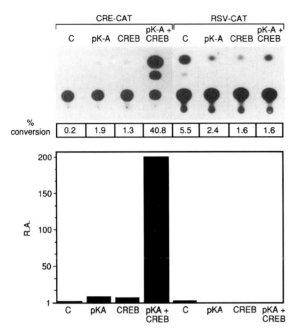

Figure 2. Effect of CREB and Kinase A on Δ(−71) CAT and RSV-CAT Expression in Undifferentiated F9 Teratocarcinoma Cells

Cells were transfected with either the Δ(−71) CAT of RSV-CAT reporter gene plus C-subunit (pK-A), CREB (CREB), or both expression plasmids (pK-A + CREB). Control (C) = F9 cells transfected with reporter gene alone. CAT activity is expressed as percent conversion shown below each lane. The graph below shows CAT activity relative to control values (R.A.).

from wild-type, M1, and M2 transfections, but was not observed in RNA from cells transfected with a nonexpressing pGEM-CREB vector (Figure 4b).

To demonstrate that the inactivity of M1 and M2 plasmids was not due to preferential degradation of the mutant CREB proteins, we performed immunoprecipitation experiments (Figure 4c). Cells transfected with wild-type, M1, or M2 plasmids expressed comparable levels of immunoreactive CREB products. Furthermore, indirect immunofluorescence microscopy revealed that both wild-type and mutant (M1 and M2) CREB products were appropriately targeted to nuclei of transfected F9 cells (Figure 4d). The inability of the CREB mutants M1 and M2 to stimulate transcription, therefore, would appear to arise from the deletion of the Ser-133 phosphoacceptor site and not from underexpression of the transfected plasmids.

The presence of two basic residues (Arg-135 and Lys-136) C-terminal to Ser-133 forms a potential kinase C phosphorylation motif (S-Y-R-K). Indeed, Ser-133 can also be phosphorylated by kinase C in vitro (Yamamoto et al., 1988), prompting us to investigate whether the C-subunit can activate CREB when the kinase C motif is ablated. Substituting Arg-135 and Lys-136 with Met-135 and Glu-136 in M3 (Figure 3a) destroyed the consensus kinase C motif. In vitro translation of M3 RNA (Figure 3b) showed the expected 43 kd band and a smaller 30 kd immunoreactive product, which is consistent with alternate translational initiation at Met-135. Furthermore, immunoprecipitation and immunofluorescence assays (Figures 4c and 4d)

a.

```
Construct          (130)                (136)
                        pK-A       pK-C
                   ─────────────  ──────────
Wild Type  . . . .AGG AGG CCT TCC  TAC AGG AAA. . . .
           . . . .Arg Arg Pro Ser* Tyr Arg Lys. . . .

M1         . . . .AGG AGG CCT GCC  TAC AGG AAA. . . .
           . . . .Arg Arg Pro Ala  Tyr Arg Lys. . . .

M2         . . . .AGG AGG CCT GAC  TAC AGA AAA. . . .
           . . . .Arg Arg Pro Asp  Tyr Arg Lys. . . .

M3         . . . .AGG AGG CCT TCC  TAC ATG GAA. . . .
           . . . .Arg Arg Pro Ser* Tyr Met Glu. . . .
```

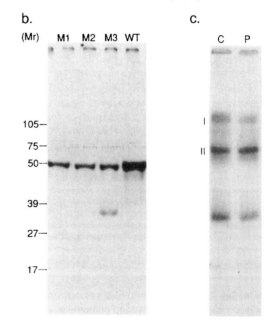

b.

c.

Figure 3. Characterization of CREB Point Mutants

(a) Nucleotides and corresponding amino acid sequences of point mutants M1, M2, and M3 compared with wild-type (WT) CREB cDNA near the Ser-133 phosphoacceptor site. Consensus kinase A (pK-A) and kinase C (pK-C) phosphorylation sites are overlined and labeled. Positions of amino acids are shown in parentheses. Serine phosphoacceptors are indicated by asterisks. Mutated residues are underlined and in bold.

(b) Immunoprecipitation of ^{35}S-labeled wild-type and mutant CREB proteins expressed in vitro. Mr = molecular weight standards (in thousands). M1, M2, M3, and WT lanes correspond to mutant and wild-type CREB proteins.

(c) Effect of C-subunit phosphorylation on CRE binding activity of bacterial CREB fusion protein in vitro. Gel shift assay of extract using ^{32}P-labeled double-stranded CRE oligonucleotide probes. I and II = putative dimer and monomer forms of CREB, respectively. C = control; P = CREB phosphorylated by the C-subunit in vitro.

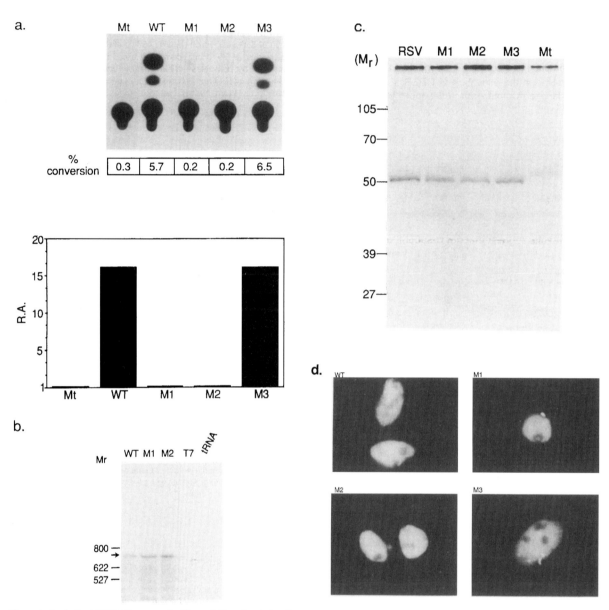

Figure 4. Analysis of CREB Mutants at the Kinase A Phosphorylation Site

(a) Analysis of wild-type and mutant CREB proteins by transient expression assay. Representative CAT assays of F9 cells transfected with wild-type (WT) or mutant (Mt, M1, M2, and M3) RSV-CREB plasmids. The mutant Mt, which contains a nonsense mutation at residue 131 and does not express immunopreciptable CREB protein, was used to control for nonspecific effects of the RSV-CREB plasmid on transcription. Each CREB plasmid was cotransfected with $\Delta(-71)$ CAT reporter, C-subunit vector MtC, and RSV–β-galactosidase as described in Figure 2. Percent conversion is indicated and relative activity (R.A.) is shown graphically below. Assays were normalized for β-galactosidase activity. Each mutant was tested in at least three separate assays.

(b) RNase protection of total RNA from F9 cells transfected with wild-type (WT) or mutant (M1, M2) CREB plasmids. Mr = molecular weight standards (in bases). Arrow points to 745 nucleotide RNAase-protected fragment. T7 = RNA from cells transfected with pGEM-CREB plasmid DNA. tRNA = control assay using tRNA only. To prepare ^{32}P-labeled antisense CREB RNA probe, pGEM-CREB plasmid, linearized with AatlI, was transcribed in vitro (Chomcyzynski and Sacchi, 1987) using T7 RNA polymerase. The 800 nucleotide probe was resolved on a 6% urea–polyacrylamide gel, removed, and electroeluted. Total RNA was prepared from transfected cells (Chromcyzynski and Sacchi, 1987) and treated with DNAase I to remove transfected plasmid DNAs. Equal amounts of RNA (6 μg, verified by agarose gel electrophoresis) were analyzed by RNAase protection of the CREB RNA probe according to Zinn et al. (1983).

(c) Immunoprecipitation of ^{35}S-labeled CREB products from F9 cells transfected with CREB plasmids. F9 cells were transfected with wild-type (WT), mutant (M1, M2, M3), and truncated (Mt) plasmids as indicated. After 24hr in medium containing [^{32}S]methionine, cells were harvested and immunoprecipitated with affinity-purified W39 CREB antiserum. Labeled products were resolved by SDS–PAGE. Mr = relative molecular weight in thousands.

(d) Immunofluorescence analysis of F9 cells following transfection with wild-type (WT) or mutant (M1, M2, M3) RSV-CREB expression plasmids. Transfected CREB protein was detected using affinity-purified W39 CREB antiserum. No immunoreactive products were detectable in control (untransfected) cells. In each case, CREB proteins were appropriately targeted to the nuclear compartment.

showed that the M3 protein was expressed and targeted to nuclei of transfected cells. Likewise, M3 retained wild-type activity when cotransfected with the C-subunit expression plasmid (Figure 4a), suggesting that kinase C is not involved in the activation of CREB at Ser-133.

The kinase A motif does not appear to participate in DNA binding. We have previously demonstrated, for example, that phosphorylation of purified CREB by kinase A in vitro has no effect on DNAase I footprinting or gel-shifting activities (Montminy and Bilezikjian, 1987; Yamamoto et al., 1988). Furthermore, a bacterial β-galactosidase–CREB fusion protein containing the C-terminal 317 amino acids of CREB shows no change in DNA binding activity upon phosphorylation by the C-subunit in vitro (Figure 3c).

Our results suggest that the C-subunit of kinase A may directly phosphorylate CREB and thereby activate transcription upon stimulation with cAMP. Indeed, treatment of cells with cAMP appears to induce transport of the C-subunit to the nucleus (Nigg et al., 1985). Furthermore, microinjection of the C-subunit into cells can directly activate transcription of cAMP-responsive genes c-fos and vasoactive intestinal peptide (Riabowol et al., 1988). Our results demonstrate that CREB contains a classic kinase A motif that is phosphorylated in response to cAMP and that is absolutely required for activation by the C-subunit in F9 cells.

How does phosphorylation activate CREB? Although a number of nuclear factors have been shown to stimulate transcription through domains that are largely acidic in character (Ptashne, 1988), our results suggest that phosphorylation of CREB does not stimulate transcription simply by providing negative charge. It is tempting to speculate that, by analogy with other substrates that are allosterically regulated by kinase A phosphorylation, CREB may also undergo a conformational change that allows a site distal to the kinase A motif to interact with the transcription apparatus. Indeed, the inability of the CREB mutant M2 to stimulate transcription, either constitutively or upon cotransfection with the C-subunit, would argue in favor of this model. Further mutagenesis studies will help to define other structural determinants involved in the activation process.

Experimental Procedures

Immunoprecipitations and Phosphopeptide Mapping
All immunoprecipitations were performed using CREB antiserum W39 as described previously (Gonzalez et al., 1989). PC12 and A126-B2 cells were labeled with inorganic ^{32}P for 5 hr and then treated with either 10 mM forskolin or ethanol vehicle (control) for 20 min. ^{32}P-labeled cell extracts were immunoprecipitated with W39 antiserum and were then fractionated by SDS–PAGE. ^{32}P-labeled CREB bands, visualized by autoradiography, were cut out and placed in elution buffer. After recovery from gel slices, equal amounts (in cpm) of ^{32}P-labeled CREB were digested with trypsin and analyzed by two-dimensional mapping as described previously (Yamamoto et al., 1988). Purified CREB, labeled in vitro with kinase A and digested with trypsin, was analyzed in parallel with test samples to confirm identity of spots observed on chromatograms.

Plasmid Constructions
The RSV-CREB expression plasmid was constructed by isolating a 1.2 kb fragment of the CREB cDNA containing the entire coding region

(1023 bp) plus 100 bp of 5′ and 52 bp of 3′ untranslated sequences. This CREB fragment was inserted into the RSV expression vector RSV-SG (gift of S. Gould), which contains 450 bp of the RSV long terminal region plus an SV40 polyadenylation site. CREB mutants were constructed by polymerase chain reaction amplification of a 200 bp StuI–KpnI cDNA fragment using 30 base oligonucleotide primers containing mutations indicated above (Figure 3a). Amplified fragments were reinserted into the CREB cDNA and then sequenced on both strands of the DNA. Mutant and wild-type constructs were then inserted into the RSV-SG and pGEM vectors. Bacterial expression plasmid KY-CREB was constructed by the insertion of a 1 kb cDNA fragment encoding the C-terminal 317 amino acids of CREB into pGEM.

Cell Lines, Transfections, and Immunofluorescence Studies
PC12 and F9 teratocarcinoma cells were maintained in Dulbecco's modified Eagle's medium (DMEM) supplemented with 10% fetal calf serum plus 5% horse serum.

Transfections were performed as described previously (Yamamoto et al.,1988), using 5 μg of reporter ($\Delta[-71]$ CAT or RSV-CAT), 2.5 μg of RSV–β-galactosidase plasmid, and when indicated, 5 μg of C-subunit plasmid MtC (kinase A) (Mellon et al.,1989), and 7.5 μg of RSV-CREB expression plasmid. For each transfection, pGEM plasmid was included to maintain a total of 20 μg of plasmid DNA. Cells were assayed for CAT activity after normalizing to β-galactosidase activity. All assays were performed at least three times.

For immunofluorescence analysis, cells were fixed in 3% formaldehyde solution 7 hr after transfection. Affinity-purified W39 CREB antiserum was added to cells at a 1:30 dilution for 15 min. Fluoresceinated goat anti-rabbit antibody was then added at a 1:50 dilution and allowed to incubate for 10 min. After each antibody, cells were washed 10 times in phosphate-buffered saline (pH 7.4). Cells were viewed and photographed on a Leitz microscope.

In Vitro Translations and Bacterial Extracts
Mutant and wild-type CREB cDNAs in pGEM were transcribed in vitro with T7 RNA polymerase, and transcripts were translated in reticulocyte lysate extracts using [^{35}S]methionine. Translation products were then immunoprecipitated with W39 CREB antiserum and resolved by SDS–PAGE. Bacterial extract preparation, C-subunit phosphorylation in vitro, and gel shift assays were performed as previously described (Gonzalez et al., 1989).

Acknowledgments

We thank K. Yamamoto for RNAase protection data, S. McKnight for the gift of MtC plasmid, M. Uhler for the gift of MaCK72M plasmid, and W. Vale for advice and support. We also thank members of the Max Planck Lab (Salk Institute) for oligonucleotide synthesis. This work was supported by United States Public Health Service grant GM 37828 and conducted in part by the Clayton Foundation for Research, California Division. M. R. M. is a Clayton Foundation Investigator.

The costs of publication of this article were defrayed in part by the payment of page charges. This article must therefore be hereby marked "*advertisement*" in accordance with 18 U.S.C. Section 1734 solely to indicate this fact.

Received July 26, 1989; revised September 7, 1989.

References

Chomczynski, R. and Sacchi, N. (1987). Single-step method of RNA isolation by acid guanidinium thiocyanate-phenol-chloroform extraction. Anal. Biochem *162*, 156–199.

Comb, M., Birnberg, N. C., Seascholtz, A., Herbert, E., and Goodman, H. M. (1986). A cyclic-AMP– and phorbol ester–inducible DNA element. Nature *323*, 353–356.

Gonzalez, G. A., Yamamoto, K. K., Fischer, W. H., Karr, D., Menzel, P,. Biggs, W., III, Vale, W. W., and Montminy, M. R. (1989). A cluster of phosphorylation sites on the cyclic AMP–regulated nuclear factor CREB predicted by its sequence. Nature *337*, 749–752.

Lewis, E. J., Harrington, C. A., and Chikaraishi, D. M. (1987). Transcriptional regulation of the tyrosine hydroxylase gene by glucocorticoid and cyclic AMR. Proc. Natl. Acad. Sci. USA *84*, 3550–3554.

Mellon, P. L., Clegg, C. H., Correll, L. A., and McKnight, G. S. (1989). Regulation of transcription by cyclic AMP–dependent protein kinase. Proc. Natl. Acad. Sci. USA 86, 4887–4892.

Montminy, M. R., and Bilezikjian, L. M. (1987). Binding of a nuclear protein to the cyclic-AMP response element of the somatostatin gene. Nature 328, 175–178.

Montminy, M. R., Sevarino, K. A., Wagner, J. A., Mandel, G., and Goodman, R. H. (1986). Identification of a cyclic-AMP–responsive element within the rat somatostatin gene. Proc. Natl. Acad. Sci. USA 83, 6682–6686.

Nigg, E. A., Hilz, H., Eppenberger, H. M., and Dutly, F. (1985). Rapid and reversible translocation of the catalytic subunit of cAMP-dependent protein kinase type II from the Golgi complex to the nucleus. EMBO J. 4, 2801–2806.

Plet, A., Evain, D., and Anderson, W. B. (1982). Effect of retinoic acid treatment of F9 embryonal carcinoma cells on the activity and distribution of cyclic AMP–dependent protein kinase. J. Biol. Chem. 257, 889–893.

Ptashne, M. (1988). How eukaryotic transcriptional activators work. Nature, 335, 683–689.

Riabowol, K. T. Fink, J. S., Gilman, M. Z., Walsh, D. A., Goodman, R. H., and Feramisco, J. R. (1988). The catalytic subunit of cAMP-dependent protein kinase induces expression of genes containing cAMP-responsive enhancer elements. Nature, 316, 83–86.

Rickles, R. J., Darrow A. L., and Strickland, S. (1989). Differentiation-responsive elements in the 5′ region of the mouse tissue plasminogen activator gene confer two-stage regualtion by retinoic acid and cyclic AMP in teratocarcinoma cells. Mol. Cell. Biol. 9, 1691–1704.

Sasaki, K., Cripe, T. P., Koch, S. R., Andreone, T. L., Peterson, D. D., Beale, E. B., and Granner, K. K. (1984). Multihormonal regulation of phosphoenolpyruvate carboxykinase gene transcription. J. Biol. Chem. 259, 15242–15251.

Yamamoto, K. R., Gonzalez, G. A., Biggs, W. H., III, and Montminy, M. R. (1988). Phosphorylation-induced binding and transcriptional efficacy of nuclear factor CREB. Nature 334, 494–498.

Zinn, K., DiMaio, D., and Maniatis, T. (1983). Identification of two distinct regulatory regions adjacent to the human β-interferon gene. Cell 34, 865–879.

Baeuerle & Baltimore (1988) Science **242**, 540–546

IκB: A Specific Inhibitor of the NF-κB Transcription Factor

PATRICK A. BAEUERLE AND DAVID BALTIMORE

In cells that do not express immunoglobulin kappa light chain genes, the kappa enhancer binding protein NF-κB is found in cytosolic fractions and exhibits DNA binding activity only in the presence of a dissociating agent such as sodium deoxycholate. The dependence on deoxycholate is shown to result from association of NF-κB with a 60- to 70-kilodalton inhibitory protein (IκB). The fractionated inhibitor can inactivate NF-κB from various sources — including the nuclei of phorbol ester–treated cells — in a specific, saturable, and reversible manner. The cytoplasmic localization of the complex of NF-κB and IκB was supported by enucleation experiments. An active phorbol ester must therefore, presumably by activation of protein kinase C, cause dissociation of a cytoplasmic complex of NF-κB and IκB by modifying IκB. This releases active NF-κB which can translocate into the nucleus to activate target enhancers. The data show the existence of a phorbol ester-responsive regulatory protein that acts by controlling the DNA binding activity and subcellular localization of a transcription factor.

IN EUKARYOTIC CELLS, THE RATE OF TRANSCRIPTION OF MANY genes is altered in response to extracellular stimuli. Changes in expression of genes transcribed by RNA polymerase II in response to such agents as steroid hormones, growth factors, interferon, tumor promoters, heavy metal ions and heat shock are mediated through distinct cis-acting DNA-sequence elements (1). Most important are those called enhancers (2), which display great positional flexibility with respect to the gene they control (3), and promoters, which are confined to the 5′ noncoding region of the gene (4). Both cis-acting elements contain multiple binding sites for sequence specific DNA-binding proteins (1, 5). The demonstration of protein-DNA interaction in vivo (6), competition experiments in vitro (7) and in vivo (8), and the definition of protein binding sites by mutational alteration of regulatory DNA sequences (9, 10) suggested that occupation of cis-acting elements by trans-acting factors is crucial for the transcriptional activity of constitutive and inducible genes. There is increasing evidence that inducible transcription of genes is mediated through induction of the activity of

The authors are at the Whitehead Institute for Biomedical Research, Nine Cambridge Center, Cambridge, MA 02142, and at the Department of Biology, Massachusetts Institute of Technology, Cambridge, MA 02139.

trans-acting protein factors (11-14). Possible mechanisms include the transcriptional activation of genes encoding trans-acting factors, the activation of DNA binding activity from an inactive precursor, inducible nuclear translocation, and modification of an already DNA-associated factor to acquire transcription activating competence.

The NF-κB transcription factor provides a model system to study the activation of a phorbol ester-inducible transcription factor. Binding of NF-κB to the κ light chain enhancer, the human immunodeficiency virus (HIV) enhancer, and an upstream sequence of the interleukin-2 receptor α-chain gene has been shown to confer transcriptional activity and phorbol ester inducibility to genes controlled by these cis-acting elements (9, 15). Mutations in the binding sites for NF-κB abolish these effects. Oligonucleotides representing NF-κB binding sites inserted in either upstream or

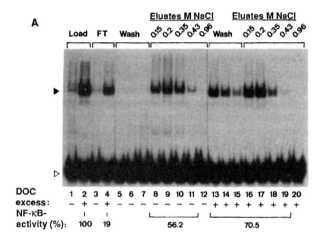

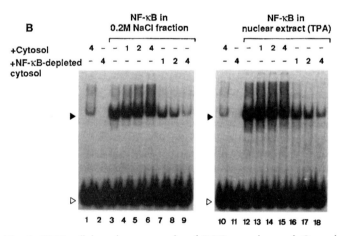

Fig. 1. DNA-cellulose chromatography of DOC-treated cytosol. Cytosol was prepared from unstimulated 70Z/3 pre-B cells (19) and protein concentrations were determined (40). In the fluorograms of native gels shown, the filled arrowheads indicate the position of the NF-κB–κ enhancer fragment complex (19), and the open arrowheads indicate the position of unbound DNA probe. (**A**) Release of DOC-independent NF-κB activity. Equal proportions of load, flow-through (FT), washings, and eluates (41) were analyzed by EMSA (18, 19) with (+) or without (−) excess DOC (41). The ³²P radioactivity in the NF-κB–DNA complexes was counted by liquid scintillation and the percentage of NF-κB activity recovered in the various fractions was calculated. (**B**) Release of an inhibitory activity. NF-κB contained in the 0.2M NaCl fraction (31 ng of protein) or NF-κB in a nuclear extract from TPA-treated 70Z/3 cells (1.1 μg of protein) was incubated under nondissociating conditions (41) with the indicated amounts (in microliters) of either cytosol which was DOC-treated but not passed over DNA-cellulose (lanes 4 to 6 and 13 to 15) or the flow-through fraction (referred to as NF-κB-depleted cytosol; lanes 7 to 9 and 16 to 18).

downstream enhancer (16) or promoter positions (17) are sufficient to confer phorbol ester-inducible and cell stage-specific transcriptional activity to test genes. In most cell lines, except for those derived from mature B cells, DNA binding activity of NF-κB is not evident in nuclear extracts unless cells have been treated with the phorbol ester TPA (12-O-tetradecanoylphorbol 13-acetate) (11, 18). This activation of NF-κB is independent of new protein synthesis, suggesting that NF-κB is present in most cell types as an inactive precursor before TPA stimulation.

We showed earlier that the NF-κB transcription factor can be detected in two forms (19). One form, in nuclear extracts from TPA-stimulated cells, does not require dissociating agents for DNA binding activity. The other form fractionates into the cytosol of unstimulated cells and is only active in the presence of dissociating agents such as sodium deoxycholate (DOC) or after denaturation and electrophoretic fractionation. The conversion of inactive into active NF-κB by TPA in vivo or by dissociating agents in vitro and the different subcellular fractionation of the two forms of NF-κB suggested that the activation of NF-κB takes place in at least two steps, namely, derepression of DNA binding activiy and nuclear translocation.

To investigate whether the DOC-dependence of cytosolic NF-κB results from its association with an inhibitor, we probed for an activity in cytosolic fractions that would specifically prevent DNA binding to NF-κB in electrophoretic mobility shift assays (EMSA). Here we demonstrate a protein inhibitor (called IκB) in cytosolic fractions of unstimulated pre-B cells that can convert NF-κB into an inactive DOC-dependent form by a reversible, saturable, and specific reaction. The inhibitory activity becomes evident only after selective removal of the endogenous cytosolic NF-κB under dissociating conditions, suggesting that NF-κB and IκB were present in a stoichiometric complex. Enucleation experiments showed that the complex of NF-κB and IκB is truly cytoplasmic. Our data are consistent with a molecular mechanism of inducible gene expression by which a cytoplasmic transcription factor–inhibitor complex is dissociated by the action of TPA presumably through activation of protein kinase C. The dissociation event results in activation and apparent nuclear translocation of the transcription factor. It would appear that IκB is the target for the TPA-induced dissociation reaction.

Separation of an inhibitor from NF-κB. We examined cytosolic fractions from unstimulated 70Z/3 pre-B cells for an activity that would impair the DNA binding activity of added NF-κB in an EMSA (18, 19). Increasing amounts of cytosol from unstimulated cells did not significantly influence the formation of a protein-DNA complex between NF-κB and a κ enhancer fragment (Fig. 1B, lanes 13 to 15). This indicated the absence of free inhibitor, presumably because all of it is complexed with endogenous NF-κB. In an attempt to liberate the inhibitor, we used DNA-cellulose to selectively remove the endogenous NF-κB from DOC-treated cytosol. Almost all NF-κB was present in a DOC-dependent form prior to DOC activation and chromatography (Fig. 1A, lanes 1 and 2). In the presence of excess DOC, about 80 percent of the NF-κB activity was retained by DNA-cellulose (Fig. 1A, compare lanes 2 and 4) most of which eluted from the DNA-cellulose between 0.15 and 0.35M NaCl (Fig. 1A, lanes 8 to 10 and 16 to 18). The NF-κB activity eluting at high salt concentration was detectable in mobility shift assays in the absence of excess DOC (Fig. 1A, lanes 8 to 11), indicating that NF-κB had been separated from an activity that caused its DOC-dependent DNA binding activity. In contrast, the small percentage of NF-κB activity contained in the washings was still dependent on DOC (Fig. 1A, compare lanes 5 to 7 and 13 to 15). These results show that aff niy chromatography is sufficient to convert DOC-dependent NF-κB precursor into DOC-independent active NF-κB, similar to that found in nuclear extracts from TPA-stimulated cells.

The flow-through fraction from the DNA-cellulose was assayed for an activity that, after neutralization of DOC by non-ionic detergent, would inactivate added NF-κB from the 0.2M NaCl fraction from nuclear extracts of TPA-stimulated cells. Increasing amounts of cytosol from which the endogenous NF-κB was removed inhibited the formation of an NF-κB–DNA complex as

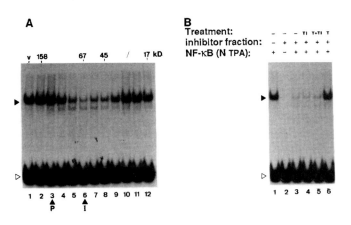

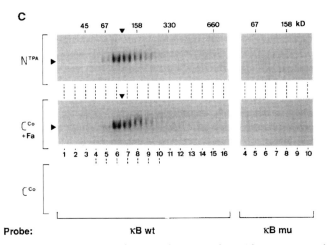

Fig. 2. Characterization of IκB and its complex with NF-κB. In the fluorograms shown, the filled arrowheads indicate the position of the NF-κB–DNA complex and the open arrowheads the position of free DNA probe. (**A**) For size determination of IκB, the flow-through from the DNA-cellulose column (41) was passed over a G-200 Sephadex column (42). Portions of fractions were incubated with NF-κB contained in nuclear extracts from TPA-stimulated 70Z/3 cells (N TPA) (43), and analyzed by EMSA [v, void volume; P, fraction where remaining NF-κB precursor (Fig. 1A, lane 4) peaked after gel filtration as assayed with excess DOC in the absence of added NF-κB; I, fraction where the inhibiting activity peaked]. (**B**) The effect of trypsin treatment (45) on the inhibiting activity of IκB. NF-κB in a nuclear extract (lane 1) was incubated with a fraction containing inhibitor (lane 2) without any addition (–; lane 3) or with bovine pancreas trypsin inhibitor (TI; lane 4), trypsin that had been incubated with BPTI (T + TI; lane 5), or with trypsin alone (T; lane 6). Samples were then used in the inhibitor assay (43). (**C**) Glycerol gradient sedimentation of NF-κB and its complex with IκB. Nuclear extract from TPA-stimulated 70Z/3 cells (N TPA) and cytosol from unstimulated cells (C Co) were subjected to sedimentation through a glycerol gradient (46). Cosedimented size markers were ovalbumin (45 kD), BSA (67 kD), immunoglobulin G (158 kD) and thyroglobulin (330 and 660 kD). NF-κB activity was detected in the fractions by EMSA with a wild-type κ enhancer fragment (κB wt, left panels). The specificity was tested with a mutant fragment (9) (κB mu, right panels). The inactive cytosolic NF-κB precursor (lower panel) was activated by formamide treatment (19) (Fa; middle panel).

monitored by EMSA (Fig. 1B, lanes 7 to 9 and 16 to 18). The DOC-treated cytosol that was not passed over DNA-cellulose had no effect (Fig. 1B, lanes 4 to 6 and 13 to 15) even if cells had been treated with TPA (20). Because, after DNA-cellulose chromatography of DOC-treated cytosol, we found both DOC-independent NF-κB activity and an inhibitory activity, we believe that we separated NF-κB from an inhibitor. We refer to this inhibitor as IκB.

IκB fractionates as a 60- to 70-kD protein. The flow-through fraction from the DNA-cellulose column was subjected to gel filtration through G-200 Sephadex, and the fractions were assayed for an activity that would interfere with the DNA binding activity of added NF-κB contained in a nuclear extract from TPA-stimulated 70Z/3 cells (Fig. 2A). The 67-kD fraction had the highest activity; it virtually completely prevented interaction of NF-κB and DNA (Fig. 2A, lane 6). In fractions from a G-75 Sephadex column, no additional inhibitor of low molecular size was detectable (20), an indication that NF-κB was inactivated by a macromolecule of defined size. No significant inhibitory activity could be demonstrated after gel filtration of a DNA-cellulose flow-through of DOC-treated cytosol from TPA-stimulated 70Z/3 cells (20), implying that TPA treatment of cells inactivated IκB.

The inhibitor fraction was treated with trypsin to test whether IκB is a protein (Fig. 2B). Tryptic digestion was stopped by the addition of bovine pancreas trypsin inhibitor (BPTI), and samples were analyzed for NF-κB inhibition. Trypsin treatment interfered with the activity of IκB, as shown by the complete inability of the treated sample to inhibit NF-κB activity (Fig. 2B, compare lanes 1 and 6). Trypsin that had been treated with BPTI had no effect (Fig. 2B, lane 5) demonstrating that the inactivation of IκB was specifically caused by the proteolytic activity of trypsin. It appears that IκB requires an intact polypeptide structure for its activity.

The cytosolic complex of IκB and NF-κB showed an apparent size of about 120 to 130 kD after both gel filtration (Fig. 2A, lane 3) and sedimentation through a glycerol gradient (Fig. 2C, lanes 6 and 7). For both methods, cytosol from unstimulated cells was analyzed under nondissociating conditions. The NF-κB was activated in fractions by either DOC (Fig. 2A) or formamide (19) (Fig. 2C, middle panel) prior to analysis by EMSA. The specificity of complexes was tested with a mutant DNA probe (9) (Fig. 2C, right panels). The apparent release of a 60- to 70-kD inhibitory protein from the cytosolic NF-κB precursor, its sedimentation velocity in glycerol gradients, and its size seen by gel filtration suggest that the inactive NF-κB precursor is a heterodimer composed of a 55- to 62-kD NF-κB molecule (19) and a 60- to 70-kD IκB molecule. Nuclear NF-κB was found to cosediment with the cytosolic complex of IκB and NF-κB (Fig. 2C, upper panel). Native gel electrophoresis, a method that allows resolution of size differences of protein-DNA complexes (21), provided evidence that the 120 kD form of nuclear NF-κB seen in glycerol gradients comes from the formation of a homodimer (22). By these interpretations, activation of NF-κB would include an additional step, that is, formation of a NF-κB homodimer. This is consistent with the observation that the protein-DNA complexes formed with in vitro-activated NF-κB have the same mobility in native gels as those formed with nuclear NF-κB (19).

Inactivation of NP-κB by IκB is reversible, saturable, and specific. Incubation with the inhibitor fraction can inhibit the DNA binding activity of NF-κB by more than 90 percent (Fig. 3A, lanes 1 and 3). Treatment of a duplicate sample with DOC after the inhibition reaction reactivated 66 percent of the added NF-κB activity (Fig. 3A; compare lanes 3, 4, and 6). This showed that a DOC-dependent form of NF-κB can be reconstituted in vitro by the addition of a fraction containing IκB to nuclear NF-κB. The incomplete activation of NF-κB by DOC might be due to the DOC-

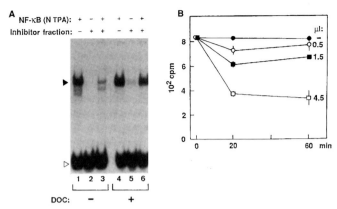

Fig. 3. Reversibility and kinetics of the inactivation of NF-κB. (**A**) The effect of DOC treatment on in vitro inactivated NF-κB. The NF-κB contained in nuclear extracts from TPA-stimulated 70Z/3 cells (N TPA; 1.1 μg of protein) was inactivated by addition of a gel filtration fraction containing IκB (2.5 μg of protein). A duplicate sample was treated after the inhibition reaction with 0.8 percent DOC followed by addition of DNA binding reaction mixture (43) containing 0.7 percent NP-40. Samples were analyzed by EMSA. In the fluorograms shown, the filled arrowhead indicates the position of the NF-κB–DNA complex, and the open arrowhead the position of unbound DNA probe. (**B**) A titration and kinetic analysis of the in vitro inactivation of NF-κB. The NF-κB contained in nuclear extracts from TPA-treated 70Z/3 cells (2.2 μg of protein) was incubated with increasing amounts (0.25 to 2.25 μg of protein) of a gel filtration fraction containing IκB. After the DNA binding reaction, samples were analyzed by EMSA. The ^{32}p radioactivity in the NF-κB–DNA complexes visualized by fluorography was determined by liquid scintillation counting. All reactions were performed in triplicates. The bars represent standard deviations.

neutralizing effect of non-ionic detergent that was still present in the sample from the preceding inhibition reaction.

A titration and kinetic analysis showed that IκB stoichiometrically interacts with NF-κB (Fig. 3B). Increasing amounts of inhibitor fraction were added to an excess amount of NF-κB and incubated for 20 or 60 minutes. After the DNA binding reaction, NF-κB-DNA complexes were separated on native gels and quantified by liquid scintillation counting. The relation between amount of IκB fraction added and extent of inhibition was linear. The amount of NF-κB inactivated after 20 minutes of incubation was not increased after 60 minutes (Fig. 3B). These kinetics were probably not the result of a rapid decay of a catalytically active inhibitor because the fractions were incubated before the reaction. Our data are consistent with rapid formation of an inactive complex by addition of IκB to NF-κB. The fraction containing IκB does not appear to catalytically or covalently inactivate NF-κB: neither the reversibility nor the kinetics support such a model.

IκB was tested for its influence on the DNA binding activity of other defined nuclear factors (Fig. 4A). These factors were contained in nuclear extracts that had essentially no active NF-κB (19), which otherwise could have inactivated IκB by complex formation. The DNA binding activity of H2TF1, a transcription factor thought to be related to NF-κB (23) was not affected by the inhibitor fraction. Ubiquitous and lymphoid-specific octamer-binding proteins (OCTA) (24, 25) were unaffected in their DNA binding activities as were two E-box binding factors, NF-μE1 (26) and NF-κE2 (9), interacting with μ heavy chain and κ light chain enhancers, respectively. Another TPA-inducible transcription factor, AP-1 (13), also showed equal complex formation after incubation in the presence and absence of the inhibitor fraction. Furthermore, none of the undefined DNA binding activities seen in the EMSA showed any inactivation by IκB (Fig. 4A). These results show that IκB is a specific inhibitor of the DNA binding activity of NF-κB.

In vivo activated NF-κB is responsive to IκB. IκB prepared

from the mouse pre-B cell line 70Z/3 was tested for inactivation of NF-κB contained in nuclear extracts from other cell lines. Human NF-κB contained in nuclear extracts from TPA-stimulated HeLa cells and H-9 T lymphoma cells was efficiently inactivated (Fig. 4B). When excess amounts of the various NF-κB activities were used in the inhibitor assay, the extent of reduction of NF-κB activities by a fixed amount of IκB was very similar as determined by liquid scintillation counting (20). Also NF-κB from nuclear extracts of TPA-stimulated Madin-Darby bovine kidney (MDBK) cells was inactivated (20), suggesting that the control of NF-κB activity by IκB is conserved among different mammalian species.

NF-κB is constitutively active in cell lines derived from mature B cells (18). We tested nuclear extracts from the mouse B cell line WEHI 231 in the inhibitor assay to examine whether NF-κB has undergone a modification in those cell lines that prevented its inactivation by IκB. The NF-κB from B cells was as efficiently inactivated as NF-κB from pre-B cells (Fig. 4B), suggesting that NF-κB is not stably modified in B cells (or in other cells after TPA stimulation) in such a way that it cannot respond to inactivation by IκB.

The NF-κB–IκB complex is present in enucleated cells. The NF-κB–IκB complex shows a cytosolic localization on subcellular fractionation (19). This procedure may, however, cause artifacts. Hypotonic lysis of cells may result in partitioning of nuclear proteins into the cytosol, especially when they are not tightly associated with nuclear components (27). We tried therefore to detect the complex of IκB and NF-κB in enucleated cells. Enucleation is performed with living cells at 37 °C (28) and should therefore not interfere with active nuclear import of proteins, which is adenosine triphosphate-dependent and blocked at low temperature (29).

Using cytochalasin B-treated HeLa cells, we obtained an enucleation efficiency of about 90 percent (Fig. 5A). Enucleated and cytochalasin B-treated complete cells were incubated in the absence and presence of TPA and solubilized by detergent; and the proteins were extracted with high salt. Because of the small number of cells analyzed, this procedure is different from our standard one. Total cell extracts were analyzed for NF-κB-specific DNA binding activity by EMSA (Fig. 5B). In both enucleated and complete cells, similar amounts of NF-κB activity were found after TPA stimulation (Fig. 5B, lanes 1 to 4). The activity was specific for NF-κB because it was not observed with a mutant κ enhancer fragment (9, 20). These results suggest that TPA-inducible NF-κB in HeLa cells is predominantly cytoplasmic because it was still present in enucleated cells. The NF-κB activity seen under control conditions (Fig. 5B, lanes 1 and 3) was most likely activated by the lysis conditions used because the activity was also observed in extracts from HeLa cells that were not treated with cytochalasin B (20) but not in fractions obtained after hypotonic lysis (19). It was still evident, however, that TPA could activate NF-κB in enucleated cells (Fig. 5B, lanes 3 and 4).

After treatment with DOC, total extracts from complete and enucleated control cells showed about a twofold increase in the amount of NF-κB activity (Fig. 5B, compare lanes 1 and 3 with 5 and 7). The demonstration of DOC-activatable NF-κB in enucleated cells as well as the presence of similar amounts of total NF-κB in enucleated and complete cells (Fig. 5B, lanes 5 to 8) shows that a substantial amount of the total cellular NF-κB–IκB complex was cytoplasmic. In contrast to NF-κB, most of the DNA binding activity of AP-1, a bona fide nuclear protein (13), was apparently lost by enucleation of cells (Fig. 5B, lanes 9 to 12).

Mechanism of NP-κB activation. We have shown that the NF-κB nuclear transcription factor exists in unstimulated pre-B cells in a cytoplasmic complex with a specific inhibitory protein, IκB. In this complex, NF-κB does not exhibit DNA binding activity in EMSA and partitions, upon subcellular fractionation, into the cytosol. The complex is apparently a heterodimer consisting of an NF-κB molecule of about 60 kD and an IκB molecule of 60 to 70 kD. Upon TPA stimulation of cells or after treatment with dissociating agents in vitro, the NF-κB–IκB complex dissociates. This releases NF-κB, which appears now to form a homodimer and can translocate into the nucleus. Whether dimerization is required for activation of NF-κB is not known.

The inhibitory effect of IκB on the DNA binding activity and nuclear localization properties of NF-κB appears to arise from a simple physical affinity of the two proteins. The complex freely dissociates and the components readily associate under in vitro conditions. Even in vivo, dissociation by short-term TPA treatment and reassociation after long-term TPA treatment (30) is evident. The latter presumably results from the degradation of protein kinase C

after TPA activation and implies that NF-κB can move back to the cytoplasm after being active in the nucleus.

The effect of TPA appears to involve an alteration of IκB but not of NF-κB. After TPA stimulation, no active IκB was found, implying its alteration, while the nuclear NF-κB remained sensitive to unmodified IκB when tested in vitro. Whether inactive IκB can be regenerated is unclear; in experiments with cycloheximide (30), irreversible loss of IκB activity was the only demonstrable effect after 8 hours of TPA treatment. In that TPA can activate protein kinase C, it is a reasonable hypothesis that direct or indirect phosphorylation of IκB results in its dissociation from NF-κB; however, we are at present unable to directly test this suggestion.

We previously found that the NF-κB–IκB complex is recovered in the cytosol (19). Here we show directly that it is not removed from the cell by enucleation and therefore is truly cytoplasmic (31). Because active protein kinase C is bound to the plasma membrane (32), it becomes increasingly attractive to suggest that the cytoplasmic

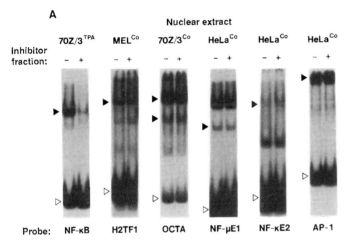

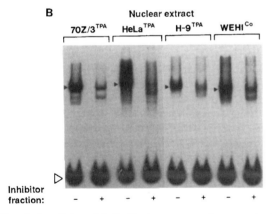

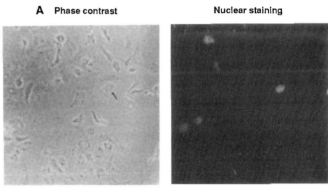

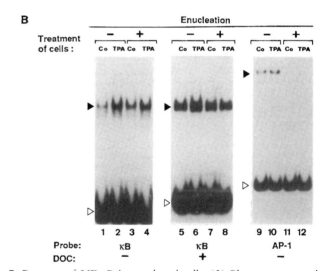

Fig. 4. Specificity of IκB. Nuclear extracts from unstimulated (Co) or TPA treated cells were incubated with 5 μl of buffer G (−) (41) or with 5 μl of a gel filtration fraction containing IκB (+) (A, in the presence of 150 mM NaCl). After DNA binding reactions, samples were analyzed by EMSA. (**A**) Influence of IκB on the DNA binding activity of various nuclear factors. The probes were: NF-κB (18); H2TF1, an oligonucleotide subcloned into pUC containing the H2TF1 binding site from the H-2 promoter (23); OCTA, an oligonucleotide subcloned into pUC containing the common binding site for the ubiquitous (upper filled arrowhead) and lymphoid-specific (lower filled arrowhead) octamer-binding proteins (25), NF-μE1 (26); NF-κE2 (9); and AP-1, Eco RI-Hind III fragment of the yeast HIS4 promoter (47) containing three binding sites recognized by mammalian AP-1/jun (48). In the fluorograms shown, filled arrowheads indicate the positions of specific protein-DNA complexes. Open arrowheads indicate the positions of uncomplexed DNA fragments. (**B**) Interaction of IκB with NF-κB from different cell lines. The filled arrowheads indicate the positions of the NF-κB-DNA complexes from the various cell lines and the open arrowhead indicates the position of uncomplexed DNA probe.

Fig. 5. Presence of NF-κB in enucleated cells. (**A**) Phase contrast and fluorescence microscopy of enucleated HeLa cells (49). From 612 cells counted on photographic prints, 63 showed nuclear staining. A representative micrograph is shown. The arrow indicates a cell that retained its nucleus. (**B**) Analysis of complete and enucleated cells for NF-κB activity. Total cell extracts (1.2 μg of protein) from control (Co) and TPA-treated complete and enucleated cells were analyzed by EMSA with a labeled κ enhancer fragment (κB) or HIS4 promoter fragment (AP-1) (47, 48), 3 μg of poly(dI-dC), 1 μg of BSA, 1.2 percent NP-40 and the binding buffer (19) in a final volume of 20 μl. In lanes 5 to 8, extracts were treated with DOC followed by the addition of the DNA binding mixture to give final concentrations of 0.8 percent DOC and 1.2 percent NP-40. Samples were analyzed by EMSA. In the fluorograms shown, the filled arrowheads indicate the positions of specific protein-DNA complexes and the open arrowheads indicate the positions of uncomplexed DNA probes.

complex interacts in the cytoplasm (perhaps near the plasma membrane) with protein kinase C, and the liberated NF-κB carries the signal from cytoplasm to nucleus. Under a number of conditions, active NF-κB is found in the cytoplasm (30): this and the reversibility of NF-κB activation in vivo (30) suggests that the protein may freely move in and out of the nucleus, bringing to the nucleus information reflecting the cytoplasmic activation state of protein kinase C and possibly of other signaling systems.

The response of NF-κB to activated protein kinase C occurs apparently indirectly through modification and subsequent release of associated IκB. The inducibility of NF-κB by TPA is thus dependent on the presence and state of activity of IκB. Changes in amount or activity of IκB should therefore influence the TPA inducibility of NF-κB. The NF-κB can indeed exist not only in TPA-inducible but also in constitutively active form, for example, in mature B cells (18). Because constitutive NF-κB from B cells is still responsive to IκB in vitro, we suspect that IκB and not NF-κB is altered during differentiation of pre-B into B cells.

IκB is apparently unstable when not complexed with NF-κB. This is suggested by the absence of excess active inhibitor in the cytosol from unstimulated cells. In a situation where the production of new inhibitor is impaired, the decay of occasionally released inhibitor could activate NF-κB. This would explain the partial activation of NF-κB seen after treatment with the protein synthesis inhibitors cycloheximide and anisomycin (11). The demonstration of a specific inhibitory protein of NF-κB and the interpretation that cycloheximide treatment can activate NF-κB presumably because cells become depleted of inhibitor suggest that IκB is the putative labile repressor of κ gene expression (33) and of NF-κB activity (11).

Similarity to the activation of glucocorticoid receptor. Inactive glucocorticoid receptor (GR) is localized in the cytosol (34) in complex with two molecules of the 90-kD heat shock protein (hsp90) (35). This complex is thought to dissociate upon binding of steroid hormone to the receptor (36, 37), thereby releasing active GR that is able to migrate into the nucleus to bind to glucocorticoid responsive elements (38). Despite a similar function, we consider it unlikely that IκB is related to hsp90. The molecular size, specificity, responsiveness to TPA, and apparent low abundance and lability distinguish IκB from hsp90. Unlike the GR-hsp90 complex (36), the NF-κB–IκB complex is stable to warming, gel filtration, dilution, and elevated ionic strength (19, and our data presented above) suggesting a high affinity interaction between IκB and NF-κB. In the case of the GR-hsp90 complex, there is no requirement for a high affinity interaction because hsp90, as a very abundant protein (39), is in great excess over the receptor. Furthermore, modification of hsp90 is apparently not required for dissociation of the complex. It is, rather, released by a conformational change of the receptor upon hormone binding (37).

REFERENCES AND NOTES

1. A. K. Hatzopoulos, U. Schlokat, P. Gruss, in *Eucaryotic RNA Synthesis and Processing*, B. D. Hames and D. M. Glover, Eds. (IRL Press, Washington, DC, in press).
2. U. Schlokat and P. Gruss, in *Oncogenes and Growth Control*, P. Kahn and T. Graf, Eds. (Springer-Verlag, Berlin, 1986), p. 226.
3. J. Banerji, S. Rusconi, W. Schaffner, *Cell* 27, 299 (1981); P. Moreau, R. Hen, B. Wasylyk, R. Everett, M. P. Gaub, P. Chambon, *Nudeic Acids Res.* 9, 6047 (1981); M. Fromm and P. Berg, *Mol. Cell. Biol.* 3, 991 (1983).
4. W. S. Dynan and R. Tjian, *Nature* 316, 774 (1985); S. McKnight and R. Tjian, *Cell* 46, 795 (1986).
5. E. Wingender, *Nucleic Acids Res.* 16, 1879 (1988).
6. A. Ephrussi, G. M. Church, S. Tonegawa, W. Gilbert, *Science* 227, 134 (1985).
7. H. R. Schoeler and P. Gruss, *Cell* 36, 403 (1984).
8. H. Schoeler and P. Gruss, *EMBO J.* 4, 3005 (1984); M. Mercola, J. Goverman, C. Mirell, K. Calame, *Science* 227, 266 (1985).
9. M. Lenardo, J. W. Pierce, D. Baltimore, *Science* 236, 1573 (1987).
10. M. E. Greenberg, Z. Siegfried, E. B. Ziff, *Mol. Cell. Biol.* 7, 1217 (1987).
11. R. Sen and D. Baltimore, *Cell* 47, 921 (1987).
12. R. Prywes and R. Roeder, *ibid.*, p. 777.
13. W. Lee, P. Mitchell, R. Tjian, *ibid.* 49, 741 (1987); P. Angel et al., *ibid.* 49, 729 (1987).
14. P. J. Godowski, S. Rusconi, R. Miesfeld, K. R. Yamamoto, *Nature* 325, 365 (1987).
15. G. Nabel and D. Baltimore, *ibid.* 326, 711 (1987); E. Boehnlein et al., *Cell* 53, 827 (1988); K. Leung and G. J. Nabel, *Nature* 333, 776 (1988).
16. J. W. Pierce, M. Lenardo, D. Baltimore, *Proc. Natl. Acad. Sci. U.S.A.* 85, 1482 (1988).
17. T. Wirth and D. Baltimore, *EMBO J.*, in press.
18. R. Sen and D. Baltimore, *Cell* 46, 705 (1986).
19. P. A. Baeuerle and D. Baltimore, *ibid.* 53, 211 (1988).
20. ———, unpublished observation.
21. I. A. Hope and K. Struhl, *EMBO J.* 6, 2781 (1987).
22. The mobility of the protein-DNA complex formed with highly homogeneous NF-κB that was renatured from a single spot of a two-dimensional reducing SDS-gel is the same as that seen with crude NF-κB from nuclear extracts used for glycerol gradient centrifugation (P. A. Baeuerle, M. Lenardo, D. Baltimore, unpublished observation).
23. A. S. Baldwin and P. A. Sharp, *Proc. Natl. Acad. Sci. U.S.A.* 85, 723 (1988).
24. H. L. Sive and R. G. Roeder, *ibid.* 83, 6382 (1986).
25. L. M. Staudt et al., *Nature* 323, 640 (1986).
26. J. Weinberger, D. Baltimore, P. A. Sharp, *ibid.* 322, 846 (1986).
27. J. J. Li and T. J. Kelly, *Proc. Natl. Acad. Sci. U.S.A.* 81, 6973 (1984).
28. D. M. Prescott and J. B. Kirkpatrick, *Methods Cell Biol.* 7, 189 (1973).
29. D. D. Newmeyer and D. J. Forbes, *Cell* 52, 641 (1987); W. D. Richardson, A. D Mills, S. M. Dilworth, R. A. Laskey D. Dingwall, *ibid.*, p. 655.
30. P. A. Baeuerle, M. Lenardo, J. W. Pierce, D. Baltimore, *Cold Spring Harbor Symp. Quant. Biol.*, in press.
31. The estrogen receptor that normally fractionates into the cytosol was shown to be nuclear by its absence from enucleated cells; W. V. Welshons et al., *Nature* 307, 747 (1984).
32. A. S. Kraft, W. B. Anderson, L. Cooper J. J. Sando, *J. Biol. Chem.* 257, 13193 (1983); M. Wolf, H. LeVine III, S. May Jr., P. Cuatrecasas, N. Sahyoun, *Nature* 317, 546 (1985); U. Kikkawa and Y. Nishizuka, *Annu. Rev. Cell. Biol.* 2, 149 (1986).
33. R. Wall et al., *Proc. Natl. Acad. Sci. U.S.A.* 83, 295 (1986).
34. D. Picard and K. R. Yamamoto, *EMBO J.* 6, 3333 (1987).
35. E. R. Sanchez, D. O. Toft, M. J. Schlesinger, W. B. Pratt, *J. Biol. Chem.* 260 12398 (1985).
36. E. R. Sanchez et al., *ibid.* 262, 6986 (1987).
37. W. B. Pratt et al., *ibid.* 263, 267 (1988); M. Denis, L. Poellinger, A. -C. Wikstroem J.-A. Gustafsson, *Nature* 333, 686 (1988).
38. K. R. Yamamoto, *Annu. Rev. Genet.* 19, 209 (1985).
39. P. M. Kelley and M. J. Schlesinger, *Mol. Cell. Biol.* 2, 267 (1982).
40. P. K. Smith et al., *Anal. Biochem.* 150, 76 (1985).
41. Cytosol from unstimulated 70Z/3 pre-B cells in buffer A (19, 44) was adjusted to a final concentration of 50 mM NaCl, 20 mM Hepes (pH 7.9), 1.5 mM EDTA, 5 percent glycerol and 0.2 percent NP-40. Cytosolic protein (45 mg) was mixed to a final volume of 4 ml with 0.6 percent DOC, 0.75 g of calf thymus (wet weight) DNA-cellulose [Sigma; equilibrated in buffer G: 10 mM tris-HCl (pH 7.5), 50 mM NaCl, 1 mM EDTA, 1 mM dithiothreitol (DTT), 5 percent glycerol, 0.2 percent DOC, 0.2 percent NP-40, and 0.5 mM phenylmethyl sulfonylfluoride (PMSF)] and 1.2% NP-40. The suspension was incubated in a mini column for 1 hour at room temperature on a rotary shaker. The flow-through fraction was used for gel filtration. DNA-cellulose was washed with buffer G and eluted with a NaCl step gradient in buffer G. Equal proportions of fractions were assayed by EMSA (18, 19) at a final concentration of 1.2 percent NP-40 in the presence of either 0.03 percent DOC (nondissociating condition) or 0.6 percent DOC (dissociating condition) and with 10 μg of bovine serum albumin (BSA) as carrier.
42. The flow-through fraction from the DNA-cellulose column (1.55 mg of protein in 250 μl) was subjected to a G-200 Sephadex column (280 by 7 mm) with a flow rate of 0.15 ml/min in buffer G at room temperature. A mixture of size markers (dextran blue; immunoglobulin G, 158 kD; BSA, 67 kD; ovalbumin, 45 kD; myoglobin, 17 kD; Bio-Rad) was run on the column before samples were analyzed. Markers were detected in fractions by their color and by SDS–polyacrylamide gel electrophoresis (SDS-PAGE), which was followed by Coomassie blue staining.
43. To detect inhibiting activity, portions of fractions (5μl; in buffer G) were mixed with 1 μl of nuclear extracts [in buffer D(+) (19, 44)] and 0.5 ml 10 percent NP-40. After 30 minutes at room temperature, the reaction volume was brought to 20 μl by the addition of a DNA binding reaction mixture containing 3.2 μg of poly(dI-dC) (Pharmacia), 1 to 4 fmole of ^{32}P–end labeled K enhancer fragment (18), 75 mM NaCl, 15 mM tris-HCl (pH 7.5), 1.5 mM EDTA, 1.5 mM DTT, 7.5 percent glycerol, 0.3 percent NP-40, and 20 μg of BSA. After a 20-minute DNA binding reaction, samples were analyzed by EMSA (18, 19).
44. J. P. Dignam, R. M. Lebovitz, R. G. Roeder, *Nucleic Acids Res.* 11, 1475 (1983).
45. Gel filtration fractions containing IκB (25 μg of protein) were incubated for 1 hour at room temperature in buffer G without any addition or with 2μg of TPCK-treated trypsin (Sigma), 8 μg of BPTI (Sigma), or with 2 μg of trypsin that had been incubated with 8 μg of BPTI. Tryptic digestion was stopped by a 10-minute incubation with 8 μg of BPTI and samples analyzed as described (43).
46. Nuclear extract from TPA-stimulated 70Z/3 cells and cytosol from untreated cells (both 220 μg of protein) were secEmented through 5 ml of a continuous 10 to 30 percent glycerol gradient in buffer D(+) at 150,000g (SW 50.1 rotor; Beckman) for 20 hours at 4°C. Cosedimented size markers were detected in fractions by SDS-PAGE and Coomassie blue staining. Portions of glycerol gradient fractions (4 μl) were analyzed by EMSA (19) with 10 μg of BSA as carrier and 0.5 μg of poly(dI-

dC). The NF-κB precursor was activated by treating 4 μl of fractions with 1.5 μI of formamide before the DNA binding reaction mixture was added.

47. T. F. Donahue, R. S. Daves, G. Lucchini, G. R. Fink, *Cell* **32**, 89 (1983).
48. K. Struhl, *ibid*. **50**, 841 (1987).
49. HeLa cells were grown in Eagle's minimum essential medium supplemented with 10 percent horse serum, penicillin (50 IU/ml), and streptomycin (50 μg/ml) (referred to as supplemented MEM) on disks (1.8 cm in diameter) cut from cell culture plastic ware. For enucleation, disks were placed upside down into centrifuge tubes filled with 10 ml of supplemented MEM at 37°C containing cytochalasin B (10 μg/ml; Sigma). Cells were enucleated for 20 minutes by centrifugation at 17,500g (JS-13 rotor; Beckrnan) at 37°C. Control cells were also treated with cytochalasin B (10 μg/ml) and held for the same time in the incubator. To estimate the enucleation efficiency, enucleated cells on one disk were fixed with formaldehyde (3.7 percent) in phosphate-buffered saline (PBS) for 20 minutes, stained for 4 minutes with 4′,6-diamidino-2-phenylindole (DAPI, 1 μg/ml; Sigma) in PBS, and washed in PBS. Fluorescence microscopy under ultraviolet and phase contrast microscopy were performed with a Zeiss Photomicroscope III. Control and enucleated cells were allowed to recover in cytochalasin B—free supplemented MEM for 30 minutes before a 2-hour incubation in the absence or ptesence of TPA (50 ng/ml). Cells were then washed in ice cold PBS, scraped off the disks in 100 μl of a buffer containing 20 mM Hepes (pH 7.9), 0. 35M NaCl, 20 percent glycerol, l percent NP-40, 1 mM MgCl$_2$, 1 mM DTT, 0.5 mM EDTA, 0.1 mM EGTA, 1 percent aprotinin (Sigma) and 1 mM PMSF. After lysis and extraction for 10 minutes on ice, particulate material was removed by centrifugation (Microfuge) for 15 minutes at 4°C, and the resulting supernatants were analyzed by EMSA.

50. We thank K. Arndt, A. Baldwin, J. LeBowitz, and C. Murre for providing labeled DNA probes to detect various DNA binding activities and for a nuclear extract from MEL cells; M. Lenardo for plasmids containing wild-type and mutant κ enhancer and for a nuclear extract from MDBK cells; R. Van Etten and T. Orr-Weaver for help with fluorescence microscopy; L. Sraudt for stimulating discussion; M. Smith for carefully reading the manuscript; and Owl Scientific (Cambridge, MA) for preparation of disks from cell culture plastic ware. Supported by a grant from the American Cancer Society (D.B.) and a fellowship of the Deutsche Forschungsgemeinschaft (P.A.B.).

27 July 1988; accepted 26 September 1988

Functional role of transcription factors

13

In many cases, specific transcription factors were identified on the basis that they bound to a particular DNA sequence element that had been shown to be involved in a particular pattern of gene expression. Thus, as discussed in Section 6, the heat-shock factor was identified on the basis of its ability to bind to the heat-shock element, which confers enhanced gene expression in response to elevated temperature. Similarly, NF-κB discussed in Section 12 binds to a DNA sequence within the immunoglobulin κ light chain gene regulatory region that is critical for its B-cell-specific pattern of gene expression. In these cases, therefore, the function of the transcription factor in particular regulatory processes was clear from the nature of the DNA sequence to which it bound and which led to its identification.

In other cases, however, transcription factors were not identified simply on the basis of their ability to bind to a particular DNA sequence. Rather, when a particular property of biological significance was analysed in detail, it was found that the protein critical for this process was in fact a transcription factor, which acted by modulating the activity of specific target genes in order to produce the observed functional effect. Cases of this type have often led to the identification of an entirely new area in which transcription factors play a critical role, and two papers that presented the first analysis of such cases are therefore presented here.

The first of these, by Desplan et al., deals with the engrailed protein of the fruit fly *Drosophila melanogaster*. Mutations in the gene encoding the engrailed protein result in a fruit fly which lacks posterior regions. The *engrailed* gene is thus one of the homoeotic genes of *Drosophila* which play a critical role in the development of the *Drosophila* embryo and whose mutation results in the abnormal development of particular regions of the fly (for reviews, see Scott and Carroll, 1987; Ingham, 1988). Like the other homoeotic genes, therefore, the *engrailed* gene was originally identified simply on the basis that mutations within it produced dramatic effects. It was therefore clear that its protein product was likely to play a critical role in the fly, but it was unclear what the precise mechanism of action of this protein might be.

Early evidence suggested that the homoeotic genes might encode DNA binding transcription factors. Thus when these genes were cloned, they were shown to contain a common 60-amino-acid sequence which was flanked on either side by sequences that differed dramatically between the different homoeotic genes. This sequence was therefore named the homoeobox or homoeodomain (for reviews, see Scott et al., 1989; Kornberg, 1993; Gehring et al., 1994). Interestingly, this homoeobox also appeared in the yeast a and α mating-type proteins, which had been suggested to act as transcription factors regulating gene expression in yeast (Shepherd et al., 1984). Similarly, predictions of the structure of the homoeodomain on the basis of its protein sequence revealed that (as discussed in Section 8) it had some similarities with the known structures of bacterial DNA binding proteins.

These observations suggested, therefore, that the homoeobox might act as a DNA binding region capable of binding to the regulatory elements of target genes, thereby allowing the homoeotic gene products to modulate transcription. In turn, the failure to regulate such target genes correctly when the homoeobox-containing genes were mutated would explain the aberrant patterns of development produced. However, although this was an attractive theory, there was no direct evidence that the *Drosophila* homoeobox-containing proteins actually bound to DNA. The significance of the paper by Desplan et al. is that it provided this evidence.

In the experiment presented in Fig. 2 of the paper, Desplan et al. showed that the engrailed protein was capable of binding with low affinity to all DNA sequences. They then reasoned that, if the protein was also able to bind to some specific sequences with high affinity, such target sites would be present in large pieces of DNA simply by random chance. Indeed, in the experiment shown in Fig. 3 of the paper, they were able to demonstrate that certain restriction enzyme fragments of bacteriophage DNA were able to bind the engrailed protein with high affinity. This therefore established that the engrailed protein was capable of binding to all DNA with low affinity, but of binding to specific sequences with high affinity. This is characteristic of DNA binding transcription factors, and was noted for the mammalian Sp1 factor in Section 7.

Most importantly, Desplan et al. were also able to show that high-affinity binding sites for the engrailed protein were present in the regulatory regions of the *engrailed* gene itself and in another *Drosophila* gene, *fushi tarazu*, as illustrated in Fig. 4 of the paper. This finding therefore established that potential target genes for regulation by the engrailed protein actually contained DNA binding sites for it. Hence these findings established the engrailed protein as a DNA binding transcription factor that was able to bind with high affinity to specific sequences that were contained within target genes. In addition, by expressing either the homoeodomain-containing region of engrailed or the remainder of the protein lacking the homoeodomain, Desplan et al. were able to show that the DNA binding activity of the intact protein could be reproduced with a fragment of the protein containing the homoeodomain. In contrast, no binding was observed when the region of the protein lacking the homoeodomain was tested in these assays. This demonstrated that, as predicted from sequence comparisons with yeast and bacterial transcription factors, it is the homoeodomain region of the protein that actually binds to DNA. This finding therefore both identified the homoeodomain as the DNA binding domain and also paved the way for the detailed structural studies of its binding to DNA that were described in Section 8.

The importance of the paper by Desplan et al. is that a gene, the mutation of which had previously been shown to dramatically disrupt *Drosophila* development, was identified as encoding a DNA binding transcription factor. This immediately showed that this factor played a critical role in *Drosophila* development. As a number of other regulatory proteins containing homoeoboxes had been identified at this time, it also indicated the existence of a large class of DNA binding transcription factors that played a critical role in *Drosophila* development. Subsequent studies have identified such homoeobox-containing genes in other organisms, including mammals, where they play a similar critical role (for reviews, see Kenyon, 1994; Krumlauf, 1994).

In contrast with the Desplan et al. paper, the second paper presented in this section, by Davis et al., directly identified a transcription factor based on its functional activity when overexpressed in a cell type that normally lacks it. It had been shown previously that treatment of the undifferentiated mouse fibroblast cell line 10T½ with 5-azacytidine resulted in its differentiation into proliferating skeletal muscle cell precursors or myoblasts (Taylor and Jones, 1982). Subsequent removal of serum from the growth medium of these cells resulted in them forming fully differentiated non-dividing skeletal muscle cells (Figure 13.1). As 5-azacytidine acts by altering the methylation of target genes and thereby switching on their expression (see Section 5, and Latchman, 1995a, for a review), it was therefore suggested that it acts by stimulating the expression of a specific regulatory gene in the 10T½ cells. The protein product of this regulatory gene would then activate a number of other genes required for differentiation into myoblasts (Konieczny and Emerson, 1984).

The paper by Davis et al. reported the cloning of this regulatory gene. To isolate the regulatory gene, Davis et al. reasoned that it would be expressed in the myoblast cells derived from the 10T½ cells, but not in undifferentiated 10T½ cells. Similarly, it would also be expressed in other myoblast cell lines. They therefore carried out a subtractive hybridization in which RNA sequences common to the undifferentiated cells and the myoblast cells were removed, leaving only RNA sequences that were specific to the myoblast cells. When these myoblast-specific sequences were used to screen a cDNA library (as illustrated in Figure 1 of the paper), 92 cDNA clones were isolated which appeared to represent mRNAs enriched in the myoblast cells. Critical to the subsequent success of Davis et al. was their decision to remove from this pool clones that did not fit the precise criteria for encoding the regulatory gene. Thus, for example, they removed sequences which were more strongly expressed in fully differentiated muscle cells than in proliferating myoblasts and which would, therefore, represent markers of terminal differentiation. Similarly, they removed sequences whose expression was induced by azacytidine in variant cell lines in which azacytidine did not induce myotube differentiation (Figure 13.2).

This stringent procedure resulted in only three clones, known as MyoA, MyoD and MyoH, remaining. All these three clones showed enhanced expression in the myoblast cells compared with undifferentiated cells. They did not, however, show further increased expression in fully differentiated muscle cells, as illustrated in Figure 2 of the paper. Similarly, they did not show greatly enhanced expression in non-differentiating variants of the 10T½ cells that did not differentiate in response to azacytidine (see Figure 3 of the paper).

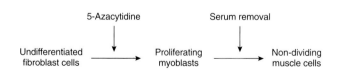

Figure 13.1 Differentiation of 10T½ cells. Undifferentiated fibroblasts can be induced to form proliferating myoblasts by treatment with 5-azacytidine. Subsequent removal of serum results in the differentiation of the myoblasts into non-dividing muscle cells.

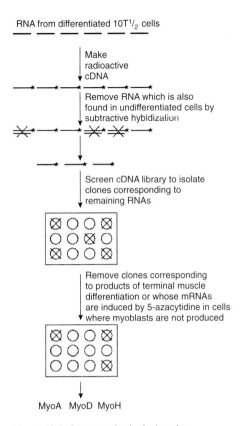

RNA from differentiated 10T$^1/_2$ cells

Make radioactive cDNA

Remove RNA which is also found in undifferentiated cells by subtractive hybridization

Screen cDNA library to isolate clones corresponding to remaining RNAs

Remove clones corresponding to products of terminal muscle differentiation or whose mRNAs are induced by 5-azacytidine in cells where myoblasts are not produced

MyoA MyoD MyoH

Figure 13.2 Strategy for isolating the master regulatory locus expressed in 10T$^1\!/_2$ cells after but not before treatment with 5-azacytidine. Clones corresponding to terminal differentiation products of muscle and to RNAs induced in non-muscle-producing cells by 5-azacytidine were removed, resulting in three candidate clones, i.e. MyoA, MyoD and MyoH.

Most critically, Davis et al. now tested the effect of overexpressing each of these genes in the undifferentiated 10T^1/$_2$ cells. As illustrated in Figure 4 of the paper, only transfection of MyoD was able to induce these cells to differentiate into myoblasts which expressed, for example, the myosin heavy chain gene. This most dramatic result therefore indicated that the overexpression of this single factor was capable of mimicking the effects of 5-azacytidine and inducing the 10T^1/$_2$ cells to differentiate into myoblasts. MyoD is therefore likely to be the regulatory factor that was originally suggested to be activated by azacytidine treatment and to induce differentiation. Subsequent DNA sequence analysis of the MyoD cDNA illustrated in Figure 8 of the paper revealed similarities with transcription factors such as the achaete-scute proteins of *Drosophila*, and it was shown subsequently that MyoD achieves its effect by activating the expression of specific target genes required for muscle differentiation (for reviews, see Edmondson and Olson, 1993; Olson and Klein, 1994).

Hence identification of the *MyoD* gene by the functional assay used by Davis et al. illustrates yet another facet of transcription factor activity. Expression of the single MyoD factor in a cell which does not normally contain it is sufficient to convert that cell into a myoblast. As expected from this, and as illustrated in Figure 7 of the paper, MyoD is only expressed in skeletal muscle cells and not in other cell types, including cardiac and smooth muscle. In contrast, MyoA and MyoH, which proved not to encode the muscle-inducing factor, were shown to have a wider expression pattern.

Thus MyoD constitutes a transcription factor that is regulated at the level of synthesis, being absent from most cell types and present only in the skeletal muscle cell lineage. It therefore illustrates the regulation of transcription factor activity by regulation of its synthesis, as discussed in Section 12. Interestingly, however, MyoD also illustrates the regulation of transcription factor activity which was also discussed in Section 12. Thus the differentiation of myoblasts into fully differentiated muscle cells which occurs upon serum removal (Figure 13.1) involves the activation of several genes whose expression is dependent upon MyoD. Yet no increase in MyoD levels is observed during this second step of the differentiation event. This effect is explained by the existence of an inhibitory protein, Id (Benezra et al., 1990), which dimerizes with MyoD and prevents it binding to DNA. Upon differentiation of myoblast cells into fully differentiated muscle cells, the levels of Id fall dramatically and MyoD is therefore capable of binding to DNA. Hence the MyoD system illustrates the regulation both of transcription factor synthesis and of transcription factor activity which were discussed in the previous section.

More importantly, the identification by Davis et al. of MyoD established the critical role of this transcription factor in muscle development, and also provided the first example of a single factor whose overexpression can induce a particular differentiated phenotype. When taken together with the experiments of Desplan et al., which showed that mutations in genes encoding transcription factors could produce dramatic changes in development, these experiments established the critical importance of transcription factors in controlling complex processes such as development. They therefore paved the way for the eventual understanding of *Drosophila* and mammalian development in terms of the interactions between multiple different transcription factors.

Desplan et al. (1985) Nature (London) **318**, 630–635

The *Drosophila* developmental gene, *engrailed*, encodes a sequence-specific DNA binding activity

Claude Desplan, James Theis & Patrick H. O'Farrell

Department of Biochemistry and Biophysics, University of California, San Francisco, San Francisco, California 94143, USA

Plasmid expression vectors carrying either the entire engrailed *coding region or a subfragment including the homoeo box, produce protein fusions having sequence-specific DNA binding activity.*

MUTATIONS in *Drosophila* have identified genes that control major steps in development[1-3]. Some of these mutants, the segmentation mutants, are defective in the processes that subdivide the embryo into the segmented body plan[4-7] while others, the homoeotic mutants, improperly specify the developmental fate of particular regions of the fly[1]. Garcia-Bellido[8] suggested that these mutations affect 'selector genes' that act, in those cells in which they are expressed, to select the developmental pathway. It was proposed that they function by controlling 'cytodifferentiation genes'[8].

A segmentation gene

Each morphologically obvious segment is composed of cells from two distinct lineages termed anterior and posterior compartments[9,10]. Genetic analysis suggests that the *engrailed* gene product is required to specify cells as members of posterior compartments[11-14]. As anticipated by the selector gene hypothesis[8], by 3.5 h of development *engrailed* gene product accumulates in narrow bands corresponding to the primordia of the posterior part of each segment[15-17]. Apparently, the

engrailed regulatory activity acts wherever it is expressed to direct cells along a pathway of development suited to cells of posterior compartments[8]. In *engrailed* mutants, the segmental fusions and the failure to specify cells of posterior compartments are thought to result from the absence of this regulator or from alterations in its expression.

Selector genes interact

In addition to *engrailed*, several other *Drosophila* developmental genes are expressed in spatially restricted patterns consistent with their apparent roles in directing particular portions of the developmental programme[15-19]. These studies have focused interest on two related issues. How are these genes regulated to achieve the appropriate pattern of expression, and how do they regulate subsequent development? Recent studies suggest that the selector genes interact in a complex regulatory network. Based on phenotypes of double mutant combinations, Struhl[20] argued that the *Ubx* gene product represses *Scr* expression in the mesothorax. Molecular studies[21] have offered further suggestions of interactions. Regulatory interactions among six different

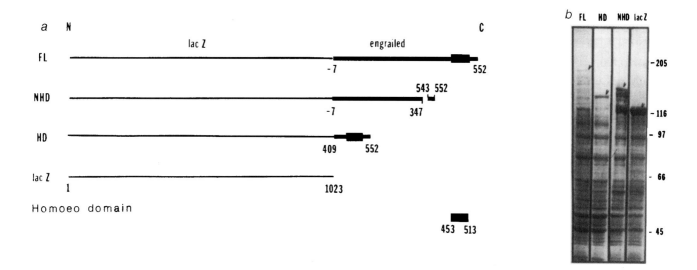

Fig. 1 Construction of *lacZ–engrailed* fusions and expression in *Escherichia coli. a*, Gene fusions; *b*, polyacrylamide gel electrophoresis of bacterial extracts; FL, full-length fusion; HD, homoeo domain fusion; NHD, non-homoeo domain fusion; *lacZ*, β-galactosidase.
Methods: *a*, DNA fragments derived from the *engrailed* locus[17,35,38] were inserted in the polylinker of expression vector plasmids pUR 290, 291 or 292 (ref. 47). These different plasmids allow inserted DNA to be expressed as a C-terminal extension of β-galactosidase. In the full-length fusion protein (FL) the extension includes 7 amino acids that precede the first methionine of the *engrailed* protein as well as the entire *engrailed* protein. In the homoeo domain construct (HD), a *Bam*HI fragment was spliced out of the FL construction, keeping only the last quarter of the *engrailed* coding sequence from amino acid 409 to the end. This fusion protein contains the entire homoeo domain (amino acids 453–513). Amino acids 347–542, containing the homoeo domain, have been deleted in the non-homoeo domain protein (NHD) by removing a *Xho*I fragment from the FL construct. In the absence of insertion the *lacZ* gene is expressed as the complete β-galactosidase. *b*, Polyacrylamide gel electrophoresis of bacterial extracts. The expression of the fusion proteins from chimaeric plasmids is under the control of the *lac* promoter in a *lacI* overproducing strain DG101 (a gift from D. Gelfand, Cetus Corporation). Bacteria in exponential growth were induced with isopropyl β-thiogalactoside at an absorbance of 0.5 (at 600 nm) and collected 2 h later. Cells were resuspended in about 0.005 culture volume of 25% sucrose, 0.2 mM EDTA, 40 mM Tris-HCl *p*H 7.5 and 1 mM dithiothreitol (DTT). Lysis and protein solubilization were achieved by lysozyme treatment (0.4 mg ml⁻¹ for 1 h at 0°C followed by addition of urea to 4 M and further incubation at 0°C for 1h). After centrifugation at 20,000 r.p.m. for 1h the urea was removed from the supernatant by dialysis first against 10 mM Tris-HCl *p*H 7.5, 25 mM NaCl, 1 mM EDTA, 0.1% Triton X-100, 1 mM DTT, 10% glycerol, 1 mM phenylmethylsulphonyl fluoride, and 0.1 mM benzamidine containing 2 M urea and subsequently against the same buffer without urea. Glycerol was added to the extracts to a final concentration of 50% and they were stored at −20°C. When analysed by SDS-gel electrophoresis the fusion proteins are seen as abundant high relatibe molecular mass proteins (arrowheads). The larger fusion protein is present in lower amounts and we detect numerous minor bands that are presumed to result from degradation of the fusion proteins.

homoeotic loci appear to coordinate their spatial patterns of expression (ref. 22 and C. Wedeen and M. Levine, personal communication).

Recent molecular analyses suggest that the segmentation genes interact to control the expression of one another. Immuno-fluorescent staining revealed that *engrailed* protein appears first in alternate segments and only later in every segment[17]. This led to the suggestion that *engrailed* expression is regulated by the products of another class of segmentation genes, the pair-rule genes[17,23], which are expressed in alternate segments[19,24]. Alterations in the pattern of *engrailed* expression in pair-rule mutants have confirmed this prediction (S. DiNardo and P.H.O'F., unpublished observations; K. Howard and P. Ingham, personal communication). Similar analyses of *fushi tarazu* expression suggest that its expression is also influenced by several other segmentation genes (S. Carroll and M. Scott, personal communication).

These observations suggest an extensive network of regulatory interactions among the *Drosophila* developmental genes, and imply that selector genes are themselves targets for the regulators they encode[17,23].

The homoeo box

The products of a number of the developmental genes include a conserved protein domain of 60 amino acids, the homoeo domain. Related sequences have been identified in species from human to annelids[25–30]. This remarkable conservation suggests that the homoeo domain has common physical interactions in all these organisms. A portion of this domain is found in two yeast proteins, a1 and α2, that determine the cell fate (mating type) via transcriptional regulation[31,32]. Because of this homology it has been proposed that the developmental genes of *Drosophila* might function similarly, by interacting with DNA[27,33]. Further, it has been noted that sequences within the homoeo domains are compatible with a protein structural motif that characterizes bacterial sequence-specific DNA binding proteins[27,34].

Here we address three issues. Does a homoeo-domain-containing protein bind to DNA? Is the binding specific? Is the homoeo domain responsible for binding?

Construction of *engrailed* fusion proteins

To study how the *engrailed* protein product might function as a regulator, we constructed bacterial expression vectors that encoded the *engrailed* protein as carboxy-terminal extensions of β-galactosidase. In order to test for possible autonomous functions of the homoeo domain, we constructed three fusion proteins (Fig. 1). The full-length fusion contained the sequence encoding the entire 552 amino acids of the *engrailed* protein. This *engrailed* protein sequence has been derived from an open reading frame in the *engrailed* cDNA sequence[35]. The pattern of evolutionary sequence conservation of this open reading frame suggests that it encodes protein (J. Kassis, D. K. Wright, and P.H.O'F., unpublished observations). We think that this predicted protein is made *in vivo* because two antisera directed against different domains of this predicted sequence detect expression of these domains in the posterior compartments of segments[17]. The 'homoeo domain fusion' includes only the terminal quarter of the protein coding sequence, encompassing the 60-amino-acid homoeo domain and an additional 44 amino acids on the N-terminal side plus 39 amino acids on the C-terminal side. The 'non-homoeo domain fusion' is deleted for a 196-amino-acid region and lacks the homoeo domain (Fig. 1).

Non-specific DNA binding

We first tested whether the fusion proteins would bind DNA non-specifically. We mixed the fusion proteins with labelled restriction fragments of DNA and then, using an antibody directed against bacterial β-galactosidase and fixed *Staphylococcus aureus* as an immunoadsorbent, we precipitated the fusion protein along with bound DNA fragments[31,36]. The bound DNA fragments were then separated by electrophoresis and detected

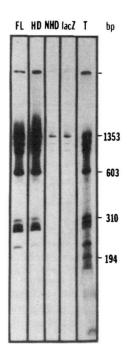

Fig. 2 Nonspecific binding of the fusion proteins to DNA. Bacteriophage ΦX174 DNA was cleaved by *Hae*III and end-labelled using T4 polymerase[48]. The labelled DNA (about 30 ng) was incubated for 30 min at 0°C in 25 μl of binding buffer (BB) (50 mM NaCl, 20 mM Tris pH 7.6, 0.25 mM EDTA, 1 mM DTT, 10% glycerol) in the presence of a bacterial extract containing the fusion protein (about 10 μg of total protein extract, see Fig. 1 legend). The complexes formed between DNA and the fusion proteins were then immunoprecipitated by 30 min further incubation at 0°C with an anti-β-galactosidase monoclonal antibody (provided by Tom Mason and Judy Partaledis, courtesy of Mike Hall) preabsorbed on cross-linked *Staphylococcus* (Pansorbin, Calbiochem). The pellet was washed twice in BB, phenol extracted and the DNA ethanol precipitated before polyacrylamide gel electrophoresis and autoradiography. The autoradiogram shows the results of precipitations performed in the presence of extracts containing the full-length fusion (FL), the homoeo domain fusion (HD), the non-homoeo domain fusion (NHD) or β-galactosidase (*lacZ*). Lane T shows the labelled digest before immunoprecipitation, representing 25% of the counts added to the incubation mixture in the immunoprecipitation experiments.

by autoradiography. At low salt concentration (50 mM NaCl) and in the absence of carrier DNA, the full-length fusion protein and the homoeo domain fusion protein bound all of the *Hae*III restriction fragments of bacteriophage ΦX174 DNA, whereas neither β-galactosidase alone, nor the non-homoeo domain fusion protein bound significant amounts of DNA (Fig. 2). Though these observations are consistent with the hypothesis that the homoeo domain would function in DNA binding, the results could potentially be due to simple ionic interactions. Sequence-specific binding would suggest that the fusion protein has a DNA binding domain.

Sequence-specific DNA binding

Since we had no knowledge of what the target sequences for binding of *engrailed* protein might be, we sought a generalized approach to detect sequence specificity. Sequence-specific DNA binding proteins recognize degenerate versions of a consensus binding site. Sufficiently complex DNA should contain, by chance, sequences recognized by the protein. For example, Ross and Landy[37] identified the sequence of several binding sites for λ integrase (Int protein) in pBR322 DNA.

To pursue this approach, we digested λ-phage DNA to produce more than 100 fragments and labelled their 3' ends. This DNA was used in the assay described above, except that we added increasing amounts of salt or carrier DNA so that only fragments that bound to the fusion proteins with higher affinities would appear in the precipitate. Figure 3 shows that at low concentrations of carrier DNA all the λ DNA fragments are bound nonselectively (lanes 2, 5), but as the stringency of the

Desplan et al. (1985) Nature (London) 318, 630–635

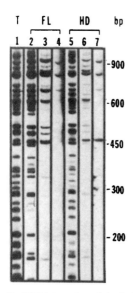

Fig. 3 Sequence-specific DNA interaction of the fusion protein with bacteriophage λ DNA fragments. Bacteriophage λ DNA was restricted with Sau3A and labelled using T4 polymerase[48]. Binding assays were performed as described in Fig. 2 legend with the addition of carrier DNA as indicated. Fragments were separated on a 5% polyacrylamide gel. Lane 1 shows the total digest. The amount of DNA loaded in lane 1 is one-quarter of the input amount used in the binding assays (10 ng; 0.4 μg ml^{-1}). Lanes 2, 3 and 4 show immunoprecipitates obtained following addition of the full-length fusion extract in the presence of 0, 4 and 40 μg ml^{-1} of carrier DNA (fragmented calf thymus DNA), respectively. Lanes 5, 6 and 7 show the results of a similar experiment using the homoeo domain fusion extract. Lanes 4 and 7 were exposed four times longer because of the reduced recovery of bands at high levels of carrier.

binding conditions is increased the binding becomes selective. For example, at high concentration of carrier DNA (Fig. 3, lanes 4, 7), only 4 fragments out of 115 DNA fragments are retained. These experiments demonstrate the specificity of *engrailed* fusion protein binding to DNA.

Whether the homoeo domain fusion or the full-length fusion is used in binding assays, the same fragments are recovered in immunoprecipitates at comparable efficiencies (Fig. 3). Thus, the two fusions bind with the same specificity and generally exhibit similar relative affinities for these fragments. The 143 amino acids of the *engrailed* sequence present in the homoeo domain fusion must include a domain competent in specific binding. At least under the conditions of our assay, the additional 409 amino acids of the full-length fusion protein make little or no contribution to the specificity of binding.

To estimate the minimal binding constant of the binding interaction, we assume that all of the fusion protein is active[31,36]. When our binding assay contains less than 10^{-8} M fusion protein and less than 10^{-11} M DNA fragments, recovery of specific DNA fragments in the immunoprecipitate exceeds 50% (for example Fig. 4A, *engrailed* fragments f and k). Accordingly, the binding constant must exceed 5×10^{7} mol^{-1} (at 170 mM NaCl). Furthermore, preliminary evidence indicating that only a fraction of the fusion protein is active (J.T., unpublished observations) suggests that the binding constant must be higher, and may well be comparable to the binding constants of other sequence-specific DNA binding proteins[34].

The binding behaviour of the fusion protein described here may not accurately reproduce the behaviour and specificities of the natural *engrailed* protein. The binding specificity might be influenced by interactions that the fusion protein cannot reproduce or by modifications that would be missing in a protein produced in *Escherichia coli*. Furthermore, our simple *in vitro* binding assay may lack accessory factors that influence *in vivo* binding of the *engrailed* protein to DNA. Nonetheless, because other work using fusion proteins or proteolytic fragments[31,34] suggests that DNA binding domains can function relatively autonomously, we believe that the results reported here are likely to reflect at least a subset of the activities of the normal *engrailed* protein.

Specific binding to *Drosophila* DNA

If *engrailed* and other selector genes act as pleiotropic regulators of transcription, we might expect their protein products to interact with DNA near the promoters of a number of target genes. Can we identify any plausible candidates for target genes? There is considerable evidence that selector genes regulate each other's expression (summarized above). Thus, we envision that the developmental genes will include regulatory sites that are targets for interaction with the products of other developmental genes. Because of the high degree of relatedness of the developmental genes, the various target sites might also be homologous. Because it is a member of this group of related proteins, perhaps the *engrailed* fusion protein will exhibit site-specific interaction with all or a subset of the related regulatory sites.

Following this line of logic, we decided to look for *engrailed* fusion protein binding adjacent to cloned selector genes. Because a detailed analysis would require DNA sequence information, we chose to examine the *engrailed* locus itself, for which we have 1.2 kilobases (kb) of upstream sequence (unpublished data), and the *fushi tarazu* locus that had been sequenced by Laughon and Scott[27].

We looked for *engrailed* fusion protein binding to a 4.9-kb *Eco*RI fragment that includes 2.6 kb of *engrailed* coding sequence and 2.3 kb of upstream sequences[38] and to a 3.2-kb fragment that includes the *fushi tarazu* coding sequence and flanking sequences[49]. Figure 4A shows that both the cloned *engrailed* sequences and cloned *fushi tarazu* sequences contain fragments that bind to the *engrailed* fusion protein under stringent assay conditions. In fact, a number of binding fragments are detected (Fig. 4A, C). The positions of binding fragments are indicated in Fig. 4B.

We purified the subfragments indicated by the hatched lines in Fig. 4B and used these to map more precisely the binding interactions upstream of the *engrailed* and *fushi tarazu* coding regions. Secondary digests of these fragments were tested for interactions with the *engrailed* fusion protein (Fig. 5). These analyses localized three binding sites within the 900-base-pair (bp) region 5′ to the *engrailed* cDNA. The higher resolution and sensitivity of these experiments showed that the binding fragment k (Fig. 4A, B) actually contained two binding sites (sites a and b in Fig. 5) and that fragment d, though not detected as a binding fragment in experiments using the whole plasmid, contains a weak binding site (site c in Fig. 5). The analysis of the *fushi tarazu* subfragment did not reveal any new binding sites but did contribute to more accurate localization of the upstream site (Fig. 5). At present the accuracy of localization of the sites does not allow us to identify a consensus binding site unambiguously.

Binding sites

Without a functional assay we cannot directly assess the importance of the binding sites detected in cloned *Drosophila* sequences. However, we can test whether affinities, frequency of occurrence, clustering and location of binding sites differ from fortuitous sites.

If the frequency of fortuitous sites were extremely low (less than 1 per 1,000 kb), the presence of a cluster of binding sites within a few kilobases of *engrailed* DNA would be highly significant. This is not the case. Although the frequency of binding sites on a 4.9-kb fragment of *engrailed* DNA is higher than the density of λ DNA, the difference was not very large—about 10-fold (Fig. 4C).

Fortuitous binding sites should have a wide range of affinities depending on their similarity to an optimal site. Higher-affinity fortuitous sites should be less frequent (chance might produce an optimal binding site but should do so less frequently than imprecise approximations of this sequence). We tested the relative affinity of the *engrailed* fusion protein for various binding sites. We used conditions where a number of labelled restriction fragments are bound selectively. Addition of cold competitor DNA displaced bacteriophage λ DNA fragments with different

Fig. 4 Sequence-specific interaction of the homoeo domain fusion protein with restriction fragments of cloned *fushi tarazu* and *engrailed* sequences. The *fushi tarazu* clone (p6-3, derived from clone pDmA439, a gift from Matt Scott)[49] contains 900 bp of upstream sequence, the entire coding sequence and 800 bp of downstream sequence. The *engrailed* clone (p615) contains 2.3 kb of upstream sequence, the complete first exon and most of the first intron[35,38] (panel *B*). Restriction fragments of p6–3, p615 of λ DNA were end-labelled and tested for fusion protein binding as described in the legend for Fig. 2, except that a higher salt concentration (170 mM NaCl) was used to diminish nonspecific interactions. Fragments were separated on a 5% acrylamide gel. In the example shown in *A*, the *fushi tarazu* plasmid (ftz) DNA was digested with *Rsa*I and the *engrailed* plasmid (*en*) was digested with *Hin*fI and *Cla*I. Lanes marked T show the total fragment pattern and lanes marked I show the fragments recovered in the immunoprecipitate. For both plasmids the amount of DNA loaded in T is one-quarter of the amount subjected to immunoprecipitation and displayed in I. Bands derived totally from plasmid sequences are identified with a 'p'. Bands derived at least in part from insert sequences are designated as **a** to **j** (*ftz*) or **a** to **o** (*en*) and their postions within the cloned sequence are indicated in *B*. Similar experiments using *Alu*I, *Hae*III, *Bst*NI, *Dde*I or *Hpa*II to digest these plasmids gave comparable results (not shown). *B* shows a map of the inserts in p6-3 (*ftz*) and p615 (*en*). The orientation of the coding region within the insert is indicated. Except for the dashed portion of the *engrailed* insert, the sequence is known (refs 27, 35, and J. Kassis, D. K. Wright and C. D., unpublished). The positions of restriction enzyme sites (*Rsa*I sites in *ftz* and *Hin*fI and *Cla*I sites in *en*) are indicated and all fragments detectable in the separation shown in panel *A* are lettered. Those fragments detected as bound by the homoeo domain fusion protein are indicated as +. Because they were not detected, a few small restriction fragments (unlettered) could not be scored in the binding experiment (*A*) but analysis of these DNAs cleaved with other restriction enzymes failed to detect additional binding sites. Exons of the transcribed region are indicated with a bold line and introns with sloping lines. The hatched region indicates the position of a subfragment that was purified and used for additional binding studies (Fig. 5). Note that the region responsible for binding of *ftz*, fragment a, was further localized by analysis of *Alu*1, *Hae*III, *Bst*NI, *Dde*I, *Hpa*II and *Hin*fI digests that defined only one site; this site corresponds to that mapped in more detail by analysis of the purified subfragment (Fig. 5). *C* Compares the efficiency of homoeo domain fusion protein binding to *engrailed* or λ sites. The binding assays were performed as in *A* except that increasing amounts of unlabelled DNA (*Hin*fI-restricted p615) was added as competitor. Lanes *T* represents 10% of the counts of the total digest submitted to immunoprecipitation in lanes 1, 2, 3 and 4. In lane 1, no unlabelled DNA was added while in lanes 2, 3 and 4, 30 ng, 250 ng and 4 μg, respectively, were added. *en* lanes: plasmid p615 was restricted with *Cla*I and *Hin*fI, labelled and submitted to immunoprecipitation. Lanes 1, 2, 3 and 4 represent immunoprecipitated DNA from ~4 ng of labelled DNA (~1 fmol). Obviously the four bands retained in *A* behave differently when unlabelled competitor DNA is added. Three bands (arrows) corresponding to fragments f, k and h in *A*, are detected in the immunoprecipitate in the presence of 1,000-fold excess of unlabelled DNA. λ lanes: λ DNA was restricted with *Cla*I and *Hin*fI, labelled and processed under the same conditions as for *en* DNA. An equimolar amount of DNA (1 fmol) was used in this experiment (20 ng). Two bands were retained at the highest stringency (arrowheads). *en* + λ lanes: T is an equimolar mixture of *en* and λ DNA restricted with *Cla*I and *Hin*fI. The bands that are retained are the same as those seen in the experiments testing *en* and λ DNAs individually. Five bands are retained in the presence of the highest concentration of competitor (arrows and arrowheads).

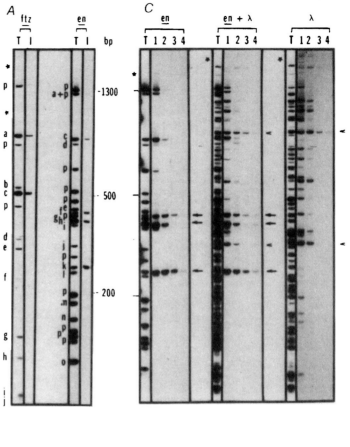

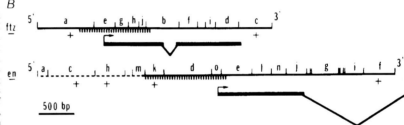

efficiencies (Fig. 4*C*). Thus, as expected, the binding sites in λ DNA have a range of affinities and there are few high-affinity sites. Assuming that the binding sites in λ DNA occur by chance, the specific binding of 14 restriction fragments at intermediate stringencies suggests that the sequence recognized by the *engrailed* fusion protein is relatively short (about 6 bp) or substantially degenerate.

For some sequence-specific DNA binding proteins (such as *lac* repressor), fortuitous occurrence of high-affinity binding sites is extremely unlikely. For these a site of functional interaction (the *lac* operator[39]) has a distinctively high affinity. For other sequence-specific interactions (for example λ integrase[37,40]) the affinities of fortuitous and functional sites overlap. Depending on the characteristics of the *engrailed* fusion protein, functionally relevant sites might have distinctively high affinities. We therefore examined the relative binding affinity of *engrailed* and bacteriophage λ DNA fragments. We observe differing affinities for the *engrailed* fusion protein interaction with various sites on *engrailed* DNA (Fig. 4*C*). Using our present assay, the ranges of affinities seen for A and *engrailed* sites overlap (Fig. 4*C*). Three *engrailed* fragments bind with

particularly high affinity (arrows) compared with two λ fragments (arrowheads).

The binding data provide no support for suggestions that the binding sites in *Drosophila* DNA are functional. It should, however, be made clear that the opposing conclusion also cannot be reached from these data; that is, the binding sites in *engrailed* DNA cannot be dismissed as nonfunctional because they have properties similar to fortuitous binding sites. Thus, the issue of function remains open.

Although the functional importance of the binding sites still requires experimental test, we propose that the binding sites we have detected in *Drosophila* DNA function *in vivo* as targets for interaction with either the *engrailed* protein or closely related gene products (that is, other selector genes). We make this suggestion on the basis of the location, clustering and conservation of the sites. The positions of binding sites in relation to the *fushi tarazu* and the *engrailed* coding regions are reminiscent of the positions of enhancer elements in other systems[41–43]. The clustering of binding sites is unlikely to be coincidental. Such clustering of binding sites for regulatory proteins is fairly common (for example, refs 44,45). Finally, if functional, the binding

Desplan et al. (1985) Nature (London) **318**, 630–635

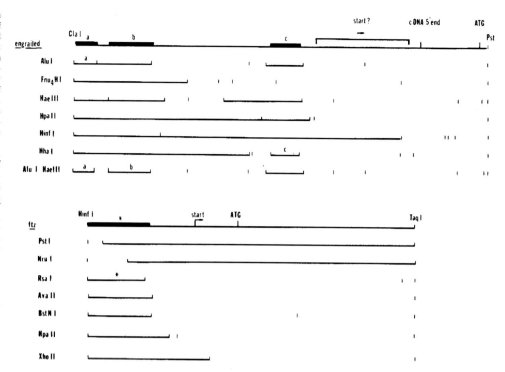

Fig. 5 Localization of binding sites in the 5′ regions of *engrailed* and *fushi tarazu*. In digests of whole plasmid DNA were identified fragments that were bound by the homoeo domain fusion protein (Fig. 4). To localize more precisely the binding sites immediately upstream of the *engrailed* and *fushi tarazu* coding regions, we purified a 1,180-bp *ClaI-PstI* fragment from the *engrailed* clone and a 939-bp *HinfI-TaqI* fragment from the *fushi tarazu* clone (Fig. 4B). These purified fragments were digested with additional enzymes as indicated and the binding of subfragments assayed as before (Fig. 4). In each of the restriction digests shown here the binding fragments are indicated as solid lines. Although the overlap of binding sites: a *ClaI-AluI* fragment (a, the binding regions, this could be misleading if the binding sites were complex. Therefore, in the summary diagrams above the restriction patterns, we have indicated the precision of the localization of the binding sites based on the minimal fragment showing binding (bold line). For *engrailed*, three fragments were identified as containing binding sites: a, *ClaI-AluI* fragment (a, 67 bp), a *HaeIII-AluI* fragment (b, 100 bp) and a *HhaI* fragment (c, 88 bp). For *ftz* a single site was localized to a 165-bp *HinfI-RsaI* fragment (*).

sites will be conserved in evolution. In the absence of a functional test, we believe that the best way to distinguish fortuitous and functional binding sites is to see whether protein binding occurs at analogous positions in distantly related genomes. We have cloned the *engrailed* gene of a distantly related *Drosophila* species, *D. virilis*[46]. A preliminary analysis indicates that the fusion protein binds to fragments upstream of the *D. virilis engrailed* gene (D. Wright, unpublished data).

Binding specificity

Together with previous arguments[27], our results predict that the homoeo domain imparts a sequence-specific DNA binding activity to the protein. Accordingly, other homoeo domain-containing proteins should also bind DNA in a sequence-specific manner and such proteins having closely homologous homoeo domains should have similar sequence specificity. We presently recognize two classes of homoeo domain sequences. Class I is comprised of seven genes which have highly homologous homoeo domains and are located in two clusters of developmental genes (the bithorax complex and the Antennapedia complex)[22,25]. Class II is comprised of the *engrailed* homoeo domain and the highly homologous homoeo domain of the *engrailed* related gene[35]. The homoeo domains of different classes have lower homology (Fig. 6). As noted previously, the regions of sequence identity suggest that class I homoeo domains might specify binding to the same sequence[27]. The differences between class I and class II homoeo domains might include an alteration of the sequence specificity.

Evolution of proteins

Duplication and divergence of a primordial gene encoding a DNA binding protein might lead to a family of interacting regulators. If the primordial protein included sequences for dimerization and for DNA binding, newly duplicated coding regions would have common binding specificities and could form heterotypic dimers. The interactions between products of duplicated genes would persist if the dimerization function and DNA binding function were conserved. We suggest that continued duplication and divergence can result in a family of DNA binding proteins that interact physically by forming heterotypic associations and interact functionally by competition for binding

to related DNA binding sites. Such interactions would link the various genes in a regulatory network. The evolution of the members of the family would be coupled because of the importance of maintaining the interactions among the members of this regulatory network. Since coordinate change of many genes is an extraordinarily unlikely event, such a network of interaction

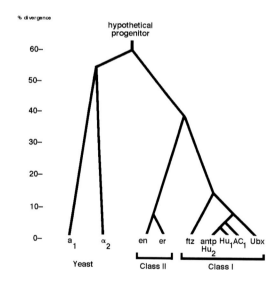

Fig. 6 Family tree of relatedness of homoeo domains. Pairwise comparisons of the protein sequences of the homoeo domains encoded by *Drosophila* genes (*Antp*, *Ubx*, *ftx*, *en*, and *er*), yeast genes (a1 and α2) and sequences isolated by homology from humans (Hu$_1$ and Hu$_2$) and frogs (AC$_1$) were used to score divergence. The comparison was confined to residues 29–58 because this region is well conserved among all of these sequences. To produce homology scores we gave one point for amino-acid identity and half a point for similarity. The data are assembled into a family tree by indicating the divergence of each sequence that approximates all the pairwise comparisons. The sequence data are taken from Levine *et al.*[28] (Hu$_1$ and Hu$_2$), Scott and Weiner[26] (*Ubx* and *ftz*), McGinnis *et al.*[25] (*Antp*), Poole *et al.*[35] (*en* and *er*), Carraso *et al.*[30] (AC$_1$), and Tatchell *et al.*[32] (α2 and a1). Note that the discrepancies in the published sequence[25,26] for *Ubx* are significant in comparison to the divergence among class I homoeo domains.

may contribute to the extraordinary conservation of homoeo domain sequences. It should be possible to test the predictions of this rationale using approaches similar to those used here to show that the *engrailed* gene encodes a sequence-specific DNA binding activity.

We thank our colleagues for discussions and experimental assistance, particularly Steve DiNardo, Mike Hall, Sandy Johnson, Judy Kassis, Jerry Kuner, Roger Miesfield, Sandro Rusconi, Elizabeth Sher and Deann Wright. We thank Steve Poole for the gift of *en* cDNA and sequence information before publication, Matt Scott for the *fushi tarazu* clone and for encouragement, and Sandy Johnson and Keith Yamamoto for their comments on the manuscript. This work was funded by NSF grant PCM-8418263 and by NIH grant GM 31286. C.D. was supported by a Fogarty fellowship and by ARC, and J.T. by an NIH training grant.

Received 2 July; accepted 5 November 1985

1. Lewis, E. B. *Nature* **276**, 565–570 (1978).
2. Kaufman, T. C., Lewis, R. & Wakimoto, B. *Genetics* **94**, 115–133 (1980).
3. Garcia-Bellido, A. & Santamaria, P. *Genetics* **72**, 87–104 (1972).
4. Nusslein-Volhard, C. & Wieschaus, E. *Nature* **287**, 795–801 (1980).
5. Nusslein-Volhard, C., Wieschaus, E & Kluding, H . *Wilhelm Roux Arch. dev. Biol.* **193**, 267–282 (1984).
6. Weischaus, E., Nusslein-Volhard, C. & Jurgens, G. *William Roux Arch. dev. Biol.* **193**, 296–307 (1984).
7. Jurgens, G., Wieschaus, E., Nusslein-Volhard, C. & Kluding, H. *Willhelm Roux Arch. dev. Biol.* **193**, 283–295 (1984).
8. Garcia-Bellido, A. *CIBA Fdn Symp* **29**, 161–182 (1975).
9. Garcia-Bellido, A., Ripoli, P. & Morata, G. *Nature new Biol.* **245**, 251–253 (1973); *Devl Biol.* **48**, 132–147 (1976).
10. Crick, F. H. C. & Lawrence, P. A. *Science* **189**, 340–347 (1975).
11. Morata, G. & Lawrence, P. A. *Nature* **255**, 608–617 (1975).
12. Lawrence, P. A. & Morata, G. *Wilhelm Roux Arch. dev. Biol.* **187**, 375–379 (1979).
13. Struhl, G. *Devl. Biol.* **84**, 372–385 (1981).
14. Kornberg, T. *Proc. natn. Acad. Sci. U.S.A.* **78**, 1095–1099; *Devl Biol* **86**, 363–372 (1981).
15. Kornberg, T., Siden, I., O'Farrell, P. & Simon, M. *Cell* **40**, 45–53 (1985).
16. Fjose, A., McGinnis, W. J. & Gehring, W. J. *Nature* **313**, 284–289 (1985).
17. DiNardo, S., Kuner, J., Theis, J. & O'Farrell, P. H. *Cell* (in the press).
18. Akam, M. E. *EMBO J.* **2**, 2075–2084 (1983).
19. Hafen, E., Kuroiwa, A. & Gehring, W. J. *Cell* **37**, 833–841 (1984).
20. Struhl, G. *Proc. natn. Acad. Sci. U.S.A.* **79**, 7380–7384 (1982).
21. Hafen, E., Levine, M. & Gehring, W. J. *Nature* **307**, 287–289 (1984).
22. Harding, K., Wedeen, C., McGinnis, W., Levine, M. *Science* (in the press).
23. O'Farrell, P. H. *et al. UCLA Symp. molec. cell. Biol., new Ser.* **31** (1985).
24. Wakimoto, B. T., Turner, R. F. & Kaufman, T. C. *Devl. Biol.* **102**, 147–172 (1984).
25. McGinnis, W., Garber, R. L., Wirz, J., Kuroiwa, A. & Gehring, W. J. *Cell* **37**, 403–408 (1984).
26. Scott, M. P. & Weiner, A.J. *Proc. natn. Acad. Sci. U.S.A.* **81**, 4115–4119 (1984).
27. Laughon, A. & Scott, M. P. *Nature* **310**, 25–31 (1984).
28. Levine, M., Rubin, G. & Tijan, R. *Cell* **38**, 667–673 (1984).
29. McGinnis, W., Hart, C. P., Gehring, W. J. & Ruddle, F. *Cell* **38**, 675–680 (1984).
30. Carrasco, A. E., McGinnis, W., Gehring, W. J. & DeRobertis, E. M. *Cell* **37**, 409–414 (1984).
31. Johnson, A. & Herskowitz, I. *Cell* **42**, 237–247 (1985).
32. Tatchell, K., Nasmyth, K., Hall, B., Astell, C. & Smith, M. *Cell* **27**, 25–35 (1981).
33. Shepherd, J. C. W., McGinnis, W., Carrasco, A. E., DeRobertis, E. M. & Gehring, W. J. *Nature* **310**, 70–71 (1984).
34. Pabo, C. O. & Sauer, R. T. A. *Rev. Biochem.* **53**, 293–321 (1984).
35. Poole, S. J., Kauvar, L. M., Drees, B. & Kornberg, T. *Cell* **40**, 37–43 (1985).
36. McKay, R. *J. molec. Biol.* **145**, 471–488 (1981).
37. Ross, W. & Landy, A. *Proc. natn. Acad. Sci. U.S.A.* **79**, 7724–7728 (1982).
38. Kuner, J. M. *et al. Cell* **42**, 309–315 (1985).
39. Lin, S.-Y. & Riggs, A. D. *J. molec. Biol.* **72**, 671–690 (1972).
40. Better, M., Lu, C., Williams, R.C. & Echols, H. *Proc. natn Acad. Sci. U.S.A.* **79**, 5837–5841 (1982).
41. Banerji, J., Rusconi, S. & Schaffner, W. *Cell* **27**, 299–308 (1981).
42. Gillies, S. D., Morrison, S. L., Oi, V. T. & Tonegawa, S. *Cell* **33**, 717–728 (1983).
43. Stuart, G.W., Searle, P. F., Chen, H.Y., Brinster, R. L. & Palmiter, R. D. *Proc. natn. Acad. Sci. U.S.A.* **81**, 7318–7322 (1984).
44. Miller, A. M., MacKay, V. L. & Nasmyth, K. A. *Nature* **314**, 598–603 (1985).
45. Dynan, W. S., & Tijan, R. *Nature* **316**, 774–778 (1985).
46. Kassis, J., Wong, M.-L.. & O'Farrell, P. H. *Molec cell. Biol.* **5**, 3600–3609 (1985).
47. Ruther, U. & Muller-Hill, B. *EMBO J.* **2**, 1791–1794 (1983).
48. O'Farrell, P. *Focus* **3**, 1–3 (1981).
49. Weiner, A. J., Scott, M. P. & Kaufman, T. C. *Cell* **37**, 843–851 (1984).

Davis et al. (1987) Cell 51, 987–1000

Expression of a Single Transfected cDNA Converts Fibroblasts to Myoblasts

Robert L. Davis,*† Harold Weintraub,*
and Andrew B. Lassar*

* Department of Genetics
Hutchinson Cancer Research Center
1124 Columbia Street
Seattle, Washington 98104
† Department of Pathology
School of Medicine
University of Washington
Seattle, Washington 98195

Summary

5-azacytidine treatment of mouse C3H10T1/2 embryonic fibroblasts converts them to myoblasts at a frequency suggesting alteration of one or only a few closely linked regulatory loci. Assuming such loci to be differentially expressed as poly(A)$^+$ RNA in proliferating myoblasts, we prepared proliferating myoblast-specific, subtracted cDNA probes to screen a myocyte cDNA library. Based on a number of criteria, three cDNAs were selected and characterized. We show that expression of one of these cDNAs transfected into C3H10T1/2 fibroblasts, where it is not normally expressed, is sufficient to convert them to stable myoblasts. Myogenesis also occurs, but to a lesser extent, when this cDNA is expressed in a number of other cell lines. The major open reading frame encoded by this cDNA contains a short protein segment similar to a sequence present in the *myc* protein family.

Introduction

Jones and colleagues originally showed that myogenic, adipogenic, and chondrogenic clones can be derived from the embryonic mouse fibroblast line C3H10T1/2 (10T1/2 cells) following a brief treatment with 5-azacytidine (Taylor and Jones, 1979). Experiments using other cytidine analogs, and the correlation of this phenomenon with incorporation of 5-azacytidine into DNA, point to DNA demethylation as having a significant role (Jones and Taylor, 1980). Moreover, based on the high frequency of myogenic conversion (25%–50%), it appears that demethylation of only one or a few closely linked loci are necessary to convert 10T1/2 cells into myoblasts (Konieczny and Emerson, 1984). This possibility has been substantiated by the observation that transfected DNA from 5-azacytidine-derived, myogenic 10T1/2 clones can convert normal 10T1/2 cells to myoblasts at a frequency (10^{-4}) consistent with the transfer of one or only a few closely linked demethylated loci, while transfected DNA from normal 10T1/2 cells has no effect (Lassar et al., 1986; see also Konieczny et al., 1981). Myogenic lines, derived after 5-azacytidine treatment or genomic myoblast DNA transfection of 10T1/2

cells, express a repertoire of muscle-specific genes when these cells are induced to differentiate.

We wish to identify the loci involved in converting 10T1/2 cells to determined myoblasts. Both immunological and biochemical studies have indicated that proliferating myoblasts express lineage-specific markers prior to terminal differentiation (Walsh and Ritter, 1981; Konieczny and Emerson, 1984; Sasse et al., 1984; Kaufman et al., 1985). We hypothesized that, unlike uncommitted 10T1/2 cells, proliferating myoblasts derived from them continually express, as poly(A)$^+$ RNA, regulatory loci that both activate the expression of lineage-specific markers in proliferating myoblasts and make these cells competent to express other muscle-specific genes when induced to differentiate. We have employed subtracted cDNA hybridization between proliferating myoblast cDNA and proliferating 10T1/2 poly(A)$^+$ RNA to prepare probes for screening a myocyte cDNA library. A number of cDNA clones have thus been isolated and tested for expression specific to myogenic cells and muscle tissue. Use of several assumptions regarding the expression of myogenic determination loci restricted further analysis to three of these clones. We show that expression of one of these cDNAs in transfected 10T1/2 cells is sufficient to convert them at high frequency to stable myoblasts. Myogenesis is also induced, but to a lesser extent, in mouse 3T3 embryonic fibroblast lines and in a variety of other cell lines transfected with this cDNA. This gene, which we call MyoD, is normally expressed only in mouse skeletal muscle in vivo and myogenic cell lines in vitro; it is not detectable in other tested adult or newborn mouse tissue, nor in several nondifferentiating variant myogenic cell lines. The single major open reading frame encoded by this cDNA contains a region with sequence similarity to a region of the *myc* protein family. This region has been shown to be essential for c-*myc* transforming activity in vitro (Stone et al., 1987). A similar sequence is shared by two transcription units of the Drosophila *achaete-scute* complex (Villares and Cabrera, 1987), which is involved in neuroblast commitment.

Results

Isolation of Myoblast-Specific cDNAs

To prepare subtracted probes enriched for putative myogenic regulatory sequences, ^{32}P-labeled cDNA was prepared using poly(A)$^+$ RNA from proliferating myoblasts of two types of myogenic cell lines. One line was derived from 5-azacytidine treated 10T1/2 cells (F3 aza-myoblasts); the other was the normal diploid mouse myogenic line C2C12 (Blau et al., 1983). In preparing these probes, we wanted to select against 5-azacytidine-induced sequences irrelevant to myogenic determination, since 5-azacytidine induces the expression of a variety of genes in various cell types (Groudine et al., 1981; Hsiao et al., 1986). We suspected that mRNAs which were not common to both aza-myoblasts and C2C12 cells might represent such 5-aza-

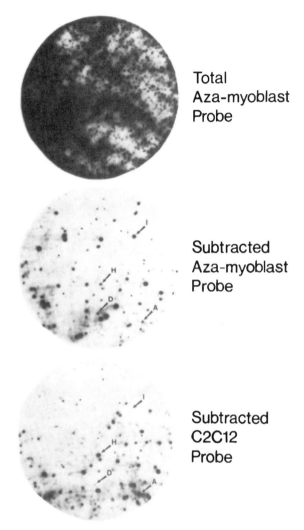

Total
Aza-myoblast
Probe

Subtracted
Aza-myoblast
Probe

Subtracted
C2C12
Probe

Figure 1. Screening a Total Aza-Myoblast/Myotube cDNA Library with Myoblast-Specific Probes

Autoradiographs of approximately 10^4 clones of an aza-myoblast/myotube cDNA library screened with ^{32}P-labeled total myoblast (overnight exposure) or subtracted, myoblast-specific probes (8 day exposure). The bottom two autoradiographs are duplicate filters. Representative phages containing MyoD (D), MyoA (A), MyoH (H), and intracisternal type A particle (I) cDNAs are indicated.

cytidine-induced messages as well as other sequences irrelevant to myogenesis. We prepared subtracted probes such that isolated cDNA clones would represent sequences that are absent in proliferating 10T1/2 cells and present in both myogenic cell lines. Aza-myoblast and C2C12 ^{32}P-cDNAs were each exhaustively hybridized ($C_0t \geq 3000$) twice to 10T1/2 poly(A)$^+$ RNA, with the myoblast-specific single-stranded material isolated each time after passage over hydroxyapatite. The myoblast-specific cDNA probes were then positively selected by hybridizing the remaining aza-myoblast cDNA to C2C12 poly(A)$^+$ RNA, and the C2C12 cDNA to aza-myoblast poly(A)$^+$ RNA, followed by recovery of double-stranded material (see Experimental Procedures for details).

The probes in final form were used to screen a λgt10 myo-

blast/myotube cDNA library prepared using cDNA synthesized from a 1:1 mixture of proliferating myoblast and differentiated myotube poly(A)$^+$ RNA of the F3 aza-myoblast line. Approximately 1% of the recombinant phage hybridized with the subtracted probes. Figure 1 shows a comparison of autoradiographs from plaque hybridizations of the two subtracted probes versus an unsubtracted aza-myoblast cDNA probe. The positive plaques from this initial screening were divided into three groups: those showing stronger hybridization to the aza-myoblast probe; those showing stronger hybridization to the C2C12 probe; or those showing equal hybridization to both probes. Plaques from each group were picked from the primary filters and rescreened (after clonal gridding) on secondary filters. From 10^4 cDNA clones screened, 92 cDNA clones were isolated, which represent 26 different sequences.

We used the following criteria to select cDNA clones for further analysis: complete lack of expression in proliferating 10T1/2 cells; little or no increase in expression in differentiated myotubes versus proliferating myoblasts to exclude markers characteristic of terminally differentiated myotubes; and reduced to absent expression in non-differentiating variants derived from an aza-myoblast line. The validity of employing this last criterion is based upon several characteristics of the variant lines, which are outlined below. Classes of isolated clones failing our criteria included intracisternal type A particle RNAs, whose expression is induced by 5-azacytidine but is not diminished in non-differentiating variants (data not shown). In addition, since a small percentage of cells in the proliferating myoblast cultures used to prepare poly(A)$^+$ RNA do differentiate, we expected to, and did, isolate some clones with expression characteristic of terminal differentiation markers (three of which were desmin, myosin heavy chain, and myosin light chain 2). Eventually, the focus narrowed to the three cDNAs described below (see Experimental Procedures for a detailed account of how the analysis was simplified).

Characterization of Selected cDNA Clones

The three cDNA clones are referred to as MyoA, MyoD, and MyoH, representing RNAs of approximately 5.0, 2.0, and 2.4 kb, respectively. Their representations in the F3 aza-myoblast/myotube library are 29 for MyoA, 20 for MyoD, and 488 for MyoH, per 10^5 cDNAs. For comparison, we screened the F3 aza-myoblast/myotube library with probes to genes thought to represent the low abundance class of mRNAs. These included c-*myc*, at 4 per 10^5 cDNAs and hypoxanthine-guanine phosphoribosyl transferase, at 3 per 10^5 cDNAs. The lowest abundance cDNA we isolated with the subtracted probe has a frequency of 4 per 10^5 cDNAs. We feel that, while we did not perform a saturating screen (since we screened only 10^4 cDNAs and a typical somatic cell is thought to contain about 10^5 mRNAs, half or more of which are of the low abundance class; Hedrick et al., 1984), we were able to detect and isolate some cDNAs in the 10^4 clones screened that represent low abundance poly(A)$^+$ RNAs. Representative phage containing the Myo cDNAs in a

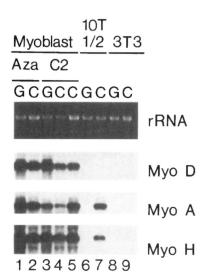

Figure 2. Expression of Myo mRNAs in Myoblasts, 10T1/2 Cells, and Swiss 3T3 Cells

Northern analysis (total RNA) of MyoD, MyoA, and MyoH expression in growing (G) and confluent (C) P2 aza-myoblasts (lanes 1 and 2), C2C12 myoblasts (lanes 3, 4, and 5), 10T1/2 cells (lanes 6 and 7), and Swiss 3T3 cells (lanes 8 ancl 9). Confluent myoblast cultures were composed ot a mixture of differentiated myotubes and undifferentiated cells. In lane 5, C2C12 myotubes were purified by incubating the culture in differentiation medium (2% horse serum) plus 10^{-5} M cytosine arabinoside for 4 days. The top panel shows an ethidium bromide stain (of rRNA) of the RNA sample used for Northern analysis.

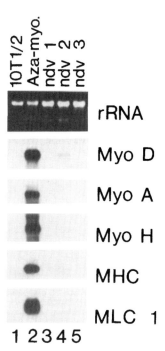

Figure 3. Expression of Myo mRNAs and Muscle Differentiation Markers in Nondifferentiating Aza-Myoblast Variants

Northern analysis (total RNA) of MyoD, MyoA, MyoH, myosin heavy chain (MHC), and myosin light chain 1 (MLC1) in 10T1/2 cells (lane 1), P2 aza-myoblasts (lane 2), and in three lines of nondifferentiating variants (ndv) derived from P2 aza-myoblasts (lanes 3–5). Levels of the Myo mRNAs were analyzed in proliferating cells whereas MHC and MLC1 levels were analyzed in cells placed in differentiation medium (2% horse serum) for 4 days. The top panel shows an ethidium bromide stain (of rRNA) of the RNA samples (from proliferating cells) used for Northern analysis of the Myo mRNAs.

screen of the F3 aza-myoblast/myotube library are indicated in Figure 1.

Figure 2 shows the expression of the three corresponding RNAs by Northern analysis. All three messages are expressed in proliferating myoblasts and differentiated myotubes (lanes 1–5). As expected, the RNAs are not expressed in proliferating 10T1/2 cells (lane 6). Whereas MyoD is not expressed in confluent, growth-arrested 10T1/2 cells, MyoA and MyoH are expressed in these cells (lane 7). In separate experiments, we have also shown that induction of MyoA and MyoH occurs in nonconfluent 10T1/2 cells when serum is removed. MyoA and MyoH are not genes generally responsive to growth arrest in other cell types, as shown t)y the lack of induction in growth-arrested Swiss 3T3 fibroblasts (lanes 8 and 9), as well as a variety of other fibroblast and adipoblast cell lines tested (data not shown).

Figure 3 shows expression of the Myo RNAs in nondifferentiating variants derived from the P2 aza-myoblast line shown in Figure 2. These nondifferentiating lines, which fail to express a number of muscle differentiation markers (see Figure 3), were derived by continual passage of proliferating cells after most of the cells on a plate had terminally withdrawn from the cell cycle and had differentiated. Cells able to proliferate were passaged several times and then cloned (see Experimental Procedures for a more detailed description of their derivation). We reasoned that these lines had either lost their ability to differentiate or in fact were no longer determined myoblasts. That expression of the three Myo RNAs is either

absent (variants 1 and 3) or highly attenuated (variant 2) in these cells (when proliferating) suggests that these lines lack a factor(s) required for expression of muscle-specific markers prior to differentiation. The absence of MyoA and MyoH expression in confluent variants (data not shown) suggests that these cells are not "simple" revertants to the 10T1/2 phenotype. Perhaps the variant lines receive a continuous growth signal even in the absence of serum factors. In contrast to the Myo RNAs, intracisternal type A particle (IAP) RNA, whose expression is induced by 5-azacytidine, remains at high levels in the nondifferentiating variants (data not shown). In addition to IAP RNA, we isolated five other cDNAs whose expression was maintained in the nondifferentiating variants. Although it is possible that these sequences play a role in myogenesis, we decided to initially concentrate on those muscle-specific cDNAs (MyoA, D, and H) whose expression is absent or highly attenuated in these variant lines.

Transfection of 10T1/2 Cells with a MyoD Expression Vector Converts Them to Stable Myoblasts

Having obtained only three cDNAs that satisfied our screening criteria, we focused our efforts on identifying phenotypes associated with their expression in proliferating 10T1/2 cells, where they are not normally expressed.

Crystal Violet

Myosin Immunostain
high magnification

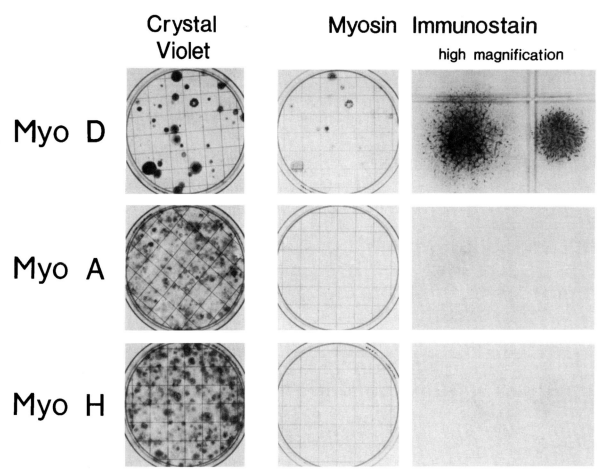

Myo D

Myo A

Myo H

Figure 4. Transfection of 10T1/2 Cells with Myo cDNA Expression Vehicles

10T1/2 cells were transfected with 10 μg of the expression vehicles (derived from pEMSVscribe), containing the Myo cDNAs in the sense orientation, 0.1 μg of pLNSAL (a plasmid encoding G418 resistance), and 10 μg of genomic 10T1/2 DNA. The left-hand panels show a crystal violet stain of the resultant G418 resistant colonies, the middle panels show the same dishes as on the left stained for myosin heavy chain, and the right-hand panels show a high magnification (12×) micrograph of myosin-stained colonies on these dishes.

The cDNAs were cloned into an expression vector under control of the Moloney sarcoma virus LTR with an SV40 poly(A) addition signal (see Experimental Procedures). Whereas both MyoD and MyoH cDNAs were nearly full length, the MyoA cDNA represented only 2 kb out of a 5 kb message.

Using the "sense" expression orientation (determined by probing Northerns with strand-specific probes), these vectors were cotransfected into 10T1/2 cells at approximately a 100:1 molar ratio with a selection vector encoding the bacterial neomycin phosphotransferase gene (neo). The cells were selected for G418 resistance 24 hr after transfection. Cotransfection with the MyoD expression vector gave a noticeable reduction in the number of G418 resistant colonies (stained with crystal violet) compared to the number of colonies resulting after cotransfection with MyoA or MyoH (Figure 4, left). The transfected cultures were fixed and scored for myosin heavy chain expression using alkaline phosphatase-linked immunostaining (Figure 4, middle and right). We observed that about 50% (an average of several transfections) of those colonies

obtained from transfection with MyoD demonstrated myosin-positive, multinucleate syncitia, morphologically indistinguishable from myogenic colonies derived by 5-aza-cytidine treatment of 10T1/2 cells. No notable phenotypic changes were observed using MyoA or MyoH vectors. Figure 4 (top right) shows representative myogenic G418 resistant colonies derived by cotransfection with the MyoD expression vector and immunostained for myosin heavy chain. Fusion or cells to form syncitia was coincident with the presence of the myosin heavy chain gene product, showing coordination of the muscle differentiation program. In addition, we have determined that these cells also express myosin alkali light chain 1,3 by immunostaining (data not shown) and myosin light chain 2 by Northern analysis (see below). Control transfections of 10T1/2 cells with MyoD in a promoterless plasmid or in the antisense orientation in the same expression vector showed no effect, either on colony formation or myogenesis (data not shown).

The reduction in number of G418 resistant colonies on MyoD transfection of 10T1/2 cells is a reproducible pheno-

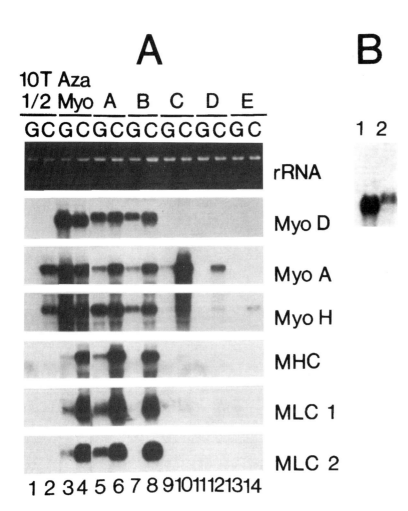

A

10T Aza
1/2 Myo A B C D E

G C G C G C G C G C G C G C

rRNA

Myo D

Myo A

Myo H

MHC

MLC 1

MLC 2

1 2 3 4 5 6 7 8 9 10 11 12 13 14

B

1 2

Figure 5. Expression of the Myo RNAs and Muscle Differentiation Markers in 10T1/2 Clones Transfected with the MyoD Expression Vehicle

(A) Northern analysis (total RNA) of MyoD, MyoA, MyoH, myosin heavy chain (MHC), myosin light chain 1 (MLC1), and myosin light chain 2 (MLC2) in growing (G) and confluent (C) 10T1/2 cells (lanes 1 and 2), P2 aza-myoblasts (lanes 3 and 4), and five clones (A, B. C, D, and E) of 10T1/2 cells transfected with the MyoD expression vehicle, picked from a parallel plate as that shown in Figure 4 (lanes 5–14). Clones A and B show extensive fusion when cultured in differentiation medium, clone C initially fused but lost this capacity during subsequent passage, and clones D and E never displayed any fusion. The top panel shows an ethidium bromide stain (of rRNA) of the RNA sample used for Northern analysis.
(B) Northern analysis of MyoD expression in proliferating P2 aza-myoblasts (lane 1) and myogenic transfectant clone A cells (lane 2).

type that becomes noticeable when the input ratio of MyoD expression vector to selection vector exceeds 10:1. One attractive explanation is that beyond a threshold level of MyoD expression some transfected cells differentiate early and fail to proliferate, thereby not contributing to the final colony scoring. As the colonies were growing, we often observed isolated rnultinucleate cells, or cells in small clusters, that had the appearance of myotubes, although it is generally difficult to discriminate between them and the morphological effects of G418 selection. However, after immunostaining plates with fully grown colonies, we again saw isolated multinucleate cells strongly positive for myosin heavy chain, which suggested that these isolated clusters of cells were, in fact, prematurely differentiating transfectants. This point is discussed further below.

About half of the G418 resistant clones cotransfected with MyoD differentiate into muscle. Several differentiating and non-differentiating clones were isolated and the expression of MyoD in each was examined by Northern analysis. Figure 5A shows that clones A and B. which express muscle structural genes, express MyoD at levels comparable to aza-myoblasts (lanes 3–8). In contrast clones D and E, which do not differentiate, lack MyoD expression (lanes 11–14). Clone C was derived from a fused, apparently myogenic colony, Which after subsequent passage lost its capacity to fuse. Muscle differentiation mark-

ers and MyoD are both expressed at only trace levels in this clone (lanes 9 and 10). For each clone, the expression of MyoD and the ability to differentiate are correlated with expression of myosin heavy chain, myosin light chain 1, and myosin light chain 2 by Northern analysis. Using low percentage gels, the exogenous and endogenous MyoD transcripts can be distinguished, due to a slight difference in electrophoretic mobility. We can detect only the exogenous transcript of MyoD in transfected myogenic colonies (Figure 5B), which suggests that the endogenous gene cannot be transactivated by its own gene product. The expression of MyoA and MyoH in proliferating cells, which is characteristic of myoblasts, is evident in transfected clones that express MyoD (clones A and B, and to a lesser extent clone C; Figure 5A). This suggests that MyoD expression alters the growth-arrest restricted expression of these genes characteristic of parental 10T1/2 cells.

Expression of MyoD in Other Cell Lines Induces Myogenesis

Myogenesis has been reported to occur in 5-azacytidine-treated Swiss 3T3 cells and CVP3SC6, a mouse prostate cell line (Taylor and Jones, 1982). However, the frequency of myogenesis in 5-azacytidine-treated 10T1/2 cells is at least two orders of magnitude greater than in Swiss 3T3 cells (Taylor and Jones, 1979). This observation led us to

Table 1. Frequency of Myogenic Colonies in Different Cell Lines Transfected with MyoD

Transfected Cell Line	# of Myogenic Colonies[a] per Total G418[R] Colonies Transfected with Expression Vehicle (without insert)	# of Myogenic Colonies[b] per Total G418[R] Colonies Transfected with MyoD Expression Vehicle	Range of % of Myosin-Positive Cells per Myogenic Colony	Extent of Myogenesis per Transfection with MyoD Expression Vehicle[c]
(A) C3H10T1/2 (mouse fibroblast)	0/456	81/154 (53%)	1%–50%	high
(B) NIH 3T3 (mouse fibroblast)	0/720	23/250 (9%)	0.1%–1%	low
(C) Swiss 3T3 (mouse fibroblast)	0/48	17/38 (45%)	1%–10%	medium
(D) Swiss 3T3 Clone 2 (C2) (mouse fibroblast)	0/61	20/50 (40%)	1%–10%	medium
(E) L Cell (mouse fibroblast)	0/900	23/800 (3%)	1%–10%	low
(F) 3T3-L1 (Swiss 3T3–derived adipoblast)	0/118	37/120 (31%)	1%–50%	high
(G) 3T3-F442A (Swiss 3T3–derived adipoblast)	0/61	30/63 (48%)	1%–10%	medium
(H) TA1 (C3H10T1/2-derived adipoblast)	15/118[d] (13%)	104/160 (65%)	1%–50%	high
(I) CV1 (monkey kidney cell)	0/190	0/186 (0%)	—	none

[a] Cells were transfected with 0.2 μg pCMV neo, 2.0 μg pEMSVscribe, and 20.0 μg C3H10T1/2 high molecular weight carrier DNA.
[b] Cells were transfected with 0.2 μg pCMVneo, 2.0 μg pEMSVscribe-MyoD, and 20.0 μg C3H10T1/2 high molecular weight carrier DNA.
[c] Determined by noting the percentage of myosin-positive cells per polyclonal culture.
[d] TA1 cells show the same level of spontaneous myogenesis in the absence of DNA transfection. This frequency of myogenic conversion is increased by transfection with the MyoD expression vehicle.

think that MyoD may operate in a permissive environment of 10T1/2 cells such that it activates a predetermined state (as if 10T1/2 cells are only one step away from myogenic determination). If so, one would not expect MyoD expression alone in other lines to result in myogenic determination. Alternatively, the low frequency of myogenesis in 5-azacytidine-treated Swiss 3T3 cells, for example, may be due to a deficiency of factors necessary for MyoD expression itself, rendering the locus relatively insensitive (directly or indirectly) to genomic demethylation. In this case, MyoD expression in Swiss 3T3 cells, driven by a viral LTR, might efficiently convert these cells into myoblasts. We cotransfected the MyoD expression vehicle with a vector encoding *neo* into the cell types listed in Table 1 and selected for G418 resistance. Colonies were immunostained for myosin heavy chain (Figure 6). With the exception of CV1 cells, myosin heavy chain synthesis and cell fusion were detected in all cell lines transfected with the MyoD expression vehicle. However, the frequency of myogenic colonies and the percentage of myosin-positive cells within these colonies varied with each line. Whereas 53% of transfected 10T1/2 colonies went myogenic, 3%, 9%, and 45% of the G418 resistant L cell, NIH3T3, and Swiss 3T3 colonies (respectively) were myogenic (see Table 1). In the latter three cell lines, the percentage of myosin-positive cells per myogenic colony is considerably

less than in myogenic 10T1/2 colonies. The low percentage of differentiated cells in these colonies probably accounts for the low degree of cell fusion (due to the lack of adjacent fusion partners). Myogenesis also occurs in BALB/C 3T3 cells transfected with the MyoD expression vehicle (data not shown). No myosin heavy chain was detectable in the above cell lines transfected with the parental expression vehicle lacking the MyoD cDNA insert.

In addition to transfected "fibroblast" lines, MyoD expression also induces myogenesis in adipoblast cell lines derived from either Swiss 3T3 (3T3–L1 and 3T3–F442A) or 5-azacytidine-treated 10T1/2 cells (TA1). Some of the 3T3–L1 and TA1 myogenic colonies show as high a degree of differentiation as transfected 10T1/2 colonies (Figures 6F and 6H, respectively). We are currently examining whether the adipocyte differentiation program is concomitantly expressed in these apparently myogenic colonies. In addition to myosin heavy chain, myosin light chain 1,3 has also been detected by immunostaining in BALB/C 3T3, 3T3–L1, and 3T3–F442A cells transfected with the MyoD expression vehicle. CV1, a monkey kidney cell line, is the only line that did not display detectable myogenesis after MyoD transfection. It is unclear if the lack of myogenesis in this cell line is due to the epithelial origin of CV1 cells or to a species incompatibility of mouse MyoD in a monkey cell. Northern analysis revealed that G418 resis-

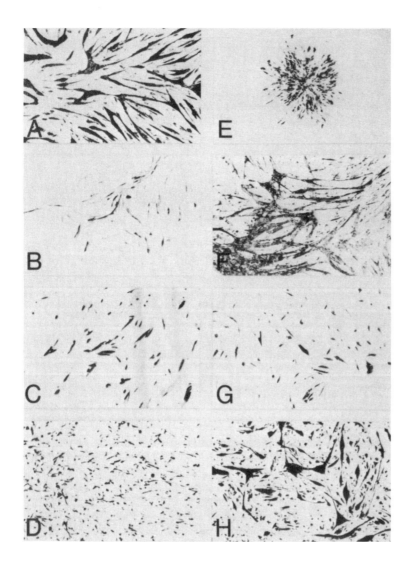

Figure 6 Myosin Staining of Different Cell Lines Transfected with the MyoD Expression Vehicle

Immunostained myosin heavy chain–positive colonies (shown at 50× magnification) resulting after transfection of 2.0 μg of the MyoD expression vehicle (pEMSVscribe MyoD) 0.2 μg of pCMVneo (a plasmid encoding G418 resistance), and 20 μg of genomic 10T1/2 DNA into the following cell lines: C3H10T1/2 (A), NIH 3T3 (B), Swiss 3T3 (C), Swiss 3T3-clone 2 (D), L cells (E), 3T3-L1 adipocytes (F), 3T3-F442A adipocytes (G), and TA1 adipocytes (H).

tant polyclones of transfected CV1 cells were expressing MyoD RNA at levels comparable to transfected 10T1/2 polyclones (data not shown). The effects of expressing MyoD in more defined differentiated cells (e.g., hepatoma cells, neuroblastomas, and myelomas), as well as in non-differentiating muscle variants, is currently under study.

Expression of MyoD, MyoA, and MyoH in Mouse Tissues

We examined the expression of MyoD, MyoA, and MyoH RNAs in various mouse adult tissues and some neonatal tissues by Northern analysis. Figure 7 shows that MyoD RNA is expressed in neonatal and adult skeletal muscle (lanes 2, 3, 6, and 7), although not in neonatal or adult cardiac muscle (lanes 4 and 8), nor in any other tissue tested. In addition, MyoD RNA is not detectable in the mouse smooth muscle cell line, BC3H1 (data not shown). The level of expression of MyoD RNA in skeletal muscle tissue is about 10% of that in skeletal muscle cell lines. It may be significant that in chicken primary cultures MyoD expression is highly reduced in myotubes as compared with proliferating myoblasts (data not shown).

MyoA RNA is expressed in neonatal and adult skeletal and cardiac muscle, as well as adult kidney, lung, brain, ovary, and stomach (trace). MyoH RNA is expressed in neonatal and adult skeletal and cardiac muscle, as well as neonatal liver, and adult lung, ovary, and stomach (trace). The expression pattern for MyoH RNA is reminiscent of a previously characterized gene, H19, described by Tilghman and colleagues in their studies of loci under control of the *raf* and *Rif* loci in developing mouse liver (Pachnis et al., 1984). In fact, MyoH cross-hybridizes to an H19 cDNA generously provided by S. Tilghman (data not shown). The function of H19 in skeletal and cardiac muscle, where it is expressed constitutively (Pachnis et al., 1984; this study), is unknown. We can also detect transcripts homologous to MyoD, MyoA, and MyoH in human and chicken myogenic cell cultures (data not shown).

Sequence of MyoD

We have sequenced MyoD by generating nested deletions (Henikoff, 1987) and using dideoxy sequencing procedures (Sanger et al., 1977). This sequence is presented in Figure 8. The sense orientation encodes only

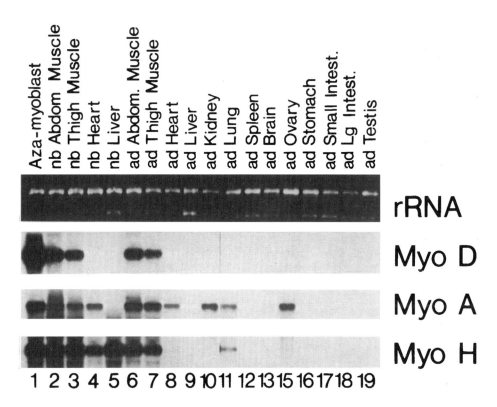

Lanes labeled above each lane: Aza-myoblast, nb Abdom Muscle, nb Thigh Muscle, nb Heart, nb Liver, ad Abdom. Muscle, ad Thigh Muscle, ad Heart, ad Liver, ad Kidney, ad Lung, ad Spleen, ad Brain, ad Ovary, ad Stomach, ad Small Intest., ad Lg Intest., ad Testis

Panels: rRNA, Myo D, Myo A, Myo H

Lane numbers: 1 2 3 4 5 6 7 8 9 10 11 12 13 15 16 17 18 19

Figure 7. Expression of Myo mRNAs in Various Mouse Tissues

Northern analysis (total RNA) of MyoD, MyoA, and MyoH expression in the P2 aza-myoblast line (lane 1) and various newborn (nb; lanes 2–5) and adult (ad; lanes 6–19) mouse tissues (indicated above each lane). The top panel shows an ethidium bromide stain (of rRNA) of the RNA sample used for Northern analysis.

one open reading frame that is greater than 100 residues and has a good initiation sequence consensus (GAT*ATG*G). This 318 amino acid reading frame is also consistent with the observation that a shorter MyoD cDNA clone (a 300 bp 3′ deletion) which contains this reading frame intact also converts 10T1/2 cells to myoblasts (data not shown). The segment of the open reading frame between residues 62 and 101, containing several clustered histidines and cysteines, is similar to the putative metal binding domains of the "zinc-finger" class (Berg, 1986), al-

```
ACACCTCTGACAGGACAGGACAGGGAGGAGGGGTAGAGGACAGCCGGTGTGCATTCCAACCCACAGAACCTTTGTCATTGTACTGTTGGGGTTCCGGAGTGGCAGAAAGTTAAGACGACT                1 2 0

CTCACGGCTTGGGTTGAGGCTGGACCCAGGAACTGGGATATGGAGCTTCTATCGCCGCCACTCCGGGACATAGACTTGACAGGCCCCGACGGCTCTCTCTGCTCCTTTGAGACAGCAGAC                2 4 0
                                        MetGluLeuLeuSerProProLeuArgAspIleAspLeuThrGlyProAspGlySerLeuCysSerPheGluThrAlaAsp               2 7

GACTTCTATGATGATCCGTGTTTCGACTCACCAGACCTGCGCTTTTTTGAGGACCTGGACCCCGCGCCTGGTGCACGTGGGAGCCCTCCTGAAACCGGAGGAGCACGCACACTTCTCTACT                3 6 0
AspPheTyrAspAspProCysPheAspSerProAspLeuArgPhePheGluAspLeuAspProArgLeuValHisValGlyAlaLeuLeuLysProGluGluHisAlaHisPheSerThr               6 7

GCGGTGCACCCAGGCCCAGGCGCTCGTGAGGATGAGCATGTGCGCGCGCCCAGCGGGCACCACCAGGCGGGTCGCTGCTTGCTGTGGGCCTGCAAGGCGTGCAAGCGCAAGACCACCAAC                4 8 0
AlaValHisProGlyProGlyAlaArgGluAspGluHisValArgAlaProSerGlyHisHisGlnAlaGlyArgCysLeuLeuTrpAlaCysLysAlaCysLysArgLysThrThrAsn              1 0 7

GCTGATCGCCGCAAGGCCGCCACCATGCGCGAGCGCCGCCGCCTGAGCAAAGTGAATGAGGCCTTCGAGACGCTCAAGCGCTGCACGTCCAGCAACCCGAACCAGCGGCTACCCAAGGTG                6 0 0
AlaAspArgArgLysAlaAlaThrMetArgGluArgArgArgLeuSerLysValAsnGluAlaPheGluThrLeuLysArgCysThrSerSerAsnProAsnGlnArgLeuProLysVal              1 4 7

GAGATCCTGCGCAACGCCATCCGCTACATCGAAGGTCTGCAGGCTCTGCTGCGCGACCAGGACGCCGCGCCCCCTGGCGCCGCTGCCTTCTACGCACCTGGACCGCTGCCCCCAGGCCGT                7 2 0
GluIleLeuArgAsnAlaIleArgTyrIleGluGlyLeuGlnAlaLeuLeuArgAspGlnAspAlaAlaProProGlyAlaAlaAlaPheTyrAlaProGlyProLeuProProGlyArg              1 8 7

GGCAGCGAGCACTACAGTGGCGACTCAGACGCGTCCAGCCCGCGCTCCAACTGCTCTGATGGCATGATGGATTACAGCGGCCCCCCAAGCGGCCCCCGGCGGCAGAATGGCTACGACACC                8 4 0
GlySerGluHisTyrSerGlyAspSerAspAlaSerSerProArgSerAsnCysSerAspGlyMetMetAspTyrSerGlyProProSerGlyProArgArgGlnAsnGlyTyrAspThr              2 2 7

GCCTACTACAGTGAGGCGGTGCGCGAGTCCAGGCCAGGGAAGAGTGCGGCTGTGTCGAGCCTCGACTGCCTGTCCAGCATAGTGGAGCGCATCTCCACAGACAGCCCCGCTGCGCCTGCG                9 6 0
AlaTyrTyrSerGluAlaValArgGluSerArgProGlyLysSerAlaAlaValSerSerLeuAspCysLeuSerSerIleValGluArgIleSerThrAspSerProAlaAlaProAla              2 6 7

CTGCTTTTGGCAGATGCACCACCAGAGTCGCCTCCGGGTCCGCCAGAGGGGGCATCCCTAAGCGACACAGAACAGGGAACCCAGACCCCGTCTCCCGACGCCGCCCCTCAGTGTCCTGCA             1 0 8 0
LeuLeuLeuAlaAspAlaProProGluSerProProGlyProProGluGlyAlaSerLeuSerAspThrGluGlnGlyThrGlnThrProSerProAspAlaAlaProGlnCysProAla              3 0 7

GGGCTCAAACCCCAATGCGATTTATCAGGTGCTTTGAGAGATCGACTGCAGCAGCAGAGGGCGCACCACCGTAGGCACTCCTGGGGATGGTGCCCCTGGTTCTTCACGCCCAAAAGATGAA             1 2 0 0
GlySerAsnProAsnAlaIleTyrGlnValLeu                                                                                                       3 1 8

GCTTAAATGACACTCTTCCCAACTGTCCTTTCGAAGCCGTTCTTCCGAGAGGGAAGGGAAGAGCAGAAGTCTGTCCTAGATCCAGCCCCAAAGAAAGGACATAGTCCTTTTTGTTGTTGTT             1 3 2 0

GTTGTAGTCCTTCAGTTGTTTGTTTGTTTTTTCATGCGGCTCACAGCGAAGGCCACTTGCACTCTGGCTGCACCTCACTGGGCCAGAGCTGATCCTTGAGTGGCCAGGCGCTCTTCCTTT             1 4 4 0

CCTCATAGCACAGGGGTGAGCCTTGCACACCTAAGCCCTGCCCTCCACATCCTTTTGTTTGTCACTTTCTGGAGCCCTCCTGGCACCCACTTTTCCCCACAGCTTGCGGAGGCCACTCAG             1 5 6 0

GTCTCAGGTGTAACAGGTGTAACCATACCCCACTCTCCCCCTTCCCGCGGTTCAGGACCACTTATTTTTTTATATAAGACTTTTGTAATCTATTCGTGTAAATAAGAGTTGCTTGGCCAG             1 6 8 0

AGCGGGAGCCCCTTGGGCTATATTTATCTCCCAGGCATGCTGTGTAGTGCAACAAAAACTTTGTATGTTTATTCCTCAAGCGGGCGAGTCAGGTGTTGGAAATCC                           1 7 8 5
```

Figure 8. Nucleotide Sequence of the MyoD1 cDNA and Its Predicted Amino Acid Sequence

Mouse C-Myc	N	E	K	A	P	K	V	V	I	L	K	K	A	T	A	Y	I	L	S	I	Q	A
Mouse N-Myc	N	E	K	A	A	K	V	V	I	L	K	K	A	T	E	Y	V	H	A	L	Q	A
Avian V-Myc	N	E	K	A	P	K	V	V	I	L	K	K	A	T	E	Y	V	L	S	L	Q	S
MyoD1	N	Q	R	L	P	K	V	E	I	L	R	N	A	I	R	Y	I	E	G	L	Q	A
T4 AS-C	H	K	K	I	S	K	V	D	T	L	R	I	A	V	E	Y	I	R	S	L	Q	D
T5 AS-C	N	K	K	L	S	K	V	S	T	L	K	M	A	V	E	Y	I	R	R	L	Q	K

Figure 9. Amino Acid Similarities in MyoD1, Myc, and *achaete-scute* Genes

Amino acids shared by MyoD1 (aa. 141–162) mouse c-*myc* (aa. 387–408, Bernard et al., 1983), mouse N-*myc* (aa. 412–433; DePinho et al., 1986), avian V-*myc* (aa. 373–394; Alitalo et al., 1983), and genes in the *achaete-scute* locus (aa. 143–164 in T4, aa. 71–92 in T5; Villares and Cabrera, 1987) are indicated. Identical amino acids shared by MyoD1 and at least one *myc* or *achaete-scute* gene are indicated by two dashes; conservative amino acid substitutions in MyoD1 are indicated by a single dash.

though it lacks certain conserved residues. This region is followed by a highly basic region (residues 102–124). In a homology search of the Protein Identification Resource (National Biomedical Research Foundation at Georgetown University Medical Center, Washington, D.C.), it was found that the basic region is just amino-terminal to a short segment (22 residues) that shows strong similarity to a conserved segment of mouse, chicken, and human c-*myc* proteins, as well a s avian v-*myc* and mouse N-*myc*. Figure 9 outlines this segment. This region of the human c-*myc* protein has been shown, by insertion and deletion mutagenesis, to be critical to its transforming potential in vitro (Stone et al., 1987). We also noticed that this same segment is shared by two recently sequenced homologous transcription units of the Drosophila *achaete-scute* complex (Villares and Cabrera, 1987), whose expression is correlated with neurogenic lineage development.

Discussion

MyoD1, a Gene that Converts Fibroblasts to Myoblasts

Using a subtracted cDNA screening procedure, we isolated a single myoblast-specific cDNA whose expression in transfected fibroblast-like 10T1/2 cells converts them to stable myoblasts. We would like to name the gene encoding this sequence MyoD1, for myoblast determination gene number 1. Myogenesis is also evident, although to a lesser extent, in several other fibroblast and adipoblast cell lines when transfected with MyoD1. MyoD1 RNA is expressed in normal skeletal muscle and has not been detected in other tissues of the mouse. Hence, we believe (but have not proven) that MyoD1 also functions to activate normal myogenesis, where it is also known that fibroblasts and myoblasts are derived from a common precursor (Dienstman and Holtzer, 1975).

Several factors aided us in identifying MyoD1. A key assumption in our analysis was that a regulator of myogenic determination would be present in undifferentiated myoblasts. This assumption allowed us to screen out differentiation-specific muscle structural markers. In addition, the nondifferentiating myoblast variants were extremely useful for screening out intracisternal type A particle and other "nonspecific" genes activated by 5-aza-cytidine. Finally, we were fortunate that the abundance of MyoD1 mRNA in the cell allowed its detection (four times) in the 10^4 cDNA clones screened. How general can one

expect the subtracted hybridization procedure to be for identifying sequences that regulate determination or differentiation? We believe that the real limitations are biological — to find two appropriate cell types, or preferably, noninducing variants of a single cultured cell type. Clearly, in our case, the myogenic conversion of 10T1/2 cells induced by genomic myoblast DNA transfection and the high frequency of myogenesis induced by 5-azacytidine (Taylor and Jones, 1982; Konieczny and Emerson, 1984) were important clues for subsequent functional analysis, since they suggested that only one event was required. In this respect, it is of interest to point out that the comparison between 10T1/2 cells and myoblasts is very different from, for example, more differentiated cultured cell lines (e.g., MEL cells, HL60 cells, or pre-B cells) in induced versus uninduced conditions. In these instances, regulatory changes may depend more on modification of a repertoire of regulatory proteins already present (Sen and Baltimore, 1986).

MyoD1 Seems to Antagonize Cell Growth

The "phenotype" of 10T1/2 cultures cotransfected with MyoD1 presents two complications: the G418 resistance transformation frequency is about 1/10 that of controls, and only 50% of all viable clones are positive for myogenesis. The second point is best explained by variability in expression of the cotransfected MyoD1 expression vehicle since nonmyogenic transfected colonies do not express significant levels of MyoD1 mRNA while myogenic colonies do. This variability in expression is probably a consequence of both a failure of successful cotransfection and a failure to express the cotransfected vector due, for example, to chromosomal position effects. One clone that initially fused eventually lost the capacity to fuse, and, when assayed, failed to express significant levels of MyoD1. Nondifferentiating aza-myoblast variants also failed to express significant levels of MyoD1. It would seem then that MyoD1 expression is required to maintain myogenic determination potential, at least for 10T1/2 derived myoblasts.

Why is the transformation frequency reduced with MyoD1 cotransfection? We would speculate that one possibility is that MyoD1 expression rapidly induces most recipient 10T1/2 cells to commit to myogenesis. As a consequence they withdraw from the division cycle and fail to form colonies. Implicit in this explanation is that there is a competition between growth and myogenesis and that when MyoD1 is expressed at high levels, those initial transfectants differentiate, even in the presence of high serum. When MyoD1 is expressed at moderate levels, differentiation occurs only in low serum, and when expressed at low levels, myogenesis does not occur at all. In support of this notion, we observed that the level of exogenous MyoD1 mRNA in two transfected myogenic clones is nearly equivalent to the endogenous mRNA levels expressed in both aza and C2C12 myoblasts.

In preliminary experiments, we tested to see if MyoD1 expression inhibited colony formation because of early differentiation by using factors that promote continued proliferation by an independent mechanism and block myogenic differentiation (Linkhart et al., 1981; Lathrop et al., 1985; Clegg et al., 1987). Using one such reagent, fibroblast growth factor (FGF), Wf3 were able to increase the transformation frequency to control levels. Also, in parallel experiments, the number of myogenic colonies increased. That some of the MyoD1 transfected 10T1/2 cells differentiated early implies that in the absence of FGF we should be able to detect them as small clusters of cells expressing muscle markers. Indeed, isolated clusters of differentiated cells were observed, and their frequency was roughly what was expected based on the transformation frequency using control DNA. Finally, we performed transient transfections of 10T1/2 cells with MyoD1 and used immunofluorescence to detect a small percentage of cells expressing desmin and myosin heavy chain.

Interestingly, MyoD1 transfection activates the continued expression of two mRNAs (MyoA and MyoH) in proliferating myoblasts that are absent in proliferating 10T1/2 cells and are otherwise expressed only in growth-arrested 10T1/2 cells. The expression in proliferating myoblasts of these growth regulated mRNAs further suggests that MyoD1 may antagonize or dampen mitogenic signals (see below).

MyoD1 Shares a Region of Amino Acid Similarity with *myc*

The similarity of a MyoD1 coding region segment with *myc* proteins may be important in ultimately explaining the mutual antagonism between growth and differentiation illustrated in this system. It has been observed that c-*myc* is induced in a variety of cell types when they enter the cell cycle from a quiescent state (Kelly et al., 1983; Thompson et al., 1985). In addition, c-myc is often down-regulated as cells terminally differentiate in vitro (Westin et al., 1982; Lachman and Skoultchi, 1984; Dony et al., 1985; Sejersen et al., 1985; Endo and Nadal-Ginard, 1986) and in vivo (Conklin and Groudine, 1986; Schneider et al., 1986). Recent experiments employing a transfected, constitutively expressed *myc* gene suggest that this down-regulation may be a necessary factor in differentiation (Coppola and Cole, 1986; Dmitrovsky et al., 1986; Prochownik and Kukowska,1986). Additionally, v-*myc* expression in MC29 transformed quail myoblasts inhibits myogenic differentiation (Falcone et al., 1985). Insertion and deletion mutation analysis of the region of human c-*myc* protein containing the shared sequences with MyoD1 has revealed that the loss of this segment or small nearby

insertions abolishes *myc* transforming activity in vitro (Stone et al., 1987). While direct or indirect (i.e., via binding to a common protein or ligand) interaction of MyoD1 with c-*myc* based on a single, small conserved segment is highly speculative, it does provide one focus for further analysis of MyoD1 function by site-directed mutagenesis. It is striking that the gene products of the *achaete-scute* locus, which play a role in Drosophila neurogenic cell determination, also share the same common region of *myc* homology with MyoD1. A common amino acid sequence in determination factors of at least two cell lineages (myogenic and neurogenic) raises the possibility that other cell lineages may use analogous determination factors that share this sequence.

The Genetic Basis for Myogenic Conversion

Our present data neither support nor refute the possibility that MyoD1 is the gene directly activated by 5-azacytidine in 10T1/2 cells, and additional experiments are in progress to address this question. Some evidence comes from preliminary observations that a specific HpaII fragment detected by MyoD1 (in a Southern blot) is observed in aza-myoblasts but not in 10T1/2 cells, suggesting preferential methylation of this site(s) in 10T1/2 cells.

Is myogenesis controlled by a single master gene? At present, it is unclear if MyoD1 expression alone is sufficient to activate the entire myogenic program. Whereas MyoD1 expression induces myogenesis in several "fibroblast" and adipoblast cell lines, the extent of differentiation within a myogenic colony differs with each line and is in most cases much less than that observed in transfected 10T1/2 cells. Furthermore, a monkey kidney cell line (CV1), expressing transfected MyoD1 mRNA, shows no evidence at all of myogenic transformation. We do not understand this variability in "penetrance" between different recipient cell lines. It is possible that 10T1/2 cells already contain some additional elements that cooperate with MyoD1 for determining the myogenic phenotype. Alternatively, it is possible that other cell lines contain an inhibitor of myogenesis which results in a decreased probability of differentiation (depending on the "balance" between positive and negative factors in individual cells of a colony) and that the argument for several positive myogenic conversion factors is incorrect. Heterokaryon experiments in which myotubes were fused with a variety of cell types have provided evidence for both positive factors in muscle cells (Blau et al., 1983; Wright, 1984a; Blau et al., 1985) as well EIS inhibitory factors in nonmyogenic cells (Wright and Aronoff, 1983; Lawrence and Coleman, 1984; for a discussion of positive and negative differentiation factors see Wright, 1984b, 1985).

The significant reduction of MyoD1 expression in *all* non-differentiating aza-myoblast variants suggests that MyoD1 expression (or factors that regulate it) may alone be necessary for myogenic determination of these cells. Alternatively, this locus and/or one controlling it may be unusually sensitive to inactivation, either because it is the most distal (and therefore dependent) element in a series of operations or because it is simply more sensitive due to its molecular properties. We are currently researching if reexpression of this gene in the variant lines will reactivate the muscle program. In so far as MyoD1 does not promote increased adipogenesis in transfected 10T1/2 cells (data not shown), it is unlikely that MyoD1 is a general differentiation agent. That MyoD11 expression is limited (as far as we can detect) to skeletal muscle and myoblasts supports this notion. It is possible, however, that MyoD1 is expressed transiently in many lineages during development, yet its continued expression is detectable only in skeletal muscle, which is known to retain muscle stem cells or satellite cells.

Previously, we suggested the possibility that 10T1/2 cells may be determined myoblasts that have lost their ability to differentiate as a result of the activation of an oncogene (Lassar et al., 1986). In this particular instance, myogenesis in 10T1/2 cells could result from expression of a suppressor of such an activated oncogene. Although we cannot rule out this scenario, both the restriction of MyoD1 expression to skeletal muscle and its ability to induce myogenesis in a number of different cell lines argue against this gene being simply an "anti-oncogene."

There is a great deal of evidence in C. elegans and Drosophila that regulatory genes are expressed in several cell lineages, supporting current and classical notions that cell type would be encoded by a subset of combinations from a pool of genomic regulatory genes. There is less evidence for cell type-specific "determination genes" being expressed in only one lineage or cell type and also being cell autonomous in that lineage. On the other hand, current DNA binding studies do detect DNA binding activities that are present only in specific differentiated cells. There are also a few "determination" genes that do seem to affect only a single type of cell (e.g., mec3 in C. elegans, *achaete-scute* and sevenless in Drosophila, and genes involved in germ-line sex determination). Possibly, combinatorial schemes (as exemplified in Drosophila segmentation) and cell lineage (as illustrated in C. elegans) are both used to generate positional identity (for example, see Wolpert and Stein, 1984) in the embryo for each individual cell. Eventually, this information, together with input from local cues, could focus down to the activation of only one or a few specific genes that are ultimately responsible for determining a specific cell type.

Experimental Procedures

Cells

C3H10Tt/2, clone 8, Swiss 3T3, and Swiss 3T3-L1 adipocytes (Green and Kehinde, 1974) were from the American Type Culture Collection. 5-azacytidine-derived myoblast lines were generated by exposing C3H10T1/2 cells to 3 μM 5-azacytidine for 24 hr (Lassar et al., 1986). Resulting myogenic colonies were subcloned five times, and two independently isolated aza-myoblast lines (F3 and P2) were established. C2C12 myoblasts (Blau et al., 1983) were provided by Helen Blau. 3T3 Swiss-C2 and 3T3-F442A cells (Green and Kehinde, 1976) were provided by Howard Green. L cells (TK$^-$, APRT$^-$) were provided by Richard Axel. TA1 cells (Chapman et al., 1984) were obtained from Gordon Ringold. Other cell lines are described in the figure legends. All cell lines were grown in Dulbecco's modified Eagle's medium (DMEM) supplemented with 15% fetal calf serum (Hyclone Laboratories). Myogenic differentiation was induced by incubation in DMEM with 2% heat-inactivated horse serum (differentiation medium).

Subtracted cDNA Probe Preparation

Myoblast cDNA was prepared using poly(A)$^+$ RNA isolated from cloned myoblast lines. Synthesis included ^{32}P-dCTP (800Ci/mmol, NEN) and AMV reverse transcriptase (Siekagaku America, Inc.) as described (Davis et al., 1984). Initial specific activities of these probes were approximately 3×10^8 dpm/μg. Subtracted hybridizations were performed (following the protocol of Davis et al., 1984) in 8 μl of 0.5 M NaH$_2$PO$_4$/Na$_2$HPO$_4$ (pH 7.0), 0.1% SDS, 5 mM EDTA at 68°C under paraffin oil. Reactions achieved C$_0$t values of 3000 or greater. Selection for single-stranded material was performed over hydroxyapatite equilibrated in 0.12 M NaH$_2$PO$_4$/Na$_2$HPO$_4$ (pH 7.0), 0.1% SDS at 60°C in a water-jacketed column (Bio-Rad). Selection for double-stranded material in the positive selection was performed after binding to hydroxyapatite by elution with 0.5 M NaH$_2$PO$_4$/Na$_2$HPO$_4$ (pH 7.0), 0.1% SDS at 60°C. The percentage of remaining material from the initial screaming (twice subtracted with C3H10T1/2 poly(A)$^+$ RNA and once positively selected with myoblast poly(A)$^+$ RNA) was 1.15% for the aza-myoblast probe and 2.40% for the C2C12 myoblast probe.

Library Screening

The F3 aza-myoblast cDNA library was prepared by standard methods in λgt10 (Gubler and Hoffman, 1983; Huynh et al., 1985). Plaque hybridizations were performed using replica nitrocellulose filters from plates containing $1-1.5 \times 10^4$ cDNA clones. Approximately $1 \times 10^5-3 \times 10^5$ dpm of subtracted or unsubtracted probes were utilized. The hybridizations were carried out for three days at 42°C in 25 mM NaH$_2$PO$_2$/Na$_2$HPO$_4$ (pH 7.0), 5× SSC (1× SSC is 150 mM NaCl, 15 mM Na-citrate), 1 × Denhardt's solution (0.02% Ficoll, 0.02% polyvinylpyrrolidone, 0.02% BSA), 250 μg/ml total Torula RNA, 50% Formamide, with 10% dextran sulfate and 2% SDS. Washings were performed in 2× SSC, 0.1% SDS for 45 min at 60°C and in 0.4× NPE (1× NPE is 150 mM NaCl, 50 mM NaH$_2$PO$_4$/Na$_2$HPO$_4$ [pH 7.0], 1 mM EDTA), 0.1% SDS for 45 min at 60°C, followed by autoradiography.

cDNA Clone Selection

Ninety-two cDNA clones, representing about 10^4 cDNA clones plated, were initially picked. The plaque density of the initial filters required a secondary screen to isolate individual clones. The secondary screen involved a subtracted "plus/minus" type of procedure. The 92 initial phage isolates were individually gridded (75 times each) and screened on triplicate nitrocellulose filters with either ^{32}P-labefed aza-myoblast cDNA subtracted once with 10T1/2 poly(A)$^+$ RNA, ^{32}P-labeled C2C12 cDNA subtracted similarly (plus probes), or ^{32}P-labeled 10T1/2 cDNA (minus probe). These probes were prepared as described above. Phage DNA was purified from clones showing myoblast-specific hybridization, and the cDNA inserts isolated by EcoRI digestion and agarose gel electrophoresis, followed by binding to and elution from DE81 paper (Whatman). We used the following initial criteria in selecting cDNAs for further analysis: lack of expression in proliferating 10T1/2 cells; little to no increase in expression in differentiated myotubes versus proliferating myoblasts (which is characteristic of terminal differentiation markers); and reduced or absent expression in nondifferentiating variants of myogenic cells. The 92 cDNAs are listed by frequency of actual isolation as follows: MyoH (37); intracisternal type A particle RNA (21); MyoA (3); MyoD (4); six different clones showing expression characteristic of terminal differentiation markers (8); five different clones expressed at low levels in 10T1/2 cells, higher levels in myogenic cells, and diminishing only slightly or not at all in differentiation-defective variants (5); one clone induced with growth arrest in 10T1/2 cells and several other 3T3 lines (1); eight different clones that showed little to no differential expression on Northern analysis (9); and two different clones containing apparently repetitive sequences by Northern analysis (4).

Derivation of Nondifferentiating Variants of Aza-Myoblasts

A 5-azacytidine-derived myogenic line (P2) was allowed to reach confluence in DMEM plus 15% fetal calf serum. The media was then switched to DMEM plus 2% heat-inactivated horse serum. After 3 days, when most of the cells had terminally withdrawn from the cell cycle, fused, and fully differentiated, the cells were trypsinized and replated at low density. Cells able to proliferate were passaged several more times following the above regimen and then cloned. When maintained at confluence in mitogen-depleted media, most of these clones failed to fuse or immunostain for myosin heavy chain.

RNA Preparation

Total cellular RNA from adherent tissue culture cells was prepared by washing plates with Tris-buffered saline followed by overlaying guanidine lysis buffer of 5 M guanidine isothiocyanate, 25 mM Na-Citrate (pH 7.0), 0.1 M 2-mercaptoethanol, 0.5% Sarkosyl. Lysares were collected and centrifuged over cushions of 5.7 M CsCl (pH 7.0), 0.1 M EDTA at 25 Krpm in a Beckman SW28.1 rotor for 18 hr at 20°C. Pellets were resuspended in 10 mM Tris (pH 7.0), 0.1% SDS, S mM EDTA, extracted once with phenol:chloroform:isoamyl alcohol (25:24:1) and once with chloroform:isoamyl alcohol (24:1), ethanol-precipitated, and stored in diethylpyrocarbonate treated water at −70°C. Whole mouse tissue RNA was prepared similarly. Six-day or 9 week old BALB/C mice were sacrificed, and dissected tissues were washed in Tris-buffered saline, frozen in liquid nitrogen, and homogenized in guanidine lysis buffer. Poly(A)$^+$ RNA selection of total cellular RNA was performed by standard methods (Maniatis et al., 1982).

RNA Analysis

Five micrograms of total cell RNA were used for Northern analysis on 1.5% agarose gels containing 6.7% formaldehyde. Integrity and relative amounts of RNA were checked by ethidium-bromide staining of ribosomal RNA on parallel gels. Gels were treated with 50 mM NaOH, 10 mM NaCl for 40 min, neutralized in 100 mM Tris (pH 7.4), 20× SSC for 40 min, and transferred for 16 hr to Gene Screen (NEN Research Products) in 20× SSC. RNA was cross-linked by exposure to UV. Hybridizations and washings were performed as described for the library screening, using 5–10 × 10^5 dpm of probes prepared by random primer synthesis (Boehringer Mannheim). The mouse myosin heavy chain, myosin light chain 2, and myosin light chain 1 probes had each been isolated from the F3 aza-myoblast/myotube cDNA library. The identity of these cDNAs was determined by hybridization to rat probes kindly provided by Bernardo Nadal-Ginaid. Intracisternal type A (IAP) cDNAs were identified by cross-hybridization to an IAP probe kindly provided by Kira Lueders.

Expression of cDNAs

MyoD, MyoA, and MyoH cDNAs were isolated from λgt10 clones. EcoRI cut λDNA was subjected to low melting point agarose gel electrophoresis, followed by extraction with 1% CTAB (hexadecyltrimethylammonium bromide) in water-saturated n-butanol. Fragments were cloned into EcoRI cut pEMSVscribe (for expression) or EcoRI cut pVZ-1 (for sequencing). Orientations were determined by preparing strand-specific ribo-probes for MyoA, MyoD, and MyoH. pEMSVscribe was generated by first digesting Bluescribe M13(+) (Stratagene) with EcoRI, filling in the ends with the Klenow fragment of E. coli pol I (Boehringer Mannheim), and religating to create Bluescribe M13(+)EcoRI−. This vector was then cut with HindIII, and the HindIII fragment of EMSV-33 (Harland and Weintraub, 1985) containing the Moloney sarcoma virus LTR, and SV40 poly(A) addition signal was inserted by religation. This construct allows unique EcoRI insertion between the LTR and poly(A) signal sequence, which are themselves flanked by T7 and T3 promoters. pVZ-1 (Henikoff and Eghtedarzadeh, 1987) is a modified version of Bluescribe M13(+) employed to generate unidirectional plasmid deletions.

Transfections

pCMV-neo, 0.1–0.2 μg (Linial, 1987) or pLNSAL (Palmer et al., 1987) were used for cotransfection to G418 resistanos. One to ten micrograms of the expression vehicles or control vehicles and 10–20 μg 10T1/2 high molecular weight DNA as carrier was used. Actual ratios are described in the text. Cells, 1.5–3.0 × 10^5, were plated 2 days prior to transfection on 10 cm diameter dishes. Cells were refed with DMEM containing 10% fetal calf serum 3 hr prior to transfection. The DNAs were combined and added to cells as a Ca–PO$_4$ precipitate (Wigler et al., 1979) with 10 μM chloroquine. This DNA was removed 24 hr later with addition of fresh medium as above. Twenty-four hours after this, the cells were split 1:10 to 10 cm diameter or 15 cm diameter gelatin coated dishes with DMEM containing 15% fetal calf serum and 0.4 mg/ml G418 (Geneticin, Gibco). To score for myogenesis, after colonies of about 10^4 cells had grown, the media was switched to DMEM

containing 2% heat-inactivated horse serum with 0.2 mg/ml G418. To score for adipogenesis, the media was switched to DMEM containing 15% fetal calf serum, 1 μM Dexamethasone, 5 μg/ml insulin with 0.2 mg/ml G418.

Fixing and Immwmostaining of Colonies

To score for myogenesis, plates were fixed 3–5 days after switching to horse serum containing media. Plates were washed once with Tris-buffered saline and incubated for 1 min in AFA (70% ethanol, 3.7% formaldehyde, 5% glacial acetic acid), followed by extensive washing with phosphate-buffered saline (PBS). Plates were stored at 4°C with PBS and 0.02% sodium azide. Fixed colonies were incubated with the anti-myosin heavy chain monoclonal antibody MF-20 (Bader et al., 1982). Immunostaining was performed with the alkaline phosphatase Vectastain ABC kit (Vector Laboratories, Inc.). Plates were examined at 50× and 20× magnification. For adipogenesis scoring, plates were fixed 7 days after switching to dexamethasone insulin-containing media as above, except that formalin was used instead of AFA. Adipogenic colonies were visualized using Oil Red-O.

Sequencing

Overlapping deletions of MyoD cloned into pVZ-1 in both orientations were obtained by unidirectional digestion methods (Henikoff, 1987). Deletion clones were sized on agarose gels and selected clones were used to prepare single-stranded DNA using an M13 helper phage and the E. coli strain TG1/λ (kindly provided by T. St. John). Single-stranded DNA was used for dideoxy sequencing using the T7 Sequenase kit (United States Biochemical Corp.) and α-thiol[^{35}S]dATP (1000 Ci/mmol, NEN). Reactions were run on 6% acrylamide gradient gels, followed by autoradiography. Sequence recording and analysis was performed using a digitizer and Genepro software (Joe Brown, Oncogen, Seattle, WA, and Jirn Wallace, Fred Hutchinson Cancer Research Center, Seattle, WA; version 4.1, 1987).

Acknowledgments

We thank Mark Groudine and Richard Harland for help with constructing the cDNA library; John McKay for assistance with cloning the aza-myoblast lines; Helen Blau, Howard Green, and Gordon Ringold for cell lines; Steve Henikoff, Maxine Linial, Kira Lueders, Dusty Miller, Bernardo Nadal-Ginard, and Shirley Tilghman for plasmids; Don Fishman and Frank Stockdale for antibodies; Pei Feng Cheng for outstanding technical assistance; Helen Devitt for typing this manuscript; and our colleagues in the Department of Genetics for their thoughtful comments on the manuscript. A. B. L. expresses his appreciation to Hazel Sive and Rufus Mandelbaum for advice and encouragement. R. L. D. is supported by the National Institutes of Health Medical Scientist Training Program. During the course of this study A. B. L. was supported by postdoctoral fellowships from the Jane Coffin Childs Memorial Fund for Ivledical Research and the Muscular Dystrophy Association; he is currently a Lucille P. Markey Scholar and this work was supported in part by a grant from the Lucille P. Markey Charitable Trust. H. W. would like to dedicate this paper to B. W.; R. L. D. would like to dedicate to H. D. This work was supported by a grant to H. W. from the National Institutes of Health.

The costs of publication of this article were defrayed in part by the payment of page charges. This article must therefore be hereby marked "advertisement" in accordance with 18 U.S.C. Section 1734 solely to indicate this fact.

Received October 20, 1987.

References

Alitalo, K., Bishop, J. M., Smith, D. H., Chen, E. Y., Colby, W. W., and Levinson, A. D. (1983). Nucleotide sequence of the v-myc oncogene of avian retrovirus MC29. Proc. Natl. Acad. Sci. USA 80, 100–104.

Bader, D., Masaki, T. and Fischman, D. A. (1982). Immunochemical analysis of myosin heavy chain during avian myogenesis in vivo and in vitro. PI. Cell Biol. 95, 763–770.

Berg, J. M. (1986). Potential metal-binding domains in nucleic acid binding proteins. Science 232, 485–487.

Bernard, O., Cory, S., Gerondakis, S., Webb, E., and Adams, J. M. (1983). Sequence of the murine and human cellular *myc* oncogenes and two modes of *myc* transcription resulting from chromosome translocation in B lymphoid tumours. EMBO J. *2*, 2375–2383.

Blau, H. M., Chiu, C.-P., and Webster, C. (1983). Cytoplasmic activation of human nuclear genes in stable heterocaryons. Cell *32*, 1171–1180.

Blau, H. M., Pavlath, G. K., Hardeman, E. C., Chiu, C.-P., Silberstein, L., Webster, S. G., Miller, S. C., and Webster, C. (1985). Plasticity of the differentiated state. Science *230*, 758–766.

Chapman, A. B., Knight, D. M., Dieckmann, B. S., and Ringold, G. M. (1984). Analysis of gene expression during differentiation of adipogenic cells in culture and hormonal control of the developmental program. J. Biol. Chem. *259*, 15548–15555.

Clegg, C. H. Linkhart, T. A., Olwin, B. B., and Hauschka, S. D. (1987). Growth factor control of skeletal muscle differentiation: commitment to terminal differentiation occurs in G1 chase and is repressed by fibroblast growth factor. J. Cell Biol. *105*, 949–956.

Conklin, K. F., and Groudine, M. (1986). Varied interactions between proviruses and adjacent host chromatin. Mol. Cell. Biol. *6*, 3999–4007.

Coppola, J. A., and Cole, M. D. (1986). Constitutive c-*myc* oncogene expression blocks mouse erythroleukaemia cell differentiation but not commitment. Nature *320*, 760–763.

Davis, M. M., Cohen, D. I., Nielsen, E. A., Steinmetz, M., Paul, W. E., and Hood, L. (1984). Cell-type-specific cDNA probes and the murine I region: the localization and orientation of A_α^d. Proc. Natl. Acad. Sci. USA *81*, 2194–2198.

DePinho, R. A., Legouy, E., Feldman, L. B., Kohl, N. E., Yancopoulos, G. D., and Alt, F. W. (1986). Structure and expression of the murine N-*myc* gene. Proc. Natl. Acad. Sci. USA *83*, 1827–1831.

Dienstman, S. R., and Holtzer, H. (1975). Myogenesis: a cell lineage interpretation. In Results and Problems in Cell Differentiation, Volume 7, J. Reinert and H. Holtzer, eds. (New York: Springer-Verlag), pp.1–25.

Dmitrovsky, E., Kuehl, W. M., Hollis, G. F. Kirsch, H. R., Bender, T. P., and Segal, S. (1986). Expression of a transfected human c-*myc* oncogene inhibits differentiation of a mouse erythroleukaemia cell line. Nature *322*, 748–750.

Dony, C., Kessel, M., and Gruss, P. (1985). Post-transcriptional control of *myc* and p53 expression during differentiation of the embryonal carcinoma cell line F9. Nature *317*, 636–639.

Endo, T., and Nadal-Ginard, B. (1986). Transcriptional and posttranscriptional control of c-*myc* during myogenesis: its mRNA remains inducible in differentiated cells and does not suppress the differentiated phenotype. Mol. Cell. Biol. *6*, 1412–1421.

Falcone, G., Tato, F. and Alema, S. (1985). Distinctive effects of the viral oncogenes *myc*, *erb*, *fps*, and *src* on the differentiation program of quail myogenic cells. Proc. Natl. Acad. Sci. USA *82*, 426–430.

Green, H., and Kehinde, O. (1974). Sublines of mouse 3T3 cells that accumulate lipid. Cell *1*, 113–116.

Green, H., and Kehinde, O. (1976). Spontaneous heritable changes leading to increased adipose conversion in 3T3 cells. Cell *7*, 105–113.

Groudine, M., Eisenman, R., and Weintraub, H. (1981). Chromatin structure of endogenous retroviral genes and activation by an inhibitor of DNA methylation. Nature *292*, 311–317.

Gubler, U., and Hoffman, B. J. (1983). A simple and very efficient method for generating cDNA libraries. Gene *25*, 262–269.

Harland, R., and Weintraub, H. (1985). Translation of mRNA injected into Xenopus oocytes is specifically inhibited by antisense RNA. J. Cell Biol. *101*, 1094–1099.

Hedrick, S. M., Cohen, D. I., Nielsen, E. A., and Davis, M. M. (1984). Isolation of cDNA clones encoding T cell-specific membrane-associated proteins. Nature *308*, 149–153.

Henikoff, S. (1987). Unidirectional digestion with exonuclease III in DNA sequence analysis. Meth. Enzymol. *155*, 156–165.

Henikoff, S., and Eghtedarzadeh, M. K. (1987). Conserved arrangement of nested genes at the Drosophila *Gart* locus. Genetics, in press.

Hsiao, W.-L.W., Gattoni-Celli, S., and Weinstein, I. B. (1986). Effects of 5-azacytidine on expression of endogenous retrovirus-related sequences in C3H10T1/2 cells. J. Virol. *57*, 1119–1126.

Huynh, T. V., Young, R. A., and Davis, R. W. (1985). DNA cloning: a practical approach, Volume 1, D. M. Glover, ed. (Oxford: IRL Press), pp. 49–78.

Jones, P. A., and Taylor, S. M. (1980). Cellular differentiation, cytidine analogs and DNA methylation. Cell *20*, 85–93.

Kaufman, S. J., Foster, R. F., Haye, K. R., and Faiman, L. E. (1985). Expression of a developmentally regulated antigen on the surface of skeletal and cardiac muscle cells. J. Cell Biol. *100*, 1977–1987

Kelly, K., Cochran, B. H., Stiles, C. D., and Leder, P. (1983). Cell-specific regulation of the c-*myc* gene by lymphocyte mitogens and platelet-derived growth factor. Cell *35*, 603–610.

Konieczny, S. F., and Emerson, C. R., Jr. (1984). 5-azacytidine induction of stable mesodermal stem cell lineages from 10T1/2 cells: evidence for regulatory genes controlling determination. Cell *38*, 791–800.

Konieczny, S. F. Baldwin, A. S., and Emerson, C. R., Jr. (1985). Myogenic determination and differentiation of 10T1/2 cell lineages: evidence for a simple genetic regulatory system. In Molecular Biology of Muscle Development, UCLA Symposium on Molecular and Cellular Biology, New Series Vol. 29, C. Emerson, D. A. Fischman, B. Nadal-Ginard, and M. A. O. Siddiqui, eds. (New York: Alan R. Liss, Inc.), pp. 21–34.

Lachman, H. M., and Skoultchi, A. I. (1984). Expression of c-*myc* changes during differentiation of mouse erythroleukemia cells. Nature *310*, 592–594.

Lassar, A. B., Paterson, B. M., and Weintraub, H. (1986). Transfection of a DNA locus that mediates the conversion of 10T1/2 fibroblasts to myoblasts. Cell *47*, 649–656.

Lathrop, B., Olson, E., and Glaser, L. (1985). Control by fibroblast growth factor of differentiation in the BC_3H1 muscle cell line. J. Cell Biol. *100*, 1540–1547.

Lawrence, J. B., and Coleman, J. R. (1984). Extinction of muscle-specific properties in somatic cell heterokaryons. Dev Biol. *101*, 463–476.

Linial, M. (1987). Creation of a processed pseudogene by retroviral infection. Cell *49*, 93–102.

Linkhart, T. A., Clegg, C. H., and Hauschka, S. D. (1981). Myogenic differentiation in permanent clonal mouse myoblast cell lines: regulation by macromolecular growth factors in the culture medium. Dev. Biol. *86*, 19–30.

Maniatis, T., Fritsch, E. F., and Sambrook, J. (1982). Molecular Cloning: A Laboratory Manual (Cold Spring Harbor, New York: Cold Spring Harbor Laboratory).

Pachnis, V, Belayew, A., and Tilghman, S. M. (1984). Locus unlinked to alpha-fetoprotein under control of the murine *raf* and *Rif* genes. Proc. Natl. Acad. Sci. USA *81*, 5523–5527.

Palmer, T. D., Hock, R. A., Osborne, W. R. A., and Miller, A. D. (1987). Efficient retrovirus-mediated transfer and expression of a human adenosine deaminase gene in diploid skin fibroblasts from an adenosine deaminase-deficient human. Proc. Natl. Acad. Sci. USA *84*, 1055–1059.

Prochownik, E. V., and Kukowska, J. (1986). Deregulated expression of c-*myc* by murine erythroleukaemia cells prevents differentiation. Nature *322*, 848–850.

Sanger, F., Nicklen, S., and Coulson, A. R. (1977). DNA sequencing with chain-terminating inhibitors. Proc. Natl. Acad. Sci. USA *74*, 5463–5467.

Sasse, J., Horwitz, A., Pacifici, M., and Holtzer, H. (1984). Separation of precursor myogenic and chondrogenic cells in early limb bud mesenchyme by a monoclonal antibody. J. Cell Biol. *99*, 1856–1866.

Schneider, M. D., Payne, P. A., Ueno, H., Perryman, M. B., and Roberts, R. (1986). Dissociated expression of c-*myc* and a *fos*-related competence gene during cardiac myogenesis. Mol. Cell. Biol. *6*, 4140–4143.

Sejersen, T., Sumegi, J., and Ringertz, N. R. (1985). Density-dependent arrest of DNA replication is accompanied by decreased levels of c-*myc* mRNA in myogenic but not in differentiation-defective myoblasts. J. Cell. Physiol. *125*, 465–470.

Sen, R., and Baltimore, D. (1986). Inducibility of κ immunoglobulin enhancer-binding protein NF-κB by a posttranslational mechanism. Cell *47*, 921–928.

Stone, J., De Lange, T., Ramsay, G., Jakobovits, E., Bishop, J. M., Varmus, H., and Lee, W. (1987). Definition of regions in human c-*myc* that are involved in transformation and nuclear localization. Mol. Cell. Biol. *7*, 1697–1709.

Taylor, S. M., and Jones, P. A. (1979). Multiple new phenotypes induced in 10T1/2 and 3T3 cells treated with 5-azacytidine. Cell *17,* 771–779.

Taylor, S. M., and Jones, P. A. (1982). Changes in phenotype expression in embryonic and adult cells treated with 5-azacytidine. J. Cell. Physiol. *111*, 187–194.

Thompson, C. B., Challoner, P. B., Neiman, P. E., and Groudine, M. (1985). Levels of c-*myc* oncogene mRNA are invariant throughout the cell cycle. Nature *314*, 363–366.

Villares, R., and Cabrera, C. V. (1987). The *achaete-scute* gene complex of D. melanogaster: conserved domains in a subset of genes required for neurogenesis and their homology to *myc*. Cell *50*, 415–424.

Walsh, F. S., and Ritter, M. A. (1981) Surface antigen differentiation during human myogenesis in culture. Nature *289*, 60–64.

Westin, E. H., Wong-Staal, F., Gelmann, E. P., Dalla-Favera, R., Papas, T. S., Lautenberger, J. A., Eva, A., Reddy, E. P., Tronick, S. R., Aaronson, S. A., and Gallo, R. C. (1982). Expression of cellular homologues of retroviral *onc* genes in human hematopoietic cells. Proc. Natl. Acad. Sci. USA *79*, 2490–2494.

Wigler, M., Pellicer, A., Silverstein, S., Axel, R., Urlaub, G., and Chasin, L. (1979). DNA-mediated transfer of the adenine phosphoribosyltransferase locus into mammalian cells. Proc. Natl. Acad. Sci. USA *76*, 1373–1376.

Wolpert, L., and Stein, W. (1984). Positional information and pattern formation. In Pattern Formation, G. Malacinski and S. Bryant, eds. (New York: MacMillan), p. 1.

Wright, W. E. (1984a). Induction of muscle genes in neural cells. J. Cell Biol. *98*, 427–435.

Wright, W. E. (1984b). Control of differentiation in heterokaryons and hybrids involving differentiation-defective myoblast variants. J. Cell Biol. *98*, 436–443.

Wright, W. E. (1985). The amplified expression of factors regulating myogenesis in L6 myoblasts. J. Cell Biol. *100*, 311–316.

Wright, W. E., and Aronoff, J. (1983). The suppression of myogenic function in heterokaryons formed by fusing chick myocytes to diploid rat fibroblasts. Cell Differ. *12*, 299–306.

Note Added in Proof

We have found that transfection of the MyoD1 expression vehicle into nondifferentiating aza-myoblast variants (1 and 2) induces myogenesis in these cell lines. A recent publication of the murine L-*myc* gene sequence indicates that this gene also shares sequence similarity (a. a. 318–339) with MyoD1 (Legouy et al., 1987, EMBO J. *6*, 3359–3366).

Conclusions and future perspectives

14

The papers presented in the preceding 13 sections are intended to provide an overview of the critical steps in our current understanding of eukaryotic gene regulation. They range from early papers, which established that the RNA content of different tissues was different while the DNA content was the same, to much more recent papers dealing with the functional domains of transcription factors and the regulation of their activity. In preparing this work, no papers have been selected that appeared more recently than 1990. This is partly so that only publications whose significance has stood the test of time are included. In addition, however, it reflects my opinion that that point in time represents a convenient boundary in the study of gene regulation.

Thus, at that time, the basic mechanisms of transcriptional control had been elucidated. In fact, the nature of individual transcription factors and the manner in which they operate had been analysed in detail in many instances. In many cases the factors responsible for a particular pattern of gene expression had been identified, their functional domains had been analysed and the manner in which their activity was regulated had been elucidated.

It has become increasingly clear over the last 5 years, however, that the understanding of gene regulation necessitates more than a simple analysis of individual transcription factors. Thus in many instances the activity of an individual factor will be regulated by the presence of other factors. An early example of this was described in Section 12, involving the regulation of NF-κB activity by IκB. However, many other instances have since been described in which the activity of an individual factor is regulated positively as well as negatively by another factor. Thus such interactions can result in combinations of two factors having a much stronger effect than either alone, as well as the inhibition of the activity of one factor by a second factor (for details, see Latchman 1995a,b). Similarly, as well as stimulating or inhibiting the activity of a particular factor, such interactions can also alter the specificity of a factor, for example by changing its DNA binding specificity. Indeed, such interactions can convert a single factor from an activator of gene expression into a repressor.

It is clear, therefore, that a full understanding of gene regulation will involve an analysis of the various interactions that can occur between different factors and the manner in which they alter both factor activity and specificity. Such an analysis will require not only functional studies but also structural studies involving the X-ray crystallographic analysis of several different factors bound to DNA. The most recent paper included in the collection presented here (Kissinger et al.), which appeared in 1990, represents one of the earliest determinations of the structure of a single transcription factor bound to DNA. Since that time many such structures have appeared and have greatly aided our understanding of the manner in which transcription factors act. To date, however, very few structural analyses of multicomponent transcription factor complexes, either in solution or bound to DNA, have appeared.

A detailed analysis of such complexes and the manner in which structural interactions alter transcription factor activity will be crucial for a future understanding of gene regulation. In particular, they will allow us to understand how different signalling pathways activating distinct transcription factors can interact with one another, so that ultimately our knowledge of these pathways will be as good as our current knowledge of the metabolic pathways central to Biochemistry. Perhaps even more importantly, such analyses will allow us to understand how relatively few transcription factors interacting with one another can control the enormously complex process of vertebrate, and in particular mammalian, development. The enormous progress that has been made over the last 20 years, represented by the papers included in this collection, offers hope that this ultimate goal can also be achieved.

References

Affolter, M., Schier, A. and Gehring, W.J. (1990) Homeodomain proteins and the regulation of gene expression. *Curr. Opin. Cell Biol.* 2, 485–495

Altmann, M. and Trachsel, H. (1993) Regulation of translational initiation and modulation of cellular physiology. *Trends Biochem. Sci.* 18, 429–432

Ashburner, M. and Bonner, J.J. (1979) The induction of gene activity in Drosophila by heat shock. *Cell* 17, 241–254

Baeuerle, P.A. and Baltimore, D. (1988) Activation of DNA-binding activity in an apparently cytoplasmic precursor of the NF-κB transcription factor. *Cell* 53, 211–217

Beg, A. and Baldwin, A.S. (1993) The IκB proteins: multifunctional regulators of Rel/NFκB transcription factors. *Genes Dev.* 7, 2064–2070

Benezra, R., Davis, R.L., Lockshon, D., Turner, D.L. and Weintraub, H. (1990) The protein Id: a negative regulator of helix-loop-helix DNA binding proteins. *Cell* 61, 49–59

Berg, J.M. (1988) Proposed structure for the zinc-binding domains from transcription factor IIIA and related proteins. *Proc. Natl. Acad. Sci. U.S.A.* 85, 99–102

Bestor, T.H. and Coxon, A. (1993) The Pros and Cons of DNA methylation. *Curr. Biol.* 3, 384–386

Bingham, P.M., Chou, T.-B., Mims, I. and Zachar, Z. (1988) On/off regulation of gene expression at the level of gene splicing. *Trends Genet.* 4, 134–138

Bird, A. (1992) The essentials of DNA methylation. *Cell* 70, 5–8

Bishop, J.O., Morton, J.G., Rosbash, M. and Richardson, M. (1974) Three abundance classes in HeLa cell messenger RNA. *Nature (London)* 250, 199–204

Blank, V., Kowitsky, P. and Israel, A. (1992) NFκB and related proteins: Rel/dorsal homologies meet ankyrin-like repeats. *Trends Biochem Sci.* 17, 135–140

Breitbart, R.E., Andreadis, A. and Nadal-Ginard, B. (1987) Alternative splicing: a ubiquitous mechanism for the generation of multiple protein isoforms from different genes. *Annu. Rev. Biochem.* 56, 467–495

Brown, K., Gerstberger, S., Carlson, L., Franzoso, G. and Siebenlist, U. (1995) Control of IκBα proteolysis by site-specific, signal-induced phosphorylation. *Science* 267, 1485–1488

Cedar, H. (1988) DNA methylation and gene activity. *Cell* 53, 3–4

Chrivia, J.C., Kwok, R.P.S., Lamb, N., Hagiwara, M., Montminy, M.R. and Goodman, R.H. (1993) Phosphorylated CREB binds specifically to the nuclear protein CBP. *Nature (London)* 365, 855–859

Constantinides, P.G., Jones, P.A. and Gevers, W. (1977) Functional striated muscle cells from non-myoblast precursors following 5-azacytidine treatment. *Nature (London)* 267, 364–366

Davidson, E.H. and Britten, R.J. (1979) Regulation of gene expression: possible role of repetitive sequences. *Science* 204, 1052–1059

Diakun, G.P., Fairall, L. and Klug, A. (1986) EXAFS study of the zinc-binding sites in the protein transcription factor IIIA. *Nature (London)* 324, 698–699

Dillon, N. and Grosveld, F. (1993) Transcriptional regulation of multigene loci, multilevel control. *Trends Genet.* 9, 134–137

Edmondson, D.F. and Olson, E.N. (1993) Helix-loop-helix proteins as regulators of muscle-specific transcription. *J. Biol. Chem.* 268, 755–758

Evans, R.M. and Hollenberg, S.M. (1988) Zinc fingers: gilt by association. *Cell* 52, 1–3

Galas, D. and Schmitz, A. (1978) DNase footprinting: a simple method for the detection of protein-DNA binding specificity. *Nucleic Acids Res.* 5, 3157–3170

Garcia, J.V., Bich-Thay, L. and Stafford, J. (1986) Synergism between immunoglobulin enhancers and promoters. *Nature (London)* 322, 383–385

Gehring, W.J. (1987) Homeoboxes in the study of development. *Science* 236, 1245–1252

Gehring, W.J., Qian, Y.Q., Billeter, M., Furukubo-Tukunagu, K., Schier, A.F., Resendez-Perez, D., Affolter, M., Otting, G. and Wuthrich, K. (1994) Homeodomain-DNA recognition. *Cell* 78, 211–223

Gellert, M. (1992) Molecular analysis of V (D) J recombination. *Annu. Rev. Genet.* 22, 425–446

Gillies, S.D., Morrison, S.L., Oi, V.T. and Tonegawa, S. (1983) A tissue specific enhancer is located downstream of the joining region in immunoglobulin heavy chain genes. *Cell* **33**, 729–740

Gilmour, R.S., Harrison, P.R., Windass, J.D., Affara, N.A. and Paul, J. (1974) Globin messenger RNA synthesis and processing during haemoglobin induction in Friend cells. 1. Evidence for transcriptional control in clone M2. *Cell Differ.* **3**, 9–22

Ginsberg, A.M., King, B.O. and Roeder, R.G. (1984) Xenopus 5S gene transcription factor, TFIIIA: characterization of a cDNA clone and measurement of RNA levels throughout development. *Cell* **39**, 479–489

Glover, J.N. and Harrison, S.C. (1995) Crystal structure of the heterodimeric bZip transcription factor c-Fos–c-Jun bound to DNA. *Nature (London)* **373**, 257–261

Graves, R.A., Pandey, N.B., Chodchoy, N. and Marzluff, W.F. (1987) Translation is required for regulation of histone mRNA degradation. *Cell* **48**, 615–626

Grimm, S. and Baeuerle, P.A. (1993) The inducible transcription factor NFκB: structure–function relationship of its protein subunits. *Biochem. J.* **290**, 297–308

Gross, D.S. and Garrard, W.T. (1988) Nuclease hypersensitive sites in chromatin. *Annu. Rev. Biochem.* **57**, 159–197

Grosschedl, R. and Birnsteil, M.L. (1980) Spacer DNA sequences upstream of the TATAAATA sequence are essential for promotion of histone H2A transcription *in vivo. Proc. Natl. Acad. Sci. U.S.A.* **77**, 7102–7106

Grosveld, F., Blom Van Assendelft, G., Greaves, D.R. and Kollias, G. (1987) Position-independent, high-level expression of the human β-globin gene in transgenic mice. *Cell* **51**, 975–985

Hatzopoulos, A.K., Schlokat, U. and Grass, P. (1988) Enhancers and other cis-acting sequences. In *Transcription and Splicing* (Hames, B.D. and Glover, D.M., eds.), pp. 43–96, IRL Press, Oxford

Hurst, H.C. (1994) bZip proteins. *Protein Profile* **1**, 123–168

Igo-Kemenes, T., Horz, W. and Zachau, H.G. (1982) Chromatin. *Annu. Rev. Biochem.* **51**, 89–121

Ingham, P.W. (1988) The molecular genetics of embryonic pattern formation in *Drosophila. Nature (London)* **335**, 25–34

Jones, N. (1990) Transcriptional regulation by dimerization: two sides to an incestuous relationship. *Cell* **61**, 9–11

Kawai, Y., Takami, K., Shiosaka, S., Emson, P.C., Hillyard, C.J., Girgis, S., MacIntyre, I. and Tohyama, M. (1985) Topographic localization of calcitonin gene-related peptide in the rat brain: an immunohistochemical analysis. *Neuroscience* **15**, 747–763

Kaye, J.S., Bellard, M., Dretzen, G., Bellard, F. and Chambon, P. (1984) A close association between sites of DNase I hypersensitivity and sites of enhanced cleavage by micrococcal nuclease in the 5' flanking region of the actively transcribed ovalbumin gene. *EMBO J.* **3**, 1137–1144

Kelly, K., Cochran, B.H., Stiles, C.D. and Leder, P. (1983) Cell-specific regulation of the c-*myc* gene by lymphocyte mitogens and platelet derived growth factor. *Cell* **35**, 603–610

Kenyon, C. (1994) If birds can fly, why can't we? *Cell* **78**, 175–180

Kerppola, T. and Curran, T. (1995) Zen and the art of Fos and Jun. *Nature (London)* **373**, 199–200

Klausner, R.D., Rouault, T.A. and Harford, J.B. (1993) Regulating the fate of cellular iron metabolism. *Cell* **72**, 19–28

Klug, A. and Rhodes, D. (1987) Zinc fingers: a novel protein motif for nucleic acid recognition. *Trends Biochem. Sci.* **12**, 464–469

Klug, A. and Schwabe, J.W.R. (1995) Zinc fingers. *FASEB J.* **9**, 597–604

Konieczny, S.F. and Emerson, Jr., C.P. (1984) 5-azacytidine induction of stable mesodermal lineages from 10T1/2 cells: evidence for regulatory genes controlling determination. *Cell* **38**, 791–800

Kornberg, T.B. (1993) Understanding the homeodomain. *J. Biol. Chem.* **268**, 26813–26816

Krumlauf, R. (1994) Hox genes in vertebrate development. *Cell* **78**, 191–201

Kwok, R.P.S., Lundblad, J.R., Chrivia, J.C., Richards, J.P., Bachinger, H.P., Brennan, R.G., Roberts, S.G.E., Green, M.R. and Goodman, R.H. (1994) Nuclear protein CBP is a coactivator for the transcription factor CREB. *Nature (London)* **370**, 223–226

Lalli, E. and Sassone-Corsi, P. (1994) Signal transduction and gene regulation: the nuclear response to cAMP. *J. Biol. Chem.* **269**, 17359–17362

Lamb, P. and McKnight, S.L. (1991) Diversity and specificity in transcriptional regulation: the benefits of heterotypic dimerization. *Trends Biochem. Sci.* **16**, 417–422

Lamond, A.I. (1991a) Nuclear RNA processing. *Curr. Opin. Cell Biol.* **3**, 493–500

Lamond, A.I. (1991b) ASF/SF2, a splice selector. *Trends Biochem. Sci.* **16**, 452–453

Landshulz, W.H., Johnson, P.F., Adashi, E.Y., Graves, B.J. and McKnight, S.L. (1988) Isolation of a recombinant copy of the gene encoding C/EBP. *Genes Dev.* **2**, 786–800

Latchman, D.S. (1990) Cell-type specific splicing factors and the regulation of alternative RNA splicing. *New Biol.* **2**, 297–303

Latchman, D.S. (1993) *Transcription Factors: A Practical Approach*, IRL Press/Oxford University Press, Oxford, New York, Tokyo

Latchman, D.S. (1995a) *Gene Regulation: A Eukaryotic Perspective*, Chapman and Hall, London

Latchman, D.S. (1995b) *Eukaryotic Transcription Factors*, Academic Press, London and San Diego

Lewin, B. (1991) Oncogenic conversion by regulatory changes in transcription factors. *Cell* **64**, 303–312

Li, E., Bestor, T.H. and Jaenisch, R. (1992) Targeted mutation of the DNA methyltransferase gene results in embryonic lethality. *Cell* **69**, 915–926

Littlewood, T. and Evan, G. (1994) Helix-loop-helix. *Protein Profile* **1**, 639–709

Lowenhaupt, K., Trent, C. and Lingrel, J.B. (1978) Mechanisms for accumulation of globin mRNA during dimethyl sulfoxide induction of mouse erythroleukaemia cells: synthesis of precursors and mature mRNA. *Dev. Biol.* **63**, 441–454

Maniatis, T., Kee, S.G., Efstratiadis, A. and Kafatos, F.C. (1976) Amplification and characterization of a β-globin gene synthesized *in vitro*. *Cell* **8**, 163–182

McKeown, M. (1995) Alternative mRNA splicing. *Annu. Rev. Cell Biol.* **8**, 133–155

Miller, D.M., Turner, P., Nienhuis, A.W., Axelrod, D.E. and Gopalakrishnan, T.V. (1978) Active conformation of the globin genes in uninduced and induced mouse erythroleukemia cells. *Cell* **14**, 511–521

Montminy, M.R., Sevarion, K.A., Wagner, J.A., Mandel, G. and Goodman, R.H. (1986) Identification of a cyclic AMP response element within the rat somatostatin gene. *Proc. Natl. Acad. Sci. U.S.A.* **83**, 6682–6686

Morimoto, R. (1993) Cells in stress: transcriptional activation of heat shock genes. *Science* **259**, 1409–1410

Morse, R.H. and Simpson, R.T. (1988) DNA in the nucleosome. *Cell* **54**, 285–287

Muller, M.M., Gerster, T. and Schaffner, W. (1988) Enhancer sequences and the regulation of gene transcription. *Eur. J. Biochem.* **176**, 485–495

Neuhaus, D., NaKaseko, Y., Nagai, A. and Klug, A. (1990) Sequence specific ¹H NMR resonance assignments and secondary structure identification for 1- and 2-zinc finger constructs from SWI5. *FEBS Lett.* **262**, 179–184

Olson, E.N. and Klein, W.H. (1994) bHLH factors in muscle development: dead lines and commitments, what to leave in and what to leave out. *Genes Dev.* **8**, 1–8

Owen, D. and Kühn, L.C. (1987) Noncoding 3′ sequences of the transferrin receptor gene are required for mRNA regulation by iron. *EMBO J.* **6**, 1287–1293

Parker, C.S. and Topol, J. (1984) A *Drosophila* RNA polymerase II transcription factor contains a promoter-region specific DNA-binding activity. *Cell* **36**, 357–369

Peltz, S.W. and Jacobson, A. (1992) mRNA stability in *trans*-it. *Curr. Opin. Cell Biol.* **4**, 979–983

Peltz, S.W., Brewer, G., Bernstein, P. and Ross, J. (1991) Regulation of mRNA turnover in eukaryotic cells. *Crit. Rev. Eukaryotic Gene Expression* **1**, 99–126

Quian, Y.Q., Billeter, M., Otting, G., Miller, M., Gehring, W.J. and Wuthrich, K. (1989) The structure of the antennapaedia homeodomain determined by NMR spectroscopy in solution: comparison with prokaryotic repressors. *Cell* **59**, 573–580

Razin, A. and Cedar, H. (1991) DNA methylation and gene expression. *Microbiol. Rev.* **55**, 451–458

Reeves, R. and Cserjesi, P. (1979) Sodium butyrate induces new gene expression in Friend erythroleukemic cells. *J. Biol. Chem.* **254**, 4283–4290

Rhodes, D. and Klug, A. (1993) Zinc finger structure. *Sci. Am.* **268**, 32–39

Rosenfeld, M.G., Lin, C.R., Amara, S.G., Stolarsky, L., Roos, B.A., Ong, E.S. and Evans, R.M. (1982) Calcitonin mRNA polymorphism: peptide switching associated with alternative RNA splicing events. *Proc. Natl. Acad. Sci. U.S.A.* **79**, 1717–1721

Ruiz-I-Altaba, A., Perry-O'Keefe, H. and Melton, D.A. (1987) Xfin: an embryonic gene encoding a multi-fingered protein in *Xenopus*. *EMBO J.* **6**, 3065–3070

Sachs, A.B. (1993) Messenger RNA degradation in eukaryotes. *Cell* **74**, 413–421

Sassone-Corsi, P. (1994) Goals for signal transduction pathways: linking up with transcriptional regulation. *EMBO J.* **13**, 4628–4717

Scott, M.P. and Carroll, S.B. (1987) The segmentation and homeotic gene network in early *Drosophila* development. *Cell* **51**, 689–698

Scott, M.P., Tamkin, J.W. and Hartzell, G.W. (1989) Structure and function of the homeodomain. *Biochim. Biophys. Acta* **989**, 25–48

Sentenac, A. (1985) Eukaryotic RNA polymerases. *CRC Crit. Rev. Biochem.* **1**, 31–90

Sharp, P.A. (1987) Splicing of messenger RNA precursors. *Science* **235**, 766–771

Shepherd, J.C.W., McGinnis, W., Carrasco, A.E., DeRoberts, E.M. and Gehring, W.J. (1984) Fly and frog homoeodomains show homologies with yeast mating type regulatory loci. *Nature (London)* **310**, 70–71

Stalder, J., Groudin, M., Dodgson, J.B., Engel, J.D. and Weintraub, H. (1980) Hb switching in chickens. *Cell* **19**, 973–980

Struhl, K. (1989) Helix-turn-helix, zinc finger and leucine zipper motifs for eukaryotic transcriptional regulatory proteins. *Trends Biochem. Sci.* **14**, 137–140

Swaneck, G.E., Nordstrom, J.L., Kreuzaler, F., Tsai, M.-J. and O'Malley, B.W. (1979) Effect of oestrogen on gene expression in chicken oviduct: evidence for transcriptional control of the ovalbumin gene. *Proc. Natl. Acad, Sci. U.S.A.* **76**, 1049–1053

Taylor, S.M. and Jones, P.A. (1982) Changes in phenotypic expression in embryonic and adult cells treated with 5-azacytidine. *J. Cell. Physiol.* **111**, 187–194

Thanos, D. and Maniatis, T. (1995) NFκB: a lesson in family values. *Cell* **80**, 529–532

Thompson, C.C. and McKnight, S.L. (1992) Anatomy of an enhancer. *Trends Genet.* **8**, 232–236

Townes, T.M. and Behringer, R.R. (1990) Human globin locus activation region (LAR): role in temporal control. *Trends Genet.* **6**, 219–223

Travers, A. (1993) *DNA–Protein Interactions*, Chapman and Hall, London

Treisman, J., Gonczy, P., Vashishtha, M., Harris, E. and Desplan, C. (1990) A single amino acid can determine the DNA binding specificity of homeodomain proteins. *Cell* **59**, 553–562

Turner, B.M. (1993) Decoding the nucleosome. *Cell* **75**, 5–8

Vinson, C.R., Sigler, P.B. and McKnight, S.L. (1989) Scissors-grip model for DNA recognition by a family of leucine zipper proteins. *Science* **246**, 911–916

Waalwijk, C. and Flavell, R.A. (1978) MspI, an isoschizomer of HpaII which cleaves both unmethylated and methylated HpaII sites. *Nucleic Acids Res.* **5**, 3231–3236

Walker, M.D., Edlund, T., Boulet, A.M. and Rutter, W.J. (1983) Cell specific expression controlled by the 5′ flanking region of the insulin and chymotrypsin genes. *Nature (London)* **306**, 557–561

Weintraub, H., Larsen, A. and Groudine, M. (1981) Alpha globin gene switching during the development of chicken embryos: expression and chromosome structure. *Cell* **24**, 333–344

Wu, C., Bingham, P.M., Livak, K.J., Holmgren, R. and Elgin, S.C.R. (1979) The chromatin structure of specific genes: I. Evidence of higher order domains of defined DNA sequence. *Cell* **16**, 797–806

Yen, T.J., Gay, D.A., Pachter, J.S. and Cleveland, D.W. (1988) Autoregulated changes in stability of polyribosome-bound β-tubulin mRNAs are specified by the first 13 translated nucleotides. *Mol. Cell. Biol.* **8**, 1224–1235

Zimarino, V. and Wu, C. (1987) Induction of sequence-specific binding of *Drosophila* heat-shock activator protein without protein synthesis. *Nature (London)* **327**, 727–730

Subject index

Landmarks in Intracellular Signalling

by R D Burgoyne and O H Petersen

The intracellular signalling pathways that control cell function have been, and still are, one of the most intensively studied aspects of biology. In recent years the detailed characterization of the multiple cell-signalling pathways by many laboratories has resulted in a bewildering increase in knowledge in this field. For this reason, students and others learning about this topic for the first time are increasingly overwhelmed by the mass of information and frequently are unable to find time to read and digest the key original papers.

The idea behind **Landmarks in Intracellular Signalling** is to provide full reproductions of a set of key papers which have been chosen as landmark papers in the various aspects of intracellular signalling. The selected papers have all resulted in significant advances in one or other aspect of intracellular signalling. Readers of **Landmarks in Intracellular Signalling** will now have easy, ready available access to the original literature from one source. The papers are accompanied by commentaries that describe why the papers are significant, how the work came about and summarize the advances that have been made up to the present time as a consequence of the original paper. The commentaries will also serve as mini-reviews of many aspects of cell regulation and can be read on their own.

The area of intracellular signalling is relevant to many areas of biology and the basic principles need to be understood by undergraduates in many disciplines. Background knowledge of this area is also important for postgraduate students in many fields as well as more senior research workers and academics.

Contents:

1 85578 101 8   June 1997   Paperback 250 pages   £20.00/US$34.00

Please supply me with ☐ copies of Landmarks in Intracellular Signalling @ £20.00/US$34.00 each

Name:

Address:

I enclose a payment of £ _____

☐ Cheque payable to Portland Press   ☐ Proforma Invoice   ☐ Visa/Mastercard/Access/AmEx

Card No. _____   Exp. Date _____

Signature _____   Date _____

Send orders to: Portland Press, Commerce Way, Colchester, CO2 8HP
Tel: 01206 796351 Fax: 01206 799331 email: sales@portlandpress.co.uk.
Postage: UK: please add £2.50 per book to a maximum of £7.50

AWW/0697/A